Lecture Notes in Computer Science   **10439**

Commenced Publication in 1973
Founding and Former Series Editors:
Gerhard Goos, Juris Hartmanis, and Jan van Leeuwen

More information about this series at http://www.springer.com/series/7409

Djamal Benslimane · Ernesto Damiani
William I. Grosky · Abdelkader Hameurlain
Amit Sheth · Roland R. Wagner (Eds.)

Database and Expert Systems Applications

28th International Conference, DEXA 2017
Lyon, France, August 28–31, 2017
Proceedings, Part II

 Springer

Editors
Djamal Benslimane (iD)
University of Lyon
Villeurbanne
France

Ernesto Damiani
University of Milan
Milan
Italy

William I. Grosky (iD)
University of Michigan
Dearborn, MI
USA

Abdelkader Hameurlain
Paul Sabatier University
Toulouse
France

Amit Sheth
Wright State University
Dayton, OH
USA

Roland R. Wagner
Johannes Kepler University
Linz
Austria

ISSN 0302-9743   ISSN 1611-3349   (electronic)
Lecture Notes in Computer Science
ISBN 978-3-319-64470-7   ISBN 978-3-319-64471-4   (eBook)
DOI 10.1007/978-3-319-64471-4

Library of Congress Control Number: 2017947505

LNCS Sublibrary: SL3 – Information Systems and Applications, incl. Internet/Web, and HCI

Printed on acid-free paper

This Springer imprint is published by Springer Nature
The registered company is Springer International Publishing AG
The registered company address is: Gewerbestrasse 11, 6330 Cham, Switzerland

Preface

The well-established International Conference on Database and Expert Systems Applications – DEXA — provides a forum to bring together researchers and practitioners who are actively engaged both in theoretical and practical aspects of database, information, and knowledge systems. It allows participants to exchange ideas, up-to-date information, and experiences in database and knowledge systems and to debate issues and directions for further research and development.

This volume contains papers selected for presentation at the 28th International Conference on Database and Expert Systems Applications (DEXA 2017), which took place in Lyon, France, during August 28–31, 2017.

DEXA 2017 attracted 166 submissions from all over the world. Decision on acceptance or rejection was based on at least three reviews for each submitted paper. After a thorough review process by the Program Committee members, to whom we owe our acknowledgment and special thanks for the time and effort they invested in reviewing papers, the DEXA 2017 chairs accepted 37 full research papers and 40 short research papers yielding an acceptance rate of 22% and 24%, respectively. Full papers were given a maximum of 15 pages in this volume and short papers were given an eight-page limit. Authors of selected papers presented at the conference will be invited to submit extended versions of their papers for publication in the Springer journal *Transactions on Large-Scale Data- and Knowledge-Centered Systems* (TLDKS). The submitted extended versions will undergo a further review process.

Two high-quality keynote presentations on "Structural and Semantic Summarization of RDF Graphs" given by Ioana Manolescu, Senior Researcher, Inria Saclay and Ecole Polytechnique, France, and "Omnipresent Multimedia – Pain and Gain of the Always Connected Paradigm" given by Gabriele Anderst-Kotsis, Johannes Kepler University Linz, Austria, were also featured in the scientific program of DEXA 2017.

This edition of DEXA also featured five international workshops covering a variety of specialized topics:

- AICTSS 2017: First International Workshop on Advanced ICT Technologies for Secure Societies
- BDMICS 2017: Second International Workshop on Big Data Management in Cloud Systems
- BIOKDD 2017: 8th International Workshop on Biological Knowledge Discovery from Data
- TIR 2017: 14th International Workshop on Technologies for Information Retrieval
- UCC 2017: First International Workshop on Uncertainty in Cloud Computing

The success of DEXA 2017 would not have been possible without the hard work and dedication of many people including Gabriela Wagner as manager of the DEXA organization for her highly skillful management and efficient assistance, Chirine Ghedira and Mahmoud Barhamgi as local Organizing Committee chairs for tackling different aspects of the local organization and their dedication and commitment to this event, Karim Benouaret, Caroline Wintergerst, Christophe Gravier, Omar Boussaid, Fadila Bentayeb, Nadia Kabachi, Nabila Benharkat, Nadia Bennani, Faty Berkaï, and Claire Petrel as local Organizing Committee members for supporting us all the way through, and Vladimir Marik as publication chair for the preparation of the proceedings volumes. Our special thanks and gratitude also go to the general chairs for their continuous encouragement and great support: Abdelkader Hameurlain (IRIT, Paul Sabatier University, Toulouse, France), Amit Sheth (Kno.e.sis - Wright State University, USA), and Roland R. Wagner (Johannes Kepler University, Linz, Austria).

DEXA 2017 received support from the following institutions: Lyon 1 University, Lyon 2 University, Lyon 3 University, University of Lyon, INSA of Lyon, LIRIS Lab, ERIC Lab, CNRS, FIL (Fédération Informatique Lyonnaise), AMIES Labex, and FAW in Austria. We gratefully thank them for their commitment to supporting this scientific event.

Last but not least, we want to thank the international community, including all the authors, the Program Committee members, and the external reviewers, for making and keeping DEXA a nice avenue and a well-established conference in its domain.

For readers of this volume, we hope you will find it both interesting and informative. We also hope it will inspire and embolden you to greater achievement and to look further into the challenges that are still ahead in our digital society.

June 2017

Djamal Benslimane
Ernesto Damiani
William I. Grosky

Organization

General Chair

Abdelkader Hameurlain   IRIT, Paul Sabatier University Toulouse, France
Amit Sheth   Kno.e.sis - Wright State University, USA
Roland R. Wagner   Johannes Kepler University Linz, Austria

Program Committee Co-chairs

Djamal Benslimane   University of Lyon 1, France
Ernesto Damiani   University of Milan, Italy
William I. Grosky   University of Michigan, USA

Publication Chair

Vladimir Marik   Czech Technical University, Czech Republic

Program Committee

Slim Abdennadher   German University, Cairo, Egypt
Witold Abramowicz   The Poznan University of Economics, Poland
Hamideh Afsarmanesh   University of Amsterdam, The Netherlands
Riccardo Albertoni   Institute of Applied Mathematics and Information Technologies - Italian National Council of Research, Italy
Idir Amine Amarouche   University Houari Boumediene, Algiers, Algeria
Rachid Anane   Coventry University, UK
Annalisa Appice   Università degli Studi di Bari, Italy
Mustafa Atay   Winston-Salem State University, USA
Faten Atigui   CNAM, France
Spiridon Bakiras   Hamad bin Khalifa University, Qatar
Zhifeng Bao   National University of Singapore, Singapore
Ladjel Bellatreche   ENSMA, France
Nadia Bennani   INSA Lyon, France
Karim Benouaret   Université Claude Bernard Lyon 1, France
Morad Benyoucef   University of Ottawa, Canada
Catherine Berrut   Grenoble University, France
Athman Bouguettaya   University of Sydney, Australia
Omar Boussaid   University of Lyon, France
Stephane Bressan   National University of Singapore, Singapore
David Camacho   Autonomous University of Madrid, Spain

Chuan-Ming Liu	National Taipei University of Technology, Taiwan
Hong-Cheu Liu	University of South Australia, Australia
Jorge Lloret Gazo	University of Zaragoza, Spain
Jianguo Lu	University of Windsor, Canada
Alessandra Lumini	University of Bologna, Italy
Hui Ma	Victoria University of Wellington, New Zealand
Qiang Ma	Kyoto University, Japan
Stephane Maag	TELECOM SudParis, France
Zakaria Maamar	Zayed University, United Arab Emirates
Elio Masciari	ICAR-CNR, Università della Calabria, Italy
Brahim Medjahed	University of Michigan - Dearborn, USA
Faouzi Mhamdi	ESSTT, University of Tunis, Tunisia
Alok Mishra	Atilim University, Ankara, Turkey
Harekrishna Mishra	Institute of Rural Management Anand, India
Sanjay Misra	University of Technology, Minna, Nigeria
Jose Mocito	Brisa Innovation, Portugal
Lars Moench	University of Hagen, Germany
Riad Mokadem	IRIT, Paul Sabatier University, France
Yang-Sae Moon	Kangwon National University, South Korea
Franck Morvan	IRIT, Paul Sabatier University, France
Dariusz Mrozek	Silesian University of Technology, Poland
Francesc Munoz-Escoi	Universitat Politecnica de Valencia, Spain
Ismael Navas-Delgado	University of Málaga, Spain
Wilfred Ng	Hong Kong University of Science and Technology, Hong Kong, SAR China
Javier Nieves Acedo	University of Deusto, Spain
Mourad Oussalah	University of Nantes, France
Gultekin Ozsoyoglu	Case Western Reserve University, USA
George Pallis	University of Cyprus, Cyprus
Ingrid Pappel	Tallinn University of Technology, Estonia
Marcin Paprzycki	Polish Academy of Sciences, Warsaw Management Academy, Poland
Oscar Pastor Lopez	Universidad Politecnica de Valencia, Spain
Clara Pizzuti	Institute for High Performance Computing and Networking, ICAR, National Research Council, CNR, Italy
Pascal Poncelet	LIRMM, France
Elaheh Pourabbas	National Research Council, Italy
Jianbin Qin	University of New South Wales, Australia
Claudia Raibulet	Università degli Studi di Milano-Bicocca, Italy
Isidro Ramos	Technical University of Valencia, Spain
Praveen Rao	University of Missouri-Kansas City, USA
Manjeet Rege	University of St. Thomas, USA
Rodolfo F. Resende	Federal University of Minas Gerais, Brazil
Claudia Roncancio	Grenoble University/LIG, France
Massimo Ruffolo	ICAR-CNR, Italy

Giovanni Maria Sacco	University of Turin, Italy
Simonas Saltenis	Aalborg University, Denmark
Carlo Sansone	Università di Napoli Federico II, Italy
Igor Santos Grueiro	Deusto University, Spain
N.L. Sarda	I.I.T. Bombay, India
Marinette Savonnet	University of Burgundy, France
Klaus-Dieter Schewe	Software Competence Centre Hagenberg, Austria
Florence Sedes	IRIT, Paul Sabatier University, Toulouse, France
Nazha Selmaoui	University of New Caledonia, New Caledonia
Michael Sheng	Macquarie University, Australia
Patrick Siarry	Université Paris 12, LiSSi, France
Gheorghe Cosmin Silaghi	Babes-Bolyai University of Cluj-Napoca, Romania
Hala Skaf-Molli	Nantes University, France
Leonid Sokolinsky	South Ural State University, Russian Federation
Bala Srinivasan	Monash University, Australia
Umberto Straccia	ISTI, CNR, Italy
Raj Sunderraman	Georgia State University, USA
David Taniar	Monash University, Australia
Maguelonne Teisseire	Irstea, TETIS, France
Sergio Tessaris	Free University of Bozen-Bolzano, Italy
Olivier Teste	IRIT, University of Toulouse, France
Stephanie Teufel	University of Fribourg, Switzerland
Jukka Teuhola	University of Turku, Finland
Jean-Marc Thevenin	University of Toulouse 1 Capitole, France
A Min Tjoa	Vienna University of Technology, Austria
Vicenc Torra	University of Skövde, Sweden
Traian Marius Truta	Northern Kentucky University, USA
Theodoros Tzouramanis	University of the Aegean, Greece
Lucia Vaira	University of Salento, Italy
Ismini Vasileiou	University of Plymouth, UK
Krishnamurthy Vidyasankar	Memorial University of Newfoundland, Canada
Marco Vieira	University of Coimbra, Portugal
Junhu Wang	Griffith University, Brisbane, Australia
Wendy Hui Wang	Stevens Institute of Technology, USA
Piotr Wisniewski	Nicolaus Copernicus University, Poland
Huayu Wu	Institute for Infocomm Research, A*STAR, Singapore
Ming Hour Yang	Chung Yuan Christian University, Taiwan
Xiaochun Yang	Northeastern University, China
Junjie Yao	ECNU, China
Hongzhi Yin	The University of Queensland, Australia
Haruo Yokota	Tokyo Institute of Technology, Japan
Yanchang Zhao	IBM Australia, Australia
Qiang Zhu	The University of Michigan, USA
Yan Zhu	Southwest Jiaotong University, China
Marcin Zimniak	TU Chemnitz, Germany

Additional Reviewers

Mira Abboud	University of Nantes, France
Amine Abdaoui	LIRMM, France
Addi Ait-Mlouk	Cadi Ayyad University, Morocco
Ahmed Bahey	Nile University, Egypt
Cristóbal Barba-González	Universidad de Málaga, Spain
Nagwa M. Baz	Nile University, Egypt
Gema Bello	Autonomous University of Madrid, Spain
Kirill Borodulin	South Ural State University, Russian Federation
Yi Bu	Indiana University, USA
Stephen Carden	Georgia Southern University, USA
Loredana Caruccio	University of Salerno, Italy
Brice Chardin	LIAS/ENSMA, Poitiers, France
Arpita Chatterjee	Georgia Southern University, USA
Hongxu Chen	The University of Queensland, Australia
Weitong Chen	The University of Queensland, Australia
Van-Dat Cung	Grenoble INP, France
Sarah Dahab	Telecom SudParis, France
Matthew Damigos	NTUA, Greece
María del Carmen Rodríguez Hernández	University of Zaragoza, Spain
Yassine Djoudi	USTHB University, Algiers, Algeria
Hai Dong	RMIT University, Australia
Xingzhong Du	The University of Queensland, Australia
Daniel Ernesto Lopez Barron	University of Missouri-Kansas City, USA
Xiu Susie Fang	Macquarie University, Australia
William Ferng	Boeing Research and Technology, USA
Marco Franceschetti	Alpen Adria University Klagenfurt, Austria
Feng George Yu	Youngstown State University, USA
Azadeh Ghari-Neiat	University of Sydney, Australia
Paola Gomez	Université Grenoble-Alpes, France
Antonio Gonzalez	Autonomous University of Madrid, Spain
Senen Gonzalez	Software Competence Center Hagenberg, Austria
Wentian Guo	NUS, Singapore
Rajan Gupta	University of Delhi, India
Hsin-Yu Ha	Florida International University, USA
Ramón Hermoso	University of Zaragoza, Spain
Juan Jose Hernandez Porras	Telecom SudParis, France
Bing Huang	University of Sydney, Australia
Xin Huang	Hong Kong Baptist University, SAR China
Liliana Ibanescu	AgroParisTech and Inria, France
Angelo Impedovo	University of Bari Aldo Moro, Italy
Daniel Kadenbach	University of Applied Sciences and Arts Hannover, Germany

Eleftherios Kalogeros	Ionian University, Greece
Johannes Kastner	University of Augsburg, Germany
Anas Katib	University of Missouri-Kansas City, USA
Shih-Wen George Ke	Chung Yuan Christian University, Taiwan
Julius Köpke	Alpen Adria University Klagenfurt, Austria
Cyril Labbe	Université Grenoble-Alpes, France
Chuan-Chi Lai	National Chiao Tung University, Taiwan
Meriem Laifa	University of Bordj Bou Arreridj, Algeria
Raul Lara	Autonomous University of Madrid, Spain
Hieu Hanh Le	Tokyo Institute of Technology, Japan
Martin Ledvinka	Czech Technical University in Prague, Czech Republic
Jason Jingshi Li	IBM, Australia
Xiao Li	Dalian University of Technology, China
Corrado Loglisci	University of Bari Aldo Moro, Italy
Li Lu	Institute for Infocomm Research, Singapore
Alejandro Martín	Autonomous University of Madrid, Spain
Jorge Martinez-Gil	Software Competence Center Hagenberg, Austria
Riccardo Martoglia	University of Modena and Reggio Emilia, Italy
Amine Mesmoudi	LIAS/Poitiers University, Poitiers, France
Paolo Mignone	University of Bari Aldo Moro, Italy
Sajib Mistry	University of Sydney, Australia
Pascal Molli	University of Nantes, France
Ermelinda Oro	ICAR-CNR, Italy
Alejandra Lorena Paoletti	Software Competence Center Hagenberg, Austria
Horst Pichler	Alpen Adria University Klagenfurt, Austria
Gianvito Pio	University of Bari Aldo Moro, Italy
Valentina Indelli Pisano	University of Salerno, Italy
Samira Pouyanfar	Florida International University, USA
Gang Qian	University of Central Oklahoma, USA
Alfonso Quarati	Italian National Council of Research, Italy
Cristian Ramirez Atencia	Autonomous University of Madrid, Spain
Franck Ravat	IRIT-Université Toulouse1 Capitole, France
Leonard Renners	University of Applied Sciences and Arts Hannover, Germany
Victor Rodriguez	Autonomous University of Madrid, Spain
Lena Rudenko	University of Augsburg, Germany
Lama Saeeda	Czech Technical University in Prague, Czech Republic
Zaidi Sahnoun	Constantine University, Algeria
Arnaud Sallaberry	LIRMM, France
Mo Sha	NUS, Singapore
Xiang Shili	Institute for Infocomm Research, Singapore
Adel Smeda	University of Nantes, France
Bayu Adhi Tama	Pukyong National University, Korea
Loreadana Tec	RISC Software, Austria
Haiman Tian	Florida International University, USA
Raquel Trillo-Lado	University of Zaragoza, Spain

Sponsors of DEXA 2017

Organism	Logo
Université de Lyon	
Claude Bernard University Lyon 1	
Université Lumière Lyon 2	
Jean Moulin Lyon 3 University	
INSA Lyon (National Institute of Applied Science)	
FIL (Fédération Informatique de Lyon)	
CNRS (National Center for Scientific Research)	
AMIES (Agence pour les mathématiques en interaction avec l'entreprise et la société)	
LIRIS Lab	
ERIC Lab	
FAW Company	

Contents – Part II

Data Mining and Machine Learning

Recommender Systems and Query Recommendation

Graph Algorithms

Semantic Clustering and Data Classification

Contents – Part I

Preferences and Query Optimization

Data Integration and RDF Matching

Security and Privacy (I)

Web Search

Data Clustering

Top-K and Skyline Queries

Data Mining and Big Data

Security and Privacy (II)

Generating k-Anonymous Microdata by Fuzzy Possibilistic Clustering

Balkis Abidi$^{(\boxtimes)}$ and Sadok Ben Yahia

Faculty of Sciences of Tunis, University of Tunis El Manar,
LIPAH-LR 11ES14, 2092 Tunis, Tunisia
{abidi.balkis,sadok.benyahia}@fst.rnu.tn

Abstract. Collecting, releasing and sharing microdata about individuals is needed in some domains to support research initiatives aiming to create new valuable knowledge, by means of data mining and analysis tools. Thus, seeking individuals' anonymity is required to guarantee their privacy prior publication. The k-anonymity by microaggregation, is a widely accepted model for data anonymization. It consists in de-associating the relationship between the identity of data subjects, *i.e.* individuals, and their confidential information. However, this method shows limits when dealing with real datasets. Indeed, the latter are characterized by their large number of attributes and the presence of noisy data. Thus, decreasing the information loss during the anonymization process is a compelling task to achieve. This paper aims to deal with such challenge. Doing so, we propose a microaggregation algorithm called MICRO-PFSOM, based on fuzzy possibilitic clustering. The main thrust of this algorithm stands in applying an hybrid anonymization process.

Keywords: k-anonymity · Hybrid micoaggregation · Information loss · Fuzzy and possibilistic clustering

1 Introduction

With the growth of digital economy, data sharing has become an essential business practice. Such concept consists in collecting data to interpret, process and analyze them for a spesific purpose, then, make these data *reusable* for other purposes [6]. Sharing allows new insights from existing data and enables organizations to make full use of this core resource. However, when the collected data concern information about individual, called *microdata*, publishing and releasing them can also introduce new ethical risks. In fact, the collected data may contain confidential and sensitive information, *e.g.* visited websites of internet users, videos watched and uploded, geolocation of smartphone users, *etc.* Collecting, analyzing and sharing such information raises threat to individual privacy. Thereby, privacy concern can be considered as a major obstacle for data sharing. Data de-identification, *i.e.* subtracting explicit *identifiers*, is regarded as a priority to prevent sensitive information from being disclosed. Such process

© Springer International Publishing AG 2017
D. Benslimane et al. (Eds.): DEXA 2017, Part II, LNCS 10439, pp. 3–17, 2017.
DOI: 10.1007/978-3-319-64471-4_1

involves removing any information which is able to *uniquely* identify an individual, *e.g.* name, SSN, *etc* [14]. However, the latter solution could not ensure individual anonymity. That is, residual information can still susceptible for *identity disclosure*. Indeed, it was revealed that is possible to manipulate de-identified datasets and recover the real identity of individual, through data linkage techniques [4, 23, 26]. Such risk is called *re-identification* [25]. Although, a study presented in [26], showed that if the direct identifiers are removed from the original records, certain attributes, called quasi-identifiers, such as the birth date and zip-code can be used to uncover the real identities of the underlying records. Therefore, a large number of Privacy Preserving Data Mining (PPDM) methods have been proposed aiming at ensuring privacy of the respondents, while preserving the statistical utility of the original data [3]. The basic idea of this research area, also known as Statistical Disclosure Control, is to modify the collected data, subject to be released, in such a way to perform analysis and knowledge discovery tasks effectively without compromising the security of sensitive information contained in the data. Thus, PPDM methods aim to balance two goals inversely related, namely *utility* and *privacy* of individual data. That is, if an aggressive protection is performed on data, then a significant information loss will be generated but a low disclosure risk. However, slight data protection outputs high disclosure risk with negligible information loss. Microaggregation is a widely accepted PPDM technique for data anonymization, aiming at de-associating the relationship between the identity of data subjects and their confidential information [10]. Microaggregation technique proceeds in two stages.

– First, the set of records in a microdata is distributed into groups in such a way that: *(i)* each group contains at least k records; *(ii)* records within a group are as similar as possible. The obtained groups form a k-partition.
– Second, records within each group are replaced by a representative of the group, typically the centroid record.

Clearly, when microaggregation is applied to the projection of records on their quasi-identifier attributes, the resulting microdata fulfills the k-anonymous model [26]. That is, for any combination of values of quasi-identifier attributes in the released microdata, there are at least k records sharing that combination of values. Normally, microaggregation gathers the closest data together, in such a way that the respective distances between the data vectors and the corresponding centroids is as small as possible. However, for a multi-dimensional dataset, achieving an optimal k-partition has been shown to be NP-hard task [22]. Thus, microaggregation methods are based on heuristics to find the appropriate combination of aggregated records which increase the partition's homogeneity, while ensuring privacy, *i.e.* preventing the re-identification risk or at least possible with probability $1/k$. To achieve such a partition, microaggregation methods rely on *refinement* steps, during the partitioning process, which aim to fine tune the obtained partition, by merging or splitting its fixed size groups. However, in real datasets the poor homogeneity within the generated partition can be significant, especially in *noisy* surroundings. So, its refinement could be costly and does not necessarily converge to the optimal partition. In our opinion, the major weakness of the micoaggregation methods lies in the fact that they apply the k-partitioning process without studying the distribution of the input data and their correlation. We think that, if there is a step to add, in order to converge to the optimal k-partition, it should be applied *before* the partitioning process. This step should analyze the similarity between the input data, in order to decide which data should be gathered in a same group.

This study proposes a microaggregation algorithm, called MICRO-PFSOM, based on fuzzy possibilistic clustering [1]. The proposed algorithm aims to generate the optimal partition, *i.e,* maintains the trade-off between privacy and data utility, even when handling noisy data and outliers. Doing so, the MICRO-PFSOM algorithm splits the original microdata into a set of disjoint sub-microdata. Then, microaggregation process can be applied independently on each sub-microdata formed by similar data. Thereby, we can ensure a decrease of information loss.

The remainder of this paper is organised as follows. In Sect. 2, we review previous microaggregation methods. Section 3 thoroughly describes the Micro-PFSOM. Section 4 discusses the encouraging results of the experimental study. Finally, the conclusion and issues for future work are sketched in Sect. 5.

2 Related Work and Problem Statement

Consider a dataset with n records and d numerical attributes. That is, each record is a d-dimensional data point in a d-dimensional space. Microaggregation involves to find the k-partition, *i.e.* the appropriate combination of n data points that form g groups of at least k size, where each data point belongs to exactly one group. Then, data anonymization process consists in replacing the quasi-identifiers of each data point by those of its representative, *i.e.* the centroid of the group to which it belongs to. Consequently, gathering similar data in the same fixed-size group results low information loss. Note that, k is a given security parameter, the higher the value of k is, the larger the information loss and the lower the disclosure risk are [26]. Selecting the optimal level of anonymity, *i.e.* the choice of the parameter k, was discussed in [8].

An optimal microaggregation aims to maintain at most the homogeneity within the k-partition, which can be a compelling task to achieve, *i.e.* NP-hard problem [10]. This issue has grasped the interest of the literature and a wealthy number of algorithms exist. The MDAV algorithm [12] is the most used for microaggregartion, which operates through an iterative process. Its principle involves computing the centroid of the quasi-identifiers of all input data points. Then, the two extreme data, x_r and x_d, relative to the centroid are extracted. Where x_r is the most distant data vector to the centroid. While x_d is the most distant data vector to x_r. Then, two groups are formed with size k around x_r and x_d, respectively. One group, called G_r, contains the data point x_r and its $(k-1)$ closest data points. The other group, called G_d, which contains x_d, is similarly formed. Such process is repeated until all input data vectors of the original microdata are partitioned. Doing so, the MDAV algorithm generates a k-partition formed by groups having a same cardinality. If the number of input data is not divisible by k, the cardinality of one group, generally the last one, ranges between k and $2k - 1$. However, the obtained partition, by the latter process, may lack flexibility for adapting the group size constraint to the distribution of the data vectors. To illustrate such limit, let X, which is exposed in Table 1, be an original microdata. Each data vector $x_i \in X$ is characterized by a two-dimensional quasi-identifier set, composed by the attributes *ZIP code* and *Age*, and a set of p confidential attributes.

Table 1. The original microdata X

	Quasi-identifiers		Confidential attributes		
	ZIP code	Age	Attribute$_1$	...	Attribute$_p$
x_1	1011	25	$a_{1(1)}$	...	$a_{1(p)}$
x_2	1007	22	$a_{2(1)}$	...	$a_{2(p)}$
x_3	1025	40	$a_{3(1)}$	...	$a_{3(p)}$
x_4	1032	42	$a_{4(1)}$	...	$a_{4(p)}$
x_5	1008	23	$a_{5(1)}$	...	$a_{5(p)}$
x_6	1012	26	$a_{6(1)}$	...	$a_{6(p)}$
x_7	1010	24	$a_{7(1)}$	...	$a_{7(p)}$
x_8	1036	40	$a_{8(1)}$	...	$a_{8(p)}$
x_9	1040	43	$a_{9(1)}$	...	$a_{9(p)}$
x_{10}	1013	27	$a_{10(1)}$	...	$a_{10(p)}$
x_{11}	1050	56	$a_{10(1)}$	...	$a_{10(p)}$

By setting the parameter k equal to 3, the MDAV algorithm starts by computing the center of all input data *i.e.* $c = (1025, 37)$. The two data vectors x_{11} and x_2 having the longest squared Euclidean distance are selected. Where x_{11} corresponds to the furthest data vector to the center, while x_2 is the most distant data vector of x_{11}. Two groups of cardinality 3 are formed, around the latter pair of data vectors, gathering their $k - 1$, *i.e.* 2, nearest data. Since the cardinality of the remaining data is less than $2k$, thus they will form a same group. Thereby, the final 3-partition, given by Table 2, is composed by the following groups: $G_1 = \{x_8, x_9, x_{11}\}$, $G_2 = \{x_3, x_4, x_{10}\}$ and $G_3 = \{x_1, x_2, x_5, x_6, x_7\}$. However, the latter partition is not optimal in terms of within-groups homogeneity. For example, the two data vectors x_3 and x_4, with the quasi-identifiers (*ZIP code*, *Age*), are respectively equal to $(1025, 40)$ and $(1032, 42)$, have been gathered in a same group of data having a *ZIP code* varying between 1011 and 1013, and their *Age* attribute ranges between 25 and 27. In fact, an optimal 3-partition of X would be $G_1 = \{x_8, x_9, x_{11}\}$ and $G_2 = \{x_2, x_5, x_7\}$ and $G_3 = \{x_1, x_3, x_4, x_6, x_{10}\}$.

Table 2. The 3-partition of X obtained by the MDAV algorithm

		Quasi-identifiers		Confidential attributes		
		ZIP code	Age	Attribute$_1$	...	Attribute$_p$
G_1	x_8	**1036**	**40**	$a_{8(1)}$	...	$a_{8(p)}$
	x_9	**1040**	**43**	$a_{9(1)}$	...	$a_{9(p)}$
	x_{11}	**1050**	**56**	$a_{10(1)}$	...	$a_{10(p)}$
G_2	x_2	1007	22	$a_{2(1)}$	...	$a_{2(p)}$
	x_5	1008	23	$a_{5(1)}$	...	$a_{5(p)}$
	x_7	1010	24	$a_{7(1)}$	...	$a_{7(p)}$
G_3	x_1	1011	25	$a_{1(1)}$	...	$a_{1(p)}$
	x_3	1025	40	$a_{3(1)}$	...	$a_{3(p)}$
	x_4	1032	42	$a_{4(1)}$	...	$a_{4(p)}$
	x_6	1012	26	$a_{6(1)}$	...	$a_{6(p)}$
	x_{10}	1013	27	$a_{10(1)}$	...	$a_{10(p)}$

In order to improve the results of fixed-size heuristic, in terms of homogeneity, several methods have been proposed offering a given freedom of adapting the distribution of the input data within the k-partition. This is accomplished by allowing the cardinality of groups varying between k and $2k - 1$. In [9], the partitioning process is performed by building *one* fixed size group, at each iteration. Then, the latter group is extended by adding its closest unassigned data according to a *gain factor*. In [7,20] a two-phase partitioning process is applied. The first phase aims to build the fixed-size groups. Then, the latter are tuned in a second phase, by decomposing or merging the formed groups. However, in real-life datasets the poor homogeneity within the generated partition can be significant, especially in noisy surroundings. So, the refinement step could be costly and does not necessarily converge to the optimal partition.

In [27] the authors proposed the *Adaptive at least k fuzzy-c-means* algorithm, that introduces fuzzy clustering within a microaggregation process to build all groups simultaneously. This algorithm is an adaptive and recursive variation of the well-known FCM algorithm [16]. At a glance, it consists in applying, first, the FCM algorithm to build a fuzzy partition. Then, if the obtained clusters are too small, *i.e.* the minimal cardinality is less than k, then the adaptive step is applied again by updating both parameters of FCM. The adaptive step is applied until the smallest cardinality of clusters is reached, which is equal to k. However, this algorithm is expensive, in terms of execution time. In fact, to achieve such a stabilization the algorithm has to perform a high number iterations. Thus, the costly time complexity constitutes a serious hamper for its effective use.

To sum up, the challenge in microaggregation is to design good heuristics for the multi-dimensional microdata, where *goodness* refers to combining high group homogeneity and computational efficiency. Note that, no previous work in the literature has addressed the impact of noisy data on the microaggregation quality. In fact, real datasets are characterized by the presence of noisy data and outliers, which can directly influence the obtained anonymized microdata. The aim of this paper is to propose a new algorithm for microaggregation, based on fuzzy possibilistic clustering [1]. Our approach's aim is twofold: *(i)* build homogeneous k-partition by using a fuzzy possibilistic clustering method to reduce the influence of noisy data and outliers; *(ii)* reduce the disclosure risk, by maintaining the constraint of k-anonymity model. In the following, we detail our proposed approach aiming to cope with the above mentioned drawbacks.

3   The MICRO-PFSOM: A New Algorithm Aiming at Generating the Optimal Partition of Microaggregation

The aim of the proposed microaggregation algorithm, called MICRO-PFSOM, is to generate an anonymized microdata subject to be released for analysis purpose. Doing so, the proposed algorithm relies on fuzzy possibilistic clustering principle, in order to: *(i)* revel the similarity between the input data, *(ii)* decrease the influence of noisy data and outliers during the k-partitioning process.

3.1   General Principle of the MICRO-PFSOM Algorithm

The main idea of the MICRO-PFSOM algorithm consists in applying an *hybrid* microaggregation by splitting the original microdata X into a set of *disjoint sub-microdata, i.e.* $X = \{X_1, X_2, \ldots, X_l\}$. Then, the anonymization process can be applied independently on each sub-microdata, while maintaining the data utility and avoiding identity disclosure. In fact, cluster analysis or clustering is a useful means to meet such purpose. Clustering is a process of partitioning a set of data objects into a set of *meaningful* sub-classes, called *clusters*, based on the information describing the data or their relationships [5]. The goal is that the data in a group will be similar (or related) to one other and different from (or unrelated to) the data in other groups. By adopting such approach, we can apply microaggregation *independently* on each sub-microdata. That is, from each sub-microdata X_i, where $i = \{1, \ldots, l\}$, a k_i-partition is trained by microaggregation process. Then, the anonymous microdata is obtained from the generated k_i-partition, $i = \{1, \ldots, l\}$. Note that, applying *independently* microaggregation can reduce the information loss, since the partitioning process is applied on homogeneous records, *i.e.* sharing the same characteristics.

In a nutshell, the proposed MICRO-PFSOM algorithm, sketched in Algorithm 1, follows the following steps:

1. *Splitting step*: This step aims at dividing the original microdata into disjoint sub-microdata
2. *Microaggregation step*: This step consists in applying the partitioning process into the set of disjoint microdata.
3. *Merging step*: In this step, the generated partitions, of the previous step, are used in order to train anonymous microdata, subject to be released

---

Algorithm 1. The general principle of the MICRO-PFSOM algorithm

---

Input: X : The original microdata
Output: X' : The anonymized microdata of X
1 **Begin**
2 Split the microdata X into c disjoint sub-microdata
   $X = Xid_1 \cup Xid_2 \cup \ldots \cup Xid_c$.
3 Let $\{\bar{x}_{id(1)}, \bar{x}_{id(2)}, \ldots, \bar{x}_{id(c)}\}$ be the cluster centres of their respective
   sub-microdata $\{Xid_1, Xid_2, \ldots, Xid_c\}$.
4 **Foreach** *sub-microdata* $Xid_j \in \{Xid_1, Xid_2, \ldots, Xid_c\}$, $\forall j \in \{1, \ldots, c\}$ **do**
5 $\quad\lfloor X'_j \leftarrow \text{Microaggregation\_process}(Xid_j, \bar{x}_{id(j)})$
6 $X' = X'_1 \cup X'_2 \cup \ldots \cup X'_c$
7 **End**

Unlike the standard microaggregation methods, the MICRO-PFSOM algorithm integrates a pre-processing step before applying the k-partitioning process, aiming at discovering the data distribution (*Algorithm 1 line* 2). That is, the MICRO-PFSOM algorithm splits the original microdata into disjoint sub-microdata $\{Xid_1, Xid_2, \ldots, Xid_c\}$, while computing their appropriate cluster centres. The latter are used, in a second step, during the anonymization process. The clustering process aims to ensure that data sharing similar characteristic of quasi-identifiers are gathered in a same sub-microdata. Thus, the set of sub-microdata $\{Xid_1, Xid_2, \ldots, Xid_c\}$ corresponds to the c clusters contained in X, where the clustering process is performed according to the quasi-identifier attributes. In this way, the k-partitioning process of the microaggregation can be applied independently on each sub-microdata (*Algorithm 1 lines* 4−5), by using the MDAV algorithm due its simplicity. Accordingly, the risk of gathering dissimilar data in a same fixed-size group will be eliminated. Afterwards, the anonymized microdata X' of X, is considered simply as the union of anonymized sub-microdata X'_j obtained from each sub-microdata Xid_j (*Algorithm 1 line* 6).

Therefore, the main challenge of the MICRO-PFSOM algorithm is to find the suitable set of sub-microdata $\{Xid_1, Xid_2, \ldots, Xid_c\}$, contained in the original microdata X, and estimate rightly their centres even in noisy surroundings. In the following we present the fuzzy possibilistic clustering algorithm, called PFSOM, aiming to achieve the latter purpose.

3.2 Fuzzy Possibilistic Clustering for Microaggregation

Given a dataset X including n data vectors and c the number of clusters, the PFSOM algorithm aims to assign each data vector to its suitable cluster, in such a way that data with similar quasi-identifiers are gathered in a same cluster. Note that, the parameter c designs the optimal number of clusters, which is estimated by a multi-level approach [2].

The PFSOM algorithm relies on fuzzy possibilistic clustering in order to decrease the influence of noisy data. In fact, real datasets are characterized by the presence of noisy data and outliers, which can directly influence the obtained data clusters. Fuzzy clustering is a useful means to partition a dataset in noisy surroundings [5]. In fact, fuzzy clustering relies on fuzzy sets [28] allowing the data objects to belong to several clusters *simultaneously* with different degrees of membership [16]. Furthermore, these membership degrees offer a much finer degree of detail of the data model. In this respect, the *Fuzzy Self Organising Map* FSOM algorithm [17], is one of the popular data clustering approaches, owe to its effectiveness for clustering high dimensional datasets. However, the FSOM algorithm heavily relies on a probabilistic constraint to compute the membership values of data to the clusters [1]. That is, the membership of a data point across cluster sum to 1 [16]. However, such constraint can result membership values sensitive to noise [18]. To mitigate such effect, possibilistic clustering has been proposed [18]. Its originality consists in interpreting the membership values as degrees of the possibility belonging the data points to the clusters. That is, it reformulates the fuzzy clustering problem in such a way to generate memberships that have a typicality interpretation.

Therefore, the PFSOM algorithm aims to extend the traditional fuzzy clustering FSOM algorithm, by integrating both of the concept of typicality and membership values during the clustering process. In fact, to classify a data point, a cluster centroid has to be the closest one to the data point, and this what aims fuzzy clustering by using a probabilistic constraint, *i.e.* membership values [16]. In addition, for estimating the centroids, the possibilistic constraint, *i.e.* typicality values, is used for mitigating the undesirable effect of outliers [18].

To split a dataset into c clusters, the PFSOM algorithm adopts the partitioning process illustrated in Algorithm 2. It starts by initializing the cluster centres (Algorithm 2 *line* 2). Then, the prototypes of the latter are adjusted during a learning process. That is, the estimation of the cluster centres is achieved through an iterative process. In each iteration, the prototype of each cluster center c_j is updated according to the membership and typicality values of all data to that cluster (Algorithm 2 *line* 8). We should mention that, when the number n of data points is large, the typicality values computed will be very small. Thus, the typicality values may need to be scaled up. Doing so, the PFSOM algorithm integrates two user-predefined parameters in order to control the relative importance of fuzzy membership and typicality values during the learning process. Doing so, the proposed algorithm defines a learning rate as follows:

$$\alpha_{ij}(t) = a \times \mu_{ij}^{m_t} + b \times t_{ij}^{\eta_t} \tag{3}$$

where μ_{ij} and t_{ij} denote, respectively, the value of fuzzy membership and typicality of a given data x_i to a cluster center c_j. The constants a and b define the importance of, respectively, the fuzzy membership and the typicality values in the learning rate. Note that, the membership value represents the degree to which a data point x_i belongs to a given cluster c_j. Such value is measured according to distances between x_i to all cluster centres, as defined by Eq. 1 (Algorithm 2 *line* 6) [15]. However, the typicality of a data point to a given cluster represents its resemblance to the other data points belonging to the same cluster, *i.e.* internal

---

Algorithm 2. The PFSOM algorithm

Input:
  - X : the training data.
  - max_{it} : the maximal iterations number for training process.
  - m_0 and η_0 : the fuzzier parameters.
  - a and b: the parameters which control the membership and typicality values.
  - ε : error threshold.

Output: The fuzzy partition

1 **Begin**

2   Initialize the prototype of the cluster centres

3     $t = 1$

4   **While** $t \leq max_{it}$ *and stability condition is not reached* **do**

5   Input the training data $X = \{x_1, x_2; \ldots ; x_n\}$

6   Compute the membership values by using the following equation :

$$\mu_{ij} = \left(\sum_{k=1}^{c} (\frac{\|x_i - v_j\|}{\|x_i - v_k\|})^{\frac{2}{m_t - 1}} \right)^{-1} \tag{1}$$

and

$$t_{ij} = \left(\sum_{k=1}^{n} (\frac{\|x_i - c_j\|}{\|x_k - c_j\|})^{\frac{2}{\eta_t - 1}} \right)^{-1} \tag{2}$$

7   where

$$m_t = m_0 - t \times \frac{m_0 - 1}{max_{it}} \ , \quad \eta_t = \eta_0 - t \times \frac{\eta_0 - 1}{max_{it}}$$

8   Update the prototype of the cluster centres according to the following equation :

$$c_j(t) = c_j(t-1) + \frac{\sum_{i=1}^{n} (a \times \mu_{ij}^{m_t} + b \times t_{ij}^{\eta_t}) \|x_i - c_j(t-1)\|}{\sum_{i=1}^{n} (a \times \mu_{ij}^{m_t} + b \times t_{ij}^{\eta_t})}$$

9   Determine the stability condition of the cluster centres

$$max\{\|c_j(t) - c_j(t-1)\|\} < \varepsilon$$

10        $t = t + 1$

11 **End**

---

resemblance. That is, the belonging of a data point x_i to a cluster c_j, depends on the distance from x_i to c_j relative to the distances of all data to that cluster [24]. This is defined by Eq. 2 (Algorithm 2 *line* 6).

The process of updating cluster centres as well as the membership and typicality values is repeated until the stability condition is fulfilled or the predefined number of iterations is achieved. Then, the learning process comes to an end (Algorithm 2 *line* 9).

Worthy to mention that our main purpose is to split the dataset into c disjoint clusters. However, the PFSOM algorithm generates a fuzzy possibilistic partition, in which the training data belong to *all* c clusters but with different membership degrees. In order to extract disjoint clusters, *i.e.* each data should belong to a unique cluster, the PFSOM algorithm simply adopts a *defuzzification* process to convert the fuzzy possibilistic clusters into crisp ones. Doing so, the PFSOM algorithm assigns a given data object to the cluster for which the data has the largest membership value. If a data object has equal membership values to more than one cluster, then that data can be assigned to a random cluster. Perhaps, some questions may arise: *"What is the great interest of using fuzzy possibilistic clustering and followed by a defuzzification process, while the main goal consists in splitting the original data into disjoint clusters ? Why not just use crisp clustering in order to achieve our purpose ?"*. Indeed, our goal is to split the original dataset into disjoint clusters, while computing their appropriate cluster centres, which will be used in a second step for the anonymization process. We should remember that traditional microaggregation methods build a fixed-size groups of homogeneous data points. The homogeneity is measured in relation to the centroid. Thus, estimating rightly the centres of the generated groups, can decrease the information loss during the anonymization process. That is why it is preferable to compute the cluster centres by using fuzzy possibilistic clustering.

3.3   Illustrative Example

To support our idea, let's return to the previous example of microdata X, given in Table 1 (page 4). We noted that applying the k-partitioning process without evaluating the similarity of data can lead to generate a non-optimal k-partition, as illustrated in Table 2 (page 5), which requires its refinement in a further step. Then, we propose to apply the k-partitioning process in hybrid manner, *i.e.* per block of similar data, and compare the obtained k-partition with that resulted above. To discover the distribution of the quasi-identifiers of the microdata X the PFSOM algorithm is applied. Thereby, the optimal fuzzy partition of X is given in Table 3 (page 11). Once the clustering process is applied on the microdata X, the centres of the quasi-identifiers are extracted and the membership values of the input data to the latter centres are computed, which are given in Table 3 (page 11). Then, a defuzzification process is applied on the fuzzy partition, yielding to obtain a disjoint sub-microdata, as shown Fig. 1. It should be noted that, if we apply the standard microggregation algorithm MDAV, independently on each sub-microdata, such as shown in Fig. 1, we can achieve the optimal k-partition that maximises the within groups homogeneity. This is accomplished without applying any refinement process.

Table 3. The fuzzy partition of the microdata X

	Cluster centres: $c_j(ZIPcode, Age)$	
	$c_1 = (1010, 25)$	$c_2 = (1036, 44)$
	Membership values	
x_i(ZIP code, Age)	μ_{i1}	μ_{i2}
$x_1(1011, 25)$	**0.999658**	0.000342
$x_2(1007, 22)$	**0.986407**	0.013593
$x_3(1025, 40)$	0.277774	**0.722226**
$x_4(1032, 42)$	0.042399	**0.957601**
$x_5(1008, 23)$	**0.993123**	0.006877
$x_6(1012, 26)$	**0.995970**	0.004030
$x_7(1010, 24)$	**0.999324**	0.000676
$x_8(1036, 40)$	0.021639	**0.978361**
$x_9(1040, 43)$	0.006887	**0.993113**
$x_{10}(1013, 27)$	**0.987214**	0.012786
$x_{11}(1050, 56)$	0.105167	**0.894833**

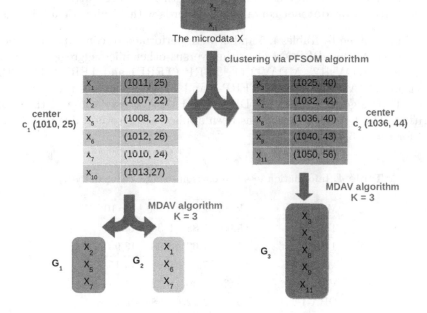

Fig. 1. The partition of X which maximises the within homogeneity

4   Experimental Results

In this section, we discuss the experimental results, of the proposed MICRO-PFSOM algorithm, and its accuracy on real-life datasets using the standard measure, $i.e.$ Information Loss (IL) [11]. In fact, the quality of a microaggregation method can be obtained from information loss, due to the anonymization of the original data. The latter measure computes the mean variation between the original and the perturbed version of a record x_i, by the following formula:

$$IL = \sum_{i=1}^{n} \left(\frac{1}{q} \times \sum_{j=1}^{q} \frac{|x_{ij} - x'_{ij}|}{\sqrt{2}S_j} \right) \tag{4}$$

where S_j is the standard deviation of the jth variable in the original data. The lower the IL is, the lower is information loss and the higher is the utility of the anonymized data.

We adopt in our experiments the following real datasets, which have been used as benchmarks in previous studies to evaluate various microaggregation methods.

– **The Census dataset** contains 1080 records with 13 numeric attributes.
– **The EIA dataset** contains 4092 records with 11 numeric attributes.
– **The Tarragona dataset** contains 834 records with 13 numeric attributes.

Experiments, given by Tables 4, 5 and 6, were performed to compare the performance of the MICRO-PFSOM algorithm versus other microaggregation methods, namely MDAV [12], MDAV-2 [9], TFRP (TERP1 and TERP2) [7], DBA (DBA1 and DBA2) [20]. Where, TFRP-1 and TFRP-2, denotes the two stages of the TFRP algorithm. As well, DBA-1 and DBA-2 refer those of the DBA algorithm. The k-anonymization has been applied using several different values for k (for $k = 3, 5, 10$).

Table 4. Information loss comparison using Census dataset.

Method	k = 3	k = 4	k = 5	k = 10
TFRP-1	5.93	7.88	9.35	14.44
TFRP-2	5.80	7.63	8.98	13.95
MDAV	5.69	7.49	9.08	14.15
VMDAV	5.65	7.40	9.12	13.94
DBA-1	6.14	9.12	10.84	15.78
DBA-2	5.58	7.59	9.04	13.52
MICRO-PFSOM	**5.42**	**6.78**	**8.54**	**13.39**

Table 5. Information loss comparison using EIA dataset.

Method	k = 3	k = 4	k = 5	k = 10
TFRP-1	0.53	0.66	1.65	3.24
TFRP-2	0.43	0.59	0.91	2.59
MDAV	0.48	0.67	1.66	2.59
VMDAV	0.41	0.58	0.94	3.83
DBA-1	1.09	0.84	1.89	4.26
DBA-2	0.42	0.55	0.81	2.08
MICRO-PFSOM	**0.35**	**0.42**	**0.72**	**1.82**

Table 6. Information loss comparison using Tarragona dataset.

Method	k = 3	k = 4	k = 5	k = 10
TRFP-1	17.22	19.39	22.11	33.18
TRFP-2	16.88	19.18	21.84	33.08
MDAV	16.93	19.54	22.46	33.19
VMDAV	16.68	19.01	22.07	33.17
DBA-1	20.69	23.82	26.00	35.39
DBA-2	16.15	22.67	25.45	34.80
MICRO-PFSOM	**15.99**	**18.02**	**20.03**	**29.82**

Tables 4, 5 and 6 show that MICRO-PFSOM works particularly well for Census, EIA and Tarragona datasets, with either the lowest or almost the lowest information loss. As it is expected, the larger the k value is, the larger the IL metric. For example, by increasing the value of k from 3 to 10, the IL of the protected Census dataset degrades from 5.42 to 13.39. Thus, the difference between the original and the protected dataset increases with larger k. In principle, when the IL value increases, the statistical utility of the protected dataset decreases, concurrently it is more difficult for an intruder to link the protected values with the original ones. This is confirmed on all dataset, regardless the microaggregation algorithm.

By examining the performance of the microaggregation methods, we can notice that applying an hybrid microaggregation can be beneficial to decrease the information loss. For example on EIA dataset, by setting the privacy parameter equal to 5, the information loss generated by the MDAV algorithm is equal to 1.66. By allowing a variable size of groups, the VMDAV algorithm improves the homogeneity within the k-partition. In fact, the information loss generated by the VMDAV is equal to 0.94 (almost equals to that MDAV algorithm). The two-phases algorithms TRFP and DBA can decrease the information loss through the second phase (TRFP-2 and DBA-2), which is respectively equal to 0.91 and 0.81. While, our proposed algorithm was able to improve the homogeneity within the fixed-size groups without any refinement steps. In fact the information loss is decreased to 0.72. This is particularly noticed regardless the training dataset and the value of the privacy parameter k.

5   Conclusion

In this paper, we introduced a new microaggregation method based on fuzzy possibilistic clustering, called MICRO-PFSOM. The main thrust of this algorithm stands in its decreasing the information loss during the anonymization process, by adopting an hybrid microaggregation. Experiments carried out on real-life datasets have shown very encouraging results, in terms of information loss. However, this methods may have some drawback on the disclosure limitation side. There is a lack of protection against *attribute disclosure, i.e.* confidential information can also be revealed. That is, if all the individuals within a k-anonymous group share the same value for a confidential attribute, an intruder can learn the confidential attribute, even without re-identification. Some refinements to the basic k-anonymity model have been proposed to improve the protection against attribute disclosure, the well known methods are l-diversity [21], t-closeness [19] and differential privacy [13]. Therefore, extending our proposed algorithm to avoid attribute disclosure risk is commonplace and essential.

References

1. Abidi, B., Yahia, S.B.: Multi-pfkcn: a fuzzy possibilistic clustering algorithm based on neural network. In: Proceedings of International Conference on Fuzzy Systems (FUZZ-IEEE 2013), Hyderabad, India, 7–10 July 2013, pp. 1–8. IEEE (2013)
2. Abidi, B., Yahia, S.B., Bouzeghoub, A.: A new algorithm for fuzzy clustering able to find the optimal number of clusters. In: Proceedings of 24th International Conference on Tools with Artificial Intelligence, ICTAI 2012, Athens, Greece, November 7–9 2012, pp. 806–813. IEEE (2012)
3. Aggarwal, C.C., Yu, P.S.: An introduction to privacy-preserving data mining. In: Aggarwal, C.C., Yu, P.S. (eds.) Privacy-Preserving Data Mining - Models and Algorithms. Advances in Database Systems, vol. 34, pp. 1–9. Springer, Boston (2008)
4. Bacher, J., Brand, R., Bender, S.: Re-identifying register data by survey data using cluster analysis: an empirical study. Int. J. Uncertainty Fuzz. Knowl. Based Syst. **10**(5), 589–607 (2002)
5. Berkhin, P., Dhillon, I.S.: Knowledge discovery: clustering. In: Meyers, R.A. (ed.) Encyclopedia of Complexity and Systems Science, pp. 5051–5064. Springer, New York (2009)
6. Borgman, C.L.: The conundrum of sharing research data. J. Am. Soc. Inf. Sci. Technol. (JASIST) **63**(6), 1059–1078 (2012)
7. Chang, C.C., Li, Y.C., Huang, W.H.: Tfrp: An efficient microaggregation algorithm for statistical disclosure control. J. Syst. Softw. **80**(11), 1866–1878 (2007)
8. Dewri, R., Ray, I., Ray, I., Whitley, D.: On the optimal selection of k in the k-anonymity problem. In: Proceedings of the 24th International Conference on Data Engineering, ICDE 7–12 2008, Cancún, México, pp. 1364–1366. IEEE Computer Society, April 2008

9. Domigo-Ferrer, J., Solanas, A., Martínez-Ballesté, A.: Privacy in statistical databases: k-anonymity through microaggregation. In: Proceedings oh the IEEE International Conference on Granular Computing, GrC 2006, Atlanta, Georgia, USA, 10–12 May 2006, pp. 774–777 (2006)

10. Domingo-Ferrer, J., Martínez-Ballesté, A., Mateo-Sanz, J.M., Sebé, F.: Efficient multivariate data-oriented microaggregation. VLDB J. **15**(4), 355–369 (2006)

11. Domingo-Ferrer, J., Torra, V.: Disclosure risk assessment in statistical data protection. J. Comput. Appl. Math. **164–165**(1), 285–293 (2004)

12. Domingo-Ferrer, J., Torra, V.: Ordinal, continuous and heterogeneous k-anonymity through microaggregation. Data Min. Knowl. Disc. **11**(2), 195–212 (2005)

13. Dwork, C.: Differential privacy. In: Bugliesi, M., Preneel, B., Sassone, V., Wegener, I. (eds.) ICALP 2006. LNCS, vol. 4052, pp. 1–12. Springer, Heidelberg (2006). doi:10.1007/11787006_1

14. Simson, L.: Garfinkel. De-identification of personal information. Technical report, National Institute of Standards and Technologie (2015)

15. Hu, W., Xie, D., Tan, T., Maybank, S.: Learning activity patterns using fuzzy self-organizing neural network. Syst. Man Cybern. Part B **34**(3), 1618–1626 (2004)

16. Ehrlich, R., Bezdek, J., Full, W.: FCM: the fuzzy c-means clustering algorithm. Comput. Geosci. **10**(2–3), 191–203 (1984)

17. Kohonen, T., Schroeder, M.R., Huang, T.S.: Self-Organizing Maps, Chap. 3. Springer, Heidelberg (2001)

18. Krishnapuram, R., Keller, J.M.: A possibilistic approach to clustering. IEEE Trans. Fuzzy Syst. **1**(2), 98–110 (1993)

19. Li, N., Li, T., Venkatasubramanian, S.: t-closeness: privacy beyond k-anonymity and l-diversity. In: Proceedings of the 23rd International Conference on Data Engineering, ICDE 2007, The Marmara Hotel, Istanbul, Turkey, 15–20 April 2007, pp. 106–115. IEEE (2007)

20. Lin, J.L., Wen, T.H., Hsieh, J.C., Chang, P.C.: Density-based microaggregation for statistical disclosure control. Expert Syst. Appl. **37**(4), 3256–3263 (2010)

21. Machanavajjhala, A., Kifer, D., Gehrke, J., Venkitasubramaniam, M.: L-diversity: privacy beyond k-anonymity. ACM Trans. Knowl. Disc. Data (TKDD) **1**(1), 3 (2007)

22. Oganian, A., Domingo-Ferrer, J.: On the complexity of optimal microaggregation for statistical disclosure control. Stat. J. United Nations Econ. Comission Eur. **18**, 345–354 (2001)

23. Ohm, P.: Broken promises of privacy: responding to the surprising failure of anonymization. UCLA Law Rev. **57**(6), 1701–1777 (2010)

24. Pal, N.R., Pal, K., Keller, J.M., Bezdek, J.C.: A possibilistic fuzzy c-means clustering algorithm. IEEE Trans. Fuzzy Syst. **13**(4), 517–530 (2005)

25. Ramachandran, A., Singh, L., Porter, E., Nagle, F.: Exploring re-identification risks in public domains. In: Proceedings of the Tenth Annual International Conference on Privacy, Security and Trust, PST 2012, Paris, France, 16–18 July 2012, pp. 35–42. IEEE (2012)

26. Sweeney, L.: K-anonymity: a model for protecting privacy. Int. J. Uncertainty Fuzz. Knowl. Based Syst. **10**(5), 557–570 (2002)

27. Torra, V., Miyamoto, S.: Evaluating fuzzy clustering algorithms for microdata protection. In: Domingo-Ferrer, J., Torra, V. (eds.) PSD 2004. LNCS, vol. 3050, pp. 175–186. Springer, Heidelberg (2004). doi:10.1007/978-3-540-25955-8_14

28. Zadeh, L.A.: Fuzzy sets. Inf. Control **8**, 338–353 (1965)

Lightweight Privacy-Preserving Task Assignment in Skill-Aware Crowdsourcing

Louis Béziaud[1,3](✉), Tristan Allard[1,2], and David Gross-Amblard[1,2]

[1] Univ. Rennes 1, Rennes, France
[2] IRISA, Rennes, France
{tristan.allard,david.gross-amblard}@irisa.fr
[3] ENS Rennes, Rennes, France
louis.beziaud@ens-rennes.fr

Abstract. Crowdsourcing platforms dedicated to work are used by a growing number of individuals and organizations, for tasks that are more and more diverse, complex, and that require very specific skills. These highly detailed worker profiles enable high-quality task assignments but may disclose a large amount of personal information to the central platform (*e.g.*, personal preferences, availabilities, wealth, occupations), jeopardizing the privacy of workers. In this paper, we propose a lightweight approach to protect workers privacy against the platform along the current crowdsourcing task assignment process. Our approach (1) satisfies differential privacy by letting each worker perturb locally her profile before sending it to the platform, and (2) copes with the resulting perturbation by leveraging a taxonomy defined on workers profiles. We overview this approach below, explaining the lightweight upgrades to be brought to the participants. We have also shown (full version of this paper [1]) formally that our approach satisfies differential privacy, and empirically, through experiments performed on various synthetic datasets, that it is a promising research track for coping with realistic cost and quality requirements.

Keywords: Crowdsourcing · Task assignment · Differential privacy · Randomized response

1 Introduction

Crowdsourcing platforms are disrupting traditional work marketplaces. Their ability to compute high-quality matchings between tasks and workers, instantly and worldwide, for paid or voluntary work, has made them unavoidable actors of the 21st century economy. Early crowdsourcing platforms did not (and still do not) require strong and specific skills; they include for example Amazon Mechanical Turk[1] (for online micro-tasks), Uber[2] (for car-driving tasks), or TaskRabbit[3]

---

[1] https://www.mturk.com/.
[2] https://www.uber.com/.
[3] https://www.taskrabbit.com/.

© Springer International Publishing AG 2017
D. Benslimane et al. (Eds.): DEXA 2017, Part II, LNCS 10439, pp. 18–26, 2017.
DOI: 10.1007/978-3-319-64471-4_2

(for simple home-related tasks—*e.g.*, cleaning, repairing). Today's crowdsourcing platforms now go one step further by addressing skill-intensive contexts (*e.g.*, general team building[4], collaborative engineering[5]) through the collection and use of fine-grained worker profiles. Such platforms carry the promise to facilitate, fasten, and spread innovation at an unprecedented scale.

However abusive behaviors from crowdsourcing platforms against workers are frequently reported in the news or on dedicated websites, whether performed willingly or not (see, *e.g.*, the privacy scandals due to illegitimate accesses to the geolocation data of a well-known drivers-riders company[6], or the large-scale exposure of workers' identifying and sensitive information—*e.g.*, real name, book reviews, or wish-list— through Amazon Mechanical Turk IDs [8]). The problem is even more pregnant with skill-intensive crowdsourcing platforms since they collect detailed workers' profiles for computing highly accurate matchings (*e.g.*, demographics, encompassive set of skills, detailed past experiences, personal preferences, daily availabilities, tools possessed). We advocate thus for a sound protection of workers' profiles against illegitimate uses: in addition to the necessary compliance with fundamental rights to privacy, it is a precondition for a wide adoption of crowdsourcing platforms by individuals.

Computing the assignment of tasks to workers is the fundamental role of the platform (or at least facilitating it). This paper considers precisely the problem of computing a high-quality matching between skill-intensive tasks and workers while preserving workers' privacy. To the best of our knowledge, this problem has only been addressed by a single recent work [6]. However, this work is based on costly homomorphism encryption primitives which strongly hamper its performances and prevent it to reason about skills within the assignment algorithm (*e.g.*, no use of semantic proximity).

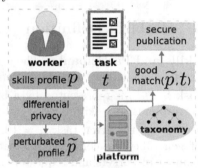

Fig. 1. Our approach to privacy-preserving task assignment

We propose an approach (see Fig. 1) that addresses these issues by making the following contributions:

1. A simple skills model for a worker's profile: a bit vector and a taxonomy.
2. An algorithm run independently by each worker for perturbing her profile locally before sending it to the platform. By building on the proven *randomized response* mechanism [3, 11], this algorithm is privacy-preserving (provides sound *differential privacy* guarantees [5]), and lightweight (no cryptography, no distributed computation, only bitwise operations).
3. A suite of weight functions to be plugged in a traditional assignment algorithm run by the platform and dedicated to increase the quality of matchings

---

[4] https://tara.ai/.
[5] https://makake.co/.
[6] https://tinyurl.com/wp-priv.

performed over perturbed profiles. Our weight functions reduce the impact of the perturbation by leveraging the skills taxonomy, vertically and horizontally, averaging the skills according to their semantic proximity in order to reduce the variance of the differentially-private perturbation. The variance reduction is mathematically sound and does not jeopardize privacy.

4. An experimental study (see the full version [1]), over a synthetic taxonomy and various synthetic datasets, that shows promising preliminary results about the practical adequacy of our approach from the sides of performance and quality.

For space reasons, we give in this paper an overview of our approach. We refer the interested reader to the full version of our work [1] that describes our approach in details, presents its experimental results, and positions it with respect to related work.

The rest of the paper is organized as follows. Section 2 introduces the notions used in our approach and defines more precisely the problem we tackle. We overview our algorithms in Sect. 3 and conclude in Sect. 4 outlining interesting future works.

2   Problem Definition

Skills and Participants. The set of skills that can be possessed by a worker (*resp.* requested by a task) is denoted S. A worker's profile $p_i \in \mathcal{P}$ (*resp.* a task $t_i \in \mathcal{T}$) is represented by a bit vector, *i.e.*, $p_i = \{0,1\}^{|S|}$, where each bit corresponds to a skill $s_j \in S$ and is set to 1 if the given worker has the given skill (*resp.* the given task $t_i = \{0,1\}^{|S|}$ requests the given skill). Without loss of generality, we consider that each requester has a single task and that the number of workers and requesters is the same (*i.e.*, $|\mathcal{P}| = |\mathcal{T}|$). Furthermore, we assume that a skills taxonomy S_T exists[7] [9], structuring the skills according to their semantic proximity, and is such that the skills in S are the leaves of S_T (*i.e.*, no non-leaf node can be possessed nor requested). The non-leaf nodes of the taxonomy are called *super-skills* (Fig. 2).

The platform is essentially in charge of intermediating between workers and requesters. The workers' profiles are considered private while the requesters' tasks are not. The platform holds the set of workers' profiles, *perturbed* to satisfy differential privacy (defined below) and denoted $\widetilde{\mathcal{P}}$, as well as the exact set of requesters' tasks $\mathcal{T}$. All participants, *i.e.*, workers, requesters, and the platform, are considered to be *honest-but-curious*. This means that they participate in the protocol without deviating from its execution sequence (*e.g.*, no message tampering, no data forging) but they will try to infer anything that is computationally-feasible to infer about private data (*i.e.*, the set of workers' non-perturbed profiles $\mathcal{P}$).

---

[7] In practice, skills taxonomies concerning numerous real-life contexts exist today (see, *e.g.*, the Skill-Project http://en.skill-project.org/skills/, or Wand's taxonomies http://www.wandinc.com/wand-skills-taxonomy.aspx).

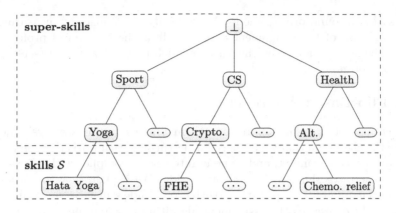

Fig. 2. Example of a skill taxonomy $\mathcal{S}_T$

The Traditional Tasks-to-Workers Assignment Problem. In a traditional context, where workers' profiles are not considered private, the objective of the crowdsourcing platform is to assign a worker to each task such that the overall expected quality is maximized. This well-known combinatorial optimization problem is referred as the assignment problem and can be expressed as a standard linear problem [7] (assuming $|\mathcal{T}| = |\mathcal{P}|$). Assignment algorithms rely on a *weight function* $C \colon \mathcal{T} \times \mathcal{P} \to \mathbb{R}$ in charge of defining the cost of each assignment, *i.e.*, the divergence between the requirements vector of a task and the skills vector of a worker. Common weight functions include the usual distance metrics (*e.g.*, Hamming distance) or disimilarities (*e.g.*, cosine distance). Since our approach is independent from the algorithm, we simply use the HUNGARIAN method [7], a standard academic choice.

Security. We say that our approach is secure against honest-but-curious participants if and only if no participant learns information about the set of non-perturbed profiles $\mathcal{P}$ that has not been perturbed by a differentially-private mechanism, where differential privacy [4] - the current *de facto* standard model for disclosing personal information while satisfying sound privacy guarantees - is defined below. Differential privacy is self-composable [10] and secure under post-processing [5].

Definition 1 (Differential Privacy [4]). *A randomized mechanism* M *satisfies* ϵ-*differential privacy with* $\epsilon > 0$ *if for any possible set of workers' profiles* $\mathcal{P}$ *and* $\mathcal{P}'$ *such that* $\mathcal{P}'$ *is* $\mathcal{P}$ *with one additional profile (or one profile less), and any possible set of output* $O \subseteq Range(M)$,

$$\Pr[M(\mathcal{P}) \in O] \le e^{\epsilon} \times \Pr[M(\mathcal{P}') \in O]. \tag{1}$$

Quality. The inherent information loss due to the differentially-private perturbation impacts the quality of the worker-to-task assignment. This is the price to pay to satisfy a sound privacy model. We quantify this impact by measuring the relative increase of the assignment cost as well as the fraction of profiles that have *all* the skills required by the task to which they are assigned (see the full version for formal definitions [1]).

3 A Flip-Based Approach

This section overviews our approach. We first focus on the workers' side: we describe the algorithm that we propose for perturbing each worker's profile, show its adequacy to our context, and demonstrate that it complies with our security model. Second, we shift to the platform's side. We explain how to reduce the impact of the differentially private perturbation (while still satisfying differential privacy) and we describe the assignment algorithm based on perturbed profiles. Finally, we overview technical means for letting workers fetch their assignment in a secure way in order to complete it. We refer the interested reader to the full version [1] for more details (including the formal proofs).

3.1 At a Worker's Side: Local Perturbation

Building Block: Randomized Response. Randomized response [11] is a simple though powerful perturbation mechanism shown to satisfy differential privacy (see below). Basically, it inputs a single bit (*e.g.*, the answer of an individual to a sensitive boolean question) and flips it randomly according to a well-chosen distribution probability. We describe below the variant called *innocuous question* that we use in this paper and show that it satisfies differential privacy[8]. Let $x \in \{0, 1\}$ be a private value. The randomized response mechanism simply outputs the perturbed value of x, denoted $\tilde{x}$, as follows:

$$\tilde{x} = \begin{cases} x & \text{with probability } 1 - \Pr_{flip} \\ 1 & \text{with probability } \Pr_{flip} \times \Pr_{inno} \\ 0 & \text{with probability } \Pr_{flip} \times (1 - \Pr_{inno}) \end{cases}$$

where Pr_{flip} depends on ϵ (see below) and $Pr_{inno} \in [0, 1]$. We use $\Pr_{inno} = 0.5$ in the rest of the paper, since it minimizes the variation of the estimated value of x after perturbation [3].

Claim. For a given differential privacy parameter $\epsilon > 0$, and a worker's profile made of a single bit to be flipped, the innocuous question randomized response scheme satisfies ϵ-differential privacy if $\Pr_{flip} = \frac{2}{1+e^\epsilon}$ (see [1] for the proof).

Flip Mechanism. Our FLIP mechanism (Algorithm 1) essentially consists in applying the randomized response mechanism to each binary skill of a worker's profile before sending it to the platform and inherits thus its high efficiency. The self-composability properties of differential privacy allow that by distributing ϵ over the bits of the skills vector.

Claim. The FLIP mechanism satisfies ϵ-differential privacy (see [1] for the proof).

---

[8] Any other variant could have been used, provided that it satisfies differential privacy.

Algorithm 1. FLIP (run by each Worker)

Input: The original profile $p = \langle p[1], \dots, p[l] \rangle$, the differential privacy budget
  $\epsilon > 0$.

1 Let Pr_{flip} be the flipping probability: $\text{Pr}_{flip} \leftarrow \frac{2}{1+e^{\epsilon/l}}$.

2 Initiate the perturbed profile: $\widetilde{p} \leftarrow \langle \widetilde{p}[1] = 0, \dots, \widetilde{p}[l] = 0 \rangle$.

3 **for** $1 \le i \le l$ **do**

4 $\quad$ $\lfloor$ $\widetilde{p}[i] \leftarrow \text{RandomizedResponse}(p[i], \text{Pr}_{flip})$.

5 **Return** *The perturbed profile* $\widetilde{p}$

3.2   At the Platform's Side: Task Assignment

Efficient traditional assignment algorithms do not need any modification to work with perturbed profiles, which are bit vectors, exactly as non-perturbed profiles are. The main question is the impact on quality due to our perturbation, and this impact is naturally related to the weight function $\texttt{C} \colon \mathcal{T} \times \mathcal{P} \to \mathbb{R}$ on which assignment algorithms rely. As there is no clear consensus on what is a good weight function for task assignment, in the sequel we recall several reasonable functions, ignoring or using the skill taxonomy. We also propose new weight functions and explain how they could cope with the differentially-private perturbation.

Existing Weight Functions. Numerous weight functions have been proposed as metrics over skills. The *Hamming distance* is a common choice to compute dissimilarity between two vectors of bits but it does not capture the semantics needed for crowdsourcing (*e.g.*, a worker possessing all the skills has a high Hamming distance from a task requiring only one skill, although he is perfectly able to perform it). The weight function proposed in [9] adresses this problem based on a taxonomy. We slightly adapt it and call it the $\texttt{Ancestors}$ weight function (AWF for short).

Definition 2 ($\texttt{Ancestors}$ Weight Function (adapted from [9])). *Let d_{max} be the maximum depth of the taxonomy $\mathcal{S}^T$. Let $\text{lca}(s, s') \in \mathcal{S}^T$ be the lowest common ancestor of skills s and s' in the taxonomy.*

$$\texttt{AWF}(t, \widetilde{p}) = \sum_{s_i \in \widetilde{p}} \min_{s_j \in t} \left(\frac{d_{max} - \text{depth}(\text{lca}(s_i, s_j))}{d_{max}} \right) \qquad (2)$$

Naive Skill-Level Weight Functions. The $\texttt{Missing}$ weight function (MWF for short) between a worker and a task revisits the Hamming distance. It settles a task-to-worker assignment cost that is intuitive in a crowdsourcing context: it is defined as the fraction of skills required by the task that the worker does not have (see Definition 3).

Definition 3 (Missing **Weight Function** (MWF)). MWF: $\mathcal{T} \times \widetilde{\mathcal{P}} \to \mathbb{R}$ *is defined as follows:*

$$\text{MWF}(t, \widetilde{p}) = \sum_{\forall i} t[i] \wedge \neg \widetilde{p}[i] \tag{3}$$

Leveraging the Taxonomy. In realistic contexts, the differentially private perturbation may overwhelm the information contained in the original profiles and make the perturbed profiles be close to uniformly random bit vectors. We cope with this challenging issue by building on the taxonomy $\mathcal{S}_T$. Indeed, the taxonomy allows to group large numbers of skills according to their semantic proximity and to reduce the variance of the perturbation by using group averages [2].

Climbing Weight Function. The Climbing weight function (CWF for short) leverages *the vertical relationship* given by the taxonomy by averaging, for each profile, the skills along the root-to-leaf paths. In other words, before performing the assignment, the platform converts each perturbed profile into a tree, *i.e.,* the same as the taxonomy S_T, and for each node n of the tree, it computes the mean of the skills that appear below n (interpreting the boolean values 1 and 0 as integers). For a given node, this mean is actually a rough estimator of the fraction of descendant skills possessed. We call it *score* below. During an assignment, given a task and a perturbed profile, the Climbing weight function consists essentially in computing the distance between the scores vector of the task and the scores vector of the profile at each level. Definition 4 formalizes the Climbing weight function.

Definition 4 (Climbing **Weight Function** (CWF)). *Let v_i (resp. u_i) denote the scores vector at level i in the tree corresponding to the profile $\widetilde{p}$ (resp. to the task t), and $\mathbf{d}: \mathbb{R}^n \times \mathbb{R}^n \to \mathbb{R}$ be a classical distance function on real-valued vectors (e.g., **Cosine**). Then, CWF: $\mathcal{T} \times \widetilde{\mathcal{P}} \to \mathbb{R}$ is defined as follows:*

$$\text{CWF}(t, \widetilde{p}) = \sum_{\forall i} i \times \mathbf{d}(u_i, v_i) \tag{4}$$

Touring Weight Function. The Touring weight function (TWF for short) leverages *the horizontal relationship* given by the taxonomy, *i.e.,* the neighbouring proximity degree between skills. As described in Definition 5, it returns the average path-length—according to the taxonomy $\mathcal{S}_T$—between the skills required by the task and the skills of the worker's profile. The expected variance reduction comes from the average path-length of the full cartesian product between the skills required by a task and the skills set to 1 in a perturbed profile. The reduction depends on the taxonomy (similarly to Climbing) and on the number of skills averaged.

Definition 5 (Touring **Weight Function** (TWF)). *Let $\rightsquigarrow$ denote the path-length operator between two skills in the taxonomy $\mathcal{S}_T$. Then, TWF: $\mathcal{T} \times \widetilde{\mathcal{P}} \to \mathbb{R}$ is defined as follows:*

$$\text{TWF}(t, \widetilde{p}) = \frac{\sum_{\forall i} \sum_{\forall j} (t[i] \wedge \widetilde{p}[j]) \times (s_i \rightsquigarrow s_j)}{\sum_{\forall i} \widetilde{p}[i] \times \sum_{\forall i} t[i]} \tag{5}$$

3.3 Post-assignment Phase

Workers need a secure way to fetch their own assignment. This can be solved easily by well-known technical means. For example, the platform could post the assignments on the Web (*e.g.,* to a dedicated webpage for each perturbed profile) so that each worker would then access it through a secure web browser (*e.g.,* TOR[9]).

4 Conclusion

We have overviewed in this paper a lightweight privacy-preserving approach to the problem of assigning tasks to workers. Our approach allows each worker to perturb her skill profile locally in order to satisfy the stringent differential privacy model without any need for additional communication or computation cost. We have proposed novel weight functions that can be easily plugged in traditional centralized assignment algorithms, and that are able to cope with the differentially private perturbation by leveraging the presence of a skill taxonomy. Additionally, promising preliminary results of experiments performed over a synthetic taxonomy and synthetic datasets are presented in the full version of the paper [1]. Future works include consolidating experiments (*e.g.,* more profiles and tasks, alternative quality measures), use the taxonomy during the FLIP mechanism, collecting a large-scale skill dataset, and continue exploring the performance/quality tradeoff by designing other profile perturbation strategies (*e.g.,* collaborative perturbation protocols designed to minimize the perturbation).

References

1. Béziaud, L., Allard, T., Gross-Amblard, D.: Lightweight Privacy-Preserving Task Assignment in Skill-Aware Crowdsourcing (Full Version) (2017). https://hal.inria. fr/hal-01534682
2. Bienaymé, I.-J.: Considérations à l'appui de la découverte de Laplace sur la loi de probabilité dans la méthode des moindres carrés. Mallet-Bachelier, Imprim (1853)
3. Blair, G., Imai, K., Zhou, Y.-Y.: Design and analysis of the randomized response technique. J. Am. Stat. Assoc. **110**(511), 1304–1319 (2015)
4. Dwork, C.: Differential privacy. In: Bugliesi, M., Preneel, B., Sassone, V., Wegener, I. (eds.) ICALP 2006. LNCS, vol. 4052, pp. 1–12. Springer, Heidelberg (2006). doi:10.1007/11787006_1
5. Dwork, C., Roth, A.: The algorithmic foundations of differential privacy. Found. Trends Theor. Comput. Sci. **9**(3–4), 211–407 (2014)
6. Kajino, H.: Privacy-Preserving Crowdsourcing. Ph.D. thesis, University of Tokyo (2016)

---

[9] https://www.torproject.org/.

7. Kuhn, H.W.: The Hungarian method for the assignment problem. Naval Res. Logistics Q. **2**(1–2), 83–97 (1955)
8. Lease, M., Hullman, J., Bigham, J.P., Bernstein, M.S., Kim, J., Lasecki, W., Bakhshi, S., Mitra, T., Miller, R.C.: Mechanical turk is not anonymous. SSRN Electron. J. (2013)
9. Mavridis, P., Gross-Amblard, D., Miklós, Z.: Using hierarchical skills for optimized task assignment in knowledge-intensive crowdsourcing. In: Proceedings of WWW 2016, pp. 843–853 (2016)
10. McSherry, F.: Privacy integrated queries: an extensible platform for privacy-preserving data analysis. In: Proceedings of ACM SIGMOD 2009, pp. 19–30 (2009)
11. Warner, S.L.: Randomized response: a survey technique for eliminating evasive answer bias. J. Am. Stat. Assoc. **60**(309), 63–69 (1965)

Clustering Heuristics for Efficient t-closeness Anonymisation

Anne V.D.M. Kayem[✉] and Christoph Meinel

Faculty of Digital Engineering, Hasso-Plattner-Institute for Digital Engineering
GmbH, University of Potsdam, Prof.-Dr.-Helmert Str. 2-3, 14440 Potsdam, Germany
anne@mykayem.org

Abstract. Anonymisation based on t-closeness is a privacy-preserving method of publishing micro-data that is safe from skewness, and similarity attacks. The t-closeness privacy requirement for publishing micro-data requires that the distance between the distribution of a sensitive attribute in an equivalence class, and the distribution of sensitive attributes in the whole micro-data set, be no greater than a threshold value of t. An equivalence class is a set records that are similar with respect to certain identifying attributes (quasi-identifiers), and a micro-data set is said to be t-close when all such equivalence classes satisfy t-closeness. However, the t-closeness anonymisation problem is NP-Hard. As a performance efficient alternative, we propose a t-clustering algorithm with an average time complexity of $O(m^2 \log n)$ where n and m are the number of tuples and attributes, respectively. We address privacy disclosures by using heuristics based on noise additions to distort the anonymised datasets, while minimising information loss. Our experiments indicate that our proposed algorithm is time efficient and practically scalable.

Keywords: Anonymisation · t-closeness · Privacy · Performance

1 Introduction

Published data, such as medical and crime data facilitates data analytics for improved service delivery. Such data releases must be protected to prevent sensitive personal information disclosures due to privacy subversion attacks. There are two categories of privacy subversion attacks, namely, identity and attribute disclosures. Identity disclosures, occur when a person can be uniquely linked to a specific data item in the released dataset. While attribute disclosures occur when the released dataset is combined with other publicly released data sources to uniquely identify individuals based on specific attribute values. Frequently, such vulnerabilities can be exploited to trigger a chain reaction of privacy disclosure attacks. For example, Machanavajjhala et al. [1,2] demonstrated that in a released medical data set, knowledge of the fact that heart attacks are rare among the Japanese could be used to reduce the range of sensitive attributes required to infer a patient's disease.

© Springer International Publishing AG 2017
D. Benslimane et al. (Eds.): DEXA 2017, Part II, LNCS 10439, pp. 27–34, 2017.
DOI: 10.1007/978-3-319-64471-4_3

In t-closeness anonymisation, data privacy is ensured by building on previous anonymisation algorithms namely, k-anonymisation [4–7], and l-diversity [1,8,9] to maximise data utility and at the same time protecting against skewness and similarity attacks [3]. t-closeness addresses both attacks by taking into account global background knowledge as well as possible disclosure levels, based on the distribution of sensitive attributes in both the equivalence classes and the entire dataset. However, as Liang and Yuan [22] have shown, the t-closeness problem is NP-Hard. Having a more efficient solution is practical for real world applications involving large datasets, and low-powered, low-processing devices. Application scenarios emerge on lossy networks and the Internet-of-things (such as opportunistic, and Fog computing networks), where personal data is collected over several devices, may need to be anonymised before it is transferred to a forwarding device.

In this paper, we propose a performance efficient algorithm based on clustering as a classification heuristic to ensure that the distance between sensitive attributes and the cluster centroid is no more than a threshold value of t. The degree of similarity between a cluster and a sensitive attribute is computed by using a combination of severity rankings (cost to privacy due to attribute exposure), the Jaccard coefficient for categorical attributes, and a Euclidean distance for numerical attributes in the quasi-identifier. A high degree of similarity, is captured by a smaller distance from the cluster centroid, and the reverse is true for a low similarity degree. Using these criteria minimises the amount of information that an observer can infer from the published data, based on background knowledge. Our proposed algorithm has an average time complexity of $O(m^2 \log n)$ where n and m are the number of tuples and attributes, respectively.

The rest of the paper is structured as follows. In Sect. 2 we present related work on the general topic of syntactic anonymisation approaches. We proceed in Sect. 3 with a description of our proposed approach to clustering supported t-closeness anonymisation. In Sect. 4, we discuss the performance complexity of our proposed t-closeness clustering algorithm. We follow this in Sect. 5, with some experimental results based on the UCI Adult dataset. In Sect. 6, we offer concluding remarks.

2   Related Work

Sweeney's work [4] on sharing personal data without revealing sensitive information, by k-anonymising the data prior to publication, has triggered a plethora of algorithms aimed at circumventing deanonymisation attacks while at the same time ensuring that the data remains usable for operations such as querying [1–3,5–24]. Privacy preserving data publishing algorithms can basically be classified into two categories namely, syntactic and semantic approaches. Syntactic approaches work well with both categorical and numerical data, and have a well defined data output format. This property allows for confirmation of privacy traits of the data by visual inspections. Adversarial models for constructing deanonymisation attacks are based for the most part on generally available information and inferences drawn from the syntactic and semantic meaning of the underlying data. Examples of algorithms that fall under this category include, k-anonymity [4], l-diversity [1], and t-closeness [3].

The t-closeness algorithm was proposed to alleviate vulnerabilities to skewness and similarity attacks [3] to which both k-anonymisation and l-diversity algorithms are vulnerable. In t-closeness the idea is to structure the anonymised dataset to ensure that the distance between the distribution of a sensitive attribute in a given equivalence class, and the distribution of sensitive attributes in the entire dataset is no more than a threshold value of t [3]. This approach to anonymisation overcomes the limitations of l-diversity in preventing attribute disclosure, and those of k-anonymisation by preventing inference of sensitive attributes, in addition to protecting against background attacks. However, t-closeness anonymisation is performance intensive in the average performance case, and as is the case with k-anonymisation [5,11,19], and l-diversity [20], achieving optimal t-closeness is an NP-Hard problem [22]. Using heuristics, is one method of obtaining near-optimal results.

Anonymisation by clustering has been studied as an approach to improving the performance of k-anonymisation by alleviating the cost of information loss [23,24]. The idea behind these clustering schemes is to cluster quasi-identifiers in equivalence classes of size k, and to avoid using generalisation hierarchies when this impacts negatively on information loss. This property of clustering lends itself well to t-closeness anonymisation as an approach to alleviating the performance demands of anonymising large datasets, particularly when this is done on low-powered, low-processing devices. In the next section we describe our proposed clustering algorithm.

3 t-closeness Clustering

Before we discuss our t-clustering algorithm we first consider aspects such as information loss and sensitive attribute severity weightings which are important in achieving a tradeoff between data utility and privacy. In order to determine information loss, we use a generalisation hierarchy denoted $T(a)$, where $T(a)_{max}$ is the root node or maximum numerical value for an attribute, and $T(a)_{min}$ a leaf node or minimum numerical value. In $T(a)$, P is the set of parent nodes, $T(a)_p$ is the subtree rooted at node $p \in P$, and $T(a)_{tot,p}$ is total number of leaf nodes in the subtree rooted at node $p \in P$. We handle NULL values by classifying them as categorical values.

To calculate information loss for categorical attributes, we consider the proportion of leaf nodes that are transformed to the parent node in the the subtree rooted at p in comparison to the total number of parent nodes P in $T(a)$ excluding the root node. Information loss as $IL(a)$ for categorical attributes is computed with

$$IL(a) = \frac{T(a)_{tot,p} - 1}{P - 1}.$$

and for numerical attributes

$$IL(a) = \frac{T(a)_{max,p} - T(a)_{min,p}}{T(a)_{max} - T(a)_{min}}$$

is used to compare the loss incurred within the subtree in which the value falls, to maximum and minimum values both in the subtree and the entire hierarchy, $T(a)$. Finally, we express the combined information loss over both categorical and numerical attributes for the entire dataset is computed using $IL_{tot} = \sum_{t \in D} \sum_{a \in A} IL_t(a)$.

We introduce a severity weighting scheme to determine the level of loss of privacy due to classifying a tuple with a given sensitive attribute in one cluster over another. For example, a severe illness like "stomach cancer" carries a higher risk of privacy loss than "flu". We denote the sensitive attribute severity weight as $S(s)$ where $s \in S(a)$ and $S(\cdot)$ maps the sensitive attribute to its weight. In this case, the weight is a guideline for the duration, severity of the illness, and/or the likelihood of stigmatisation in the case of exposure. For instance, on a scale of $1 - 10$, S(Cancer) $= 10$, while S(Allergy) $= 4$.

In order to cluster data to ensure t-closeness anonymity with clustering, it is important to determine the minimum size of a cluster required to guarantee a global minimum level of t-closeness that all clusters must adhere to. The clustering algorithm uses a value k_{min} as the minimum cluster size and moves tuples into appropriate clusters, based on both the severity weighting and the distance from the cluster centroid. We define k_{min} as follows: $k_{min} = Max(k_{cons}, min(S_D(\cdot)))$ where k_{cons} is a pre-defined minimum cluster size and $S_D(\cdot))$ represents the set of all sensitive attribute severities for D.

Based on the cluster size, we must determine which tuples to either include or exclude from a cluster. As a first step, we use the relative distance between tuples to decide which tuples to classify in the same cluster. The inter-tuple distance is computed based on both categorical and numerical attributes. The distance between categorical attributes is measured using the Jaccard's coefficient [17], as a similarity measure that is easy to interpret and works well for large datasets with a proportionately small number of NULL or missing values. We define the Jaccard coefficient for our t-clustering algorithm using

$$sim_{t_i,t_j} = \frac{Q_{t_i} \cap Q_{t_j}}{Q_{t_i} \cup Q_{t_j}}$$

where Q_{t_i} and Q_{t_i} are the quasi-identifiers for t_i and t_j, respectively. t_j is the centroid of the cluster that t_i is classified in. The value of sim_{t_i,t_j} varies between 0 and 1, 1 indicates a strong similarity between the tuples, and 0 a strong dissimilarity, based on the quasi-identifier attributes.

To reduce the rate of information loss due to tuple suppressions, we also compute the Euclidean distance between numerical attributes with an n-dimensional space function which is represented as follows:

$$Dist(t_i, t_j) = \sqrt{\left((t_i(a_1) - t_j(a_1))^2 + \ldots + (t_i(a_m) - t_j(a_m))^2 \right)}$$

where a_i is an attribute in Q. Tuples separated by a small Euclidean distance are classified in the same cluster.

Next we consider the sensitive attribute severity weightings and compute the average severity weighting AS_D for D as well as the average severity weighting AS_e for e for a given cluster (equivalence class). The AS_e serves to evaluate the distribution of sensitive attributes in e, while AS_D does this for the entire dataset D, which is similar to how t-closeness decides on tuple classifications based on statistical distributions of sensitive attributes, and also to prevent skewness as well as similarity attacks. The AS_D is used to start the anonymisation process and is computed as follows:

$$AS_D = \frac{\sum S_{t_i}(a)}{\|D\|}$$

where $S_{t_i}(a)$ is the severity weight of sensitive attribute $a \in t_i$. A high AS_D indicates a high level of diversity in the entire dataset. In a similar manner, we compute AS_e as follows:

$$AS_e = \frac{\sum S_{t_i}(a)}{\|e\|}$$

In line with using the t parameter in the t-closeness scheme as a method of optimising dataset utility, we evaluate the level of loss of privacy with respect to information loss in forming the clusters, using a fitness function that is expressed as follows:

$$t = \frac{1}{Max\,(AS_D, IL_{tot})}.$$

Expressing the fitness function in this way captures the fact that when t is low a high degree of loss of either privacy or information is likely to occur, while a high value indicates a good balance between privacy and data utility.

Finally, the Kullback-Leibler distance between AS_e and AS_D is used to determine the level of diversity of sensitive attributes in e with respect to D. Using the Kullback-Leibler distance ($\sum AS_e \log \frac{AS_e}{AS_D}$) serves as an entropy-based measure to quantify the distribution of sensitive attributes both in e and D; and is computed as follows: $Dist\,(AS_e, AS_D) = \sum AS_e \log \frac{AS_e}{AS_D} \leq t$ where $\sum AS_e \log \frac{AS_e}{AS_D} = H(AS_e) - H(AS_e, AS_D)$ such that $H(AS_e) = \sum AS_e \log AS_e$ is the entropy of AS_e and $H(AS_e, AS_D)$ is the cross entropy of AS_e as well as AS_D. When $Dist\,(AS_e, AS_D) \leq t$ the anonymised dataset mimics t-closeness by ensuring that sensitive attributes are classified according to severity of exposure. When $Dist\,(AS_e, AS_D) \nleq t$, we must rerun the whole algorithm to re-compute cluster structures to ensure privacy. In the next section we provide a complexity analysis of the average case running time for our proposed scheme.

4   Complexity Analysis

We know from Liang and Yuan's work [22] that the t-closeness anonymization problem is NP-Hard. With respect to t-clustering, we know that clustering problems are in general NP-Hard. However, with our heuristics we are able to drop the performance cost to $O(n^2 \log m)$ where n and m represent the tuples and attributes in the dataset D. We achieve this by dividing up D into at most n clusters, computations required for classification are in $O(n)$ and the fraction of attributes that are critical for classification are in $O(\log m)$, which results in a total time complexity of $O(n^2 \log m)$.

5   Experiments and Results

In this section we present some results of experiments that we conducted to evaluate the performance of our proposed t-clustering anonymisation scheme. We applied our proposed scheme to the Adult Database from the UCI Machine Learning Repository [25]. We modified the table to include 12 attributes namely: *Age, Race, Gender, Salary, Marital Status, Occupation, Education, Employer, Number of Years of Education, Workclass, Relationship, Native Country.* We included 3 quasi-identifiers, and 2 sensitive attributes. From the base original table (45222 tuples), we extracted dataset sizes to experiment with, and randomly generated an additional 20000 tuples to observe the behaviour of the proposed scheme on larger dataset sizes. With respect to the anonymisation process, we used the following parameters - cluster size: 2, 3, 9, 10, 11, 15, 16, 17, 18; maximum suppression allowed: 0%, 1%; $0.017 \leq t \leq 0.2$. From the table above, we observe that the Kullback-Leibler distance ($Dist\,(AS_e, AS_D)$) between the severity weightings both within the clusters and the dataset are relatively low which indicates a high level of privacy in terms of protection against background knowledge attacks such as skewness and similarity attacks. In this way our proposed scheme inherits the privacy properties of the t-closeness anonymisation algorithm. In terms of performance of our proposed scheme, in line with the theoretical performance discussed in Sect. 4, we note that the time required for clustering grows linearly with the size of the dataset. Finally, the percentage information loss falls between 9% and 25% depending on the number of clusters formed, the cluster size, and the dataset size. Lower information loss percentages occur when smaller and more clusters are formed for a dataset and the reverse happens when larger clusters are formed. The trade-off however, is that smaller clusters result in a higher risk of privacy loss while larger clusters reduce the privacy risk (Table 1).

Table 1. Classification time with respect to dataset size

Dataset size	Cluster size	$Dist\,(AS_e, AS_D)$	Time (ms)
30000	9	0,0015	74
35000	17	0,00117	83
40000	16	0,0015	92
45000	16	0,00035	90
50000	16	0,002	98

6   Conclusion

In this paper we presented a clustering scheme to alleviate the performance cost of t-closeness anonymisation. Basically, what we do is to rank sensitive attributes by a severity weighting and classify tuples to minimise the risk of privacy disclosure of tuples containing high severity weight sensitive attributes. Clustering has the advantage of reducing the need for extensive attribute generalisation in order to classify tuples based on similarity. This is good, in addition, because it reduces the cost of information loss. As we have mentioned earlier, high levels of information loss make datasets unusable in practical situations. By considering severity weightings both for individual clusters and the entire dataset, we mimic the t-closeness principle, of seeking to distribute tuples in ways that ensure that the difference in distributions both within the equivalence classes and the entire dataset, does not surpass a threshold value of t. In this way, our proposed scheme also offers protection against skewness and similarity attacks. Finally, a further benefit of our scheme is that because it is not performance intensive, it can be used on low-powered, low-processing networks for guaranteeing privacy of data under data forwarding schemes.

References

1. Machanavajjhala, A., Kifer, D., Gehrke, J., Venkitasubramaniam, M.: l-diversity: privacy beyond k-anonymity. ACM Trans. Knowl. Discov. Data **1**(1), 1–52 (2007). Article 3
2. Kifer, D., Machanavajjhala, A.: No free lunch in data privacy. In: Proceedings of the 2011 ACM SIGMOD International Conference on Management of Data, SIGMOD 2011, pp. 193–204. ACM, New York (2011)
3. Li, N., Li, T., Venkitasubramaniam, S.: t-closeness: privacy beyond k-anonymity and l-diversity. In: Proceedings of the 23rd International Conference on Data Engineering, pp. 106–115 (2007)
4. Sweeney, L.: K-anonymity: a model for protecting privacy. Int. J. Uncertainty Fuzziness Knowl. Based Syst. **10**(5), 557–570 (2002)
5. Aggarwal, C.: On k-anonymity and the curse of dimensionality. In: Proceedings of the 31st International Conference on Very Large Databases, VLDB 2005, pp. 901–909. VLDB Endowment (2005)
6. Bayardo, R.J., Agrawal, R.: Data privacy through optimal k-anonymization. In: Proceedings of the 21st International Conference on Data Engineering, ICDE 2005, pp. 217–228. IEEE (2005)
7. Liu, K., Giannella, C., Kargupta, H.: A survey of attack techniques on privacy-preserving data perturbation methods. In: Aggarwal, C.C., Yu, P.S. (eds.) Privacy-Preserving Data Mining. Advances in Database Systems, vol. 34, pp. 359–381. Springer, Boston (2008). doi:10.1007/978-0-387-70992-5_15
8. Shmueli, E., Tassa, T.: Privacy by diversity in sequential releases of databases. Inf. Sci. **298**, 344–372 (2015)
9. Xiao, X., Yi, K., Tao., Y.: The hardness of approximation algorithms for l-diversity. In: Proceedings of the 13th International Conference on Extending Database Technology, EDBT 2010, pp. 135–146. ACM, New York (2010)

10. Iyengar, V.S.: Transforming data to satisfy privacy constraints. In: Proceedings of the 8th ACM SIGKDD International Conference on Knowledge Discovery and Data Mining, KDD 2002, pp. 279–288. ACM, New York (2002)
11. Aggarwal, G., Feder, T., Kenthapadi, K., Motwani, R., Panigrahy, R., Thomas, D., Zhu, A.: Anonymizing tables. In: Eiter, T., Libkin, L. (eds.) ICDT 2005. LNCS, vol. 3363, pp. 246–258. Springer, Heidelberg (2004). doi:10.1007/978-3-540-30570-5_17
12. Ciriani, V., Tassa, T., De Capitani Di Vimercati, S., Foresti, S., Samarati, P.: Privacy by diversity in sequential releases of databases. Inf. Sci. **298**, 344–372 (2015)
13. Aggarwal, C.C.: On unifying privacy and uncertain data models. In: Proceedings of the 2008 IEEE 24th International Conference on Data Engineering, ICDE 2008, pp. 386–395. IEEE, Washingtion, D.C. (2008)
14. Aggarwal, C.C., Yu, P.S.: Privacy-Preserving Data Mining: Models and Algorithms, 1st edn. Springer Publishing Company Incorporated, New York (2008)
15. Lin, J.-L., Wei, M.-C.: Genetic algorithm-based clustering approach for k-anonymization. Expert Syst. Appl. **36**(6), 9784–9792 (2009)
16. Shmueli, E., Tassa, T., Wasserstein, R., Shapira, B., Rokach, L.: Limiting disclosure of sensitive data in sequential releases of databases. Inf. Sci. **191**, 98–127 (2012)
17. Aggarwal, C.C.: Data Mining: The Textbook. Springer, Cham (2015)
18. Xiao, Q., Reiter, K., Zhang, Y.: Mitigating storage side channels using statistical privacy mechanisms. In: Proceedings of 22nd ACM SIGSAC Conference on Computer Communications Security, CCS 2015, pp. 1582–1594. ACM, New York (2015)
19. Meyerson, A., Williams, R.: On the complexity of optimal k-anonymity. In: Proceedings of the 23rd ACM SIGMOD-SIGACT-SIGART Symposium on the Principles of Database Systems, PODS 2004, pp. 223–228. ACM, New York (2004)
20. Dondi, R., Mauri, G., Zoppis, I.: On the complexity of the l-diversity problem. In: Murlak, F., Sankowski, P. (eds.) MFCS 2011. LNCS, vol. 6907, pp. 266–277. Springer, Heidelberg (2011). doi:10.1007/978-3-642-22993-0_26
21. Ciglic, M., Eder, J., Koncilia, C.: k-anonymity of microdata with NULL values. In: Decker, H., Lhotská, L., Link, S., Spies, M., Wagner, R.R. (eds.) DEXA 2014. LNCS, vol. 8644, pp. 328–342. Springer, Cham (2014). doi:10.1007/978-3-319-10073-9_27
22. Liang, H., Yuan, H.: On the complexity of t-closeness anonymization and related problems. In: Meng, W., Feng, L., Bressan, S., Winiwarter, W., Song, W. (eds.) DASFAA 2013. LNCS, vol. 7825, pp. 331–345. Springer, Heidelberg (2013). doi:10.1007/978-3-642-37487-6_26
23. Kabir, M.E., Wang, H., Bertino, E., Chi, Y.: Systematic clustering method for l-diversity model. In: Proceedings of the Twenty-First Australasian Conference on Database Technologies, ADC 2010, Brisbane, Australia, vol. 104, pp. 93–102 (2010)
24. Aggarwal, G., Panigrahy, R., Feder, T., Thomas, D., Kenthapadi, K., Khuller, S., Zhu, A.: Achieving anonymity via clustering. ACM Trans. Algorithms **6**(3), 1–19 (2010). ACM, New York
25. Frank, A., Asuncion, A.: UCI machine learning repository (2010). http://archive.ics.uci.edu/ml

Service Computing

A QoS-Aware Web Service Composition Approach Based on Genetic Programming and Graph Databases

Alexandre Sawczuk da Silva[1](✉), Ewan Moshi[1], Hui Ma[1], and Sven Hartmann[2]

[1] School of Engineering and Computer Science, Victoria University of Wellington,
PO Box 600, Wellington 6140, New Zealand
{sawczualex,ewan.moshi,hui.ma}@ecs.vuw.ac.nz
[2] Department of Informatics, Clausthal University of Technology,
Julius-Albert-Strasse 4, 38678 Clausthal-zellerfeld, Germany
sven.hartmann@tu-clausthal.de

Abstract. A Web service can be thought of as a software module designed to accomplish specific tasks over the Internet. Web services are very popular, as they encourage code reuse as opposed to re-implementing already existing functionality. The process of combining multiple Web services is known as Web service composition. Previous attempts at automatically generating compositions have made use of genetic programming to optimize compositions, or introduced databases to keep track of relationships between services. This paper presents an approach that combines these two ideas, generating new compositions based on information stored in a graph database and then optimising their quality using genetic programming. Experiments were conducted comparing the performance of the newly proposed approach against that of existing works. Results show that the new approach executes faster than the previously proposed works, though it does not always reach the same solution quality as the compositions produced by them. Despite this, the experiments demonstrate that the fundamental idea of combining graph databases and genetic programming for Web service composition is feasible and a promising area of investigation.

1 Introduction

A Web service is a software module designed to accomplish specific tasks over the Internet [8], with the capability of being executed in different platforms [3]. Each Web service requires a set of inputs to be provided and produces a set of outputs. Web services by their very nature promote code reuse and simplify the process of information sharing, as developers can simply invoke the Web service as opposed to rewriting the functionality from scratch [7]. There are problems that a single Web service cannot address, and as a result, multiple Web services are required. This process of combining multiple Web services to solve more complex requests is known as Web service composition [6].

© Springer International Publishing AG 2017
D. Benslimane et al. (Eds.): DEXA 2017, Part II, LNCS 10439, pp. 37–44, 2017.
DOI: 10.1007/978-3-319-64471-4_4

Much research has gone into developing approaches that automatically yield the best possible composition for the given task. Some of these approaches utilise a directed acyclic graph (DAG) to represent the interaction between the Web services with regards to the required inputs and produced outputs [4,8,9]. Before they can find suitable compositions, these methods must load a repository of Web services and discover the relationships between the inputs and outputs of different services. Naturally, repeating this process every time a composition is to be created becomes quite onerous. Genetic Programming (GP) approaches have also been applied for composition, though they present limitations such as producing solutions that are not guaranteed to be functionally correct (i.e. solutions that are not fully executable) [4], employing fitness functions that do not take into consideration the Quality of Service (QoS) of the generated solution [8], and containing redundancies in the GP trees, namely, the reappearance of Web services at different levels of the tree [9]. Similarly, there have been previous attempts at using databases (relational and graph databases) to solve the Web service composition problem [5,12]. These approaches are generally efficient, as they store service repositories and dependencies in the database. However, these approaches are typically not effective at performing global QoS optimization.

Given these limitations, the objective of this paper is to propose a QoS-aware Web service composition approach that combines GP with a graph database. The key idea is to manage services and their connections using the database, then use that information to create and optimize solutions in GP. The advantage of combining these two components is that together they prevent the repository from being repeatedly loaded, while also improving the QoS of the solutions produced during the composition process.

2   Background and Related Work

A typical example of an automated Web service composition is the travel planning scenario [10]. In this scenario, the solution can automatically book hotels and flights according to the customer's request. A *composition request R* is provided to the composition system, specifying I_R input information (such as the destination, departure date, and duration of stay) and the desired O_R output information (such as the return ticket and hotel booking). The system then creates a Web service composition that satisfies this request by combining services from the repository. This Web service composition is illustrated in Fig. 1.

In this work, four QoS attributes are considered [11]: availability (A), which is the probability that a service will be available when a request is sent, reliability (R), which is the probability that a service will respond appropriately and timely, cost (C), which is the financial cost associated with invoking the service, and time (T), which indicates the length of time needed for a service to respond to a request. For availability and reliability, higher values indicate better quality, whereas lower values indicate better quality for cost and time. These quality attributes are associated with each atomic service in the repository, and overall QoS attributes can be completed for a composition by aggregating individual QoS values according to the constructs included in the composition [1]:

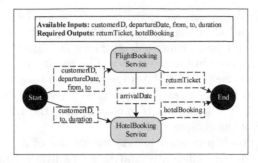

Fig. 1. A Web service composition for the travel planning scenario.

- **Sequence construct:** Web services are connected sequentially, so that some outputs of the first service are used as the inputs of the subsequent one. To calculate the aggregate availability and reliability for this construct, values of all the individual services in the composition are multiplied. To calculate the aggregate cost and time, the values of each individual service are added.
- **Parallel construct:** Web services are independently executed in parallel, so that the inputs and outputs of each Web service are produced independently. The same calculations used for the sequential construct are employed here to calculate the availability, reliability, and cost. However, in the case of total time, we must select the Web service with the highest execution time, as the other services in the construct will be executed in parallel.

QoS values are normalised to ensure their contribution to the fitness calculation is proportional. The lower bounds are obtained from the service in the repository with the smallest value for the attribute in question (except for A_{min} and R_{min}, which are both 0), and the upper bounds are obtained from the service with the largest value (for T_{max} and C_{max}, values are multiplied by the size of the repository as an estimate of the largest possible solution). T and C are offset during the normalisation so that higher values denote better quality.

2.1 Related Work

The work presented in [4] automatically produces compositions by building a GP tree using workflow constructs as the non-terminal nodes and atomic Web services as the terminal nodes. One of the underlying problems of this approach is the random initialisation of the population, which results in compositions that are not guaranteed to be *functionally correct* (i.e. a composition that is fully executable given the available input). The work in [8] represents its candidates as a graph, which reduces the associated complexity from the verification of node dependencies while ensuring functional correctness of solutions. The problem with this approach is that it cannot handle QoS-aware Web service compositions, meaning that the fitness function does not consider the QoS of candidates.

The authors of [5] propose an approach that uses relational databases to store Web services and their dependencies. This approach utilises a database to prevent loading data into memory. It employs an algorithm that creates a path between Web services in the repository. The results are then stored in the relational database for use to satisfy future requests. An issue with this approach is that paths need to be regenerated if a new Web service is introduced in the repository. Additionally, the approach only considers the time property when measuring the QoS of the compositions.

Finally, the approach presented in [12] makes use of a graph database to store services dependencies efficiently and in an easily modifiable way. When receiving a composition request, a subgraph of relevant nodes is filtered from the original dependency graph. Then, a number of functionally correct composition candidates are generated from within this subgraph, and the composition with the best overall QoS is chosen as the solution. Despite being an improvement on the relational database strategy described above, this work does not optimise the QoS attributes of the composition solution.

3   Graph Database and GP Approach

The approach presented in this paper extends the graph database work done in [12], introducing genetic programming as a means of improving the QoS of composition solutions. This process is summarised in Algorithm 1, and the core elements are discussed in the following subsections.

ALGORITHM 1. Steps in the proposed graph database and GP composition approach.

Data: Composition request $R(I_R, O_R)$
Result: Best composition solution found
1 Create a graph database for all available Web services in the repository;
2 From the initial graph database, create a reduced graph database which contains only the Web services related to the given task's inputs and outputs;
3 Initialise a population by creating compositions that can satisfy the requested task from the reduced graph database, transforming them from DAG into a GP tree form;
4 Evaluate the initial population according to the fitness function;
5 **while** *Stopping criteria not met* **do**
6 | Select the fittest individuals for mating;
7 | Apply crossover and mutation operators to individuals in order to generate offspring;
8 | Evaluate the fitness of the newly created individuals;
9 | Replace the least fit individuals in the population with the newly created offspring;
10 **return** Fittest solution encountered

3.1 Fitness Function

The fitness function used to evaluate candidates is based on the one presented in [9] but without penalties, as our DAG-to-tree algorithm ensures that all trees are functionally correct. The values produced by the fitness function range from 0 to 1, with 1 being the best possible solution and 0 being the worst. The fitness function is $Fitness_i = w_1A_i + w_2R_i + w_3T_i + w_4C_i$, where A_i, R_i, T_i and C_i denote the normalized *availability, reliability, execution cost,* and *response time* of the candidate i, and the weights w_i are rational non-negative numbers where $w_1 + w_2 + w_3 + w_4 = 1$. Each weight is associated with a quality attribute, and their values are configured by users to reflect the importance of each attribute.

3.2 Initialisation

The initialisation begins by obtaining a composition in the form of a DAG with a single start node, which is done by querying the graph database [12]. Before the evolutionary process begins the DAG for each candidate is transformed into a corresponding GP tree. We begin by dividing the DAG into layers, which are identified by traversing every node in the DAG and finding the longest path from the start to each node. The length of this path is the node's level (layer number). Figure 2a shows an example of this process. The layer information is then used to build a GP tree. This is done by Algorithm 2, which traverses the layers in reverse and creates sequence, parallel, and terminal nodes depending on the situation. Figure 2b shows a tree version of Fig. 2a.

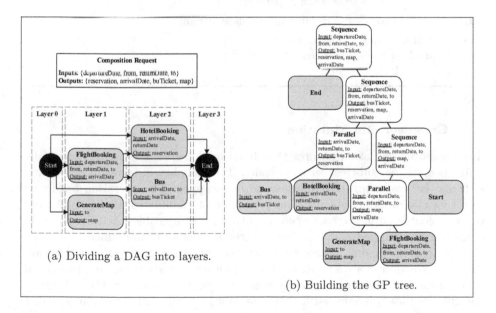

(a) Dividing a DAG into layers.

(b) Building the GP tree.

Fig. 2. Initialisation process.

ALGORITHM 2. Converting the DAG to a tree.

Data: NodeLayers

Result: Converted GP Tree

1 Initialize tree;

2 Create TreeNode previous = null;

3 **for** $i = |NodeLayers|$ *-1; $i >= 0$* **do**

4   Create SequenceNode sequenceCurrent with no parent;

5   **if** *previous is not null* **then**

6   **if** *node in nodelayer i is start node* **then**

7   Create TerminalNode startNode with previous as parent;

8   Add startNode to children of previous;

9   Break;

10   **else**

11   Set parent of sequenceCurrent to previous;

12   Add sequenceCurrent to children of previous;

13   **if** *number of nodes in current layer = 1* **then**

14   Create TerminalTreeNode n with the Web service's name and sequenceCurrent as parent;

15   Add n to children of sequenceCurrent;

16   **else**

17   Create ParallelNode parallel with sequenceCurrent as the parent;

18   **foreach** *Node in layer i* **do**

19   Create TerminalTreeNode n with the Web service's name and parallel as parent;

20   Add n to children of parallel;

21   Add parallel to children of sequenceCurrent;

22   Add sequenceCurrent to the tree;

23   previous = sequenceCurrent;

24   i = i - 1;

3.3 Crossover and Mutation

The crossover operation ensures that functional correctness is maintained by only allowing two subtrees with equivalent functionality in terms of inputs and outputs, i.e. two *compatible* subtrees, to be swapped. The operation begins by selecting two random subtrees from two different candidates. If the inputs and outputs contained in the root of each subtree are compatible, then the two subtrees can be swapped and functional correctness is still maintained, otherwise another pair is sought. In case there are no compatible nodes across two candidates, crossover does not occur. The mutation operation ensures functional correctness by restricting the newly generated subtree to satisfy the outputs of the subtree that was selected for the mutation. The operation randomly selects a subtree from a candidate, then randomly creates a new subtree based on the inputs and outputs of the selected node and following the initialisation principles described earlier. Finally, the selected subtree is replaced with the new subtree.

4   Experimental Design and Results

Experimental comparisons were conducted against GraphEvol [8] and a graph database approach [12]. The evaluation of the candidate solutions produced by these two approaches was performed using the fitness function proposed in this paper. WSC-2008 [2] was the benchmark dataset chosen for the evaluation. Each approach was run 30 independent times on a personal computer with an i7 CPU (3.6 GHz) and 8 GB RAM. In the case of GraphEvol, 500 individuals were evolved for 51 generations, with a crossover probability of 0.8, a mutation probability of 0.1, and a reproduction probability of 0.1. Tournament selection was used, with a tournament size of 2. For the proposed GP approach, 30 individuals were evolved for 50 generations, with a crossover probability of 0.9, a mutation probability of 0.1, and a tournament selection strategy with size 2. Finally, the graph database approach randomly produced 30 solutions for each run, and selected the one with the best quality. For all approaches, fitness function weights were set to 0.25 for each attribute. Experimental results are shown in Table 1, which contains the mean and standard deviation for the execution time and fitness of each approach. For displaying purposes, the QoS attributes of the final composition from each run were re-normalised. This was done by choosing normalisation bounds for each QoS attribute from the set of solutions produced by each approach. Wilcoxon rank-sum tests with a 0.05 significance level were conducted to detect statistically significant differences. Results show that the proposed GP approach required significantly less time to execute than the others for all datasets except Dataset 6, where runs did not conclude even after several hours. Results for those runs are not available. The GP approach produces solutions with significantly lower quality than at least one other approach for each dataset, though its fitness is often not the lowest out of the three.

Table 1. Mean and standard deviation for the execution time and solution fitness for the three approaches. Significantly lower values are indicated using ↓.

Dataset 2008	GP approach Time (ms)	Fitness	Graph database approach Time (ms)	Fitness	GraphEvol approach Time (ms)	Fitness
1	22.44 ± 0.66 ↓	0.46 ± 0.12 ↓	2197.90 ± 329	0.521 ± 0.169	4845.57 ± 315.42	0.645 ± 0.139
2	369.38 ± 231.27 ↓	0.58 ± 0.07 ↓	5347.13 ± 880	0.48 ± 0.152	3699.77 ± 364.57	0.906 ± 0
3	91.94 ± 2.99 ↓	0.36 ± 0.11 ↓	10961.53 ± 790	0.387 ± 0.1	17221.53 ± 764.85	0.176 ± 0.045
4	201.04 ± 73.11 ↓	0.43 ± 0.07 ↓	3885.70 ± 399	0.431 ± 0.066	6076.7 ± 281.58	0.305 ± 0.066
5	61.25 ± 1.68 ↓	0.37 ± 0.11 ↓	4510.6 ± 468	0.403 ± 0.128	10444.2 ± 572.59	0.164 ± 0.046
6	-	-	258503.33 ± 42324	0.407 ± 0.089	22183.53 ± 1639	0.228 ± 0.06
7	190.6 ± 26.89 ↓	0.38 ± 0.09 ↓	17839.77 ± 763	0.457 ± 0.097	20304.37 ± 1257	0.316 ± 0.039
8	1117.09 ± 219.39 ↓	0.39 ± 0.08 ↓	53003.7 ± 4465	0.468 ± 0.091	18567.03 ± 2055	0.315 ± 0.028

5   Conclusions

This paper introduced an automated QoS-aware Web service composition approach that combines graph databases and GP to produce service composition solutions. Experiments were conducted to evaluate the proposed approach against two existing methods, one that purely employs a graph database and another that purely employs an evolutionary computation technique. Results show that the GP approach requires less time to execute, though it does not always match the quality of the solutions produced by other methods. Thus, future work should investigate improvements to the GP component of this approach, in particular the genetic operators, to produce higher fitness solutions.

References

1. van der Aalst, W.M.P., Dumas, M., ter Hofstede, A.H.M.: Web service composition languages: Old wine in new bottles? In: 29th EUROMICRO 2003 Conference on New Waves in System Architecture, 3–5 September 2003, Belek-Antalya, Turkey, pp. 298–307 (2003)
2. Bansal, A., Blake, M.B., Kona, S., Bleul, S., Weise, T., Jaeger, M.C.: WSC-08: continuing the Web services challenge. In: 2008 10th IEEE Conference on E-Commerce Technology and the Fifth IEEE Conference on Enterprise Computing, E-Commerce and E-Services, pp. 351–354. IEEE (2008)
3. Hashemian, S.V., Mavaddat, F.: A graph-based approach to Web services composition. In: 2005 IEEE/IPSJ International Symposium on Applications and the Internet (SAINT 2005), Trento, Italy, 31 January–4 February 2005, pp. 183–189 (2005)
4. Lerina Aversano, M.D., Taneja, K.: A genetic programming approach to support the design of service compositions. In: First International Workshop on Engineering Service Compositions (WESC 2005) (2005)
5. Li, J., Yan, Y., Lemire, D.: Full solution indexing using database for QoS-aware Web service composition. In: IEEE International Conference on Services Computing (SCC), Anchorage, AK, USA, 27 June–2 July 2014, pp. 99–106 (2014)
6. Milanovic, N., Malek, M.: Current solutions for web service composition. IEEE Internet Comput. 8(6), 51–59 (2004)
7. Ni, J., Zhao, X., Zhu, L.: A semantic Web service-oriented architecture for enterprises. In: International Conference on Research and Practical Issues of Enterprise Information Systems (CONFENIS), Beijing, China, 14–16 October 2007, pp. 535–544 (2007)
8. da Silva, A.S., Ma, H., Zhang, M.: GraphEvol: A graph evolution technique for Web service composition. In: Chen, Q., Hameurlain, A., Toumani, F., Wagner, R., Decker, H. (eds.) DEXA 2015. LNCS, vol. 9262, pp. 134–142. Springer, Cham (2015). doi:10.1007/978-3-319-22852-5_12
9. da Silva, A.S., Ma, H., Zhang, M.: Genetic programming for QoS-aware web service composition and selection. Soft. Comput. 20(10), 3851–3867 (2016)
10. Srivastava, B., Koehler, J.: Web service composition - current solutions and open problems. In: ICAPS 2003 Workshop on Planning for Web Services, pp. 28–35 (2003)
11. Yu, Y., Ma, H., Zhang, M.: An adaptive genetic programming approach to QoS-aware Web services composition. In: IEEE Congress on Evolutionary Computation (CEC), Cancun, Mexico, 20–23 June 2013, pp. 1740–1747 (2013)
12. Zhang, Z., Ma, H.: Using Graph Databases for Automatic QoS-Aware Web Service Composition. Technical report ECSTR 16–07, Victoria University of Wellington (2016). http://ecs.victoria.ac.nz/foswiki/pub/Main/TechnicalReportSeries/ECSTR16-07.pdf

Combining Web-Service and Rule-Based Systems to Implement a Reliable DRG-Solution

Idir Amine Amarouche[1(✉)], Lydia Rabia[1(✉)], and Tayeb Kenaza[2(✉)]

[1] Department of Medical and Hospital Informatics, Central Hospital of Army,
Ain Naadja, 16005 Algiers, Algeria
i.a.amarouche@gmail.com, rabia.lydia@protonmail.com
[2] Ecole Militaire Polytechnique, BP 17, BEB, 16046 Algiers, Algeria
ken.tayeb@gmail.com

Abstract. Diagnosis-Related Groups (DRG) is a Patient Classification System that allows classifying inpatients stays among well-known groups in order to estimate the appropriate fees to refund hospitals. Each DRG solution has its own grouping approach. Using data gathered from the Hospital Information System (HIS), it assigns to an inpatient stay the appropriate group called DRG. In this paper, using both Web Service technology and Rule Based Expert System, we develop a rule-based system as a service for handling the grouping solution. This approach provides for hospitals and public/private stockholders the possibility to avoid the stringent software requirements on their local IT infrastructure by reusing a shared and distributed DRG solution. Moreover, combining these two technologies enhance knowledge maintenance and improve its reusability.

Keywords: Diagnosis-Related Groups · Rule-based system · Web Service · Drools

1 Introduction

Started in the USA in 1983, Diagnosis-Related Groups (DRG) is a Patient Classification Systems (PCS) for inpatient stays, which are commonly used as a refunding system to pay hospital activities in many countries [1]. More precisely, DRG system groups together cases that a hospital treats with respect to the used resources. Thus, all patients within the same DRG are expected to have similar resources consumption in terms of costs based on clinical condition. Routinely collected data on patient discharge are used to classify patients into a manageable number of groups that are clinically meaningful and economically homogeneous [2, 3].

Figure 1 shows the process of grouping an inpatient stay. In step (1) treating doctors fill in the patient data management (e.g. HIS) information about the patient stay, namely clinical data (i.e., diagnoses, procedures), demographic data (i.e., age, gender), and resources consumption data (i.e., Length of stay). In step (2) of the process, a clinical coder converts the patient's medical data, contained in Discharge Summary (DS), into a sequence of codes. A clinical coder assigns a sequence of diagnosis and procedure

© Springer International Publishing AG 2017
D. Benslimane et al. (Eds.): DEXA 2017, Part II, LNCS 10439, pp. 45–53, 2017.
DOI: 10.1007/978-3-319-64471-4_5

codes expressed respectively in ICD[1] and in a country-specific classification of procedures. In step (3), the DRG-system maps the sequence of codes and data to a DRG according to a pre-defined schema. The last step is performed by a subsystem called "grouping solution" or "grouper software" [4].

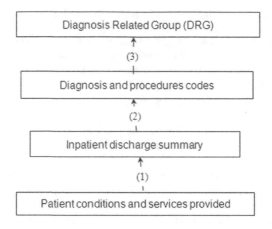

Fig. 1. Information flow for a typical inpatient stay

The grouping solution takes inpatient DS data as input and returns a corresponding DRG as output. It is based on an algorithm that consists of several consecutive predefined checking steps in order to assign to an inpatient stay the DRG that best reflects the corresponding resource consumption. The grouping solution could be represented as function G such that: *DRG = G(age; sex; diagnoses; procedures, etc.)*. This function is implemented and adapted by each health management system according to its needs and its coding systems [5, 6]. For this reason, there is no an advocated approach to design and implement a grouping solution.

Furthermore, grouping solution is subject to continuous changes in term of the algorithm's structure that would capture continuously the changing number of DRGs. According to [2], these changes are mainly influenced by the covered health care services' scope and the availability of new treatment patterns.

The remainder of this paper is structured as follows. Section 2 presents our motivations, challenges and contributions. Sections 3 and 4 demonstrate the proposed architecture and provides explanations about its implementation issues respectively. Section 5 presents the developed prototype to illustrate the capabilities of our approach. Finally, conclusions and perspectives are summarized in Sect. 6.

2   Motivations, Challenges and Contributions

Let's expose a DRG grouping case with the main data processing requirements that grouper software must meet. Let imagine a masculine patient having been admitted in

---

[1] International classification of diseases.

hospital for a dialysis session. The procedure performed is Automated Peritoneal Dialysis (APD) that involves a machine called "cycler", which is used to introduce and drain the dialysis fluid. Figure 2 presents the considered data to explain the grouping case. The set of checking is performed according to the path indicated in the flowchart that is mainly used, as a representation, to describe the grouping algorithm [2]. First, the presence of a session will be checked. Then, the Main Diagnosis (MD) is checked if it is motivated by a session and corresponds to an extra-renal purification. Finally, the procedure is checked if it is a peritoneal dialysis using the machine. If the tests are successful, the case is classified in an appropriate group.

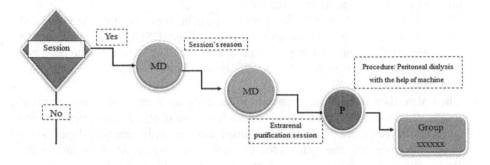

Fig. 2. Flowchart representation of the dialyze session grouping case.

This example illustrates the possibility to express grouping case, in rule-based format. Furthermore, most DRG systems contain between 650 and 2300 groups [2, 3]. Also, in each DRG system changes continually occur in terms of group's number and required data. All this enforce the need for a grouping solution that takes into consideration the following challenges:

- Expressing grouping algorithm, at a high level, by flowchart or decision tree formats help to be understandable by large health practitioners. But, coding it in procedural or Object-Oriented approach can be very difficult. For this, ensuring the evolution and maintaining the coherence of grouping algorithm have to be examined as an essential requirement for grouping solution.
- The deployment of grouping solution is performed by health government agencies or payment providers. For this, providing an executable grouping solution accessible from different stakeholder's information systems has to be addressed as another challenge.

On the basis of the above challenges, there is a real need to support grouping function by a comprehensive and evolving solution. In this paper, we propose a new approach to design a grouping solution as Rule-based system provided as a Web Service (WS).

3 The Proposed Architecture: Overview and Capabilities

In order to describe our proposed architecture, we first explain major functionalities and related modules and their capabilities. The proposed architecture, presented in Fig. 3, presents the main modules, namely, Grouping Rules (GR) editor; GR Converter; GR verifier; GR engine; GR repository and Web Services API. The definition of each module in terms of functionality is described as follow.

- **The GR editor.** It acts as a subsystem that helps to manage and to create new grouping rules which are graphically represented. The rules can be retrieved by their unique name and can be updated thereafter.
- **The GR Converter.** It is an auxiliary tool responsible for converting the visualized rule, from GR editor, into a Domain Specific Language (DSL). It allows mainly the translation of the graphical representation of rules into an executable rule language understandable by a Rule based-system or into an XML format supported by any rule-based expert system.
- **The GR verifier.** It is a module that aims to detect eventual inconsistencies occurring in a set of rules. For this, the adopted approach consists of transforming rules set into a Directed Hypergraphs [7, 8]. This formalism provides several analytical techniques in order to ensure that the updated rules do not introduce any risk of inconsistency. By inconsistency, it is meant redundant, incorrect or conflicting rules.
- **The GR engine.** It is responsible for processing the submitted case for grouping in either unitary or batch mode. Thus, for a single or several DSs, it selects the appropriate grouping rules according to the inference strategy of Rule based-system, the forward chaining in our case [9].

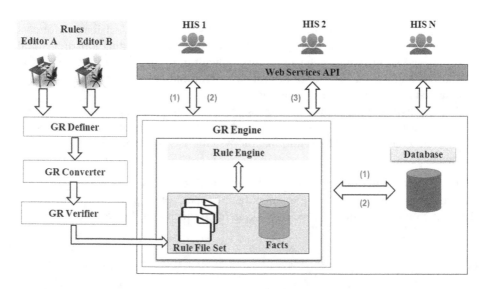

Fig. 3. The Proposed grouping solution architecture

- **The GR repository.** This module stores the rule-related information and consists mainly of "drl" files (drools file) containing the grouping rules. Each file contains a set of rules belonging to a given diagnosis category.
- **Web Services API.** This module provides a set of WSs. The following WS are implemented, namely: DS data loading and uploading, unitary and batch grouping. They are parameterized by HIS identity, DS Identifier, starting date and ending date. Thus, the proposed architecture highlights the following interactions depicted in Fig. 3 as follows: (3–1) Uploading WS provides the possibility to upload DSs from HIS database; (3–2) Grouping service, unitary or batch, invokes the "GR engine" module; (3–3) Assigning DRG group to an inpatient stay is made available for downloading by the client or the concerned HIS.

4 Implementation

4.1 Grouping Cases: Rule Representation and Verification

Rule representation and structuring. A rule is expressed using Horn clauses form, and named by corresponding DRG group. For instance, the interpretation of flowchart of Fig. 2 into rule is depicted in Fig. 4. This Figure illustrates both the rule description in pseudo-natural language, or in DSL, (Fig. 4.A) and in rule language using Drools (Fig. 4.B). The specification of the rule description language is expressed as a set of DSL definitions, which are mapped to Drools rule representation. This process is repeated for each path, from root to leaf, till achieving all grouping cases. Moreover, our proposal suggests that the tree representing all groups, in a given DRG system, will be transformed in rule base containing "n" rules that correspond to the number of DRG groups.

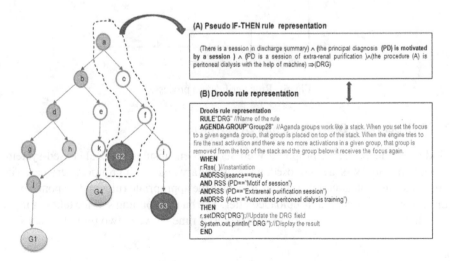

Fig. 4. Rule reprsentation in DSL and in DROOLS.

Conditions. The Left Hand Side (LHS) of the rule defines a conjunction of conditions or tests on DS data elements, that include mainly: one quantitative item (length of stay), two qualitative items (output mode of stay, patient's age) and several symbolic items (MD, associated diagnosis, medical or surgical procedures). Tests must be represented in a predefined order as shown in Fig. 4. For instance, the path "a–c–f" represents the set of conditions that should be verified to reach the DRG "G2".

Consequences. The Right Hand Side (RHS) of the rule defines the DRG code that is represented as a leaf node in grouping algorithm.

Checking consistency of grouping rules. Rules in DSL form are mainly comprehensible by experts but still too complicated for processing. Also, Drools does not support reliable verification of the Knowledge-base[2]. For instance, it accepts describing two rules having the same LHS and RHS which are named differently or two different rules, in terms of LHS and RHS, and having the same names. For this reason, we propose to add a verification module to ensure consistency of rule base in case of its updating. The adopted approach is based on the transformation of the rule base on Directed Hypergraphs [7] which is a sound method as indicated in [8]. Doing so, each operation aiming to update the knowledge (rule) base, the process depicted in Fig. 5 will be triggered. It allows checking the consistency of rules, mainly detecting structural anomalies.

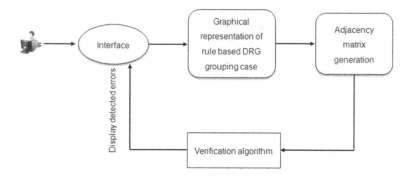

Fig. 5. Rule verification process.

4.2 Web Services: Description and Deployment

As indicated in Sect. 3, the proposed Web Services encapsulate the Rule based-system. Indeed, two main WSs are proposed, namely, unitary and batch grouping services. The former takes the uploaded DS as input, executes the appropriate rules and responds back to the client with output that is uploaded after. The batch grouping service takes as input DSs belonging to a given date or to an interval of time between two provided dates.

---

[2] "Drools Documentation" [Online] [Cited: 03. 18. 2017] https://docs.jboss.org/drools/release/6.0.1.Final/drools-docs/html_single/.

5   Experimentations

To illustrate the viability of our grouping solution, we implemented a Rule based system as a service accessible by several HISs. We tested the proposed solution that supplies grouper software or solution as a service.

5.1   Creation of Knowledge Base

The expert uses the knowledge acquisition tool, GR editor, to build the knowledge base by drawing grouping algorithm in forms of Directed Hypergraphs. Adding a rule is done by drawing the corresponding part of the DRG tree and entering the appropriate information for each node, and then the "drl" file is generated. The number of created rules is one-thousand representing the considered DRG groups in our test. Moreover, the same tool allows the expert to modify, delete, activate and deactivate rules in a simple way. All these operations are done while the implemented verification algorithm ensures the detection of eventual inconstancies. Once detected, the expert is informed about their existence and their nature. After the completion of the knowledge base acquisition, we evaluated the solution's correctness with real world cases and provide required adjustment if necessary. By correctness, we mean that the DS is grouped into the required groups according to user expert.

5.2   Experimental Settings and Some Tests' Results

The developed solution allows healthcare practitioners to perform unit grouping or batch grouping in a user-friendly manner. The user submits DS data to the Rule based-system

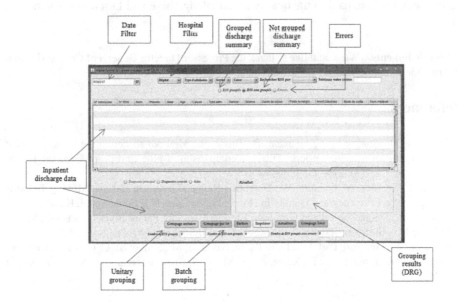

Fig. 6. Snapshots of the implemented grouping solution

via WS API. After that, the recommendations and answers generated by the inference engine are presented to the user on the screen. Figure 6 shows a snapshot of the main interface of the developed solution.

Also, we analyze the performance of the grouping solution to perform unitary or batch grouping. For this, we tested on a large sample of DS data that consists of 1000 DS, stemmed from six HISs. Tests performed highlight the following two aspects. First, it appears that the variation of the execution time is influenced mainly by the number of processed DSs. For instance, our solution can complete the grouping of 30 DSs in less than two (02) minutes. Also, the grouping execution time is influenced by the number of rules of the knowledge base and the number of antecedents of rule's LHS. For this, splitting the rule base into several files has revealed it efficiency in term of processing time. Another aspect highlighted by performed tests is that 70% of DSs were processed successfully. That means that the remaining 30% of DSs are not assigned any DRG groups. The verification done upon these cases revealed the existence of data errors on DS in terms of data incompatibilities and data incompleteness.

6   Conclusion

Providing grouping solution as a service aims to maintain and to evolve easily this function. The developed prototype is implemented, deployed and accessible from several HISs. The first phase of solution exploitation produced satisfying results and proves its practicality and its usefulness. To our knowledge, combining these technologies has not been explored till now in DRG systems. Furthermore, this work demonstrates other issues that would be highlighted as perspectives. As explained in experiment section, many Discharge Summaries have not been assigned a DRG. So, a focus should be, therefore, paid to the quality of knowledge base and Discharge Summaries data.

Acknowledgments. We would like to thank the Professor Director General of Central Hospital of Army for his support during the achievement of this work.

References

1. Schreyögg, J., Stargardt, T., Tiemann, O., Busse, R.: Methods to determine reimbursement rates for diagnosis related groups (DRG): a comparison of nine European countries. Health care Manag. sci. **9**(3), 215–223 (2006)
2. Busse, R., Geissler, A., Quentin, W.: Diagnosis-Related Groups in Europe: Moving Towards Transparency, Efficiency and Quality in Hospitals. McGraw-Hill Education, UK (2011)
3. Fetter, R.B.: Diagnosis Related Groups in Europe: Uses and Perspectives (Casas, M., Wiley, M.M. (eds.)). Springer Science & Business Media, Heidelberg (2012)
4. Preskitt, J.T.: Medicare Part A and DRG's. In: Savarise, M., Senkowsi, C. (eds.) Principles of Coding and Reimbursement for Surgeons, pp. 81–93. Springer International Publishing, Cham (2017)

5. Bellanger, M.M., Quentin, W., Tan, S.S.: Childbirth and Diagnosis Related Groups (DRGs): patient classification and hospital reimbursement in 11 European countries. Eur. J. Obstet. Gynecol. Reprod. Biol. **168**(1), 12–19 (2013)
6. Luo, W., Gallagher, M.: Unsupervised DRG upcoding detection in healthcare databases. In: 2010 IEEE International Conference on Data Mining Workshops (ICDMW), pp. 600–605 IEEE (2010)
7. Ausiello, G., Laura, L.: Directed hypergraphs: Introduction and fundamental algorithms— A survey. Theoret. Comput. Sci. **658**, 293–306 (2017)
8. Huang, H., Huang, S., Zhang, T.: A formal method for verifying production knowledge base. In: 2009 Fourth International Conference on Internet Computing for Science and Engineering (ICICSE), IEEE (2009)
9. Grosan, C., Abraham, A.: Rule-based expert systems. In: Grosan, C., Abraham, A. (eds.) Intelligent Systems, pp. 149–185. Springer, Heidelberg (2011)

Usage-Aware Service Identification for Architecture Migration of Object-Oriented Systems to SoA

Dionysis Athanasopoulos[✉]

School of Engineering and Computer Science,
Victoria University of Wellington, Wellington, New Zealand
dionysis.athanasopoulos@ecs.vuw.ac.nz

Abstract. Organisations currently migrate the architecture of their traditional systems to service-oriented systems. Since their systems usually serve a lot of clients, a diversity of the system usage may exist. The evolution of a service-oriented system is facilitated if the offered services are specific for each group of clients. However, the state-of-the-art service-identification approaches do not consider the system usage. Thus, we propose an online process that dynamically re-identifies services by the arrival of new method traces of the system usage. The preliminary evaluation of our process on real-world case-studies shows high effectiveness on identifying usage-aware services.

Keywords: Service identification · Online process · Usage-trace pattern and relatedness

1 Introduction

Organisations that have relied their function on traditional systems have now entered in the era of the digital connected world. The architecture of their systems is evolving to interconnected Service-oriented Architecture (SoA) systems [1] that usually serve a lot of clients. SoA client is a piece of software that accesses services. A high number of clients may imply diversity in the system usage, i.e. different clients may use different parts of service interfaces. In this case, (group of) clients should have been developed with respect to the (parts of) service interfaces that they use. In this way, the evolution of SoA systems is facilitated, since changes in service interfaces do not break the clients which do not depend on the changed interface parts [2]. However, the existing approaches, which support architecture migration of object-oriented (OO) systems to SoA[1], do not identify services that are specific for each group of clients. On the contrary, the approaches identify services based on internal system-artifacts exclusively.

---

[1] For a systematic literature review, the interested reader may refer to the recent survey in [3].

© Springer International Publishing AG 2017
D. Benslimane et al. (Eds.): DEXA 2017, Part II, LNCS 10439, pp. 54–64, 2017.
DOI: 10.1007/978-3-319-64471-4_6

Challenge. Since groups of clients are not available beforehand and clients of the same group do not necessarily use exactly the same part of service interfaces, we face the challenge of dynamically (re-)identifying services based on the system usage.

Contribution. To address this challenge, we propose an initial version of a usage-aware service-identification process. The process keeps histories of the method traces of the OO system usage (we concisely call them usage traces) and identifies services based on the traces that are related to each other. The process is further online, since it dynamically re-identifies services by the arrival of new usage-traces. To do so, the process includes two mechanisms. The first mechanism identifies reusable single-method services (singleton services) via determining patterns in usage traces. Since it identifies the interface and implementation of services (a.k.a. front- and back-end methods), we call it vertical (vertical lines in Fig. 3(a)). The second (horizontal) mechanism merges singleton services to form sets of related methods (horizontal lines in Fig. 3(a)), using the usage-trace relatedness metric that we propose. Finally, we evaluate the effectiveness of the mechanisms on real-world case-studies.

The rest of the paper is structured as follows. Sections 2 and 3 describe related approaches and our running example, respectively. Section 4 defines the basic notions. Section 5 specifies the mechanisms. Section 6 presents the evaluation results. Finally, Sect. 7 summarizes our approach and discusses its future research directions.

2   Related Work

The migration of OO systems to SoA typically includes the phases of the system understanding, migration feasibility, service identification, technical evolution, and service deployment [3]. Focusing on representative approaches that deal with service identification from architecture perspective, we observe that the majority of them offers guidelines to identify candidate services (e.g. [4–7]) or proposes manual processes (e.g. [8,9]). The remaining (semi-) automated approaches (e.g. [10,11]) adopt techniques, such as feature extraction [10] and dominance analysis in directed graphs (e.g. [11]).

Independently of the automation degree, all the approaches identify services by analysing internal system-artifacts, such as requirements specification, system architecture, and source code. Moreover, the approaches cannot dynamically improve the quality of identified services, since the former are off-line. On top of this, the approaches, except for the recent one in [12] and ours, rely on ad-hoc criteria for evaluating service quality. The metric adopted in [12] assesses the cohesion and coupling of services. Our metric assesses the service usage and is related to the usage-cohesion metric in [13]. However, the latter is applied only on front-end methods and assesses usage cohesion in terms of specific client groups.

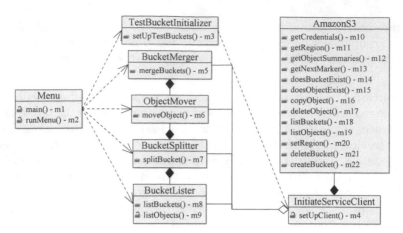

Fig. 1. The UML class-diagram (www.omg.org/spec/UML) of the example system.

3   Running Example

We exemplify our process with the OO system of Fig. 1 that organises user objects (e.g. photos) into buckets stores them on the Amazon cloud using the service S3 (aws.amazon.com/s3). To migrate the system to SoA, we initially applied the related approach in [12] that identifies services of high cohesion and low coupling. Following, we monitored the system usage and observed the following client groups: (i) bucket organization (BucketSplitter and BucketMerger); (ii) object movement (ObjectMover and BucketLister); (iii) test-bucket generation (setUpTestBuckets). Constructing the expected services based on the client groups (Fig. 2(a)), we observe that the clients do not depend on methods that they do not invoke. However, this observation does not hold in Fig. 2(b), which depicts the services identified by [12].

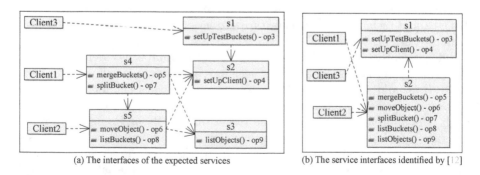

(a) The interfaces of the expected services   (b) The service interfaces identified by [12]

Fig. 2. The UML class-diagrams of the identified services for the example system.

4 Basic Notions

4.1 Representation Models of OO and SoA Systems

Independently of programming languages, our model represents an OO system as a directed tree of methods (Fig. 3(a)). A method m (Table 1(1)) is mainly characterized by the fields id, name, and isAccessorMutator[2]. The method usage is characterized by its total invocation number and the set of its directly invoked methods[3] (inv and edges in Table 1(1)). Our model does not depend on the system classes and packages.

A SoA system comprises a set of services. A service s (Table 1(2)) consists of its interface si and implementation simpl. A service interface SI (Table 1(3)) is defined as a set of operations. Each operation OP (Table 1(4)) encapsulates and has the same id, name, and inv with an OO method. A service implementation SIMPL (Table 1(5)) includes for each service operation the encapsulated OO method, along with the set of all the OO methods that are (directly or indirectly) invoked by the encapsulated method (set internalM of Table 1(5)). A service implementation further includes the invoked operations of its nested services (set exteranlOP of Table 1(5)).

4.2 Usage Trace

Definition 1 *[Usage trace]. A usage trace T is a sequence of methods* $(id_1, id_2, \ldots, id_N)$ *that corresponds to a path of the tree system-representation.* $\square$

Definition 2 *[Usage-trace history]. The history of T is defined by a discrete function h: id → inv that maps each method id of T to the method invocation number inv.* $\square$

Figure 3(a) depicts the usage-trace histories of the example system (node label includes a method id and its total invocation-number inv in parentheses). Figure 3(b) further plots the history of $(m_1, m_2, m_3, m_4, m_{10})$, in which h has a *local minimum*. According to the discrete mathematics [14], h has local minimum at id_x if its discrete derivative Δ^4 at id_{x-1} is negative ($\Delta_{id_{x-1}} h < 0$) and positive at id_x ($\Delta_{id_x} h > 0$). Checking for potential local minima of h, the following patterns can be identified.

---

[2] The usage of common accessors/mutators indicates that methods operate on the same data-models.

[3] If a method contains a condition block (e.g. while, if), the system processor assigns weights w_i (Table 1(1)) to the edges that connect a method to the inner-block methods. Each weight corresponds to an invocation probability (Fig. 3(a) depicts the weights that are smaller than 1.0). The processor further breaks invocation cycles via removing the edges that close cycles.

[4] The discrete derivative of a function f at n is given by the formula $\Delta_n f = f(n+1) - f(n)$.

Table 1. The definitions of the representation models of OO and SoA systems.

[**Method**]: $M := (id : \text{int}, \; name : \text{string}, \; isAccessorMutator : \text{boolean}, \; inv : \text{int}, edges) \mid edges = (w_i \in [0,1], \; m_i : M)$	(1)
[**Service**]: $S := (si : SI, \; simpl : SIMPL)$	(2)
[**Interface**]: $SI := \{op_i : OP\}$	(3)
[**Operation**]: $OP := (id : \text{int}, \; name : \text{string}, \; inv : \text{int})$	(4)
[**Implementation**]: $SIMPL := \{(op_i : OP, internalM, externalOP)\} \mid simpl.op_i.id = si.op_i.id \wedge internalM = \{m_k : M\} \wedge externalOP = \{op_j : OP\}$	(5)

Definition 3 [Non-service pattern]. *Let id_x be a method of a usage trace, id_x participates to a non-service pattern if h at id_x equals to the total times that a system has been used by its clients (system-usage times).* □

For instance, id_1 and id_2 in Fig. 3(b) do not belong to services, since their h values equal to 29, which is the total system-usage times in our example.

Definition 4 [Service-interface pattern]. *Let id_x be a method of a usage trace, id_x belongs to a service-interface pattern if (i) id_x is not an accessor or mutator; (ii) h has local minimum at id_x or is decreasing at id_{x-1} and non-decreasing at id_x.* □

Returning to Fig. 3(b), id_3 belongs to a service interface, since h has local minimum at id_3 and id_3 is not an accessor or mutator.

Definition 5 [Service-implementation pattern]. *Let id_x be a method of a usage trace, id_x belongs to a service-implementation pattern if h is decreasing at id_{x-1} and non-decreasing at id_x or is constant at both id_{x-1} and id_x.* □

Returning to Fig. 3(b), id_4 (resp. id_{10}) belongs to a service implementation, since h is decreasing at id_3 and non-decreasing at id_4 (resp. constant at id_4 and id_{10}).

Definition 6 [Related usage-trace histories]. *Let h_1, h_2 be the histories of T_1, T_2, h_1 and h_2 are related if (i) $\exists \; id_x \in T_1$, $id_y \in T_2$: $|h_1(id_x) - h_2(id_y)| \le \delta$; (ii) the sub-sequences $(id_x, \ldots, T_1.length)$ and $(id_y, \ldots, T_2.length)$ have common methods.* □

Returning to Fig. 3(a), since m_5 and m_7 of (m_1, m_2, m_5, m_{14}) and (m_1, m_2, m_7, m_{14}) have close invocation values ($m_5.inv = 5$ and $m_7.inv = 3$) and their sub-sequences (m_5, m_{14}) and (m_7, m_{14}) have a common method, the usage-traces are related to each other and s_4 of Fig. 2(a) is defined based on them.

4.3   Metric of Usage-Trace Relatedness

Our metric R, defined for two singleton services (s_1, s_2), includes two parts. R_1 assesses the degree to which s_1 and s_2 have been used a close number of times via calculating the percentage of the min inv divided by the max inv of the service operations, $\frac{MIN(s_1.si.op.inv,\ s_2.si.op.inv)}{MAX(s_1.si.op.inv,\ s_2.si.op.inv)}$. R_2 calculates the percentage of the common methods of the service implementations, $\frac{|s_1.simpl.internalM \cap s_2.simpl.internalM|}{MAX(|s_1.simpl.internalM|,\ |s_2.simpl.internalM|)}$. R equals to the product of R_1 and R_2, assuming that R is high when both R_1 and R_2 are high.

5   Online Usage-Aware Service-Identification Mechanisms

Vertical service-identification. The mechanism determines the patterns in usage-trace histories. Its underlying Algorithm 1 accepts the system-usage times E and the tree system-representation (rooted at m_0), and returns a list of singleton services sList. The algorithm traverses the tree paths via adopting the classical Depth-First Search (DFS) (Algorithm 1(3–23)). The algorithm constructs each tree path that corresponds to a usage trace t (Algorithm 1(6)). For each method id_x of t (Algorithm 1(9)), the algorithm checks if id_x participates to the non-service (Definition 3), service-interface (Definition 4), or service-implementation (Definition 5) patterns in Algorithm 1(10), (11–16), or (17–21), respectively.

Returning to our running example, the algorithm determines the patterns in the usage-trace histories of Fig. 3(a) and identifies the singleton services of Table 2(a).

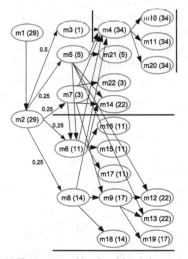

(a) The usage-trace histories of the whole system

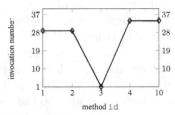

(b) Plotting the history of $(m_1, m_2, m_3, m_4, m_{10})$

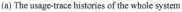

	s1	s2	s3	s4	s5	s6	s7
s1	-	0.01	0.05	0.01	0.06	0.01	0.00
s2	0.01	-	0.02	0.17	0.02	0.11	0.00
s3	0.05	0.02	-	0.11	**0.39**	0.07	0.03
s4	0.01	0.17	0.11	-	0.09	**0.32**	0.00
s5	0.06	0.02	0.39	0.09	-	0.20	0.06
s6	0.01	0.11	0.07	0.32	0.20	-	0.26
s7	0.00	0.00	0.03	0.00	0.06	0.26	-

(c) The usage-trace relatedness values of the singleton services

Fig. 3. The usage-trace histories and relatedness values for the running example.

Algorithm 1. Vertical Service-identification

Input: int E, m_0
Output: List<S> $sList$

1: $T\ t \leftarrow \emptyset$
2: $S\ currentService \leftarrow null$
3: Stack $stack \leftarrow new\ Stack(m_0)$
4: **while** $stack \neq \emptyset$ **do**
5:     M $m \leftarrow stack.\text{POP}()$
6:     $t.\text{ADD}(m)$
7:     **for all** $m_i \in m.edges$ **do** $stack.\text{PUSH}(m_i)$
8:     **if** $m.edges = \emptyset$ **then**
9:         **for all** $id_x \in t$ **do**
10:             **if** $h(id_x) = E$ **then** CONTINUE
11:             **if** $m_x.isAccessorMutator = \text{false} \mid m_x.id = id_x$ **then**
12:             **if** $\left(\Delta_{id_{x-1}}h < 0\ \&\&\ \Delta_{id_x}h > 0\right) \mid\mid \left(\Delta_{id_{x-1}}h > 0\ \&\&\ \Delta_{id_x}h \geq 0\right)$ **then**
13:                 $OP\ op \leftarrow new\ OP(m_x, h(id_x))$
14:                 $currentService \leftarrow new\ S().si.\text{ADD}(op)$
15:                 $currentService.simpl.internalM.\text{ADD}(m_x)$
16:                 $sList.\text{ADD}(currentService)$
17:             **if** $\left(\Delta_{id_{x-1}}h > 0\ \&\&\ \Delta_{id_x}h > 0\right) \mid\mid \left(h(id_{x-1}) = h(id_x)\right)$ **then**
18:                 **if** $\exists s_j \in sList : op_x \in sList.s_j.si$ **then**
19:                     $op_x.inv \leftarrow h(id_x)$
20:                     $currentService.simpl.externalOP.\text{ADD}(op_x)$
21:                 **else** $currentService.simpl.internalM.\text{ADD}(m_x)$
22:         $t \leftarrow \emptyset$
23: **end while**

Horizontal service-identification. The mechanism merges services based on related usage-trace histories. Its underlying Algorithm 2 accepts a set of singleton services sList and a threshold (the lowest acceptable usage-trace relatedness), and returns a list of (non-)singleton services. The algorithm initially calculates the usage-trace relatedness for every pair of singleton services by using the metric R (Algorithm 2(1)). To form service groups, the algorithm adopts the classical agglomerative clustering-method (Algorithm 2(2–10)) [15]. Concerning the usage-trace relatedness between two clusters (proximity), the algorithm calculates their min relatedness (Algorithm 2(4–5)), which guarantees that every service in a cluster is related to the rest services.

Table 2. The services identified by our mechanisms for the running example.

(a) Vertical mechanism

s	SI	SIMPL	
		externalOP	internalM
s_1	op_3	$\{op_4\}$	$\{m_3\}$
s_2	op_4	$\emptyset$	$\{m_4, m_{10}, m_{11}, m_{20}\}$
s_3	op_9	$\emptyset$	$\{m_9, m_{12}, m_{13}, m_{19}\}$
s_4	op_5	$\{op_4, op_6\}$	$\{m_5, m_{12}, m_{13}, m_{14}, m_{21}\}$
s_5	op_7	$\{op_4, op_9\}$	$\{m_7, m_{14}, m_{22}\}$
s_6	op_6	$\{op_4\}$	$\{m_6, m_{14}, m_{15}, m_{16}, m_{17}\}$
s_7	op_8	$\{op_4, op_9\}$	$\{m_8, m_{18}\}$

(b) Horizontal mechanism

s	SI	SIMPL	
		externalOP	internalM
s_1	op_3	$\{op_4\}$	$\{m_3\}$
s_2	op_4	$\emptyset$	$\{m_4, m_{10}, m_{11}, m_{20}\}$
s_3	op_9	$\emptyset$	$\{m_9, m_{12}, m_{13}, m_{19}\}$
s_4	op_5	$\{op_4, op_6\}$	$\{m_5, m_{12}, m_{13}, m_{14}, m_{21}\}$
	op_7	$\{op_4, op_9\}$	$\{m_7, m_{14}, m_{22}\}$
s_5	op_6	$\{op_4\}$	$\{m_6, m_{14}, m_{15}, m_{16}, m_{17}\}$
	op_8	$\{op_4, op_9\}$	$\{m_8, m_{18}\}$

Algorithm 2. Horizontal Service-identification

Input: List<S> $sList$, double $threshold$
Output: List<S> $services$
1: **for all** $(s_i, s_j) \in sList \times sList$ **do** $r[i][j] \leftarrow R(s_i, s_j)$
2: **for all** $s_i \in sList$ **do** $services.\text{ADD}(s_i)$
3: **repeat**
4:    $(min, s_i, s_j) \leftarrow \text{FINDMINPROXIMITYPAIR}(r, services)$
5:    **if** $min \geq threshold$ **then**
6:        S $s \leftarrow \text{MERGE}(s_i, s_j)$
7:        $services.\text{REMOVE}(s_i, s_j)$
8:        $services.\text{INSERT}(s)$
9:        $\text{UPDATE}(r, services, s)$
10: **until** $|services| = 1$ **or** $min < threshold$

Returning to our running example, the algorithm accepts the services of Table 2(a) and identifies those of Table 2(b), which are the same to the expected services (Fig. 2(a)). In particular, the algorithm merges the pairs (s_3, s_5) and (s_4, s_6), since they have the highest R values (Fig. 3(c)). The algorithm does not merge other services, since their R values are lower than the proximity threshold 0.2 (see Sect. 6).

6   Preliminary Evaluation

We implemented the vertical and horizontal mechanisms in Java (our research prototype is available online[5]) to evaluate their effectiveness on two real-world case-studies (see footnote 5).

Experimental setup. The first case-study is the small-sized S3 app of our running example (8 classes and 22 methods) and the second case-study is a medium-sized information system IS (30 classes and 95 methods). We executed the mechanisms for a set of monitoring usage-traces and repeated the executions for a range of proximity values. Finally, we evaluated the identified services using the following metrics.

Effectiveness metrics. We calculated the precision and recall of the identified services with respect to the expected services (F-measure [16]). We define the precision (resp. recall) as the percentage of the operations, which correctly participate in the identified services over all the operations of the identified (resp. expected) services. We further measured the interface usage-cohesion (IUC) of the identified services with respect to their clients via using the metric IUC of [13]. Since IUC is defined for front-end services exclusively, we extend it for back-end services (Eq. 6). We finally calculated the average IUC value and its standard deviation for each case-study and proximity value.

$$IUC(i) := \frac{\sum_{j=1}^{|clients|} \frac{used\_methods(j,i)}{all\_methods(i)}}{|clients|} \;\Big|\; j \text{ denotes a client of a } \textit{front-end} \text{ service } i \text{ or a } \textit{back-end}$$

service i for which there is a sequence of services whose front-end service is used by j (6)

---

[5] ecs.victoria.ac.nz/foswiki/pub/Main/DionysisAthanasopoulos/MigrationSources.zip.

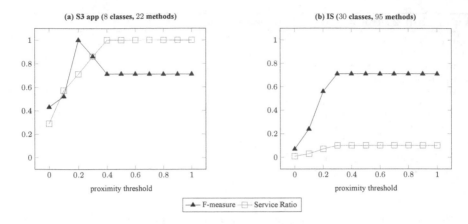

Fig. 4. The F-measure values for the services identified by our mechanisms in both case-studies.

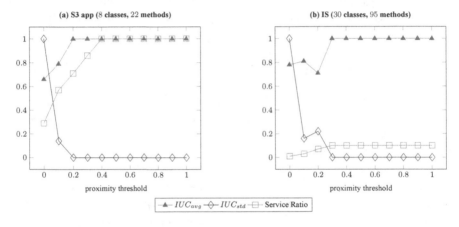

Fig. 5. The IUC values for the services identified by our mechanisms in both case-studies.

Results. We observe in s3 app (Fig. 4(a)) that our mechanisms identify the expected services, since the max F-measure is achieved for the proximity 0.2. In detail, the F-measure increases starting from a low value, then it is maximized, and finally it decreases until a medium value. To justify this behaviour, we check the ratios of the identified services (their number is divided by the total number of singleton services). We observe that when the proximity is low, then the ratios are also low, since many services are related and are finally merged (a few fat service-interfaces are identified). For medium proximity, the ratios increase, since many thin and/or singleton service-interfaces are identified. In the case of IS (Fig. 4(b)), the mechanisms do not achieve max F-measure, since they identify less services (≤10) than the 14 expected services.

Concerning the average IUC value and its standard deviation (stdev), the mechanisms achieve in the S3 app (Fig. 5(a)) the max (resp. min) average (resp. stdev of) IUC value for the proximity 0.2, as expected. In the case of IS (Fig. 5(b)), the mechanisms achieve max (resp. min) average (resp. stdev of) IUC value for the proximity 0.3–1.0. Even if the mechanisms identify less services than the expected ones, its average IUC is max, since the mechanisms identify the expected front-end services.

7   Conclusions and Future Work

We proposed the initial version of an online process that re-identifies services based on the usage of OO systems. The preliminary evaluation results of our process on real-world case-studies show high effectiveness on identifying usage-aware services. A future research direction is to extend the process with extra patterns and ways to relate usage traces. Moreover, experiments on large-scale legacy systems could be conducted.

References

1. Erl, T.: Service-Oriented Architecture: Concepts, Technology, and Design. Prentice Hall, Upper Saddle River (2005)
2. Romano, D., Raemaekers, S., Pinzger, M.: Refactoring fat interfaces using a genetic algorithm. In: International Conference on Software Maintenance & Evolution (2014)
3. Razavian, M., Lago, P.: A systematic literature review on SOA migration. J. Softw. Evol. Process **27**(5), 337–372 (2015)
4. Lewis, G.A., Morris, E.J., Smith, D.B., O'Brien, L.: Service-oriented migration and reuse technique (SMART). In: International Workshop on Software Technology and Engineering Practice, pp. 222–229 (2005)
5. Sneed, H.M.: Integrating legacy software into a service oriented architecture. In: European Conference on Software Maintenance and Reengineering, pp. 3–14 (2006)
6. Khadka, R., Reijnders, G., Saeidi, A., Jansen, S., Hage, J.: A method engineering based legacy to SOA migration method. In: International Conference on Software Maintenance, pp. 163–172 (2011)
7. Khadka, R., Saeidi, A., Jansen, S., Hage, J.: A structured legacy to SOA migration process and its evaluation in practice. In: International Symposium on the Maintenance and Evolution of Service-Oriented and Cloud-Based Systems, pp. 2–11 (2013)
8. Alahmari, S., Zaluska, E., De Roure, D.: A service identification framework for legacy system migration into SOA. In: International Conference on Services Computing, pp. 614–617 (2010)
9. Canfora, G., Fasolino, A.R., Frattolillo, G., Tramontana, P.: Migrating interactive legacy systems to web services. In: European Conference on Software Maintenance and Reengineering, pp. 24–36 (2006)
10. Aversano, L., Cerulo, L., Palumbo, C.: Mining candidate web services from legacy code. In: International Symposium on Web Systems Evolution, pp. 37–40 (2008)

11. Li, S., Tahvildari, L.: A service-oriented componentization framework for Java software systems. In: Working Conference on Reverse Engineering, pp. 115–124 (2006)
12. Adjoyan, S., Seriai, A.-D., Shatnawi, A.: Service identification based on quality metrics. In: International Conference on Software Engineering & Knowledge Engineering (2014)
13. Perepletchikov, M., Ryan, C., Tari, Z.: The impact of service cohesion on the analyzability of service-oriented software. IEEE Trans. Serv. Comput. **3**(2), 89–103 (2010)
14. Rosen, K.H.: Discrete Mathematics and its Applications. McGraw-Hill, New York (2002)
15. Maqbool, O., Babri, H.A.: Hierarchical clustering for software architecture recovery. IEEE Trans. Softw. Eng. **33**(11), 759–780 (2007)
16. Baeza-Yates, R.A., Ribeiro-Neto, B.A.: Modern Information Retrieval. ACM Press/Addison-Wesley, Boston (1999)

Continuous and Temporal Data, and Continuous Query Language

On Computing Temporal Aggregates over Null Time Intervals

Kai Cheng[✉]

Graduate School of Information Science, Kyushu Sangyo University,
2-3-1, Mtsukadai, Higashi-ku, Fukuoka 813-8503, Japan
chengk@is.kyusan-u.ac.jp
http://www.is.kyusan-u.ac.jp/~chengk/

Abstract. In a temporal database, each data tuple is accompanied by a time interval during which its attribute values are valid. In this paper, we consider the null time intervals, that is, time intervals not intersected by any time intervals in the temporal database. We deal with the problem of computing temporal aggregates over null time intervals with length constraints. By interval folding, we transform the problem into aggregates over stabbing groups, maximal stabbing interval sets. We describe the detailed algorithms and report the experimental results.

Keywords: Temporal database · Temporal aggregation · Stabbing query · Null time interval · Interval folding

1 Introduction

We consider the problem of computing temporal aggregates over null time intervals. In a temporal database, data tuples are typically accompanied by time intervals that capture the valid time of the information or facts. Consider a scheduling system where scheduled activities for individuals or groups are stored in a temporal relation. In order to create a new activity for a group of people, one has to find time intervals during which all members can participate. We call such time intervals *null time intervals*. A time interval is said to be a null time interval when no time intervals in the database intersect with it. The qualifying null time intervals should also satisfy length constraint. For example there must be at least 90 minutes free time for the new activity. Furthermore, when no qualifying null time interval is available, a partially null time interval can also be seen as a feasible choice. For example, a query for free time intervals of 10 members may accept results with 1 or 2 members absent.

To report qualifying null time intervals, it is important to compute temporal aggregates. Support for temporal aggregates is a predominant feature of many data management systems. When aggregating temporal relations, tuples are grouped according to their timestamp values. There are basically two types of temporal aggregation: instant temporal aggregation and span temporal aggregation [3,7]. *Instant temporal aggregation (ITA)* computes aggregates on each

© Springer International Publishing AG 2017
D. Benslimane et al. (Eds.): DEXA 2017, Part II, LNCS 10439, pp. 67–79, 2017.
DOI: 10.1007/978-3-319-64471-4_7

time instant and consecutive time instants with identical aggregate values are coalesced into so-called *constant intervals*, i.e., tuples over maximal time intervals during which the aggregate results are constant [4]. On the other hand, *Span temporal aggregation (STA)* allows an application to control the result size by specifying the time intervals, such as year, month, or week. For each of these intervals a result tuple is produced by aggregating over all argument tuples that overlap that interval.

(a) Event time			(b) Null time			(c) Aggregates		
	Symbol	**Event Time**		**Symbol**	**Null Time**	**Null Time**	**Group**	**CNT**
e_1	A	$[1, 4]$	r_1	A	$[5, 13]$	$[20, 21]$	$\{A\}$	1
e_2	A	$[14, 17]$	r_2	A	$[18, 24]$	$\mathbf{[22, 24]}$	$\{A, B\}$	**2**
e_3	B	$[7, 12]$	r_3	A	$[30, 50]$	$[26, 29]$	$\{B\}$	1
e_4	B	$[19, 21]$	r_4	B	$[1, 2]$	$\mathbf{[30, 35]}$	$\{A, B\}$	**2**
e_5	A	$[25, 29]$	r_5	B	$[6, 6]$			
e_6	B	$[3, 5]$	r_6	B	$[13, 18]$			
			r_7	B	$[22, 50]$			

Fig. 1. Running example of null time aggregates on $[20, 35]$

In this paper, we study the problem of temporal aggregates over null time intervals. Figure 1 gives a running example. Assume the time domain is $[1, 50]$. Event time in Fig. 1(a) is physically stored in a temporal database, in which each event symbol is accompanied by an event time. Null time in Fig. 1(b) is a derived relation during query time. For example, A has an event sequence $\{[1, 4], [14, 17], [25, 29]\}$, from which the null time $\{[5, 13], [18, 24], [30, 50]\}$ is derived. Temporal aggregates on $[20, 35]$ as shown in Fig. 1(c) are computed by grouping tuples in the query range at first and then applying aggregate functions to each group. During $[22, 25]$, $[30, 35]$, both A and B are not overlapped by any event time, we call them truly null time, whereas $[20, 21]$ and $[26, 29]$ are partially null time.

Support for null time intervals is not provided by current database products. Syntactically, all relational database management systems (RDBMS) support a representation of "missing information and inapplicable information". Null (or NULL) is a special value to indicate that a data value does not exist in the database. However, to our knowledge, there not exist database systems that support null time since it not a practical solution to explicitly store null time intervals in databases. Neither NOT IN nor NOT EXISTS is suitable for querying time intervals that are not intersected by other intervals because null time intervals depend on time domain.

Our contributions include: (1) we introduce a new operation called interval folding and transform the problem into interval stabbing problem; (2) we propose stabbing group as a new temporal grouping method to solve the interval stabbing problem. (3) We develop a balanced tree based data structure and algorithms for efficient computation of temporal aggregates over null time intervals.

The rest of paper is organized as follows. In Sect. 2, we define the problem and propose the main techniques. Section 3 describes the main techniques. Section 4 introduces the experimental results. Section 6 concludes the paper and points out some future directions.

2   Problem Definition

Let $E = \{e_1, e_2, \cdots, e_k\}$ be the set of event symbols and N be the time domain. The triplet $(e_i, s_i, f_i) \in E \times N \times N$ is an *event interval* or *real time interval*. The two time points s_i, f_i are called event times, where s_i is the starting time and f_i is the finishing time, $s_i \leq f_i$. The set of all event intervals over E is denoted by I. An *event sequence* is a series of event interval triplets $E_S = \langle (e_1, s_1, f_1), (e_2, s_2, f_2), \cdots, (e_n, s_n, f_n) \rangle$, where $s_i \leq s_{i+1}$, and $s_i < f_i$. A temporal database D is a set of records, $\{r_1, r_2, \cdots, r_m\}$, each record r_i $(1 \leq i \leq m)$ consists of a sequence-id and an event interval.

event intervals (real time intervals)

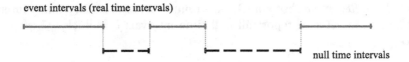

null time intervals

Fig. 2. Real time interval vs. null time interval

For $S \subseteq D$, a *null time interval* $\boldsymbol{a}$ is an interval that for any $\boldsymbol{b} \in S$, $\boldsymbol{a} \cap \boldsymbol{b} = \emptyset$. As shown in Fig. 2, during real time intervals events are valid, while events are invalid during null time intervals. We assume in this paper that only real time intervals are explicitly stored in database. Given the temporal database D and a query interval $[p, q]$ null time intervals can be derived as follows.

$$[p, q] - \bigcup_{[s,f] \in D} [s, f]$$

Given an event sequence $q = \langle (e_1, s_1, f_1), (e_2, s_2, f_2), \cdots, (e_n, s_n, f_n) \rangle$, the set $T = \{s_1, f_1, s_2, f_2, \cdots, s_i, f_i, \cdots, s_n, f_n\}$ is called a time set corresponding to sequence q where $1 \leq i \leq n$. If we order all the elements in T and eliminate redundant elements, we can derive a sequence $T_S = \langle t_1, t_2, t_3, \cdots, t_k \rangle$ where $t_i \in T$, $t_i < t_{i+1}$. T_S is called a time sequence corresponding to event sequence q.

When discussing time intervals, it is important to describe pairwise relationships between two time interval-based events. According to Allen's temporal logics [1], the basic temporal relations between any two event intervals are shown in Fig. 3. Except (g), each of (a)–(f) has its inverse relation. For example, "A before B" also means "B after A", "A contains B" implies "B is contained by A", etc. These relationships can describe any relative position of two intervals based on the arrangements of the starting and finishing time points.

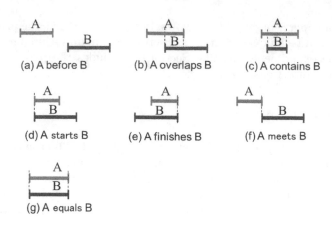

Fig. 3. Temporal relations between two intervals

Now we can formulate the problem we target as follows:

Null Time Reporting Problem: Given a temporal database D, a query interval $[p, q]$ and a parameter α, report all null time intervals $\boldsymbol{b} = [b_s, b_f] \subseteq [p, q]$ and $|\boldsymbol{b}| \geq \alpha$.

2.1   Interval Folding

We introduce *interval folding*, an operation that transforms an interval to a shorter one. For interval $\boldsymbol{b} = [b_s, b_f]$, the α-folding of $\boldsymbol{b}$ is defined as:

$$\boldsymbol{b} - \alpha = [b_s + (1 - \lambda)\alpha, b_f - \lambda\alpha]$$

where λ is a parameter, which can be any real value between 0 and 1, for example

1. $\lambda = 0$, $[s, f] \rightarrow [s + \alpha, f]$
2. $\lambda = 1$, $[s, f] \rightarrow [s, f - \alpha]$
3. $\lambda = \frac{1}{2}$, $[s, f] \rightarrow [s + \frac{\alpha}{2}, f - \frac{\alpha}{2}]$

The α-folding of an interval set S, denoted by $S - \alpha$, is defined by applying α-folding to each element interval (Fig. 4).

Lemma 1. *Let S be a set of intervals. If $\bigcap_{b \in S}(\boldsymbol{b} - \alpha) \neq \emptyset$, then $\bigcap_{b \in S}(\boldsymbol{b} - \alpha) = \bigcap_{b \in S} \boldsymbol{b} - \alpha$.*

Proof. Let $\hat{s} = \max\{b_s \mid [b_s, b_f] \in S\}$, $\hat{f} = \min\{b_f \mid [b_s, b_f] \in S\}$. Then

$$\bigcap_{b \in S} \boldsymbol{b} = [\hat{s}, \hat{f}]$$

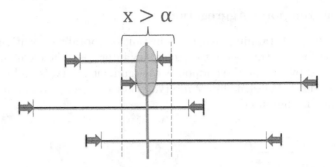

Fig. 4. Interval folding

It is obvious that $\max\{b_s+(1-\lambda)\alpha \mid [b_s, b_f] \in S\} = \hat{s}+(1-\lambda)\alpha$, $\min\{b_f-\lambda\alpha \mid [b_s, b_f] \in S\} = \hat{f} - \lambda\alpha$, that is

$$\bigcap_{b \in S}(\boldsymbol{b} - \alpha) = [\hat{s} + (1 - \lambda)\alpha, \hat{f} - \lambda\alpha] = \bigcap_{b \in S} \boldsymbol{b} - \alpha.$$

Theorem 1. *The intersection of S has a length larger than α, if and only if the intersection of $S - \alpha$ is non–empty:*

$$\bigcap_{b \in S} \boldsymbol{b} - \alpha \neq \emptyset \Leftrightarrow \bigcap_{b \in S}(\boldsymbol{b} - \alpha) \neq \emptyset$$

Proof. Let $\boldsymbol{x}$ be the intersection of S and $\boldsymbol{x} - \alpha \neq \emptyset$. As for any $\boldsymbol{b} \in S$, $\boldsymbol{b} - \alpha \supseteq \boldsymbol{x} - \alpha$, therefore

$$\bigcap_{b \in S}(\boldsymbol{b} - \alpha) \supseteq \boldsymbol{x} - \alpha \neq \emptyset$$

On the other hand, if $\bigcap_{b \in S}(\boldsymbol{b} - \alpha) \neq \emptyset$, then by Lemma 1, $\bigcap_{b \in S} \boldsymbol{b} - \alpha = \bigcap_{b \in S}(\boldsymbol{b} - \alpha) \neq \emptyset$.

Theorem 1 tells that by α-folding, a null time reporting problem can be transformed into the interval stabbing problem: given a query interval $[p, q]$, report all non–empty null time intervals in $[p, q]$.

3   Temporal Aggregates over Null Time Intervals

We now present the techniques for computing temporal aggregates over null time intervals. Given a temporal database D, a query interval $[p, q]$ and the length threshold α, the basic idea to compute the temporal aggregates is to derive null time intervals from the event times, and then by α-folding, transform the problem to interval stabbing problem. Thus, one just need to report all non–empty null time intervals contained in the query interval.

3.1   Instant Temporal Aggregation

A solution to interval stabbing problem is instant temporal aggregation. The key idea is to partition the timeline into elementary intervals. The elementary intervals are obtained by sorting the endpoints of argument intervals and consecutive two endpoints define an elementary intervals. For each elementary interval, an aggregate value is computed.

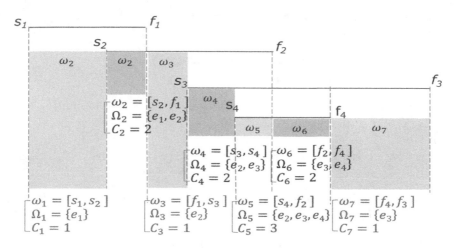

Fig. 5. Instant temporal aggregation

Figure 5 shows the temporal aggregates by instant temporal aggregation over $\{< e_1, s_1, f_1 >, < e_2, s_2, f_2 >, \cdots, < e_4, s_4, f_4 >\}$. The timeline is partitioned into a sequence of elementary intervals $\omega_1, \cdots, \omega_7$ from left to right: $\omega_1 = [s_1, s_2], \omega_2 = [s_s, f_1], \omega_3 = [f_1, s_3], \omega_4 = [s_3, s_4], \omega_5 = [s_4, f_2], \omega_6 = [f_2, f_4], \omega_7 = [f_4, f_3]$. With each elementary interval, we maintain a list of event symbols Ω_i and a count C_i.

$$\omega_i = \bigcap_{e_k \in \Omega_i} [s_k, f_k], C_i = |\Omega_i|$$

3.2   Stabbing Groups

The brute–force approach to computing instant temporal aggregates requires multiple passes to scan the argument relation. We propose a balanced tree based approach for efficient computation. The basic idea is to maintain the aggregate groups using a balanced search tree.

As shown in Fig. 6, each node of the balanced tree keeps a stabbing group. A stabbing group is a set of intervals stabbed by a common set of points. An interval b is *stabbed* by a point q is $q \in b$. The common set of points is called *group representative*, which is actually the intersection of the argument intervals.

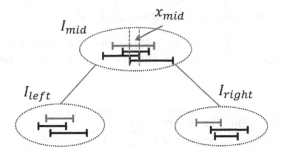

Fig. 6. Balanced tree for stabbing groups

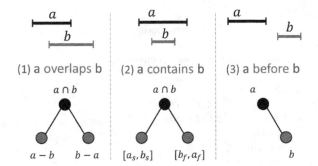

Fig. 7. Insert new intervals into the tree

In Fig. 6, the the representative of stabbing group I_{mid} is an interval x_{mid}. The two children of I_{mid}, stabbing groups I_{left} and I_{right} have their group representatives x_{left} and x_{right}, and $x_{left} < x_{mid} < x_{right}$. The $<$ relation is the same as Allen's *before* relation, that is, $a = [a_s, a_f]$, $b = [b_s, b_f]$

$$a < b \Leftrightarrow a_f < b_s$$

To insert a new interval into the balanced tree, we first do binary tree search by comparing with the group representatives on the search path. Whenever the interval to be added overlaps the group representative of a node, it is added to that node, which may cause change to the group representative. Figure 7 depicts how b is inserted into the tree. Suppose a is the group representative of the target group,

1. if a overlaps b, the representative of the a is changed to $a \cap b$. In addition, $a - b$, and $b - a$ are added recursively to the left and right subtrees.
2. if a contains b, the representative of the current group is changed to $a \cap b$. In addition, $[a_s, b_s]$, and $[b_f, a_f]$ are added recursively to the left and right subtrees.
3. if a before b, if the right child is absent, b will be added as a new right child of a.

3.3   Stabbing Temporal Aggregation (BTA)

We now describe the algorithm in detail. Since the temporal aggregates are computed based on stabbing groups, the algorithm is called *stabbing temporal aggregation* (BTA).

Algorithm 1. BTA (Building aggregation tree)

Input: α-folded null time intervals $S = \{[s_1, f_1], [s_2, f_2], \cdots, [s_n, f_n]\}$
Output: Temporal aggregates of S

1: **for** $i \leftarrow 1 \cdots n$ **do**
2:　　$b \leftarrow [s_i, f_i]$
3:　　$t \leftarrow T.find(b)$
4:　　**if** $t = nil$ **then**　　　　♯ Add a new node
5:　　　　$t.add(b)$
6:　　**else**　　　♯ Update existing nodes recursively
7:　　　　$a \leftarrow t.\omega$
8:　　　　$t.\omega \leftarrow a \cap b$
9:　　　　$t.\Omega \leftarrow t.\Omega \cup \{b\}$　　　♯ For report
10:　　　　$t.C \leftarrow t.C + 1$　　♯ For count
11:　　　　$L \leftarrow t.leftChild$
12:　　　　$R \leftarrow t.rightChild$
13:　　　　**if** a overlaps b **then**
14:　　　　　　$L.insert(a - b)$
15:　　　　　　$R.insert(b - a)$
16:　　　　**else if** a contains b **then**
17:　　　　　　$L.insert([a_s, b_s])$
18:　　　　　　$R.insert([b_f, a_f])$;
19:　　　　**end if**
20:　　**end if**
21: **end for**
22: **return** T

Given a temporal database D, the algorithm takes a query interval $[p, q]$ and the parameter α as input, and report all null time intervals that satisfy the length constraint. For count, report the number of qualifying null time intervals. The process of computing aggregates are outlined as follows.

1. Query the database D for event times that intersect $[p, q]$;
2. Derive the null time intervals from the obtained event times;
3. Apply α-folding to the null time intervals;
4. Create the balanced search tree for the α-**folded** null time intervals;
5. Traverse the tree and report groups and their representatives.

In Algorithm 1, we use an AVL tree T as the main data structure. T is built by a recursively inserting intervals.

– $T.find()$: Search the tree/subtree rooted on T for a given interval, return the first node whose representative intersecting with the argument interval
– $T.insert()$: Insert a node to the tree/subtree rooted on T
– $t.add()$: Add a new node to the current location t

The time complexity of BTA algorithm is the complexity of constructing a balanced search tree $O(n \log n)$. Notice step 2, 3 and 4 can be done simultaneously, which means the proposed method needs to scan the query result only once.

4   Experimental Analysis

To verify the proposed method, we implemented naive ITA and the proposed BTA algorithms for computing temporal aggregates over null time intervals. We evaluate algorithms in terms of response time and memory space. The response time includes database query, α-folding, and temporal aggregation. We do experiments for two types of queries: *stabbing report* and *stabbing count*. In stabbing report, the detail of stabbing groups are reported, while in stabbing count, only count for each group is maintained and output.

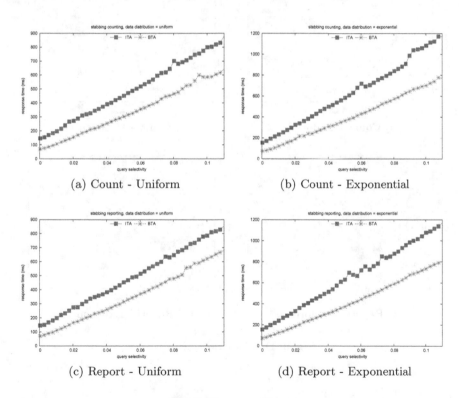

(a) Count - Uniform   (b) Count - Exponential

(c) Report - Uniform   (d) Report - Exponential

Fig. 8. Response times

The input to the algorithm consists of 1, 000 event symbols and 1, 000 event times for each event. The starting times are random numbers with uniformly distributed on $[1 \cdot \cdot 10^6]$. The length of the random intervals are either uniform or short. Uniform random intervals have a uniform distributed among $[1, 2, 000]$. Short random intervals have an exponentially distributed length with expected value 2, 000.

The results are shown in Figs. 8, 9, 10 and 11. In each of the plots, x-axis represents query intervals, ranging from 1% to 11%. First, in terms of response time, BTA outperforms naive ITA under uniform dataset as well as exponential dataset and for both count and report queries. Database query time as a part of response time varies with the query ranges but it is not a dominant part (only 10% of response time, see Fig. 9). The aggregation time is the main part of response time has a similar trend with response time (Fig. 10). The memory space for BTA is significantly smaller than naive ITA (Fig. 11). In total, BTA provides better performance for computing null time aggregates.

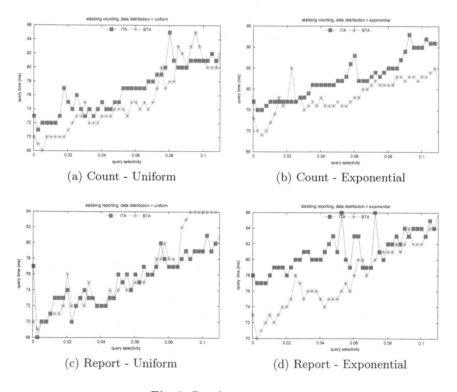

(a) Count - Uniform

(b) Count - Exponential

(c) Report - Uniform

(d) Report - Exponential

Fig. 9. Database query cost

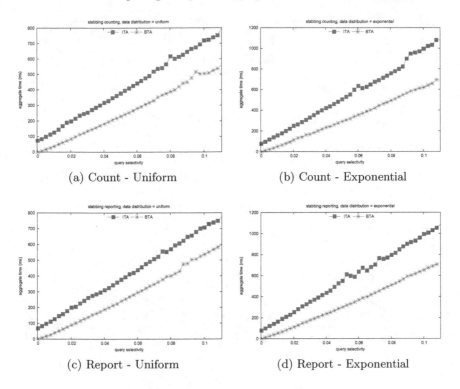

(a) Count - Uniform

(b) Count - Exponential

(c) Report - Uniform

(d) Report - Exponential

Fig. 10. Aggregate computation cost

5   Related Work

In [7], Kline and Snodgrass proposed an algorithm for computing temporal aggregation using a main memory-based data structure, called aggregation tree. It builds a tree while scanning a temporal relation. After the tree has been built, the answer to the temporal aggregation query is obtained by traversing the tree in depth-first search. It should be noted that this tree is not balanced. The worst case time to create an aggregation tree is $O(n^2)$ for n stored tuples. In an extension of his previous work, Kline proposed using a $2-3$ tree, which is a balanced tree, to compute temporal aggregates [2]. The leaf nodes of the tree store the time intervals of the aggregate results. Like the aggregation tree, this approach requires only one database scan. The running time is $O(n \log n)$ given that the database was initially sorted.

Interval stabbing problem is a well-known problem in computational geometry and there exist a number of results, for example [5,6]. However, interval stabbing in this context is aimed to preprocess the intervals into a data structure for repetitive query. The query time is at least $\Omega(1 + k)$ for output size k. The preprocessing is often expensive, requires multiple scans.

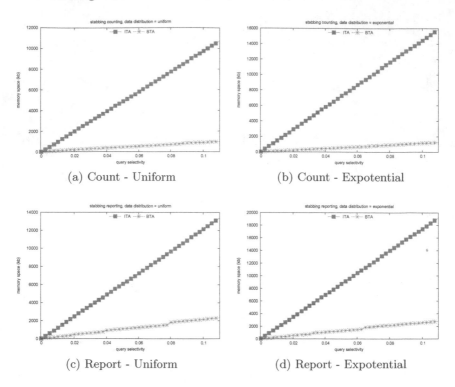

(a) Count - Uniform

(b) Count - Expotential

(c) Report - Uniform

(d) Report - Expotential

Fig. 11. Memory space

6 Conclusion

In this paper, we dealt with the problem of computing temporal aggregates over null time intervals. Null time intervals are intervals not overlapped by any event time intervals. We introduced α-folding and transformed the problem into the interval stabbing problem. To compute aggregates for stabbing groups efficiently, we proposed balanced search tree based data structure that maintains stabbing groups and their associated aggregates. The proposed algorithm requires only scan the argument intervals exactly once.

Acknowledgments. We would like to thank Prof. Yuichi Asahiro for the valuable advices that led to this paper. We also thank the anonymous reviewers for their valuable comments.

References

1. Allen, J.F.: Maintaining knowledge about temporal intervals. Commun. ACM **26**(11), 832–843 (1983)
2. Kline, N.: Aggregation in temporal databases, Ph.D. thesis. Univ. of Arizona, May 1999

3. Lopez, I.F.V., Snodgrass, R.T., Moon, B.: Spatiotemporal aggregate computation: a survey. IEEE Trans. Knowl. Data Eng. **17**(2), 271–286 (2005)
4. Böhlen, M., Gamper, J., Jensen, C.S.: Multi-dimensional aggregation for temporal data. In: Ioannidis, Y., Scholl, M.H., Schmidt, J.W., Matthes, F., Hatzopoulos, M., Boehm, K., Kemper, A., Grust, T., Boehm, C. (eds.) EDBT 2006. LNCS, vol. 3896, pp. 257–275. Springer, Heidelberg (2006). doi:10.1007/11687238_18
5. de Berg, M., Cheong, O., van Kreveld, M., Overmars, M.: Computational Geometry, 3rd revised edn. Springer, Heidelberg (2008)
6. Schmidt, J.M.: Interval stabbing problems in small integer ranges. In: Dong, Y., Du, D.-Z., Ibarra, O. (eds.) ISAAC 2009. LNCS, vol. 5878, pp. 163–172. Springer, Heidelberg (2009). doi:10.1007/978-3-642-10631-6_18
7. Kline, N., Snodgrass, R.T.: Computing temporal aggregates. In: Proceedings of International Conference on Data Engineering, pp. 222–231, March 1995

A Continuous Query Language for Stream-Based Artifacts

Maroun Abi Assaf[✉], Youakim Badr, and Youssef Amghar

University of Lyon, CNRS, INSA-Lyon, LIRIS, UMR5205, 69621 Villeurbanne, France
{maroun.abi-assaf,youakim.badr,youssef.amghar}@insa-lyon.fr

Abstract. Artifacts not only combine data and process into the same semantic entities that evolve according to precise state-based lifecycles, but they also provide casual users with a simple and intuitive framework for modeling, executing, and managing business processes. Despite the surge in interest, artifacts still lack effective languages and tools to define, manipulate and interrogate their instances. Most existing artifact languages and notations focuses only on modeling and executing artifact-based processes in which data have discrete values and fail to handle artifacts in which their data include continuous data streams. In this paper, we present the *Continuous Artifact Query Language (CAQL)* for modeling stream-based artifacts and semantically manipulating and interrogating their instances.

Keywords: Domain specific language · Artifact Query Language · Data streams · Continuous query language · Complex event processing

1 Introduction

Artifact-centric processes are introduced by IBM Research labs in 2003 [16]. The artifact-centric approach, rather than entity-relationship modeling in Databases or activity-centric modeling in Business Process Modeling, combines both data and process into self-contained entities, known as *Artifacts* that serve as the basic building blocks to model business processes. Roughly speaking, artifact-centric processes involve; a set of artifacts each of which includes data model as attribute-value pairs, a set of tasks operating on data models, and a state-based lifecycle describing the possible ways that tasks can be invoked on data by following transitions between lifecycle states. An artifact type is thus a blend of data and process acting as a dynamic entity that evolves according to a specified lifecycle to achieve a particular goal. As reported in [8, 11], the artifact-centric approach has successfully been applied to process management and case handling, and has demonstrated many advantages such as enabling a natural modularity and componentization of processes, providing a framework of varying levels of abstraction between data used in business processes. Yet another promising application of artifacts is the *Internet of Things* in which smart objects link networks of sensors and actuators. In this context, smart objects can be modeled as self-evolving artifacts gathering data streams from various sensors, detecting complex events, and performing actions on actuators. Moreover, the artifact data model should not only represent data with discrete values but also data with continuous values originating from data streams. As

© Springer International Publishing AG 2017
D. Benslimane et al. (Eds.): DEXA 2017, Part II, LNCS 10439, pp. 80–89, 2017.
DOI: 10.1007/978-3-319-64471-4_8

a result, using Artifacts in the Internet of Things requires an appropriate language to continuously process artifact data streams and handle complex events. To the best of our knowledge, current work on artifacts still lack effective languages and tools that take full advantage of their semantic nature in order to define, manipulate and interrogate their instances. Most existing languages [4, 9, 20] are graphical or textual notations that not only focus on modeling and executing traditional business processes but also neglect to provide declarative languages similar to SQL in relational databases for specifically managing artifact instances and handling their data streams. In order to define, manipulate and interrogate artifact instances, we have introduced the *Artifact Query Language* (known as *AQL*) in [3]. The *AQL* is built on top of the relational algebra and consists of statements each of which is translated into SQL queries that involve multiple tables, complex joins, and nested statements. In this paper, we extend the *AQL* to manage artifacts in which their data models contain data streams and require complex event processing. To this end, we introduce a model for stream-based artifact systems and define its *Continuous Artifact Query Language (CAQL)*. Finally, we develop a prototype for executing and testing *CAQL* queries.

The remainder of the paper is organized as follow. Section 2 describes the stream-based artifact system through a motivation case study from the *Internet of Things*. Section 3 presents the *CAQL* statements. Section 4 illustrates the prototype implementation. Section 5 presents related works and finally, Sect. 6 concludes the paper.

2 Stream-Based Artifact System Model

A stream-based artifact system W is a triplet *(C, S, R)* where C is a set of *Artifact Classes*, S is a set of *Services*, and R is a set of *Artifact Rules*.

In order to demonstrate the various components of stream-based artifact systems, we introduce an example scenario about a process automating that detects and controls fire incidents in the context of a smart home. We assume that every house is equipped with temperature and smoke sensors and is wirelessly connected to the city control center that remotely detects fire incidents and controls responses. Moreover, the city control center manages, in its databases, information about every house. The city control center detects a fire incident when house temperature sensor values become higher than 57 °C and smoke sensor levels exceed a threshold of 3 in a range of values between 0 and 5. When the fire incident occurs, the control system turns on the alarms, activates water ejectors and alerts the nearest fire station with the house address and location. In addition, the city control system informs the house habitats about the fire incidents by automatically sending short messages. When fire is extinguished and water has been depleted from ejectors, water pumps are remotely activated to refill the ejectors. Water levels are detected using water level sensors. Finally, when the house temperature becomes less than 50 °C the fire is considered to be extinguished and the alarm is then turned off. Finally, the fire incident is archived.

2.1   Artifact Classes

The purpose of every artifact is to achieve a particular goal. In the example scenario, the *Fire Control Artifact (FCA)* deals with fire detection and control. All artifact instances of one particular type (i.e., *FCA*) are described by an *Artifact Class* that includes an *Information Model* in addition to a state-based *Lifecycle* (Fig. 1). The *Information Model* holds data about physical things or entities expressed as a set of data attribute-value pairs of four categories:

1. *Simple attributes* hold one value at a time and record information about artifact instances (i.e., instance identifier, alarm status,…).
2. *Complex attributes* correspond to relations and are expressed as lists of simple attributes (i.e., *House* and *Habitats*).
3. *Reference attributes* refer to artifacts that are directly related to the main artifact in Parent/Child relationships. For example, *Fire Control Artifact* contains a reference to the smoke detector artifact.
4. *Stream attributes* represent data streams generated by streaming sources. They consist of a list of simple attributes with an additional timestamp attribute representing the creation time (i.e., the house's indoor temperature).

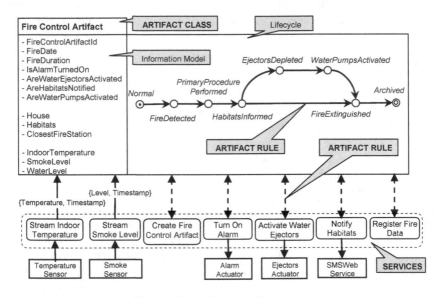

Fig. 1.   Stream-based artifact system

The state-based *Lifecycle* dictates how artifact instances evolve from their initial states to final states passing through intermediate states as triggered by *Artifact Rules*. Figure 1 illustrates the *Fire Control Artifact*'s lifecycle consisting of eight states.

We formalize an *Artifact Class* c as a tuple $(c, A, \gamma_{sim}, \gamma_{com}, \gamma_{ref}, \gamma_{str}, Q, s, F)$ where: c is the artifact name. A is the set of artifact attributes which includes four partitions; simple, complex, reference, and stream attributes. $\gamma_{sim}, \gamma_{com}, \gamma_{ref}, \gamma_{str}$ are the type functions that map attributes in *A* to their associated data types. *Q* is the set of *Lifecycle* states where *s* is its initial state and *F* are its final states.

2.2 Services

Services represent units of work that update artifact attributes and trigger state transitions according to artifact lifecycles. Similarly to Semantic Web Services, *services* are specified using Input, Output, Precondition, and Effect (IOPE). Thus, a *Service s* is the tuple *(s, C_I, C_O, P, E)* where: *s* is the service name. C_I is the set of input artifacts read by *s*. C_O is the set of output artifact modified or created by *s*. And, *P* and *E* are respectively the precondition and effects of *s*. In stream-based artifact systems, services can be of two types: (1) **Adhoc Services** are executed once when certain situations occur as specified by *Artifact Rules*. Their effect is immediate and non-lasting. *RegisterFireData* and *TurnOnAlarm* are examples of *Adhoc Services*. (2) **Stream Services** are invoked on stream attributes when artifacts are instantiated. *Stream Services* continuously read data from stream sources until the artifact instance reaches one of its final states and is archived. *StreamIndoorTemperature* is an example of stream service that reads the house's indoor temperature sensor.

2.3 Artifact Rules

Artifact Rules are variants of declarative Event-Condition-Action (ECA) rules that execute artifact processes in conformance with the artifact state-based lifecycle [6]. *Artifact Rules* fall into two categories: (1) rules that invoke services on artifacts; *"on e if ρ invoke S"*. And, (2) rules that perform state transitions; *"on e if ρ change state of c to q"*, where *e* is an external, timely, or user-generated event. *ρ* is a condition on the states and data attributes of artifact instances. *S* is a list of services to be invoked. *c* is an artifact class. And *q* is a state. *Artifact Rules* can only invoke *Adhoc Services* whereas *Stream Services* are invoked upon artifact's instantiation.

In stream-based artifact systems, *Artifact Rules* are responsible for performing Complex Event Processing (CEP). In CEP, complex events are detected using event expressions involving primitive and composite events. When complex events are detected, associated ECA Rules are invoked and, if the condition holds, the specified action is performed. In stream-based artifact systems, the states of artifact lifecycles represent the complex events. The state transitions correspond to the event expressions that detect the complex events. As a result, *Artifact Rules* that change the state of artifacts are rules that detect complex events. And, *Artifact Rules* that invoke services are rules that execute actions when complex events are detected. From the list of most common event operators described in [7], *Artifact Rules* implement the fundamental event operators; **And, Or**, and **Sequence** operators.

3   The Continuous Artifact Query Language

The *Continuous Artifact Query Language* consists of nine statements as listed in Table 1 and is divided into two sub-languages:

Table 1. *CAQL* statements

Statement	Syntax
Create Artifact	**Create Artifact** \<name\> **Attributes** \<list of attributes\> **States** \<list of states\>
Create Service	**Create Service** \<name\> **Input** \<list of artifacts\> **Output** \<list of artifacts\> **Precondition** \<expression\> **Effect** \<expression\>
Create Rule	**Create Rule** \<name\> (**On** \<event\> | **On** \<event\> **If** \<condition\> | **If** \<condition\>) (**Change State Of** \<artifact\> **To** \<state\> | **Invoke** \<list of services\>)
New	**New** \<artifact\> **Values** \<list of simple attribute values\> {\<complex attribute\> \<list of tuples\>} {\<reference attribute\> **Having** \<condition\>} {\<stream attribute\> **Using** \<stream service\>} [**Set state to** \<state\>]
Retrieve	**Retrieve** \<list of attributes\> **From** \<list of artifacts\> [**Where** \<condition\>] [**Within** \<range\>]
Update	**Update** \<artifact\> **Set** \<list of assignments\> [**Where** ' \<condition\>]
Insert	**Insert** \<attribute\> **Into** \<artifact\> \<list of tuples\> [**Where** \<condition\>]
Remove	**Remove** \<attribute\> **From** \<artifact\> [**Where** \<condition\>]
Delete	**Delete** \<artifact\> [**Where** \<condition\>]

Continuous Artifact Definition Language (CADL): The *Create Artifact* statement defines an artifact class. It consists of a list of data attributes including simple, complex, reference, and stream types, and a list of states. The query 1 in Fig. 2 depicts the definition of the *Fire Control Artifact* using the **Create Artifact** statement. *FireControlArtifactId, FireDate, FireDuration, WaterEjectorsActivated,* and *AlarmTurnedOn* are simple attributes. *House* is a complex attribute consisting of the *Address* and *Surface* of the house. The *"**As Stream**"* keywords denote that *IndoorTemperature* and *SmokeLevel* are stream attributes, consisting of pairs of simple attributes and timestamps. *Normal, FireDetected, PrimaryProcedurePerformed,* and *FireExtinguished* denote states of the artifact lifecycle in which *Normal* is the initial state, *FireExtinguished* is a final state.

```
//Query 1                                          //Query 3
Create Artifact FCA                                Create Rule r2
Attributes (                                       If state(FCA, Normal)
FireControlArtifactId :Integer,                        And FCA.IndoorTemperature.Tmp>57
FireDate : Date,                                   And FCA.SmokeLevel.Lvl>= 3
   FireDuration : Integer,                         Change State OfFCA To FireDetected
   House :{ Address : String, Surface : Real},
   Habitats : { Name : String, PhoneNum : Integer},  //Query 4
IndoorTemperature : { Tmp : Integer,               Create Service StreamIndoorTmp
Time : TimeStamp } As Stream,                      Input FCA
SmokeLevel : { Lvl :  Integer,                     OutputFCA
Time : TimeStamp } As Stream,                      Preconditionclosed(FCA.IndoorTemperature)
   WaterEjectorsActivated : Boolean,                   And defined(FCA.FireControlArtifactId)
   AlarmTurnedOn : Boolean                          And defined(FCA.House)
)                                                  Effect opened(FCA.IndoorTemperature)
States (
   Normal As Initial State,                        //Query 5
   FireDetected,                                   Create Service ActivateWaterEjectors
   PrimaryProcedurePerformed,                      Input FCA
   FireExtinguished As Final State                 Output FCA
)                                                  Precondition FCA.WaterEjectorsActivated = false
//Query 2                                              And defined(FCA.FireControlArtifactId)
Create Rule r1                                     And defined(FCA.House)
If state(FCA, FireDetected)                        EffectFCA.WaterEjectorsActivated = true
Invoke  TurnOnAlarm(FCA),  ActivateWaterEjectors(FCA)
```

Fig. 2. *CADL* query examples

The *Create Service* statement defines a service by specifying its *IOPE* as illustrated in queries 4 and 5 of Fig. 2. *Input* is a list of artifacts that are read by the service. *Output* is a list of artifacts that are modified or instantiated by the service. The *Pre-condition* is the condition that holds before the invocation of the service whereas the *Effect* is the condition that holds after the invocation of the service. Condition expressions are formed from the conjunctions of the following predicates: *new(artifact)*, *opened(stream)*, *closed(stream)*, *defined(attribute)*, *notDefined(attribute)*, and scalar comparison predicates (>, <, <=, >=, =, !=). The *new(artifact)* predicate signifies that a new instance of *artifact* is created. The *opened/closed(stream)* predicates signify that the *stream* attribute is respectively receiving or not receiving input. The *defined/notDefined(attribute)* predicates signifies that *attribute* respectively holds, or does not hold a value.

The *Create Rule* statement defines rules (queries 2 and 3 of Fig. 2). Two types of rules exist: "***If*** *condition* ***Invoke*** *services*", and "***If*** *condition* ***Change State Of*** *artifact* ***To*** *state*". An optional "***On*** *event*" clause can be appended to rules. The event in this case represents an external, timely, or user generated event, i.e. creation of a new application, submission of required documents... etc. In the case that the "***On*** *event*" clause is specified, the "***If*** *condition*" clause can be omitted. The condition expression is formed from the conjunction and disjunction of the following predicates: ***state****(artifact, state)*, ***defined****(selector)*, ***notDefined****(selector)*, and scalar comparison predicates (>, <, <=, >=, =, !=). The *selector* is a cascading reference to simple attributes inside artifacts, such as *selector = {artifact.attribute1, artifact.attribute1.attribute2}*. The ***state****(artifact, state)* predicate signifies that the state of *artifact* should be *state*.

Continuous Artifact Manipulation Language (CAML): The *New* statement instantiates an artifact instance based on its artifact class and initializes its attribute-values and states in addition to invoking the artifact's stream services. The *New* statement has several modes of usages that can be combined to initialize: (1) some or all of the simple attributes using: "**Values**(*value₁, value₂.*)". (2) state using: "**Set State To** *state*", (3) complex attributes by using name of attribute followed by a list of tuples. (4) reference attributes by using: "*attribute* **Having**(*condition*)" where the child artifact referenced by *attribute* should satisfy *condition*, The *New* statement also invokes stream services on stream attributes using: "*attribute* **Using** *service*" (see Fig. 3).

```
//Query 1                                          //Query 4
NewFCA                                             Update FCA
Values(100235)                                     Set Habitats.PhoneNum = 0033763423758
House("20 Av. Albert Einstein", 64),               Where Habitats.Name = "John"
Habitats { ("John", 00330675839457),               And FCA.FireControlArtifactId = 100325
          ("Sam", 00330625374883)}
IndoorTemperature Using StreamIndoorTemperature(this)  //Query 5
SmokeLevel Using StreamSmokeLevel(this)            Insert Habitats Into FCA
Set State To Normal                                { ("Sebastien", 0033823459876),
                                                     ("Nicole", 003357643214) }
//Query 2                                          Where FCA.FireControlArtifactId = 100325
Retrieve * From FCA
Where state(FCA, FireDetected)                     //Query 6
And FCA.IndoorTemperature.Tmp>100                  Remove Habitats From FCA
Within 10 Seconds                                  Where FCA.FireControlArtifactId = 100325
                                                   And Habitats.Name = "John"
//Query 3
Retrieve IndoorTemperature From FCA                //Query 7
Where state(FCA, Normal)                           Delete FCA
Within 3 Seconds                                   WhereFCA.FireControlArtifactId = 100325
```

Fig. 3. *CAML* query examples

The *Retrieve* statement expresses continuous queries that retrieve artifact instances (i.e., queries 2 and 3 of Fig. 3). The result of a *Retrieve* query is continuously updated to reflect changes in the source relations and/or the Sliding Window on streams. The condition of the *Retrieve* statement is specified using the "**Where** *condition*" clause. Three types of filtering conditions can be used: Conditions on simple attributes and states using scalar comparison and state predicates. Conditions on complex attributes using the *include* predicate; "*complex attribute* **Include** {*tuple list*}". And, conditions on reference attributes using the *having* predicate. Additionally, the *Retrieve* statement supports the specification of a *Sliding Window* when stream attributes are involved. The *Sliding Window* is specified using the optional "**Within** *range*" clause where range is a time interval.

The remaining statements of the *CAML* are statements that manipulate artifact instances, including *Update, Insert, Remove*, and *Delete* statements (see Fig. 3 queries 4 to 7). The *Update* statement is used to update simple attributes and states of artifacts. The *Insert* statement is used to insert tuples into complex attributes. It is also used to insert child artifact references into reference attributes. On the other hand, the *Remove* statement is used to remove tuples from complex attributes and child references from reference attributes. Finally, the *Delete* statement is used to entirely delete artifact instances including values of their complex, reference, and stream attributes. Readers are invited to refer to [3] for more information on these statements.

4 Prototyping

We develop an Eclipse RCP-based prototype with a domain-specific language for writing and executing *CAQL* queries. The prototype relies on *Odysseus* Data Stream Management System and its *Procedural Query Language (PQL)* to implement our continuous queries using its logical operators (i.e., selections and repetitions). Figure 4 illustrates the prototype graphical interface and its architecture.

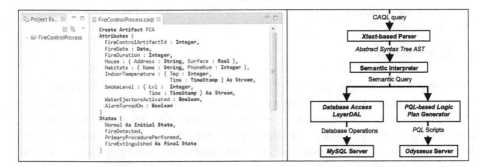

Fig. 4. Prototype graphical interface and architecture

As depicted in Fig. 4, the *Xtext*-based parser reads a *CAQL* query, performs syntax analysis, and generates an *Abstract Syntax Tree (AST)*. The semantic interpreter reads the *AST* and generates an internal representation of the *CAQL* query. The *CAQL* query is rewritten as a *PQL* logical query scripts and executed by the *Odysseus* server.

5 Related Works

Artifacts as building blocks for modeling processes was first introduced in [16]. In [20], *ArtiFlow* is introduced to model artifact processes as finite-state machines In [9], artifact processes are declaratively modeled using the *Guard-Stage-Milestone (GSM)* notations. Works in [15] have introduced the *Business Entity Definition Language (BEDL)*, an XML-based language, for modeling business artifact processes. In [4], artifact processes are defined using *Active XML*. On the other hand, *SQL for Business Artifacts (BASQL)*, introduced in [14], was a first attempt to describe SQL-like statements to define and manipulate artifact instances. In general, these languages and notations for modeling artifact processes have focused on modeling traditional business processes and fail to provide suitable statements streaming based data. The *CAQL* was first introduced in [2]. A first version of the language, referred to as *AQL*, was introduced in [3]. However, *AQL* usages were limited to supporting existing artifact processes and did not involve data streams, continuous queries and complex event processing. From the *Continuous Query*, and *Complex Event Processing (CEP)* perspective, the amount of literature is vast and divergent. The roots of continuous query go back to *Materialized Views* [10] where a view is continuously updated to reflect changes to its base relations. The concept of *Data Streams* was first introduced in [12] under the name of *Chronicles*.

Tapestry [18] was the first to introduce the notion of continuous queries. Since then, many sophisticated data stream management systems have been proposed like: *Aurora* [1], *STREAM* [5], and *Odysseus* [17]. *CEP* goes back to *Active Database* [19] where ECA rules or *Triggers*, fire when an event of interest occurs. Since events in triggers are simple update, insert, or delete operations, works in [7] focuses on specifying complex events. Recent works on *CEP* techniques [13] focuses on expressing continuous queries that detect complex events from input streams. In comparison to *CAQL*, existing continuous query and complex event languages treat low-level streams and relations, while *CAQL* deal with dynamic entities.

6   Conclusion

In this paper, we present the *CAQL* for modeling artifact processes involving data streams and complex event processing capabilities. The *CAQL* allows end-users to efficiently model high-level semantic entities, referring to smart objects or connected devices in the context of the *Internet of Things*. These entities are characterized by their data, which have discrete or continuous values. Future research perspectives include the extension of artifact rules with event operators such as *Not*, *Any*, *Aperiodic*, and *Periodic*. Another important future work consists of conducting an experimental study of the *CAQL* query performance and comparing it with existing benchmark suite in the field of existing continuous query languages.

Acknowledgment. This work is generously supported by the Rhône-Alpes Region, through the *"Big Data Platform Citizens Behaviors in Urban Worlds"* project.

References

1. Abadi, D.J., Carney, D., Çetintemel, U., Cherniack, M., Convey, C., Lee, S., Stonebraker, M., Tatbul, N., Zdonik, S.: Aurora: a new model and architecture for data stream management. Int. J. Very Large Data Bases **12**(2), 120–139 (2003)
2. Abi Assaf, M.: Towards an integration system for artifact-centric processes. In: Proceedings of the 2016 ACM SIGMOD/PODS PhD Symposium, pp. 2–6 (2016)
3. Abi Assaf, M., Badr, Y., Barbar, K., Amghar, Y.: AQL: a declarative artifact query language. In: Pokorný, J., Ivanović, M., Thalheim, B., Šaloun, P. (eds.) ADBIS 2016. LNCS, vol. 9809, pp. 119–133. Springer, Cham (2016). doi:10.1007/978-3-319-44039-2_9
4. Abiteboul, S., Bourhis, P., Galland, A., Marinoiu, B.: The AXML artifact model. In: The 16th International Symposium on Temporal Representation and Reasoning, pp. 11–17 (2009)
5. Arasu, A., Babu, S., Widom, J.: The CQL continuous query language: Semantic foundations and query execution. Int. J. Very Large Data Bases **15**(2), 121–142 (2006)
6. Bhattacharya, K., Gerede, C., Hull, R., Liu, R., Su, J.: Towards formal analysis of artifact-centric business process models. In: Alonso, G., Dadam, P., Rosemann, M. (eds.) BPM 2007. LNCS, vol. 4714, pp. 288–304. Springer, Heidelberg (2007). doi:10.1007/978-3-540-75183-0_21
7. Chakravarthy, S., Krishnaprasad, V., Anwar, E., Kim, S.K.: Composite events for active databases: semantics, contexts and detection. In: Proceedings of the Conference on VLDB, pp. 606–617 (1994)

8. Cohn, D., Hull, R.: Business artifacts: a data-centric approach to modeling business operations and processes. Bull. IEEE Comput. Soc. Tech. Committee Data Eng. **32**(3), 3–9 (2009)

9. Damaggio, E., Hull, R., Vaculín, R.: On the equivalence of incremental and fixpoint semantics for business artifacts with Guard–Stage–Milestone lifecycles. Inf. Syst. **38**(4), 561–584 (2013)

10. Gupta, A., Mumick, I.S.: Maintenance of materialized views: problems, techniques, and applications. Bull. IEEE Comput. Soc. Tech. Committee Data Eng. **18**(2), 3–18 (1995)

11. Hull, R.: Artifact-centric business process models: brief survey of research results and challenges. In: Meersman, R., Tari, Z. (eds.) OTM 2008. LNCS, vol. 5332, pp. 1152–1163. Springer, Heidelberg (2008). doi:10.1007/978-3-540-88873-4_17

12. Jagadish, H.V., Mumick, I.S., Silberschatz, A.: View maintenance issues for the chronicle data model. In: Proceedings of the 14th ACM SIGACT-SIGMOD-SIGART Symposium on Principles of Database Systems, pp. 113–124 (1995)

13. Jiang, Q., Adaikkalavan, R., Chakravarthy, S.: MavEStream: synergistic integration of stream and event processing. In: The International Conference on Digital Telecommunications, p. 29 (2007)

14. Joseph, H.R., Badr, Y.: Business artifact modeling: a framework for business artifacts in traditional database systems. In: Enterprise Systems Conference, pp. 13–18 (2014)

15. Nandi, P., Koenig, D., Moser, S., Hull, R., Klicnik, V., Claussen, S., Kloppmann, M., Vergo, J.: Data4BPM, Part 1: Introducing Business Entities and the Business Entity Definition Language (BEDL). IBM Corporation (2010)

16. Nigam, A., Caswell, N.S.: Business artifacts: an approach to operational specification. IBM Syst. J. **42**(3), 428–445 (2003)

17. Odysseus, Data Stream Management System. https://odysseus.informatik.uni-oldenburg.de

18. Terry, D.B., Goldberg, D., Nichols, D., Oki, B.M.: Continuous queries over append-only databases. In: The ACM SIGMOD International Conference on Management of Data, pp. 321–330 (1992)

19. Widom, J., Ceri, S.: Active Database Systems: Triggers and Rules for Advanced Database Processing. Morgan Kaufmann Publishers, San Francisco (1996)

20. Xu, W., Su, J., Yan, Z., Yang, J., Zhang, L.: An artifact-centric approach to dynamic modification of workflow execution. In: Meersman, R., et al. (eds.) OTM 2011. LNCS, vol. 7044, pp. 256–273. Springer, Heidelberg (2011). doi:10.1007/978-3-642-25109-2_17

Past Indeterminacy in Data Warehouse Design

Christina Khnaisser[1,2], Luc Lavoie[1], Anita Burgun[2],
and Jean-François Ethier[1,2,3(✉)]

[1] Département d'informatique, Université de Sherbrooke, Sherbrooke, Canada
{christina.khnaisser,luc.lavoie,
jf.ethier}@usherbrooke.ca
[2] INSERM, UMR 1138 team 22, Centre de Recherche des Cordeliers,
Université Paris Descartes, Paris, France
anita.burgun@aphp.fr
[3] Département de médecine, Université de Sherbrooke, Sherbrooke, Canada

Abstract. Traditional data warehouse design methods do not fully address some important challenges, particularly temporal ones. Among them past indeterminacy is not handled systematically and uniformly. Furthermore, most methods published until now present transformation approaches by providing examples rather than general and systematic transformation rules. As a result, real-world applications require manual adaptations and implementations. This hinders scalability, long-term maintenance and increases the risk of inconsistency in case of manual implementation. This article extends the Unified Bitemporal Historicization Framework with a set of specifications and a deterministic process that defines simple steps for transforming a non-historical database schema into a historical schema allowing data evolution and traceability, including past and future indeterminacy. The primary aim of this work is to help data warehouse designers to model historicized schema based on a sound theory ensuring a sound temporal semantic, data integrity and query expressiveness.

Keywords: Data warehouse design · Temporal data warehouse · Temporal indeterminacy · Missing information

1 Introduction

Historicization is the process that consists of transforming a non-historical (database) schema into a historicized one. There are complex design problems that need to be addressed [2]. A recent survey paper [4] did not identify new innovative models successfully addressing the following gaps. First, when applied to historical data warehouses (HDW), the published methods require multiple manual steps (both in terms of design and implementation) [12]. Second, the existing solutions are based on different data models (relational, entity-relationship, object-oriented, multidimensional) and use different structures and semantics. This diminishes generalizability and large-scale adoption. Finally, many solutions for temporal data warehouses do not handle missing information.

© Springer International Publishing AG 2017
D. Benslimane et al. (Eds.): DEXA 2017, Part II, LNCS 10439, pp. 90–100, 2017.
DOI: 10.1007/978-3-319-64471-4_9

This paper extends the Unified Bitemporal Historicization Framework[1] (UBHF), to cope with past indeterminacy. UBHF and it's time model were formally defined in [11].

In 2006, Rizzi et al. [13] wrote: "Though a lot has been written about how data warehouse should be designed, there is no consensus on a design method yet". According to our last survey of data warehouse design (DWD) methods [12] this is still valid.

Two major temporal models have emerged in the literature and in our domain of interest (clinical data warehousing). The first one is the Bitemporal conceptual data model (BCDM) a bitemporal model based on SQL. Snodgrass presents design "best practices" to build a bitemporal schema starting from a conceptual model (entity relationship) ending with SQL code (and TSQL2 [22] a temporal SQL extension). BCDM was initially defined in [9], a more recent presentation can be found in [14] and an extension to temporal indeterminacy in [3]. The second one is the Date-Darwen-Lorentzos Model (DDLM) which is based on the relational theory. It proposes three sub-models, two unitemporal model and one bitemporal model based on the third manifesto relational model [5] and Allen's interval logic [1]. In previous work, a comparative study [10] showed interesting similarities (and some differences) between BCDM and DDLM when using UBHF to express them [11] but none can express past indeterminacy.

Regularly, patients cannot provide exact dates regarding important health events that happened sometimes more than 40 years ago. In the event of a fracture, the exact onset moment was known at some point in time but might have been forgotten, leading to a missing date in the electronic medical record. A related but different problem occurs when a new diagnosis is made (e.g. diabetes). While we know that the disease is present at the time of the first diagnosis, for most diseases, the exact onset moment is unknown and it could be days, weeks, months or even years (e.g. hypertension) before. Currently, temporal data related to the examples above is often not included in the database. In the case where the application forces the user to enter a precise value, the clinicians will input an approximate date. This can lead to significant inconsistencies in the treatment of temporal operations during queries or the exclusion of relevant information.

2 UBHF Concepts

UBHF is a conceptual framework that defines specifications and deterministic trans-formations based on fundamental relational and temporal concepts [7, 14]. This paper, presents a synthesis of these concepts, a more elaborated version can be found in [11].

2.1 Time Model

UBHF uses a discrete time model based on points and intervals derived from the one defined by Allen in [1] and used in [7]. TIMPEPOINT and PERIOD will be used as a representative time point type and time interval type in the following sections. The

---

[1] Technical report can be found at http://griis.ca/surl/ubfh-dexa.

retained timelines are transaction time (@T) and valid time (@V) as defined in the consensus glossary [8]. In the context of a HDW, the distinction between future indeterminacy and past indeterminacy allows easier processing of the retrospective and perspective data.

Future indeterminacy (indeterminate end) concerns facts with a known beginning and an unknown ending. In BCDM, this endpoint is represented by the constant « until changed » for a valid-time period and with « forever » for a transaction-time period. These constants are encoded in the database tables as the maximal value of TIMESTAMP type (9999-12-31 when granularity is one day). In DDLM and UBHF, this is represented with the function *ufn* «until further notice» and is only used in query expression (not for storage purpose).

Past indeterminacy (indeterminate begin) concerns facts with a known ending and an unknown beginning. This indeterminacy is overlooked by most models including DDLM and BCDM. In UBHF, this is represented with the function *saw* «since a while» and is only used in query expression (not for storage purpose).

2.2 Timelines

UBHF represents a timeline by an attribute (called timeline attribute). A timeline attribute can have different types and values defined as follows:

- A period timeline attribute (be): where the beginning and the end point values are known.
- A point timeline attribute with unknown end value (bx): where the beginning point is known and the end is unknown.
- A point timeline attribute with unknown beginning value (xe): where the beginning point is unknown and the end is known.

The table below defines the notation used. Note that, for transaction timelines the xe-type is not defined, as in a DBMS the beginning point of a transaction is always known (Table 1).

Table 1. Timeline attribute definition and notation.

Notation	Definition	Timeline
@Vbe	Valid time period	Valid
@Vbx	Valid time begin point	
@Vxe	Valid time end point	
@Tbe	Transaction time period	Transaction
@Tbx	Transaction time point	

2.3 Attributes and Relations

In a non-historical schema, we conventionally distinguish between key and non-key attributes. A non-historicized relation R is denoted R(K, A), where:

- $K = \{k_1,..., k_{|K|}\}$ is the set of key attributes ($|K| \geq 1$). Without loss of generality, we consider that each relation contains only one key - although this key may have more than one attribute.
- $A = \{a_1,..., a_{|A|}\}$ is the set of non-key attributes ($|A| \geq 0$).

A historicized relation R is denoted $R'(K, B, C, D_V, D_T)$ with $A = B \cup C$ where:

- $B = \{b_1,..., b_{|B|}\}$ is the set of non-key attributes ($|B| \geq 0$) associated with a valid timeline attribute (called historicized attributes); B is a subset of A.
- $C = \{c_1,..., c_{|C|}\}$ is the set of non-key attributes ($|C| \geq 0$) **not** associated with a valid timeline attribute(called non-historicized attributes); C is a subset of A.
- $D_V = \{@V, b_1@V,..., b_{|B|}@V\}$ is the set of valid timeline attributes, with the following notations @V is associated with K, and $b_i@V$ is associated with $b_i \in B$.

$D_T = \{@T, a_1@T, ..., b_{|A|}@T\}$ is the set of transaction timeline attribute where @T is associated with K, and $a_i@T$ is associated with $a_i \in A$. In the context of R', K and "key" do not refer to the same entity. K is the key set of the original relation R where the "key" of R' is the union of K with D_V and D_T.

Keeping history changes may introduce redundancy, contradiction, circumlocution and non-denseness when attributes in the same relation are modified independently. DDLM studied these problems in detail [7] (chap. 5 and 13) and proposed constraints to avoid them. See Sect. 4 for a generalized form of constraint definitions.

3 Historicization Process

Data modifications may introduce data inconsistency as described in previous sections. Data inconsistency is usually addressed by schema normalization and constraint specification. The normalization process uses lossless relational decomposition to split a relation into smaller relations ("relparts" for short). Two types of such decomposition are used hereafter: projection-join decomposition (PJ) and the restriction-union decomposition (RU) [11]. The historicization process consists of the following activities which are presented in the remaining of the present section:

- Schema annotation (guided).
- Historical schema construction (automated).
- Temporal constraints specification (automated).

3.1 Schema Annotation

The aim of the schema annotation is to "parameterize" the algorithm of the schema transformation according to the domain needs. For each relation R(K, A) of the initial schema previously normalized in 5th normal form (5NF): Split A into B (attributes with valid time) or C (attributes without).

3.2 Modelling

For each relation R(K, B, C) the following steps are required:

1. Decompose the relation R into 6NF using the PJ decomposition[2]
   $PJ(R\_K\{K\}, R\_b_1\{K, b_1\}, ..., R\_b_n\{K, b_n\}, R\_c_1\{K, c_1\}, ..., R\_c_m\{K, c_m\})$.

   To facilitate constraint definition, these (numerous) relparts are conceptually grouped into three types of groupings [11]:

- The *K-grouping* of a relation R is the set of all the K relparts. A K-relpart, denoted R_K, is a relation with K and the associated timeline attributes only.
- A *b_i-grouping* of a relation R is the set of all b_i-relparts. Collectively, the b_i-groupings are called *B-groupings*.
- A *c_i-grouping* of a relation R is the set of all c_i-relparts. Collectively, the c_i-groupings are called *C-groupings*.

For each relation R(K, B, C) the following steps are required to produce the historical representation (i.e. R!VT(K, B, C, D_V, D_T)).

2. For each relpart in K-grouping and B-groupings: Add the timeline attribute @V.
3. For each relpart in K-grouping and B-groupings: Decompose each resulting relparts, using RU decomposition over the @V timeline attribute to separate timeline between @Vxe, @Vbe and @Vbx types, renaming the relparts accordingly.
4. For all relparts (in K-, B- and C-groupings): Add two transaction time relparts: one with @Tbx and one with @Tbe.

In the following, without loss of generality, all relations are considered bitemporal. If a relation is a transaction time relation, only the last step is required.

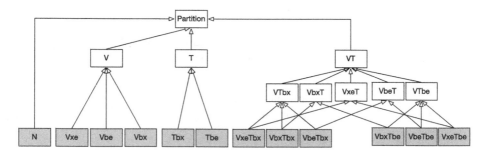

Fig. 1. Hierarchy of all potential relational partitions (Color figure online)

---

[2] A PJ decomposition of a relation R is defined by its contributive projection subsets, each of them containing the common key and the union of them being equals to the attributes of R. Here, the subsets are precisely to the 6NF relparts of R (including the key relpart).

3.3 Relational Partitions

Relparts are conceptually split into "partitions" to facilitate constraint definition and query expressiveness [11]. A partition is defined regarding the relation category and timeline category. Figure 1 shows the hierarchy of potential relational partitions that can be found in a historical schema depending on the decomposition. In UBFH, the structure and the constraints are defined over the leaf partitions (in blue).

In the following section a relational partition of R is denoted R@S where S is one of the categories illustrated in the previous figure (white and blue boxes). We may also refer to a specified grouping in a specific partition denoted R_g@S.

4 Constraints

This section presents constraint specifications that are automatically generated for a specified relation. Constraints are defined on V partitions because they can be modified by the user. No constraints are defined for T partitions because they cannot be modified by the user as they are managed by the DBMS using any of the following solutions: (a) as views when the values of the transaction timeline attribute are obtained by a function call to the DBMS journal as in DDLM or (b) as base relations (table) when the values of the transaction timeline attribute are set by the DBMS (as in SQL:2011 with SYSTEM TIME), or by triggers (as in BCDM [14]). In UBHF, all solutions are supported if they satisfy the transaction timeline semantic.

The following constraints must be defined regarding each relation of the initial schema. A constraint is defined with a unique identifier and a boolean expression.

4.1 Candidate Keys

For each grouping R_g in partition S{N, Vxe, Vbx, Tbx, VxeTbx, VbxTbx}, the key constraint is the same as the initial relation. The constraint of R_g@S is:

```
RELATION R_g@S (K, B, C, D_V, D_T)   KEY {K};
```

For historicized relparts with be-type timeline, the key constraint must be applied to each time point in the period. For each grouping R_g in partition S{Vbe, VbeTbx}, add:

```
RELATION R_g@S (K, B, C, D_V, D_T)   USING(@Vbe): KEY {K, @Vbe};
```

For each grouping R_g in partition S{Tbe, VbxTbe, VxeTbe}, add:

```
RELATION R_g@S (K, B, C, D_V, D_T)   USING(@Tbe): KEY {K, @Tbe};
```

For each grouping R_g in partition S{VbeTbe}, add:

```
RELATION R_g@S (K, B, C, D_v, D_T)  USING(@Vbe, @Tbe): KEY {K, @Vbe, @Tbe};
```

4.2   Temporal Denseness

The temporal denseness is defined over all relparts of an initial relation to ensure the history completeness. Each initial relation R has been decomposed in one K-grouping and some B- and C-groupings by historicization. Those groupings must stay consistent, i.e.: each key defined in the K-relpart must also be defined in the same period in one of its related present or missing relpart of B- and C-groupings. This constraint is inspired by the requirements 3 and 6 in [7] and by the equality dependency defined in [6].

First, a shorthand operator, **gSpace**, defined in [11], is extended to cover Vxe partition. The operator extracts the history (if any) of a specified grouping by calculating the union of all relparts of a grouping R_g with respect to the applicable partition[3] (@N or @V):

```
OPERATOR gSpace(R_g GROUPING) RETURNS RELATION;
  IF R_g is a C-grouping THEN
    R_g@N
  ELSE // R_g is a K-grouping or a B-grouping
    WITH (
      r_xe := (EXTEND R_g@Vxe : {@V:=[saw:@Vxe]}) {ALL BUT @Vxe}),
      r_bx := (EXTEND R_g@Vbx : {@V:=[@Vbx:ufn}) {ALL BUT @Vbx}),
      r_be := R_g@Vbe RENAME {@Vbe AS @V}
    ): USING (@V): r_xe UNION r_bx UNION r_be
  END IF
END OPERATOR
```

The temporal denseness may now be expressed using the two following rules: For each b_i-grouping, the constraint is defined as follows:

```
CONSTRAINT R_b_i_denseness
  USING(@V): gSpace(R_K) = gSpace(R_b_i){K,@V}
```

For each c_i-grouping, the constraint is defined as follows:

```
CONSTRAINT R_c_i_denseness
  gSpace(R_K){K} = gSpace(R_c_i){K}
```

---

[3] gSpace is defined to deal with @T and @VT partitions. In this paper only @N and @V are represented.

4.3   Foreign Keys

A foreign key is evaluated regarding related attributes in different relparts. In a historicized relation, the associated timeline attributes must be considered to guarantee that their related values are asserted at the same time (at each moment of the unpacked relation). On the one hand, the foreign key in the initial schema must be maintained. Let Rs{X}→Rd be a foreign key in the initial schema where Rs is the source relation with X being any subset of attributes of Rs equivalent to the key (K) of Rd, the destination relation. The constraint guarantees temporal referential consistency between groupings by verifying that the projection of Rs on X is included in the projection of Rd on K (with suitable renaming). Using gSpace, another shorthand operator, **gUnpack**, is defined, extracting the unpacked history of an attribute x of a specified relation R:

```
OPERATOR gUnpack(R RelationName, x AttributeName) RETURNS RELATION;
   IF (x in K of R) THEN     UNPACK (@V): (gSpace(R_K){x,@V})
   ELSIF (x in B of R) THEN UNPACK (@V): (gSpace(R_x){x,@V})
   ELSE                     UNPACK (@V): gSpace(R_x){x} JOIN gSpace(R_K)
   END IF
END OPERATOR
```

Each foreign key Rs{X}→Rd induce the following constraint:

```
CONSTRAINT Rs_Rd_@V_fk
   gUnpack(Rs, x₁) RENAME {x₁ AS k₁} ⊆ gUnpack(Rd_K, k₁)
   AND … AND
   gUnpack(Rs, xₙ) RENAME {xₙ AS kₙ} ⊆ gUnpack(Rd_K, kₙ);
```

4.4   Key History Uniqueness

The key history uniqueness is defined over the value of the timeline associated with K. It ensures non-redundancy and non-contradiction of the key attributes values over time. In other words, it ensures consistency of the history of a tuple by verifying that the same proposition is represented only once. This constraint is similar to requirement 1 in [7]. For the K-relpart, the following constraint must be applied:

```
CONSTRAINT R_K_history
   IS_EMPTY (R_K@Vbx JOIN R_K@Vbe WHERE @Vbx < POST(@Vbe)) AND
   IS_EMPTY (R_K@Vbe JOIN R_K@Vxe WHERE @Vxe > PRE(@Vbe)) AND
   IS_EMPTY (R_K@Vxe JOIN R_K@Vbx WHERE @Vxe > PRE(@Vbx));
```

4.5   Attribute History Uniqueness

The attribute history uniqueness is defined over the value of the timeline attribute associated with a b_i attribute. It ensures non-redundancy and non-contradiction of b_i values over time. This constraint is similar to requirement 4 in [7]. For each grouping R_g in {R_b₁, …, R_bₙ}, the following constraint must be applied:

```
CONSTRAINT R_g_uniqueness
  IS_EMPTY (R_g@Vbx {K, @Vbx} JOIN R_g@Vbe {K, @Vbe}
    WHERE @Vbx < POST(@Vbe)) AND
  IS_EMPTY (R_g@Vbe {K, @Vbe} JOIN R_g@Vxe {K, @Vxe}
    WHERE @Vxe > PRE(@Vbe)) AND
  IS_EMPTY (R_g@Vxe {K, @Vxe} JOIN R_g@Vbx {K, @Vbx}
    WHERE @Vxe > PRE(@Vbx));
```

4.6   Key History Non-circumlocution

The key history non-circumlocution is defined over the value of the timeline attribute associated with K. It ensures non-circumlocution of the key attributes values over time. This constraint is similar to requirement 2 in [7]. For the K-relpart, the following constraint must be applied:

```
CONSTRAINT R_K_circumlocution
  IS_EMPTY (R_K@Vbx JOIN R_K@Vbe WHERE @Vbx = POST(@Vbe)) AND
  IS_EMPTY (R_K@Vbe JOIN R_K@Vxe WHERE @Vxe = PRE(@Vbe)) AND
  IS_EMPTY (R_K@Vxe JOIN R_K@Vbx WHERE @Vxe = PRE(@Vbx));
```

The two key history constraints (uniqueness and non-circumlocution) may be combined in one.

```
CONSTRAINT R_K_key_history
  IS_EMPTY (R_K@Vbx JOIN R_K@Vbe WHERE @Vbx <= POST(@Vbe)) AND
  IS_EMPTY (R_K@Vbe JOIN R_K@Vxe WHERE @Vxe >= PRE(@Vbe)) AND
  IS_EMPTY (R_K@Vxe JOIN R_K@Vbx WHERE @Vxe >= PRE(@Vbx));
```

4.7   Attribute History Non-circumlocution

The attribute history non-circumlocution is defined over the value of the timeline attribute associated to a b_i attribute. It ensures non-circumlocution of b_i values over time. This constraint is similar to requirement 5 in [7]. For each present value relpart R_b_i, the following constraint must be applied:

```
CONSTRAINT R_bi_circumlocution
  IS_EMPTY (R_bi@Vbx {K, bi, @Vbx} JOIN R_bi@Vbe {K, bi, @Vbe}
    WHERE @Vbx = POST(@Vbe)) AND
  IS_EMPTY (R_bi@Vbe {K, bi, @Vbe} JOIN R_bi@Vxe {K, bi, @Vxe}
    WHERE @Vxe = PRE(@Vbe)) AND
  IS_EMPTY (R_bi@Vxe {K, bi, @Vxe} JOIN R_bi@Vbx {K, bi, @Vbx}
    WHERE @Vxe = PRE(@Vbx));
```

5 Conclusion

Most proposed DW design methods define transformation rules "by-example" and must largely be adapted and applied manually. More specifically, regarding past indeterminacy, some DW design methods propose "ideas" but none presents a completely integrated deterministic process. UBHF defines (a) relation, attribute and timeline categorization to provide unique semantic; (b) a unified temporal structure and general constraints to be independent of the domain and context (yet providing formal definition and superior automation capabilities); (c) historicization processes with past indeterminacy ensuring traceability over the transformation steps without losing the initial conceptual view.

A model based on UBHF satisfies (1) data integrity with the definition of the described constraints (2), sound temporal schema design using relational concepts and well-defined temporal concepts (3) data schema evolution due to the normalization in 6NF (4) schema traceability and guided automation. This paper has extended UBHF which is now suitable to guided automated historicization of a database schema including past indeterminacy. This will enable improved modelling and query possibilities in many domains, especially in healthcare.

Despite all UBHF concepts, the DW schema still not "fully" historicized. In other words, adding the past indeterminacy induces a fourth timeline category, denoted eb, that represents events that occur "certainly" at some period (from e to b) but the beginning and the end point are not known. This situation may appear when two tuples having the same a_i values and two different timelines xe and bx that merge.

Future work is required to offer a fully applicative solution:

- To add the eb-type timeline.
- To model missing information.
- To implement a tool for designing historicized schema and translating it into standard SQL code or TutorialD code so existing DBMS may be used directly.
- To evaluate UBFH approach in real applications.
- To propose a generalized set of data modification operators (insert, delete, update).

References

1. Allen, J.F.: Maintaining knowledge about temporal intervals. Commun. ACM **26**(11), 832–843 (1983)
2. Anselma, L., Piovesan, L., Terenziani, P.: A 1NF temporal relational model and algebra coping with valid-time temporal indeterminacy. J. Intell. Inf. Syst., 1–30 (2015)
3. Anselma, L., Terenziani, P., Snodgrass, R.T.: Valid-time indeterminacy in temporal relational databases: semantics and representations. IEEE Trans. Knowl. Data Eng. **25**(12), 2880–2894 (2013)
4. Arora, S.: A comparative study on temporal database models: a survey. In: 2015 International Symposium on Advanced Computing and Communication (ISACC), pp. 161–167 (2015)
5. Darwen, H., Date, C.J.: The third manifesto. SIGMOD Rec. **24**(1), 39–49 (1995)

6. Date, C.J., Darwen, H.: Database Explorations: Essays on the Third Manifesto and Related Topics. Trafford Publishing, San Francisco (2010)
7. Date, C.J., Darwen, H., Lorentzos, N.A.: Time and relational theory: temporal databases in the relational model and SQL. Morgan Kaufmann, Waltham (2014)
8. Jensen, C.S., et al.: The consensus glossary of temporal database concepts — February 1998 version. In: Etzion, O., Jajodia, S., Sripada, S. (eds.) Temporal Databases: Research and Practice. LNCS, vol. 1399, pp. 367–405. Springer, Heidelberg (1998). doi:10.1007/BFb0053710
9. Jensen, C.S., Soo, M.D., Snodgrass, R.T.: Unifying temporal data models via a conceptual model. Inf. Syst. **19**, 513–547 (1993)
10. Khnaisser, C.: Méthode de construction d'entrepôt de données temporalisé pour un système informationnel de santé. Faculté des sciences, Université de Sherbrooke (2016)
11. Khnaisser, C., Lavoie, L., Burgun, A., Ethier, J.-F.: Unified Bitemporal Historicization Framework. Université de Sherbrooke (GRIIS), Sherbrooke, Québec, Canada (2017)
12. Khnaisser, C., Lavoie, L., Diab, H., Ethier, J.-F.: Data warehouse design methods review: trends, challenges and future directions for the healthcare domain. In: Morzy, T., Valduriez, P., Bellatreche, L. (eds.) ADBIS 2015. CCIS, vol. 539, pp. 76–87. Springer, Cham (2015). doi:10.1007/978-3-319-23201-0_10
13. Rizzi, S., Abello, A., Lechtenborger, J., Trujillo, J.: Research in data warehouse modeling and design: dead or alive? In: 9th ACM International Workshop on Data Warehousing and OLAP – DOLAP 2006, pp. 3–10. Association for Computing Machinery, New York (2006)
14. Snodgrass, R.T.: Developing Time-Oriented Database Applications in SQL. Morgan Kaufmann Publishers, San Francisco (2000)

Text Processing and Semantic Search

A Case for Term Weighting
Using a Dictionary on GPUs

Toshiaki Wakatsuki$^{(\boxtimes)}$, Atsushi Keyaki, and Jun Miyazaki

Department of Computer Science, School of Computing,
Tokyo Institute of Technology, Tokyo, Japan
{wakatsuki,keyaki}@lsc.cs.titech.ac.jp, miyazaki@cs.titech.ac.jp

Abstract. This paper explains the demonstration of a fast method of
Okapi BM25 term weighting on graphics processing units (GPUs) for
information retrieval by combining a GPU-based dictionary using a succinct data structure and data parallel primitives. The main problem with
handling documents on GPUs is in processing variable length strings,
such as the documents themselves and words. Processing variable sizes
of data causes many idle cores, i.e., load imbalances in threads, due to
the single instruction multiple data (SIMD) nature of the GPU architecture. Our term weighting method is carefully composed of efficient
data parallel primitives to avoid load imbalance. Additionally, we implemented a high performance compressed dictionary on GPUs. As words
are converted into identifiers (IDs) with this dictionary, costly string
comparisons could be avoided. Our experimental results revealed that
the proposed method of term weighting on GPUs performed up to 5×
faster than the MapReduce-based one on multi-core CPUs.

Keywords: GPGPU · Term weighting · Dictionary · Parallel primitive

1 Introduction

Large numbers of documents have been created with the spread of computers
and the Web. A method of high throughput document processing is necessary
to process these large numbers of documents within a practical amount of time.
Much research has been carried out on utilizing commodity hardware to address
this issue. MapReduce is one of the most popular parallel programming models for processing large-scale data on large computer clusters [2]. MapReduce
has widely been used for processing documents, such as by term weighting and
inverted index construction [8].

GPUs, on the other hand, have widely been adopted in much research and
many applications because they offer high performance computing by utilizing
many cores and high levels of memory bandwidth. Emerging general-purpose
computing on GPUs has led to the drastic expansion of fields of application. For
example, many high performance numerical calculation libraries support GPUs

© Springer International Publishing AG 2017
D. Benslimane et al. (Eds.): DEXA 2017, Part II, LNCS 10439, pp. 103–117, 2017.
DOI: 10.1007/978-3-319-64471-4_10

for their floating point operations[1]. Furthermore, even high performance non-numerical calculation libraries have come to support GPUs. For instance, a graph processing library utilizes GPU computing primitives and optimization strategies to achieve a balance between performance and expressiveness [18]. The effective use of data parallel primitives is the key to constructing high performance programs. Although string processing has been recognized as being unsuitable for GPUs, some GPU-accelerated algorithms have been proposed such as string matching for database applications [16].

There have been some variations in MapReduce for various platforms such as GPUs [3] as well as shared memory processors [17] and clusters[2]. However, MapReduce cannot extend the potential power of GPUs due to the load imbalance among tasks.

This paper proposes an implementation of an efficient dictionary on GPUs as a new data parallel primitive for document processing, and a methodology of constructing efficient document processing by using the dictionary and existing data parallel primitives on GPUs. We then discuss its applicability to a realistic application. We particularly focus on the high performance calculations of Okapi's BM25 term weighting [15] which is a widely used term weighting scheme in information retrieval systems by using these primitives on GPUs as an example, rather than simple string match computations. Many strings need to be compared in document processing, e.g., exact matches of two words and sorting a set of words. However, comparing strings involves very costly calculations on GPUs because of their variable sizes. The dictionary is very useful when handling many strings in document processing because it efficiently converts all words in a document into corresponding integer IDs so that costly comparisons of strings are replaced with low cost comparisons of integers on GPUs.

When Web pages are processed, the size of the vocabulary increases because they contain numerous proper nouns, such as named entities and URLs. The dictionary size, however, must remain small even for large vocabularies due to the memory size limitations of GPUs. Therefore, we improved a compressed dictionary algorithm with a succinct data structure [7] to implement a GPU-accelerated dictionary, so that a large vocabulary could be handled even with the limited memory size of GPUs.

We also conducted experiments to demonstrate the power of our dictionary and the suitability of combining it with other existing data parallel primitives for calculating BM25 term weights as an example of document processing.

The rest of the paper is organized as follows. Section 2 refers to the background for this study including parallel primitives, and Sect. 3 describes an implementation of BM25 term weighting with our dictionary on GPUs. The experimental results are discussed in Sect. 4, which are followed by concluding remarks in Sect. 5.

---

[1] https://developer.nvidia.com/gpu-accelerated-libraries.
[2] http://hadoop.apache.org/.

2   Background

2.1   Notations

Let Σ be a finite set of alphabets. We denote the text of length n by $T = c_1c_2c_3 \ldots c_n$, $c_i \in \Sigma$. We have $\Sigma = \{a, b, \ldots, z\}$, $|\Sigma| = 26$ for the English text. When T is written in a natural language, it is assumed that almost all T is composed of a finite vocabulary, V_k, where $k = |V_k|$. Then, T is expressed as $T = t_1t_2t_3 \ldots t_m$, $t_i \in V_k$. Let d be the ID of a document and t be the ID of a term.

2.2   Term Weighting

The weight of term t in document d on Okapi's BM25 term weighting scheme [15] is defined by Eq. (1).

$$w_{td} = \log \frac{N}{df_t} \cdot \frac{(k_1 + 1)tf_{td}}{k_1((1 - b) + b \times (L_d/L_{ave})) + tf_{td}}, \tag{1}$$

where N is the number of documents, df_t is the number of documents that contain term t, and tf_{td} is the frequency of term t in document d. Here, L_d is the length of document d and L_{ave} is the average of all L_d. The variables, k_1 and b, are tuning parameters.

2.3   Term Weighting Method Using MapReduce

Map. Each document is assigned to a map worker. The map worker extracts terms from the document that is assigned, and then generates key-value pairs, each of which contains term t as a key and document ID d as a value. If the same term appears multiple times, pairs are generated multiple times. The generated pairs are stored, and the length of pairs is equal to that of document d.

Reduce. A reduce worker processes the group of key-value pairs that have the same term, t, as a key. The list of document ID d is obtained from aggregated values. When term t appears multiple times in document d, the same number of d exists in the list. For each d, tf_{td} is obtained by counting the number of d in the list. In addition, df_t is calculated by counting the number of unique d in the list.

We sort the list of d and use the sorted list of d to obtain tf_{td}, df_t, and w_{td}. First, the whole list is scanned to obtain df_t. Then, the list is scanned again to obtain tf_{td}, while w_{td} is calculated each time.

2.4   Dictionary

A dictionary is a data structure that handles a set of strings. A set of strings with supplementary information such as ID and description is stored in the dictionary in advance. After that, we can use this dictionary to determine whether

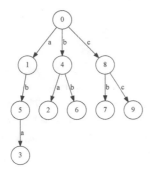

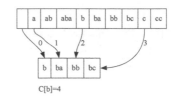

C[b]=4

Fig. 1. Example of trie.

Fig. 2. Example transition from each node by b. $B_b = 1100100010$, $C[a] = 1, C[b] = 4, C[c] = 8$.

Table 1. Example of state transition table (STT).

	0	1	2	3	4	5	6	7	8	9
a	1	−1	−1	−1	2	3	−1	−1	−1	−1
b	4	5	−1	−1	6	−1	−1	−1	7	−1
c	8	−1	−1	−1	−1	−1	−1	−1	9	−1

a certain string is included in the dictionary or not. If the string is included, its supplementary information is retrieved. The ideal dictionary algorithm in areas managing large-scale vocabulary and/or requiring limited memory capacity needs to have features that achieve a minimal memory footprint and fast lookup operation. Martínez-Prieto et al. conducted a comparative study of compressed string dictionaries from the viewpoint of both theoretical and practical performance [9].

Dictionary Encoding. Let us consider mapping that converts word w into corresponding ID t. Using this mapping, T is expressed as a dictionary-encoded sequence of integers $T = t_1 t_2 t_3 \ldots t_m$.

Instead of expressing T as a string, dictionary encoding enables an efficient comparison of words and conserves memory space in most situations.

STT. Trie is one of the data structures that support the retrieving of information using variable length keys [4]. It is represented as a tree structure. The edges are labeled with a character. Each node corresponds to a prefix of the key. The information associated with the key can be retrieved by tracing the edges from the root. The state transition table (STT) explicitly stores the next node ID for all nodes and for all characters.

For example, Fig. 1 and Table 1 outline the trie and the corresponding STT that has aba, ba, bb, cb, and cc as keys. In this example, -1 denotes that the transition is unavailable.

XBW. We applied compressed prefix matching with the XBW proposed by Hon et al. [7] to convert words on GPUs. Note that we used the algorithm without further compression because of noticeable overheads. See Hon et al. [7] for the theoretical details.

The reverse prefix is defined as the concatenation of characters from the node to the root. The ID of each node is assigned in lexicographical order of the associated reverse prefix. Note that the ID of the root node is zero. There is an example of the assigned IDs in Fig. 1.

The bit-vector, B_c, is constructed for each character c. The i-th element of the bit-vector B_c indicates node i can transition by character c if the i-th element is one. The function, $\mathrm{rank}(i, B_c)$, is defined as the number of 1s in $B_c[0, i)$. Additionally, the smallest node ID for each character c, where a node's reverse prefix begins with c is stored as $C[c]$.

The node transition by c from node x is obtained in a two-step procedure.

1. If the x-th element of B_c is zero, there is no valid transition.
2. If there is a valid transition, the next node ID is calculated with $\mathrm{rank}(x, B_c) + C[c]$.

Figure 2 outlines example transitions from each node by b. The nodes that can transition with b have an arrow labeled with a rank. Two examples are given below:

- Transition by b from root node. $\mathrm{rank}(0, B_b) + C[b] = 0 + 4 = 4$
- Transition by b from node 8. $\mathrm{rank}(8, B_b) + C[b] = 3 + 4 = 7$

Bit-Vector. RRR [14] is used for the bit-vector supporting rank in $O(1)$ time. The bit-vector is divided into blocks with bits of length t. The blocks are classified according to the number of 1s in the bits. Thus, $\binom{t}{k}$ blocks belong to class k and each block has unique index r. Therefore, the pair, (k, r), identifies a block. There are two methods of decoding from a pair to bits, i.e., using a pre-computed table or computing on the fly [12]. Additionally, a superblock groups some blocks and stores ranks and pointers at the beginning bits of the superblock.

2.5 GPU Architecture

We used a GTX 970, which is a GPU with the Maxwell architecture, for evaluation and the program was written using CUDA. See the official documents [13] for further details on the GPU architecture.

All threads within a warp perform one common instruction in a lockstep. When the threads within a warp follow different execution paths, each path is

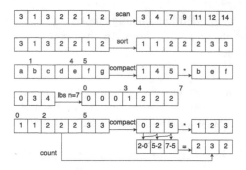

Fig. 3. Example behavior of parallel primitives.

executed one by one while the threads following the others discard the results. This is called warp divergence, which is one of the causes of low performance.

Global memory is basically accessed via 128-byte memory transactions. When all threads within a warp access a continuous 128-byte memory region, these memory accesses are coalesced. These coalesced accesses are important to maximize the performance of memory access.

2.6   Data Parallel Primitives

Data parallel primitives on GPUs are the fundamental algorithms that are used as building blocks for constructing programs. Some efficient algorithms such as sort [11], scan [6], and merge [5] have been proposed. In addition, several implementations and libraries are available. We used the ModernGPU [1] library in the research discussed in this paper.

This section describes the data parallel primitives that comprised our proposed term weighting method. Figure 3 outlines example behaviors by each primitive.

Scan. Scan, or prefix sum, takes an associative binary operator, $\oplus$, and an array, $[a_0, a_1, a_2, ..., a_{n-1}]$, as input, and then generates an array, $[a_0, (a_0 \oplus a_1), ..., (a_0 \oplus a_1 \oplus ... \oplus a_{n-1})]$. If the i-th output includes the i-th input, it is called an inclusive scan. Instead, if the i-th output does not include the i-th input, it is called an exclusive scan.

Sort. Radix sort [11] is considered to be the fastest sorting algorithm on GPUs. However, there is a limitation on the property of keys to perform radix sort. For example, radix sort cannot perform with variable length keys such as strings. In comparison with sorting algorithms, merge sort based on efficient merge [5] is competitive with radix sort. Still sorting variable length keys with merge sort yields suboptimal performance on GPUs. We used merge sort [1] as an implementation of a sort primitive.

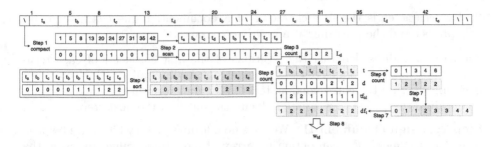

Fig. 4. Overview of proposed term weighting method using data parallel primitives.

Compact. Compact, or filter, extracts the elements that satisfy a predicate from input arrays. It consists of three steps. First, it marks the elements that satisfy a condition. Then, it calculates their indices. Finally, it generates output arrays by using the indices (Fig. 4).

Load Balancing Search. Load balancing search takes an array of length k in which elements represent the positions of the boundary of an array of length n that needs to be generated. Then, it generates an array of length n in which elements represent the indices to denote what part of an array is of length k. It can be regarded as a special case of a vectorized sorted search [1].

Count. Count is the operation that counts the numbers of unique keys in an input array. It is not so much a primitive as a composition of primitives. However, we describe it here because we employed it several times in our proposed method. First, it sorts an input array if needed. Then, it extracts the indices of boundaries using compact. The predicate for compact is whether or not the key is different to the preceding key. After that, it calculates the number of keys by subtracting the index from the succeeding index for each index. The corresponding key with the number can be retrieved by using the indices of boundaries.

3 Term Weighting Method Using Data Parallel Primitives

Assumption. We calculate all w_{td} corresponding to the pair, $\langle t, d \rangle$, when document d contains term t. A set of documents is represented as a sequence of terms separated by a space. The boundary between documents is represented as a blank line.

Step 1: Extract Word. We need to extract terms from a document to calculate term weights. As we assumed that documents would be preprocessed by the method described above, two conditions hold:

1. The character is the first character of a word, if the character is an alphabet and the preceding is a blank character.

2. The character is the first character of a document, if the character is an alphabet and the preceding is a new line character.

An array of indices denoting the positions of the first character of each term can be obtained by compact using predicate 1. Additionally, a Boolean array denoting whether the corresponding term is the first term in the document can be obtained. It is used to generate a document index in the next step.

Step 2: Assign Document ID. We have no information on which term belongs to which document. We can obtain an array of monotonic values by using the inclusive scan of sum, where terms within the same document have the same value. We used these values as document ID d.

Step 3: Calculate Document Length. The number of terms in each document is needed to calculate BM25. It can be obtained by count of an array of document IDs.

Step 4: Sort. The positions of terms and document IDs are sorted in this step by the position of the terms as the key and the document IDs as the value. The position of the terms is compared with the corresponding term by using string comparison. We used stable sort so that document IDs could preserve monotonic order within the same terms.

Step 5: Calculate tf_{td}. The number of elements that have the same term and document ID is $tf_t d$, which is obtained by count.

Step 6: Calculate df_t. The array of unique pairs of term t and document ID d was obtained in the previous step. The number of elements that have the same term is df_t in this array. In the same way, df_t is obtained by count.

Step 7: Assign df_t. We need both tf_{td} and df_t to calculate w_{td}. If one thread handles one df_t, it is easy to obtain tf_{td} from df_t by calculating offsets and iterating df_t times. However, some threads may calculate large amounts of BM25 weights in this way while others only calculate a few. Thus, significant load imbalance occurs and warp divergence adversely affects performance.

We created an expanded array of df_t whose element had a one-to-one correspondence to tf_{td} to avoid this imbalance and balance the load. This array is generated by using load balancing search.

Step 8: Calculate BM25. Since the values necessary for calculation are obtained with these seven steps, BM25 weights are calculated with Eq. (1).

3.1 Implementation of Dictionary

We implemented the dictionary with XBW, as described in Subsection 2.4. First, the block size is set to 15. Class k ranges from $[0, \ldots, 15]$. Thus, class k is represented as a 4-bit integer. The superblock groups 16 blocks. A rank, a pointer, and 16 classes are stored in a structure. The size of the structure is 16 bytes. This structure is handled as built-in vector type `uint4` so that it can be fetched in a single memory transaction.

We use the table for decoding a pair, (k, r). The number of entries is 2^{15} and the size of each entry shares 2 bytes. Thus, the size of the table is 64 KB in total. The constant, $\lceil \log \binom{t}{k} \rceil$, and offsets for accessing the table are cached in the shared memory.

Modifications to Term Weighting Method. It was assumed that the dictionary would be constructed in advance and reside in the CPU side memory. The dictionary was transferred to the GPU side memory along with input documents. We added the converting step before the sort step. This step converted the positions of terms into IDs by using the dictionary. Comparisons of the terms were replaced by those of IDs after this step.

If there is a term that is not listed in the dictionary, this term is treated as an unknown term, and the same ID is assigned to all unknown terms. Note that the existence of unknown terms never affects the term weight of other terms, although the term weight of unknown terms becomes the same value.

4   Evaluation

This section compares four methods where former two methods are comparison methods and latter two methods are proposed methods and evaluates their performance:

MapReduce Phoenix++ (MRP). This is a MapReduce based method on multi-core processors using Phoenix++ [17]. We conducted experiments using eight threads, which is the same as the number of logical cores.

MapReduce Mars revised (MRM). This is a MapReduce based method on GPUs using Mars [3]. We replaced the original sort algorithm to the merge-based sort primitive to enable a fair comparison. The merge-based sort primitive is faster than the original sort algorithm and the overhead of sorting is a bottleneck in the shuffle step.

Parallel Primitives (PP). This is a data parallel primitives based method on GPUs without a dictionary. The number of threads per block is heuristically determined for each kernel and the number of blocks is set to sufficiently large in order to exploit the latency hiding ability of GPUs.

Parallel Primitives with Dictionary (PPD). This is a data parallel primitives based method on GPUs with the dictionary to convert terms into IDs.

We measured the execution times from when all inputs resided in the CPU side memory until calculated results were stored back into the CPU side memory. Hence, the data transfer time between the CPU and the GPU was included if a GPU was used.

4.1   Setup

We collected the frequencies of terms from around 50 million English pages in the Text REtrieval Conference (TREC) ClueWeb09 Category B[3]. The terms

---

[3] http://trec.nist.gov/.

were case insensitive and only consisted of letters of the alphabet. Thus, $\Sigma = \{a, b, \ldots, z\}$ and $|\Sigma| = 26$. We used the top-k terms when the vocabulary size is k.

We used artificially created documents for evaluation. The words in the document were randomly selected by using discrete distribution based on the above frequencies of terms. The lengths of the documents were randomly determined by using a lognormal distribution whose parameters were $\mu = 6.0$ and $\sigma = 1.1$. Table 2 lists the statistics for the datasets that were generated. Here, n_w is the total number of words, n_{dl} is the number of documents, n_{tf} is that of tf_{td}, and n_{df} is that of df_t.

Table 2. Statistics on datasets.

	n_w	n_{dl}	n_{tf}	n_{df}
100 MB	17,881,505	24,411	11,029,756	447,663
200 MB	35,767,202	48,725	22,074,419	687,255
300 MB	53,651,412	73,163	33,119,396	882,477
400 MB	71,532,437	97,683	44,123,823	1,051,534
500 MB	89,409,102	122,178	55,165,752	1,203,716

The experiments were conducted on a PC with an Intel Core i7-6700K, 16 GB of DDR4 memory, and an NVIDIA GeForce GTX 970, running on Ubuntu 14.04 and CUDA 7.5.

4.2   Results

Figure 5 compares the performance of the four methods. Both the proposed methods based on parallel primitives on GPUs, PP and PPD, outperformed the multi-core CPUs. In particular, PPD, which used the dictionary, outperformed MRP by a factor of 5.0–5.1 in terms of runtime. The advantageous effects of using the dictionary were observed by comparing PPD and PP. PPD reduced the overall runtime by a factor of 3.8–4.5. In contrast, MRM, which used Mars on GPUs, failed to gain in performance against MRP. Furthermore, its memory requirements were larger those of the other methods; therefore, MRM could not run with datasets of more than 400 MB.

Figure 6 has breakdowns of the execution times for MapReduce-based MRP and MRM running on CPUs and GPUs. The shuffle step aggregates key-value pairs. This step is implemented by sort in Mars and hash tables in Phoenix++. Thus, the time for shuffle is included within map and reduce steps in Phoenix++. The merge step gathers the results of each reduce worker into one list of key-value pairs. Although the reduce step of MRM does not contain any operation handling strings, this step expends a large portion of the execution time. This is due to load imbalance within threads, which deteriorates performance.

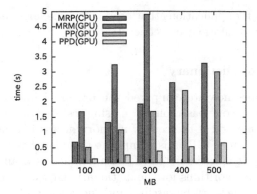

Fig. 5. Execution times for methods.

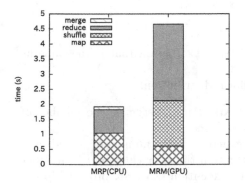

Fig. 6. Breakdown of execution time of MapReduce based methods.

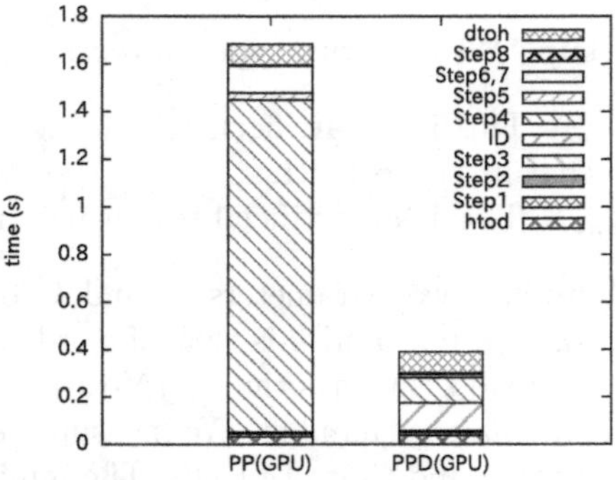

Fig. 7. Breakdown of execution time for data parallel primitive based methods. dtoh: data transfer time from GPU to CPU. htod: data transfer time from CPU to GPU. Each step corresponds to steps in Sect. 3

Figure 7 has the breakdowns for the execution times of data parallel primitives based methods. The steps in the PP method that handle variable length strings such as sort and the calculations of tf_{td} and df_t, expend a large portion of the execution time. In particular, sorting dominates the overall execution time. In contrast, the PPD method, which employs the dictionary, reduces the overheads of costly sorting and comparing of strings by converting terms into IDs in advance. Although a converting step has been added, the cost of converting and sorting strings by IDs is sufficient smaller than sorting by strings.

PPD achieves significant speedup by using the dictionary. There are two disadvantages in compensation for using the dictionary. First, an additional cost for converting terms is involved. Second, it cannot calculate weights of terms that are not listed in the dictionary. Moreover, execution without data transfer expends about 45% of the time. However, it is important to use a fast and compact dictionary for applications that need to handle large amounts of vocabulary such as those in information retrieval.

4.3 Evaluation of Dictionary

This section describes how we compared the proposed implementation of the dictionary with a naïve one based on a trie using a state transition table (STT). STT is the fastest method if the cost of memory access is constant. In reality, the memory subsystem of GPUs is complicated and varies depending on the architecture [10, 19]. It must be noted that random access is inherently unavoidable to achieve these algorithms for the dictionary. Hence, their performance is heavily influenced by memory hierarchies and the cache replacement algorithms of the hardware.

Setup. We used two types of text:

T_{freq}. This is an artificial text using discrete distribution that is the same as that in Subsect. 4.1.
T_{unif}. This is an artificial text using uniform distribution.

There are two parameters m and k for texts. Parameter m is the number of words in a text and k is that of words in the vocabulary.

Thread i converts the $i + jN$-th word, where N is the number of threads and j is the natural number. We measured execution times by changing the number of threads and that of blocks. The minimal execution times are provided in the following results.

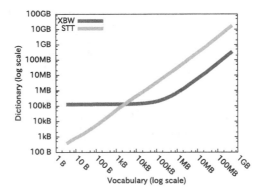

Fig. 8. Memory footprint of dictionary.

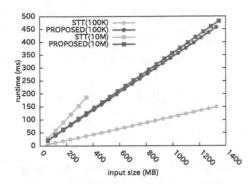

Fig. 9. Runtime in milliseconds of converting words where T_{freq}, $k = 100K$, 10M, and length of text m is varied.

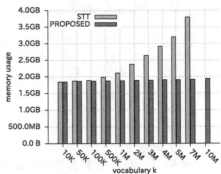

Fig. 10. Total memory usage on GPUs by converting words where T_{freq}, $m = 100M$, and number of words in vocabulary is varied.

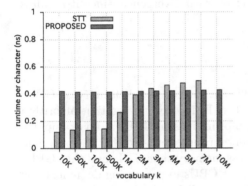

Fig. 11. Runtime per character in nanoseconds of converting words where T_{freq}, $m = 100M$, and number of words in vocabulary is varied.

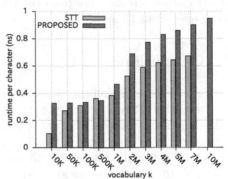

Fig. 12. Runtime per character in nanoseconds of converting words where T_{unif}, $m = 100M$, and number of words in vocabulary is varied.

Results. The memory footprint of the dictionary in our setup is plotted in Fig. 8. The XBW for a large vocabulary requires a memory footprint that is about 40× less than that of STT.

The execution time in Fig. 9 is proportional to the size of input in all cases. The performance of STT is heavily dependent on the vocabulary size. In contrast, the proposed method slightly increases the execution time.

Figure 11 has the execution time per character, while the total memory usage is shown in Fig. 10. The input text is T_{freq}, $m = 100M$, and imitates an actual text by using term frequencies. The execution time of STT increases along with vocabulary size. The proposed method outperforms STT in this scenario for a large vocabulary.

Table 3. Mean and variance in length of words in text.

Vocabulary size k		10K	100K	1M	10M
T_{freq}	Mean	4.55	4.78	4.84	4.86
	Variance	6.34	6.91	7.09	7.24
T_{unif}	Mean	6.66	7.12	7.84	9.44
	Variance	6.42	7.04	8.32	17.2

Behavior is that in Fig. 12 when the input text is T_{unif}. Both methods increase execution times along with vocabulary size. When using a uniform distribution, the characteristics of text change according to vocabulary size. The variance in the length of words increases because rare words are often longer. The mean and variance in the length of words in a text are summarized in Table 3. The variance in the length of words causes warp divergence. The proposed method suffers more from the effect of warp divergence, since it requires more operations than STT.

5   Conclusion

We proposed the implementation of an efficient compressed dictionary on GPUs and a methodology to construct a very fast BM25 term weighting algorithm with a dictionary and existing data parallel primitives on GPUs as an example of major document processing. The experimental results revealed that the proposed dictionary required a footprint with 40× less memory than STT, and that GPUs were able to achieve obvious speedup against CPUs in processing documents by properly composing the dictionary and data parallel primitives.

Although performance in converting words into integer IDs depended on the characteristics of text and vocabulary, the proposed dictionary was competitive with STT in our setup with a large vocabulary. By using the dictionary to avoid string comparisons, GPUs with the dictionary performed up to 5.1× faster than multi-core CPUs and up to 4.1× faster than GPUs without the dictionary. The importance of a compact and fast dictionary was clarified as it proved to be very effective in efficiently processing documents.

Further studies are needed to conclude whether data parallel primitives and the dictionary can be applied to other document processes. The simplest parallelization method in terms of the dictionary was used for converting words into IDs. Since the processing time for each word depended on the length of words, we considered that some sophisticated scheduling method that could rearrange words could balance the load among threads and achieve greater improvements to performance.

Acknowledgements. This work was partly supported by JSPS KAKENHI Grant Numbers 15H02701, 15K20990, 16H02908, 26540042, 26280115, 25240014 and 17K12684.

References

1. Baxter, S.: Moderngpu 2.0. https://github.com/moderngpu/moderngpu/
2. Dean, J., Ghemawat, S.: MapReduce: simplified data processing on large clusters. Commun. ACM **51**(1), 107–113 (2008)
3. Fang, W., He, B., Luo, Q., Govindaraju, N.K.: Mars: accelerating MapReduce with graphics processors. IEEE Trans. Parallel Distrib. Syst. **22**(4), 608–620 (2011)
4. Fredkin, E.: Trie memory. Commun. ACM **3**(9), 490–499 (1960)
5. Green, O., McColl, R., Bader, D.A.: GPU merge path: a GPU merging algorithm. In: Proceedings of the 26th ACM International Conference on Supercomputing, ICS 2012, pp. 331–340 (2012)
6. Harris, M., Sengupta, S., Owens, J.D.: Parallel prefix sum (scan) with CUDA. In: Nguyen, H. (ed.) GPU Gems 3. Addison Wesley, Boston (2007)
7. Hon, W.K., Ku, T.H., Shah, R., Thankachan, S.V., Vitter, J.S.: Faster compressed dictionary matching. Theoret. Comput. Sci. **475**, 113–119 (2013)
8. Lin, J., Dyer, C.: Data-Intensive Text Processing with MapReduce. Morgan and Claypool Publishers, San Rafael (2010)
9. Martínez-Prieto, M.A., Brisaboa, N., Cnovas, R., Claude, F., Navarro, G.: Practical compressed string dictionaries. Inf. Syst. **56**, 73–108 (2016)
10. Mei, X., Chu, X.: Dissecting GPU memory hierarchy through microbenchmarking. IEEE Trans. Parallel Distrib. Syst. **28**(1), 72–86 (2017)
11. Merrill, D., Grimshaw, A.: High performance and scalable radix sorting: a case study of implementing dynamic parallelism for GPU computing. Parallel Process. Lett. **21**(02), 245–272 (2011)
12. Navarro, G., Providel, E.: Fast, small, simple rank/select on bitmaps. In: Klasing, R. (ed.) SEA 2012. LNCS, vol. 7276, pp. 295–306. Springer, Heidelberg (2012). doi:10.1007/978-3-642-30850-5_26
13. NVIDIA: CUDA toolkit documentation. http://docs.nvidia.com/cuda/
14. Raman, R., Raman, V., Satti, S.R.: Succinct indexable dictionaries with applications to encoding k-ary trees, prefix sums and multisets. ACM Trans. Algorithms **3**(4), Article No. 43 (2007)
15. Robertson, S.E., Walker, S., Jones, S., Hancock-Beaulieu, M., Gatford, M.: Okapi at TREC-3. In: Proceedings of the 3rd Text REtrieval Conference, pp. 109–126 (1994)
16. Sitaridi, E.A., Ross, K.A.: GPU-accelerated string matching for database applications. VLDB J. **25**(5), 719–740 (2016)
17. Talbot, J., Yoo, R.M., Kozyrakis, C.: Phoenix++: modular MapReduce for shared-memory systems. In: Proceedings of the Second International Workshop on MapReduce and Its Applications, MapReduce 2011, pp. 9–16 (2011)
18. Wang, Y., Davidson, A., Pan, Y., Wu, Y., Riffel, A., Owens, J.D.: Gunrock: a high-performance graph processing library on the GPU. In: Proceedings of the 21st ACM SIGPLAN Symposium on Principles and Practice of Parallel Programming, PPoPP 2016, pp. 11:1–11:12 (2016)
19. Wong, H., Papadopoulou, M., Sadooghi-Alvandi, M., Moshovos, A.: Demystifying GPU microarchitecture through microbenchmarking. In: IEEE International Symposium on Performance Analysis of Systems and Software, ISPASS 2010, pp. 235–246 (2010)

Process and Tool-Support to Collaboratively Formalize Statutory Texts by Executable Models

Bernhard Waltl[(✉)], Thomas Reschenhofer, and Florian Matthes

Department of Informatics, Software Engineering for Business Information Systems,
Technische Universität München,
Boltzmannstr. 3, 85748 Garching bei München, Germany
b.waltl@tum.de, {reschenh,matthes}@in.tum.de

Abstract. This paper describes a collaborative modeling environment to support the analysis and interpretation of statutory texts, i.e., laws. The paper performs a case study on formalizing the product liability act and proposes a data-centric process that enables the formalization of laws. The implemented application integrates state-of-the-art text mining technologies to assist legal experts during the formalization of norms by automatically detecting key concepts within legal texts, e.g., legal definitions, obligations, etc. The work at hand elaborates on the implementation of data science environment and describes key requirements, a reference architecture and a collaborative user interface.

Keywords: Formalizing norms · Semantic modeling · Computable law · Data analytics · Text mining · Apache UIMA · Reference architecture

1 Introduction

Although, several legislations have the ambitious goal that laws should be understandable by everyone (see [1]), understanding and interpreting legal norms is a difficult and complex task. Civil law systems have developed complex interpretation canons, that can be used during the interpretation of legal norms to avoid misinterpretations and to unveil and determine the intrinsic uncertainty and vagueness [2,3].

Since the interpretation of laws is a repeating, data-, time-, and knowledge-intensive task, the lack of appropriate tool-support is counter-intuitive in several ways. On the one hand, providing tool-support to model and store the result of the interpretation process is a measure to unveil intrinsic vagueness, inconsistencies, complex regulations. On the other hand, the development of an application to collaboratively formalize and refine a semantic model of normative texts becomes—due to the advances in enterprise information systems engineering and because of text mining algorithms—more and more attractive [4,5]. Up to now, there is a gap between the technological possibilities and the current support of legal interpretation processes, which holds especially for the legal domain in Germany. In 2015 the dutch researcher van Engers explicitly stated: "While many

© Springer International Publishing AG 2017
D. Benslimane et al. (Eds.): DEXA 2017, Part II, LNCS 10439, pp. 118–125, 2017.
DOI: 10.1007/978-3-319-64471-4_11

attempts to automate law [...] have been made before, hardly any attention has been paid to the 'translation process' from legal rules expressed in natural language to specifications in computer executable form" [6].

It is unlikely, that the interpretation process can be fully automated through algorithms. Instead, the advances in natural language processing (NLP) and detection of patterns in legal texts are considered to be supportive to legal (data) scientists and practitioners. The paper's contribution can be stated as follows:

(i) How does a data-driven process aiming at the analysis and interpretation of normative texts look like?
(ii) How can prevalent textual representation with normative content, e.g., laws, be transformed into models?
(iii) What are requirements for an application supporting the collaborative derivation of models using NLP technologies?

2 Analysis and Interpretation of Normative Texts

The potential of tool-support during the analysis and interpretation of normative texts and the subsequent formalization in semantic models can be described by a process model shown in Fig. 1.

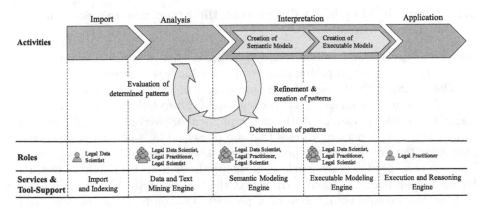

Fig. 1. Structured and data-centric process of collaboratively analyze, interpret, and model normative texts with specification of clear roles and potential tool-support.

The process is built upon the legal theory and interpretation canon proposed by Larenz and Canaris (see [2]) and consists of four steps:

1. **Import.** Integration of relevant literature into the system requires a generic data model representing the particularities of normative texts, such as structural information. Depending on the case, that a legal expert wants to deal with, an import of various literature, e.g. laws, contracts, judgments, commentaries, etc., needs to be performed.

2. **Analysis.** During the analysis step, legal experts explore and investigate the key norms and relationships between them. They implicitly determine the relevance of norms and their applicability by considering particular linguistic and semantic patterns which indicate whether a norm is relevant in a case, respectively a set of cases, or not.

3. **Interpretation.** During the interpretation step the textual representation of norms is transferred into a coherent semantic and normative model (see Figs. 2 and 3). This could be an explication of the mental model of an experience lawyer or legal scientist might has. The example in Sect. 3 illustrates how the determination of obligations, exclusions, prohibitions, etc. can be assisted during the interpretation process. It supports the legal experts to access the content of the law and prevents him of missing important norms.

4. **Application.** The models arising from the textual interpretations can be stored in order to provide access to different collaborating users. Based on the interpretation and creation of the model the system automatically generates forms which allow end-users to describe a particular case. Thereby, the user inputs the known evidence from a fact and the system automatically infers the solution represented by derived attributes.

3 The Product Liability Act

3.1 Semantic Modeling to Represent the Structure and Semantics of Legal Norms

Based on linguistic and semantic analysis of legal texts, it is possible to support the interpretation by proposing important semantic legal concepts.

The act explicitly states that there has to be a product (Sect. 2) with a defect (Sect. 3) causing a damage (Sect. 3) on a legally protected good (Sect. 1). This might be the life (in case of death) or the physical or psychological integrity (in case of injury) of a person or items that are possessed by a person. If this is the case, then the producer of the product (Sect. 4) becomes liable for the damage. The law also specifies reasons releasing the producer from his liability (Sect. 1). For sake of simplicity, we omit attributes of the types described by the legislator, such as the manufacturing date of a product. Obviously, it is possible—and depending on purpose of the modeling necessary—to deviate from the textual representation, by either modeling more or less information than provided in the text, e.g., interpretation. However, the proposed modeling approach does not make constraints regarding the quality of the model. This is intentionally left to the user.

3.2 Executable Models Representing Decision Structures and Behavioral Elements

Beside the semantic model representing types with their attributes and the relationships among them, the decision structure of norms has to be represented.

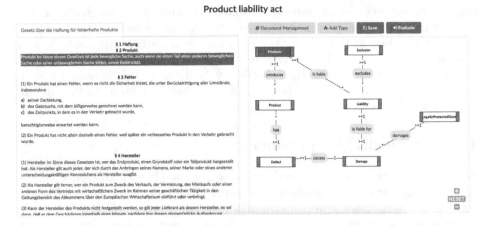

Fig. 2. Model based formalization of the product liability act in a web based environment. Types, attributes, and relations, can be linked with text, which is then highlighted.

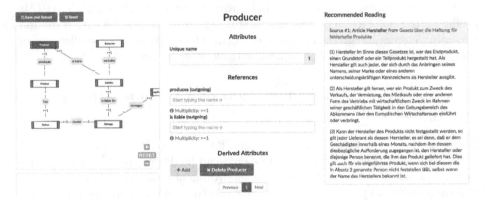

Fig. 3. The form is automatically generated and evaluated by the reasoning engine. End-users have to provided the facts, i.e., filling the form.

This so-called executable model contains the decision structures, i.e. derived attributes. Consequently, the models aim at capturing the executable (or computable) logic that can be used to decide whether someone has a claim or not. Those derived attributes are expressed in a domain specific language (MxL) and are evaluated at run time based. Those statically type-safe expressions do not only support the specification of first-order propositional logic but also higher-order logic and almost arbitrarily complex arithmetical operations.

Figure 2 shows semantic model of the product liability act. Each type in the model contains different attributes and relations to other types.

Thereby, it is necessary that the set of facts contains a product (Sect. 2) and a producer (Sect. 4) who has manufactured the product. In addition, there needs to be a defect (Sect. 3) causing a damage. In principle, producers are liable for damages that a product manufactured by him has caused. Of course, there are several exclusions that release the producer from his liability (Sect. 1). The existence or absence of those exclusions decides whether there is an effective liability or not. So if the producer is effectively liable and if there is a legally protected and damaged good, then the plaintiff has a claim according to the product liability act.

4   Technological Requirements

The requirements (listed in Table 1) served as the base line for the implementation of the data science environment. To ensure an easy extension and adaptability, the system follows the principle of high cohesion and low coupling regarding its components. Technologically, the environment was implemented as a web application and the programming language used in the back-end was Java. Elasticsearch serves as the data storage, which allows an efficient handling of a large amount of textual data. The execution and reasoning engine, which is already existing and maintained at our research group, is accessed via a REST API. It fulfills all technological requirements to store the models [7]. The execution and reasoning engine integrates a DSL, i.e. MxL, which allows the specification of complex queries as well as logical and arithmetic operations [8].

5   Concept and Implementation of a Collaborative Web Application

Based on a case study of the product liability act presented in Sect. 3 and the process shown in Fig. 1, several requirements can be derived that have to be met by a text mining and semantic modeling environment to foster collaboration. On this fundament we propose a reference architecture and an implementation.

5.1   Reference Architecture

Based on the requirements from Table 1 and on the framework proposed in [9]) it is possible to define a reference architecture focusing on the analysis of legal texts and the creation of semantic and executable models.

The data and text mining engine is the central component of the platform supporting the modeling process by unveiling linguistic and semantic patterns. Since the main task consists of the creation of semantic and executable models based on textual descriptions that are usually of normative character, the assistance during the analysis and interpretation consists in parts of the automated detection of relevant sentences and phrases (patterns).

Within the data and text mining engine, several components have to be provided, e.g., dictionaries and pattern definitions. The legal language has some particularities that make it well suited for the analysis by NLP [10], for example the usage of particular linguistic patterns (phrases or words) that indicate

Table 1. Main requirements for collaborative tool-support to model the semantics of statutory texts structured into four phases.

Import		
1	Flexible import structure	Baseline for the analysis and interpretation is the consideration of various literature (laws, judgments, contracts, commentaries, etc.) that is present in different sources (xml, html, pdf, etc.)
2	Mapping and indexing of legal data	The legal literature has to be indexed and mapped to a data model, that does not only preserve the content, i.e. text and metadata, but also structural properties, such as references and nested content
Analysis		
3	Preserving textual representation	Enabling users to access the content, i.e. legal literature. The visualizations of legal literature has to show the structural information, such as nestedness and links between articles and documents
4	Collaborative creation and maintenance of patterns	The creation, refinement and deletion of the required pattern definitions should be done collaboratively in the application, so that different users are able to share their knowledge and contributions
5	Lifecycle management of pattern descriptions	Support of the full lifecycle of the pattern specifications, namely creation, refinement, evaluation, and maintenance
6	Automated pattern detection	Automated identification of linguistic and semantic patterns through data and text mining components
7	Reuse of existing NLP components	Building of NLP pipelines, that allow the easy reuse and sharing of highly specified software components for NLP
8	Evaluation of annotation quality	Possibility to view the annotations, to examine precision and recall manually, or to export this information to compare against a manually tagged corpus
9	Manually annotating and commenting of legal texts	Users should be able to manually add relevant semantic information and comments to the legal literature
10	Storing of annotations	Storing and indexing the automatically determined and manually added annotations
Interpretation		
11	Creation of semantic and executable model elements	Step-wise definition of model elements (types, attributes, relationships, operators) for semantic and executable models
12	Lifecycle support for semantic models	Defining, maintaining and storing of static model elements, such as types, attributes, relationships
13	Lifecycle support for executable models	Defining, maintaining and storing of executable model elements, such as types, relationships, operators
14	Connecting model elements with text phrases	Creation of connections between model entities and the relevant (interpreted) text. Thereby various levels of the interpreted text should be linkable to model elements, such as words, phrases, sentences, sections, and documents
15	Domain specific language (DSL) to express semantics of operators	Specification of the operations and executable semantics of relationships with a model-based expression language
Application		
16	Access to existing models	Viewing and exploring of semantic and executable models to grasp the result of prior interpretation processes
17	Application of decision models	Executing the defined models through intelligent form-based or spreadsheet-based reasoning

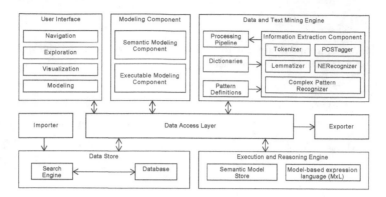

Fig. 4. The reference architecture for collaborative modeling of normative texts based on linguistic and semantic text analysis (extension of [9]).

particular legal ideas or concepts. This is what makes pattern determination valuable for the legal domain (see also [11]). In order to determine patterns which are more elaborate than common regular expressions, it is necessary to integrate a component that allows the specification of patterns which can easily nest, reuse and recombine those pattern specifications (rules). Our implementation integrates Apache Ruta (rule-based text annotation), which shows some great potential as a linguistic pattern definition technology [11] (Fig. 4).

Modeling component. The modeling engine offers the required functionality to access, create, refine, maintain, and delete models. The differentiation between the semantic model and the executable model is necessary. In contrast to the semantic model, the executable model requires the definition of the executable semantics, i.e., functions and operators, between model elements, i.e., defining derived attributes based on atomic attributes. Figure 2 shows the types which are input, namely producer, product, etc., whereas "legally protected good" is a type, which is the output (relevant intermediate result) of an operator. The semantics of these operators can be expressed using an existing model-based expression language (as described above).

The modeling engine is also capable of managing the associations between model elements and text phrases. It offers proper technical functions and services, and is also the component that observes changes in the textual representation that might lead to changes within the semantic and executable models.

Execution and reasoning engine. The execution of the model is done by an existing reasoning engine. Heart of the engine is a model-based domain specific expression language (MxL), that was developed at our research group [8]. It is a type-safe functional expression language, implemented in Java and hosted within a web application that can be accessed via a REST API. This expression language allows the specification of almost arbitrarily complex logical and numerical operations. These operations accept typed input values (integer, string, boolean, model elements, etc.) and compute a typed output. The operations are part of the interpretation process and can therefore be linked to the text.

6 Conclusion

In this paper we have developed a collaborative data-driven process that structures the analysis and interpretation of normative texts by leading to executable models of statutory texts, i.e., laws.

We have identified four phases that are required to formalized legal norms: Import, analysis, interpretation, and application. For each step we have identified and implemented tool-support.

In a case study we have shown the approach by formalizing a basic claim arising from the product liability act. Based on the proposed data- and tool-centric process we have derived key requirements. Finally, we showed how the co-existence of these semantic models and the corresponding textual representation can be implemented in a collaborative web environment.

References

1. Bundesministerium des Innern, Gemeinsame Geschäftsordnung der Bundesministerien, Berlin
2. Larenz, K., Canaris, C.-W.: Methodenlehre der Rechtswissenschaft. Springer, Berlin [u.a.] (1995)
3. Hage, J.: A theory of legal reasoning and a logic to match. Artif. Intell. Law **4**, 199–273 (1996)
4. Rotolo, A.: Legal Knowledge and Information Systems: JURIX 2015: The Twenty-Eighth Annual Conference. Frontiers in Artificial Intelligence and Applications. IOS Press, Amsterdam (2015)
5. Proceedings of the 15th International Conference on Artificial Intelligence and Law, ICAIL 2015, San Diego, California. ACM, New York (2015). ISBN 978-1-4503-3522-5
6. van Engers, T.M., van Doesburg, R.: First steps towards a formal analysis of law. In: Proceedings of eKNOW (2015)
7. Reschenhofer, T., Bhat, M., Hernandez-Mendez, A., Matthes, F.: Lessons learned in aligning data and model evolution in collaborative information systems. In: Proceedings of the International Conference on Software Engineering (2016). (accepted for publication)
8. Reschenhofer, T., Monahov, I., Matthes, F.: Type-safety in EA model analysis. In: IEEE EDOCW (2014)
9. Waltl, B., Matthes, F., Waltl, T., Grass, T.: LEXIA: a data science environment for Semantic analysis of German legal texts. In: Jusletter IT (2016)
10. Francesconi, E., Montemagni, S., Peters, W., Tiscornia, D. (eds.): Semantic Processing of Legal Texts: Where the Language of Law Meets the Law of Language. LNCS (LNAI), vol. 6036. Springer, Heidelberg (2010)
11. Grabmair, M., Ashley, K.D., Chen, R., Sureshkumar, P., Wang, C., Nyberg, E., Walker, V.R.: Introducing LUIMA: an experiment in legal conceptual retrieval of vaccine injury decisions using a UIMA type system and tools. In: ICAIL 2015: Proceedings of the 15th International Conference on Artificial Intelligence and Law, pp. 69–78. ACM, New York (2015)

A Bayesian Approach for Semantic Search Based on DAG-Shaped Ontologies

Anna Formica$^{(\boxtimes)}$, Michele Missikoff, Elaheh Pourabbas, and Francesco Taglino

National Research Council, Istituto di Analisi dei Sistemi ed Informatica
"Antonio Ruberti", Via dei Taurini 19, 00185 Roma, Italy
{anna.formica,michele.missikoff,elaheh.pourabbas,
francesco.taglino}@iasi.cnr.it

Abstract. Semantic search has a great potentiality in helping users to make choices, since it appears to outperform traditional keyword-based approaches. This paper presents an ontology-based semantic search method, referred to as *influential SemSim* (*i-SemSim*), which relies on the Bayesian probabilistic approach for weighting the reference ontology. The Bayesian approach seems promising when the reference ontology is organized according to a Directed Acyclic Graph (DAG). In particular, in the proposed method the similarity among a user request and semantically annotated resources is evaluated. The user request, as well as each annotated resource, is represented by a set of concepts of the reference ontology. The experimental results of this paper show that the adoption of the Bayesian method for weighting DAG-based reference ontologies allows *i-SemSim* to outperform the most representative methods selected in the literature.

Keywords: Semantic search · Bayesian network · Semantic annotation · Similarity reasoning · Weighted reference ontology

1 Introduction

SemSim is a semantic similarity search method that operates starting from semantic annotations associated with resources in a given search space [2]. A semantic annotation, as well as a request vector, is defined as a set of characterizing features represented by concepts of a reference ontology. In order to compute the semantic similarity between a request vector and a semantic annotation, *SemSim* is based on a Weighted Reference Ontology (WRO), where each concept is associated with a weight corresponding to the probability that, selecting a resource at random, the resource is characterized by that concept.

In a previous work [3], we have explored the use of a Bayesian approach to ontology weighting, applied to tree-shaped ontologies, that improved the performance of *SemSim*. The contribution of this paper is twofold: firstly, we show that the Bayesian approach can be effectively used also for DAG-shaped WRO; secondly, we present the *influential SemSim* (*i-SemSim*) method, which is conceived

© Springer International Publishing AG 2017
D. Benslimane et al. (Eds.): DEXA 2017, Part II, LNCS 10439, pp. 126–140, 2017.
DOI: 10.1007/978-3-319-64471-4_12

to address particular scenarios where the annotation vectors and the request vectors contain a concept that is more relevant (influential), in order to characterize the resource, than the other concepts in the vector. The underlying idea is to capture the user's mental model by identifying one feature of the resource that is dominant with respect to the others for describing it. Indeed, in the user evaluation, the similarity between an annotation vector and a request vector is usually biased by the existence of a more relevant concept in the annotation vector, independently of the other features. The experimental results of this paper show that *i-SemSim* improves the *SemSim* performance and, in most of the considered cases, it outperforms the selected similarity methods.

The paper is organized as follows. In the next Section, the Related Work is given, in Sect. 3, the notions of a WRO, request and annotation vectors are recalled. In Sect. 4, the Bayesian approach for weighting ontologies defined in [3] is extended to DAG-shaped ontologies, and in Sect. 5 the *i-SemSim* method is defined. Finally, in Sect. 6 the experimental results are given and Sect. 7 concludes.

2 Related Work

In this section, we recall some representative proposals concerning the weighting of the concepts of an ontology and the integration of Bayesian Networks (BN) and Ontologies.

In [11] a method has been proposed for measuring the information content (IC) of terms based on the assumption that the more descendants a concept has the less information it expresses. Concepts that are leaf nodes are the most specific in the taxonomy and their information content is maximal. Analogously to our proposal, in this method the information content is computed by using only the structure of the specialization hierarchy. However, it forces all the leaves to have the same IC, independently of their depth in the hierarchy.

The method for measuring semantic similarities proposed in [4] is based on edge-counting and the information content theory. In particular, different ways of weighting the shortest path length are presented, although they are essentially based on WordNet frequencies. In this work, we do not adopt the WordNet frequencies. Firstly, because we deal with specialized domains (e.g., personal computer) requiring specialized domain ontologies, and WordNet is a generic lexical ontology, secondly, because there are concepts in WordNet for which the frequency is not given.

Regarding the integration of BN with ontologies, in [12], a similarity measure for the retrieval of medical cases has been proposed. This approach is based on a BN, where the a priori probabilities are given by experts on the basis of cause-effect conditional dependencies. In our approach, the a priori probabilities are not given by experts and rely on a probabilistic-based approach.

In [6], an ontology mapping-based search methodology (OntSE) is proposed in order to evaluate the semantic similarity between user keywords and terms (concepts) stored in the ontology, using a BN. Furthermore, in [5], the authors emphasize the need of having a non-empirical mathematical method for computing conditional probabilities in order to integrate a BN in an ontology. In particular, in the proposed approach the conditional probabilities depend only on the structure of the domain ontology. However, in the last two mentioned papers, the conditional probability tables for non-root nodes are computed starting from a fixed value, namely 0.9. In line with [5], we also provide a non-empirical mathematical method for computing conditional probabilities, but our approach does not depend on a fixed value as initial assumption. In fact, in *i-SemSim* the conditional probabilities are computed on the basis of the weight w_a, which depends only on the structure of the domain ontology, i.e., the probability of the parent node divided by the number of sibling nodes.

3   Weighted Ontologies and Ontology-Based Feature Vectors

Similar to our previous work in [2,3], in this paper we adopt a simplified notion of ontology consisting of a set of concepts organized according to a specialization hierarchy. In particular, an ontology Ont is a taxonomy defined by the pair:

$$Ont = < C, ISA >$$

where $C = \{c_i\}$ is a set of concepts and ISA is the set of pairs of concepts in C that are in subsumption (subs) relation:

$$ISA = \{(c_i, c_j) \in C \times C | subs(c_i, c_j)\}$$

However, in this work, we assume that the hierarchy is a direct acyclic graph (DAG), instead of a tree.

A *Weighted Reference Ontology* (*WRO*) is then defined as follows:

$$WRO = < Ont, w >$$

where w, the concept weighting function, is a probability distribution defined on C, such that given $c \in C$, $w(c)$ is a decimal number in the interval $[0 \ldots 1]$.

With respect to the *SemSim* method proposed in [3], in this paper the weight w_a (a standing for *a priori*, as it will be explained in the next section) has been introduced by revisiting the uniform probabilistic approach, in order to take into account the DAG-shaped ontology, where a concept can have more than one parent. Figure 1 shows the *WRO* drawn upon the personal computer domain that will be used in the running example. In this figure, the weights w_a have been assigned by taking into consideration that in a DAG *multiple inheritance* is allowed. In other words, given a concept $c \in C$, there can exist more than one concept c_i such that $(c_i, c) \in ISA$, and we have:

$$w_a(c) = \min \{ \frac{w_a(c_i)}{|children(c_i)|} | c_i \in parent(c) \},$$

where $parent(c)$ is the set of parents of c in the ISA hierarchy. For instance, let us consider the concept $SmallSizeMediumResScreen$ (where Res stands for resolution). The associated w_a is 0.013 because $SmallSizeMediumResScreen$ has two parents, namely $MediumResScreen$ and $SmallSizeScreen$, which have both w_a equal to 0.04, with three and two children, respectively.

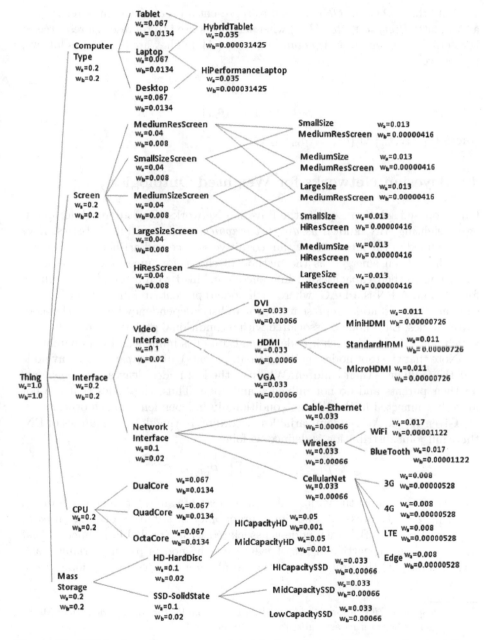

Fig. 1. The WRO of our running example with the corresponding w_a and w_b weights.

The WRO is then used to annotate each resource in the *Universe of Digital Resources* (UDR) by means of an OFV^1, which is a vector that gathers a set of concepts of the ontology *Ont*, aimed at capturing its semantic content. The same also holds for a user request, and is represented as follows:

$$ofv = (c_1, ..., c_n), \text{ where } c_i \in C, i = 1, ..., n$$

Note that, when an *OFV* is used to represent a user request, it is referred to as semantic *Request Vector* (*RV*) whereas, if it is used to represent a resource, it is referred to as semantic *Annotation Vector* (*AV*). They are denoted as follows, respectively:

$$rv = \{r_1, \ldots, r_n\},$$
$$av = \{a_1, \ldots, a_m\},$$

where $\{r_1, \ldots, r_n\} \cup \{a_1, \ldots, a_m\} \subseteq C$.

4   Bayesian Networks for Weighted Ontologies

Our proposed solution exploits the Bayesian Networks (BN) approach for ontology weighting. BN, known as *directed graphical models* within statistics, have been defined in the late 1970s within cognitive science and artificial intelligence. They have been emerged as the method of choice for uncertain reasoning [9] according to their capacities for inferences combined with a sound probabilistic foundation. A BN is a DAG, where each node represents a domain variable, and each arc between nodes represents a probabilistic dependency between the associated nodes. Each node is associated with a conditional probability distribution, which quantifies how much a node depends on or is influenced by its parents.

Note that the root nodes (nodes without parents) are *independent* of any node but they influence their children. Whereas, the leaf nodes (leaves) are *dependent* on their parents and do not influence any node. Thus, the nodes that are not directly connected in the BN are conditionally independent of each other.

Given $X_1, ..., X_n$ random variables, according to the global semantics of BN, their full joint distribution is defined as follows:

$$P(x_1, ..., x_n) = \prod_i P(x_i|pa_i)$$

where x_i is a value of the variable X_i, pa_i is a set of values for the parents of X_i, and $P(x_i|pa_i)$ denotes the conditional probability distribution of x_i given pa_i.

Therefore, in a BN, each node is associated with a probability function that takes, as input, a particular set of values for the node's parent variables, and gives (as output) the probability of the variable represented by the node. These

---

[1] The proposed *OFV* approach is based on the *Term Vector* (or *Vector Space*) Model approach, where terms are substituted by concepts [10].

probability functions are specified by means of conditional probability tables[2], one for each node of the graph.

In our approach, we build a BN isomorphic to the *ISA* hierarchy, referred to as *Onto-Bayesian Network (OBN)*. In the *OBN*, according to the BN approach, the concepts are boolean variables and the *Bayesian weight* associated with a concept c, indicated as w_b, is the probability P that the concept c is True (T), i.e.:

$$w_b(c) = P(c = T)$$

In order to compute the weights w_b, conditional probability tables are defined by using the w_a which has been proposed according to the probabilistic approach recalled in the previous section, as *a priori* weight. In particular, given a concept c, we assume that:

$$P(c = T | c_1 = T, \ldots, c_n = T) = w_a(c)$$

where $c_1, \ldots, c_n$ are the parents of c according to the *ISA* relation.

In order to illustrate the underlying idea, let us consider the concepts *Interface* and *VideoInterface* of the ontology shown in Fig. 1. The concept *Interface* (I for short) has as parent *Thing*, which is always True ($w_a(Thing) = w_b(Thing) = 1$). Consequently, the weight w_b of I coincides with the weight w_a, i.e., $w_b(I) = w_a(I)$, where $w_a(I) = P(I = T | Thing = T) = P(I = T) = 0.2$, as shown in Table 1. Now consider the concept *VideoInterface* (V), where its conditional probability depends on the True/False (T/F) values of its parent *Interface* (see Table 2), that is $w_a(V) = P(V = T | I = T) = 0.1$ and, consequently, $P(V = F | I = T) = 0.9$. The weight w_b associated with V, $w_b(V) = P(V = T)$, is computed based on the probability of its parent *Interface* by taking into account the Kolmogorov definition (given two variables A and B: $P(A, B) = P(A|B)P(B)$), as follows:

$$P(V = T) = \sum_{v \in \{T, F\}} P(V = T, I = v)$$
$$= P(V = T, I = T) + P(V = T, I = F)$$
$$= P(V = T | I = T)P(I = T) + P(V = T | I = F)P(I = F)$$
$$= 0.2 \times 0.1 + 0 \times 0.9 = 0.02$$

Therefore, the Bayesian weight $w_b(V) = P(V = T) = 0.02$, as shown in Table 3.

Now let us consider the concept *MediumSize&MediumResScreen* (MM, for short). In order to calculate $w_b(MM)$, we need the Bayesian weight of MM parents. Similar to the example above, we can compute w_b of the concepts *MediumSizeScreen* (MS), and *MediumResScreen* (MR), that is $w_b(MS) = w_b(MR) = 0.008$. Furthermore, we need $w_a(MM)$. As we can see in Fig. 1, MM is a child of

---

[2] A conditional probability table is defined for a set of (non-independent) random variables to represent the marginal probability of a single variable w.r.t. the others.

Table 1. Probability table of *Interface-I*

I	
T	*F*
0.2	0.8

Table 2. Conditional probability table of *VideoInterface-V*

	V	
I	*T*	*F*
T	0.1	0.9
F	0	1

both MS and MR. According to the uniform probabilistic approach, the weight of MM is evaluated on the basis of the weights of its parents, which are 0.013 (i.e. the a priori weight of MR, that is 0.04, divided by 3), and 0.02 (i.e. the a priori weight of MS, that is 0.04, divided by 2), respectively. Thus, according to the formula given in Sect. 3, $w_a(MM)$ is the minimum between 0.013 and 0.02.

The Bayesian weight of MM, $w_b(MM)$, is given according to the conditional probability shown in Table 4 as follows:

$$\begin{aligned}
P(MM=T) &= \sum_{v,v' \in \{T,F\}} P(MM=T, MR=v, MS=v') \\
&= \sum_{v,v' \in \{T,F\}} P(MM=T|MR=v, MS=v')P(MR=v, MS=v') \\
&= \sum_{v,v' \in \{T,F\}} P(MM=T|MR=v, MS=v') \\
&\qquad\qquad P(MR=v|MS=v')P(MS=v') \\
&= P(MM=T|MR=T, MS=T)P(MR=T|MS=T)P(MS=T) \\
&+ P(MM=T|MR=T, MS=F)P(MR=T|MS=F)P(MS=F) \\
&+ P(MM=T|MR=F, MS=T)P(MR=F|MS=T)P(MS=T) \\
&+ P(MM=T|MR=F, MS=F)P(MR=F|MS=F)P(MS=F)
\end{aligned} \tag{1}$$

In the above formula, the only non null addend is the following:
$$P(MM=T|MR=T, MS=T)P(MR=T|MS=T)P(MS=T)$$
which becomes:

$$\begin{aligned}
&P(MM=T|MR=T, MS=T)P(MR=T|S=T)P(MS=T) \\
&= w_a(MM) \times w_a(MR) \times w_b(MS) = 0.013 \times 0.04 \times 0.008 = 0.00000416
\end{aligned}$$

where S stands for *Screen* which is the parent of both MR and MS. Note that in the expression above according to the *ISA* hierarchy, $MS = T$ implies that

Table 3. Probability table of *VideoInterface-V*

V	
T	*F*
0.02	0.98

Table 4. Conditional probability table of *MidumSizeScreen&MediumResScreen-MM*

		MM	
MS	MR	T	F
T	T	0.013	0.987
T	F	0	1
F	T	0	1
F	F	0	1

$S = T$, therefore, $P(MR = T|MS = T)$ is equal to $P(MR = T|S = T)$ because the dependency of MR on MS can be re-conducted to the dependency of MR on S.

5   i-SemSim

In order to introduce *i-SemSim*, the *SemSim* method [2,3], is recalled which is the basis of the proposed approach.

The *SemSim* method has been conceived to search for the resources in the UDR that best match the *RV*, by contrasting it with the various *AV*, associated with the digital resources to be searched. This is achieved by applying the *semsim* function, which has been defined to compute the semantic similarity between *OFV*.

In *SemSim*, the weights are used to derive the information content (IC) of the concepts that, according to [8], represent the basis for computing the concept similarity. In particular, according to the information theory, the IC of a concept c is defined as:

$$IC = -log(w(c))$$

The *semsim* function is based on the notion of similarity between concepts (features), referred to as *consim*. Given two concepts c_i, c_j, it is defined as follows:

$$consim(c_i, c_j) = \frac{2 \times IC(lub(c_i, c_j))}{IC(c_i) + IC(c_j)} \qquad (2)$$

where the *lub* represents the least abstract concept of the ontology that subsumes both c_i and c_j. While in a tree the least upper bound, $lub(c_i, c_j)$, is unique, in a DAG it is not always uniquely defined. In this work, if more than a *lub* exists, the one with the greatest IC is selected.

Given an instance of RV and an instance of AV, say rv and av respectively, the *semsim* function computes the *consim* for each pair of concepts in the set formed by the Cartesian product of rv, and av. However, we focus on the pairs that exhibit high affinity. In particular, we adopt the exclusive matching philosophy, where the elements of each pair of concepts do not participate in any other pair. The method aims to identify the set of pairs of concepts of the rv and

av that maximizes the sum of the *consim* similarity values (*maximum weighted matching problem in bipartite graphs* [1]). In particular, given:

$$rv = \{r_1, ..., r_n\}$$
$$av = \{a_1, ..., a_m\}$$

as defined in Sect. 3, let S be the Cartesian Product of rv and av:

$$S = rv \times av$$

then, $\mathcal{P}(rv, av)$ is defined as follows:

$$\mathcal{P}(rv, av) = \{P \subset S | \forall (r_i, a_j), (r_h, a_k) \in P,$$
$$r_i \neq r_h, a_j \neq a_k, |P| = min\{n, m\}\}.$$

Therefore, on the basis of the *maximum weighted matching problem in bipartite graphs*, $semsim(rv, av)$ is given below:

$$semsim(rv, av) = \frac{max_{P \in \mathcal{P}(rv, av)} \{\sum_{(r_i, a_j) \in P} consim(r_i, a_j)\}}{max\{n, m\}} \quad (3)$$

As mentioned in the Introduction, *i-SemSim* is conceived in order to address particular scenarios where the annotation vector, or the request vector, contains a concept that better characterizes the resource than the other concepts. For instance, consider our running example concerning the computer domain. The first concept of each annotation vector specifies the type of computer the annotated resource belongs to (e.g., desktop, laptop). If the user is searching for a desktop computer, he/she will certainly discard laptops and tablets, even if the other requested features (e.g., type of screen and CPU) match. On the other hand, if there is not a characterizing feature, the risk is that a laptop results more similar to a desktop than to a hybrid tablet. This is the case of the *SemSim* method, as well as the other representative methods which we have selected in the experimental results Section (see Sect. 6).

The *i-SemSim* method rewards those resources that are annotated with either the most relevant concept in the request or a sub-concept of it. The related *i-semsim* function is defined as follows:

$$i\text{-}semsim(rv, av) = \begin{cases} semsim(rv, av) + (1 - \max_{av_i \in UDR} semsim(rv, av_i)), \\ \quad \text{if } av \text{ contains a concept that is subsumed} \\ \quad \text{by the most characterizing concept of } rv \\ semsim(rv, av), \qquad\qquad\qquad\qquad \text{otherwise} \end{cases} \quad (4)$$

For instance, suppose the user wants to buy a desktop and formulates his/her needs according to the *HiPerformance* request vector in Table 5. Consider now the annotation vectors av_2 and av_7 defined in Table 6 representing the former a desktop and the latter a tablet. He/she will consider the *HiPerformance* request vector more similar to av_2 than av_7 because av_2 represents the same typology of computer requested by the user. Nevertheless, if we compute the similarity between each of the two annotation vectors and the request vector by means of the majority of the representative methods defined in the literature, including *SemSim*, av_7 is more similar to the *HiPerformance* request vector than av_2. Whereas, if we apply the *i-semsim* function, we have that rv is more similar to av_2 than to av_7, as expected (see the next Section about the experimental results).

6 Experimental Results

In this section an experiment is illustrated which allows us to compare our proposal with some of the representative methods defined in the literature. In the experiment we suppose a user wants to buy a new computer and has to choose among possible computer offers. In particular, assume we have three possible kinds of customer requirements, each expressed in the form of a request vector, and ten possible computer offers, each expressed in the form of an annotation vector. The request vectors and the annotation vectors are taken from the computer ontology of Fig. 1. In the experiment we selected 22 people and we asked them to specify how much each computer offer matches each of the three kinds of customer requirements, giving a score from 1 to 4, 1 standing for strong similarity, and 4 for low or no similarity at all.

Table 5. Request vectors

LowCost = (Laptop, SmallSizeMediumResScreen, VGA, WiFi, MidCapacityHD, DualCore)
HiPerformance = (Desktop, LargeSizeHiResScreen, HDMI, Cable-Ethernet, WiFi, HiCapacityHD, OctaCore)
Nomadic = (Tablet, SmallSizeScreen, MicroHDMI, WiFi, BlueTooth, LTE, MidCapacitySSD, QuadCore)

The three kinds of customer requirements are expressed by the three request vectors given in Table 5, named *LowCost*, *HiPerformance*, and *Nomadic*, respectively. The computer offers are expressed by the ten annotation vectors av_1,..., av_{10} which are illustrated in Table 6. For instance, the *LowCost* request vector corresponds to the requirements of a customer who needs a *Laptop* having a small size and a medium resolution screen, a VGA video interface, the WiFi network interface, a medium capacity Hard Disk mass storage, and a Dual Core CPU. On the computer offer side, for instance, the offer av_1 describes a *Laptop* with a small size screen, a DVI video interface, a Wireless network interface, a medium capacity Hard Disk mass storage, and a Dual Core CPU.

Table 6. Annotation vectors

av_1 = (Laptop, SmallSizeScreen, DVI, Wireless, MidCapacityHD)
av_2 = (Desktop, LargeSizeMediumRes, StandardHDMI, Cable-Ethernet, MidCapacityHD, DualCore)
av_3 = (Tablet, SmallSizeMediumResScreen, MiniHDMI, BlueTooth, 3G, LowCapacitySSD, DualCore)
av_4 = (HybridTablet, MediumSizeHiResScreen, MiniHDMI, WiFi, 4G, MidCapacitySSD, OctaCore)
av_5 = (Desktop, LargeSizeHiResScreen, VGA, Cable-Ethernet, HiCapacitySSD, OctaCore)
av_6 = (Desktop, LargeSizeHiResScreen, MiniHDMI, WiFi, Ethernet, MidCapacityHD, QuadCore)
av_7 = (Tablet, SmallSizeHiResScreen, HDMI, WiFi, CellularNet, HiCapacitySSD, OctaCore)
av_8 = (Laptop, MediumSizeMediumResScreen, MicroHDMI, WiFi, CellularNet, QuadCore)
av_9 = (HybridTablet, MediumSizeHiResScreen, HDMI, WiFi, BlueTooth, HiCapacityHD, QuadCore)
av_{10} = (Laptop, MediumSizeHiResScreen, VGA, Cable-Ethernet, WiFi, MidCapacityHD, DualCore)

Table 7. *Pearson* correlation for *LowCost* request vector

	HJ	SemSim	i-SemSim	Dice	Jaccard	Salton	WSum
av_1	0.87	0.65	0.91	0.36	0.22	0.07	0.55
av_2	0.39	0.60	0.60	0.33	0.20	0.06	0.33
av_3	0.36	0.52	0.52	0.31	0.18	0.05	0.31
av_4	0.49	0.39	0.65	0.15	0.08	0.02	0.23
av_5	0.27	0.42	0.42	0.17	0.09	0.03	0.17
av_6	0.30	0.47	0.47	0.31	0.18	0.05	0.31
av_7	0.46	0.41	0.41	0.15	0.08	0.02	0.15
av_8	0.60	0.53	0.79	0.33	0.20	0.06	0.33
av_9	0.46	0.46	0.72	0.15	0.08	0.02	0.23
av_{10}	0.90	0.73	0.99	0.77	0.63	0.12	0.77
Corr.	**1.00**	**0.74**	**0.92**	**0.68**	**0.69**	**0.72**	**0.84**

For each of the three request vectors *LowCost*, *HiPerformance*, and *Nomadic*, the Human Judgment (HJ) has been evaluated as the average of the scores given by the 22 people that filled out the form of our experiment. Successively, the similarity between each of the three request vectors and each of the ten annotation vectors has been evaluated according to *SemSim*, *i-SemSim*, as well as

Table 8. *Spearman* correlation for *LowCost* request vector

	HJ	SemSim	i-SemSim	Dice	Jaccard	Salton	WSum
av_1	2	2	2	2	2	2	2
av_2	7	3	6	3	3	3	3
av_3	8	5	7	5	5	5	5
av_4	4	10	5	8	8	8	7
av_5	10	8	9	7	7	7	9
av_6	9	6	8	5	5	5	5
av_7	5	9	10	8	8	8	10
av_8	3	4	3	3	3	3	3
av_9	6	7	4	8	8	8	7
av_{10}	1	1	1	1	1	1	1
Corr.	**1.00**	**0.44**	**0.79**	**0.48**	**0.48**	**0.48**	**0.54**

Table 9. *Pearson* correlation for *HiPerformance* request vector

	HJ	SemSim	i-SemSim	Dice	Jaccard	Salton	WSum
av_1	0.22	0.35	0.35	0.00	0.00	0.00	0.08
av_2	0.60	0.59	0.77	0.31	0.18	0.05	0.38
av_3	0.16	0.42	0.42	0.00	0.00	0.00	0.07
av_4	0.42	0.57	0.57	0.29	0.17	0.04	0.36
av_5	0.76	0.68	0.86	0.62	0.44	0.10	0.62
av_6	0.69	0.81	0.99	0.57	0.40	0.08	0.64
av_7	0.43	0.65	0.65	0.43	0.27	0.06	0.43
av_8	0.37	0.45	0.45	0.15	0.08	0.02	0.23
av_9	0.43	0.63	0.63	0.43	0.27	0.06	0.43
av_{10}	0.24	0.61	0.61	0.29	0.17	0.04	0.29
Corr.	**1.00**	**0.77**	**0.90**	**0.86**	**0.88**	**0.88**	**0.92**

the selected representative methods *Dice*, *Jaccard*, *Salton*, and *WeightedSum* (*WSum*). For each request vector, the *Pearson* and *Spearman* correlations have been computed. We recall that the *Pearson* correlation evaluates the linear relationship between continuous variables, whereas *Spearman* correlation evaluates the monotonic relationship between continuous or ordinal variables [7].

The *Pearson* and *Spearman* correlations with HJ are shown in Tables 7 and 8, for the request vector *LowCost*, in Tables 9 and 10, for the request vector *HiPerformance*, and in Tables 11 and 12, for the request vector *Nomadic*, respectively. According to the experimental results, in the case of the *LowCost* request vector, *i-SemSim* shows the highest *Pearson* correlation (0.92) with an increment of 0.08 with respect to the *WSum* best result among the others (see Table 7), and the

Table 10. *Spearman* correlation for *HiPerformance* request vector

	HJ	SemSim	i-SemSim	Dice	Jaccard	Salton	WSum
av_1	9	10	10	9	9	9	9
av_2	3	6	3	5	5	5	5
av_3	10	9	9	9	9	9	10
av_4	6	7	7	6	6	6	6
av_5	1	2	2	1	1	1	2
av_6	2	1	1	2	2	2	1
av_7	4	3	4	3	3	3	3
av_8	7	8	8	8	8	8	8
av_9	5	4	5	3	3	3	3
av_{10}	8	5	6	6	6	6	7
Corr.	**1.00**	**0.84**	**0.94**	**0.91**	**0.91**	**0.91**	**0.92**

Table 11. *Pearson* correlation for *Nomadic* request vector

	HJ	SemSim	i-SemSim	Dice	Jaccard	Salton	WSum
av_1	0.46	0.35	0.35	0.15	0.08	0.03	0.23
av_2	0.19	0.27	0.27	0.00	0.00	0.00	0.00
av_3	0.64	0.59	0.86	0.27	0.15	0.04	0.33
av_4	0.76	0.55	0.82	0.27	0.15	0.04	0.33
av_5	0.16	0.29	0.29	0.00	0.00	0.00	0.00
av_6	0.28	0.48	0.48	0.27	0.15	0.04	0.27
av_7	0.78	0.62	0.89	0.27	0.15	0.04	0.47
av_8	0.54	0.54	0.54	0.43	0.27	0.06	0.50
av_9	0.78	0.60	0.87	0.40	0.25	0.05	0.53
av_{10}	0.33	0.37	0.37	0.13	0.07	0.02	0.13
Corr.	**1.00**	**0.89**	**0.93**	**0.75**	**0.72**	**0.72**	**0.86**

highest *Spearman* correlation (0.79), with an increment of 0.25 with respect to the *WSum* best result (see Table 8). In the case of the *Nomadic* request vector, *i-SemSim* shows the highest *Pearson* correlation (0.93) with an increment of 0.04 with respect to the *SemSim* best result among the others (see Table 11), whereas in the case of *Spearman* correlation, both *SemSim* and *i-SemSim* show the highest results (0.93), with an increment of 0.07 with respect to the *WSum* best result (see Table 12). Finally, in the case of the request vector *HiPerformance*, *i-SemSim* still shows the highest *Spearman* correlation (0.94), with an increment of 0.02 with respect to the *WSum* best result (see Table 10) whereas, in the case of *Pearson*, the highest correlation is shown by *WSum* (0.92), with an increment of 0.02 with respect to the *i-SemSim* best result (see Table 9).

Table 12. *Spearman* correlation for *Nomadic* request vector

	HJ	SemSim	i-SemSim	Dice	Jaccard	Salton	WSum
av_1	6	8	8	7	7	7	7
av_2	9	10	10	9	9	9	9
av_3	4	3	3	3	3	3	4
av_4	3	4	4	3	3	3	4
av_5	10	9	9	9	9	9	9
av_6	8	6	6	3	3	3	6
av_7	1	1	1	3	3	3	3
av_8	5	5	5	1	1	1	2
av_9	2	2	2	2	2	2	1
av_{10}	7	7	7	8	8	8	8
Corr.	**1.00**	**0.93**	**0.93**	**0.74**	**0.74**	**0.74**	**0.86**

Overall, the results of our experiment show the best *Pearson* and *Spearman* correlations of *i-SemSim* with HJ with respect to the other proposals in the cases of both the request vectors *LowCost* and *Nomadic*, with an average increment of about 0.17, in the former case, and an average increment of about 0.06, in the latter case. With regard to the request vector *HiPerformance*, *i-SemSim* still shows the best result in the case of *Spearman* correlation, with an increment of 0.02 with respect to *WSum*, whereas the opposite holds in the case of *Pearson* correlation, i.e., *WSum* shows in increment of 0.02 with respect to *i-SemSim*, and this is the only case where *i-SemSim* doesn't show the best result.

7   Conclusion

In this paper, we presented the evolution of *SemSim*, referred to as *i-SemSim*, that allows us to deal with DAG-shaped WRO. Furthermore, we improved the search mechanism by adopting, in the user request, the possibility to indicate one feature as the most influential (i.e., relevant) among the others listed in the request vector. The paper also presents the experimental results that show that the proposed method outperforms the previous versions of *SemSim* and, at the same time, the most popular similarity methods proposed in the literature. As a future work, in order to further investigate the *i-SemSim* method, we plan to address a new application domain characterized by a search space within high energy physics.

References

1. Dulmage, A., Mendelsohn, N.: Coverings of bipartite graphs. Canad. J. Math. **10**, 517–534 (1958)
2. Formica, A., Missikoff, M., Pourabbas, E., Taglino, F.: Semantic search for matching user requests with profiled enterprises. Comput. Ind. **64**(3), 191–202 (2013)
3. Formica, A., Missikoff, M., Pourabbas, E., Taglino, F.: A Bayesian approach for weighted ontologies and semantic search. In: Proceedings of IC3K, KEOD, Porto, Portugal, 9–11 November 2016, vol. 2, pp. 171–178 (2016)
4. Gao, J.B., Zhang, B.W., Chen, X.H.: A wordnet-based semantic similarity measurement combining edge-counting and information content theory. Eng. Appl. Artif. Intell. **39**, 80–88 (2015)
5. Grubisic, A., Stankov, S., Perai, I.: Ontology based approach to Bayesian student model design. Expert Syst. Appl. **40**(13), 5363–5371 (2013)
6. Jung, M., Jun, H.B., Kim, K.W., Suh, H.W.: Ontology mapping-based search with multidimensional similarity and Bayesian Network. Int. J. Adv. Manufact. Technol. **48**(1), 367–382 (2010)
7. Kendall, M.G., Stuart, A.: The Advanced Theory of Statistics. Inference and Relationship, vol. 2. Griffin, London (1973). ISBN 0-85264-215-6
8. Lin, D.: An information-theoretic definition of similarity. In: Proceedings of the 15th International Conference on Machine Learning, pp. 296–304. Morgan Kaufmann (1998)
9. Pearl, J., Russell, S.: Bayesian networks. In: Arbib, M.A. (ed.) Handbook of Brain Theory and Neural Networks, pp. 157–160. MIT Press (2001)
10. Salton, G., Wong, A., Yang, C.S.: A vector space model for automatic indexing. Commun. ACM **18**(11), 613–620 (1975)
11. Seco, N., Veale, T., Hayes, J.: An intrinsic information content metric for semantic similarity in Word-Net. In: Proceedings of ECAI, vol. 4, pp. 1089–1090 (2004)
12. Yazid, H., Kalti, K., Amara, N.E.B.: A new similarity measure based on Bayesian Network signature correspondence for braint tumors cases retrieval. Int. J. Comput. Intell. Syst. **7**(6), 1123–1136 (2014)

Indexing and Concurrency Control Methods

Indexing Multiple-Instance Objects

Linfei Zhou[1], Wei Ye[1], Zhen Wang[2], Claudia Plant[3], and Christian Böhm[1(✉)]

[1] Ludwig-Maximilians-Universität München, Munich, Germany
{zhou,ye,boehm}@dbs.ifi.lmu.de
[2] University of Electronic Science and Technology of China, Chengdu, China
zhen.wang@std.uestc.edu.cn
[3] University of Vienna, Vienna, Austria
claudia.plant@univie.ac.at

Abstract. As an actively investigated topic in machine learning, Multiple-Instance Learning (MIL) has many proposed solutions, including supervised and unsupervised methods. We introduce an indexing technique supporting efficient queries on Multiple-Instance (MI) objects. Our technique has a dynamic structure that supports efficient insertions and deletions and is based on an effective similarity measure for MI objects. Some MIL approaches have proposed their similarity measures for MI objects, but they either do not use all information or are time consuming. In this paper, we use two joint Gaussian based measures for MIL, Joint Gaussian Similarity (JGS) and Joint Gaussian Distance (JGD). They are based on intuitive definitions and take all the information into account while being robust to noise. For JGS, we propose the Instance based Index for querying MI objects. For JGD, metric trees can be directly used as the index because of its metric properties. Extensive experimental evaluations on various synthetic and real-world data sets demonstrate the effectiveness and efficiency of the similarity measures and the performance of the corresponding index structures.

1 Introduction

First motivated by the problem of drug activity predictions, Multiple-Instance Learning (MIL) deals with Multiple-Instance (MI) objects that are sets (or bags) of instances [1]. For objects with inherent structures, which are very common in real-world data, MI is a natural way to represent them. Therefore, various MIL methods have been proposed in many application domains like image classification [2], text categorization [3], activity recognition [4], etc.

With the increase of generated and stored data quantity, the efficiency of querying on MI data becomes a more and more important aspect. However, dynamic index structures for MI objects are yet to be developed and tested. A competitive candidate for such a structure has the properties to guarantee the query accuracy and to keep high efficiency in similarity calculations and pruning steps, which largely depends on the choice of similarity measures.

For MIL itself, the study of similarity measures is also the future direction. Most of MIL approaches are under a set of assumptions including the standard

© Springer International Publishing AG 2017
D. Benslimane et al. (Eds.): DEXA 2017, Part II, LNCS 10439, pp. 143–157, 2017.
DOI: 10.1007/978-3-319-64471-4_13

assumption [1] and the collective assumption [5]. For all these assumptions, each instance is assumed to have an explicit label, known or unknown, which is the same type as the label of the MI object. These assumptions work well for many applications such as drug activity predictions and content based image retrieval. However, they cannot deal with situations when the instances or bags have no labels, or there is no clear relation between instance-level labels and over-all labels. For example, the performance of an athlete is a MI object when the statistics of each match is regarded as an instance. It is impossible to obtain the learning model from instance spaces because there is a large number of instances (more than 0.7 million) need to be labeled, and even for a single instance it is difficult to label it for the evaluation of athletes. Figure 1(a) shows the Andrews plot (a smoothed version of parallel coordinate plot) of ten match logs from three NBA players, M. Jordan, K. Bryant and D. Harris. Each match log includes three statistics, minutes, field goal made and field goal attempted. As shooting guards, Jordan and Bryant have similar statistics, except that two match logs of Bryant are more like that of Harris who is a point guard. Nevertheless, due to the difference of play positions, it is not fair to label those two logs the same as the logs of Harris while label the other eight the same as the logs of Jordan. With the help of similarity measures, we can avoid these problems by taking each MI object as a whole, instead of starting with learning in instance spaces.

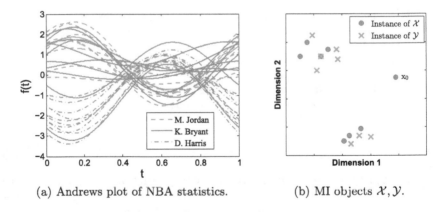

(a) Andrews plot of NBA statistics.   (b) MI objects $\mathcal{X}, \mathcal{Y}$.

Fig. 1. Demonstration of motivations.

Some MIL algorthms have introduced their similarity measures for MI objects, such as minHausdorff distance [6] and ϕ-quantile distance [7]. What is more, all distance functions that measure (dis)similarities between point sets can be used for MI objects. However, they either lose the information of MI objects or are time consuming. Take MI objects $\mathcal{X}, \mathcal{Y}$ (Fig. 1(b)) for example, they have one instance in common and the centers of instances are the same. Hausdorff distance between $\mathcal{X}$ and $\mathcal{Y}$ is highly determined by the position of instance x_0, making it extremely sensitive to outliers. Since the means of instances in $\mathcal{X}$ and

$\mathcal{Y}$ are equal, in algorithm SimpleMI, the dissimilarity of $\mathcal{X}$ and $\mathcal{Y}$ is zero although the two objects are not exactly the same.

A suitable similarity measure has competencies to be robust to noise, to be efficient in its computation, and to facilitate indexes and further analysis. As we will demonstrate, similarity measures used in this paper, Joint Gaussian Similarity (JGS) and Joint Gaussian Distance (JGD), are effective and efficient, and also support indexes for improving data retrieval operations. The main contributions of this paper are:

- We introduce Instance based Index to execute efficient queries on MI objects using JGS. Instance based Index stores a fixed number of instances in each entry, and has an effective strategy to prune non-qualified candidates.
- As a metric, JGD enables any metric tree to index MI objects. We apply VP-tree [8] on the index of MI objects using JGD.
- Experimental results show the effectiveness of JGS and JGD, and the efficiency of both indexes for MI objects.

The rest of this paper is organized as follows. In Sect. 2, we survey the previous work. Section 3 introduces JGS and JGD, and Sect. 4 describes the idea of Instance based Index for MI objects. Section 5 shows the experimental studies to verify the effectiveness of similarity measures and the efficiency of the proposed index. Finally, Sect. 6 summarizes this paper and presents some ideas for further research.

2   Related Work

In this section we give a brief survey and discussion of similarity measures for MI objects and indexes in previous work.

2.1   Similarity Measures for MI Objects

MIL algorithms can be grouped into three categories, the instance space based paradigm, the embedded space based paradigm and the bag space based paradigm [9]. For the last paradigm, the similarity measure for MI data is the essential part. In the bag space based paradigm, algorithms treat each bag as a single object and define similarity measures for bags. In this case, all distance based technologies can be used in MIL, such as k-NN, SVM, k-mediods, DBSCAN, etc.

All distance functions that measure the (dis)similarity between point sets can be used for MI objects, including Hausdorff distance [10], Sum of Minimum Distance (SMD) [11], Chamfer matching [12], EMD [13], Netflow distance [14], etc. What is more, some MIL algorithms have introduced their similarity measures for MI objects. J. Wang and J. Zucker have proposed modified Hausdorff distances for Bayesian k-NN and Citation k-NN, and concluded that the minimal Hausdorff distance slightly outperformed the maximal one [6]. It is worth noting that these two algorithms can also handle general classification tasks besides

MIL, and it is the similarity measures that solve the MIL problem. Similarly, W. Zhang et al. have applied Quantile based k-NN on MI data, defining ϕ-quantile distance [7]. X. He has proposed Probabilistic Integral Metric (PIM) on the basis of an idea that each MI object is a manifestation of some templates [15]. L. Sørensen et al. have compared BWmax and BWmean distance, and found that the latter had a better performance [16]. T. Fukui and T. Wada have introduced four similarity measures based on Diverse Density measure in their clustering algorithm, but all these measures can only handle the binary situation [17]. Metric learning has been extended to MIL [18,19], however, they actually learn a metric for instances, replacing Euclidean distances with Mahalanobis distances.

The existing similarity measures range from simple and efficient ones like Hausdorff distance to complex ones like Netflow distance and PIM, but only a few of them are metrics and take account of the information of all instances. Taking Hausdorff distance for example, it is a metric, but it only uses the distance between two single instances of bags, which makes it sensitive to noise.

2.2   Index

Index structures, besides linear scan, are essential techniques to make accesses to data more efficient.

Most of indexes are based on classical binary search algorithms, for instance, k-d tree [20], R-tree [21], etc. Spatial objects that can be treated as vectors are grouped by L-norm distance. Gauss-tree [22] and Gaussian Components based Index (GCI) [23] store objects in parameter space instead of feature spaces, and customized distance measures are used.

For general case where only a collection of objects and a function for measuring (dis)similarities are given, metric trees are introduced. However, similarity measures are required to satisfy the triangle inequality to prune candidates using the result of each similarity comparison. Metric trees includes M-tree [24], VP-tree [8], etc.

3   Joint Gaussian Based Measures

In this section, we present the ideas of JGS and JGD for MI objects. Firstly, we introduce Multiple-Instance Density and generalize it to Potential Instance Density for MI objects. Then we introduce JGS and JGD as the similarity measures.

3.1   Density of Instances

Being a finite set of instances, a MI object can be treated as a probability density function of instances. Instead of assuming that all the instances of a bag are generated from a known distribution such as a Gaussian model [25–27], we use a Gaussian distribution to represent the suppositional density around each instance. We define Potential Multiple-Instance Density as follows.

Definition 1 *(Potential Multiple-Instance Density). Given a MI object $\mathcal{X} = \{x_i\}_1^n$ where x_i is a feature vector in a space $\mathbb{R}^D$, Potential Multiple-Instance Density $f_{\mathcal{X}}(t)$ is represented by:*

$$f_{\mathcal{X}}(t) = \sum_{1 \leq i \leq n} w_i \frac{1}{\sqrt{2\pi\sigma_{\mathcal{X}}^2}} e^{-\frac{(t-x_i)^2}{2\sigma_{\mathcal{X}}^2}} \tag{1}$$

where w_i is the weight of instance x_i, $\sum_1^n w_i = 1$, and $\sigma_{\mathcal{X}}^2$ is the potential variance of $\mathcal{X}$.

3.2   Joint Gaussian Measures

Representing MI objects by their Potential Multiple-Instance Density functions, we apply JGS and JGD [28] on them. JGS is a measure of similarity for MI objects while JGD is a measure of dissimilarity and also a metric.

JGS considers all the potential positions of suppositional instances, and sums up the joint densities for two MI objects. The definition of JGS is shown as follows.

Definition 2 *(Joint Gaussian Similarity). Given two MI objects $\mathcal{X}, \mathcal{Y} \in \mathscr{P}(\mathbb{R}^D)$, JGS can be determined on the basis of Potential Multiple-Instance Density in the following way:*

$$d_{JGS}(\mathcal{X}, \mathcal{Y}) = \int_{\mathbb{R}^D} f_{\mathcal{X}}(t) f_{\mathcal{Y}}(t) \, dt$$

$$= \sum_{x \in \mathcal{X}} \sum_{y \in \mathcal{Y}} w_x w_y \frac{1}{\sqrt{2\pi(\sigma_{\mathcal{X}}^2 + \sigma_{\mathcal{Y}}^2)}} e^{-\frac{\|x-y\|^2}{2(\sigma_{\mathcal{X}}^2 + \sigma_{\mathcal{Y}}^2)}} \tag{2}$$

where $\| \cdot \|$ is Euclidean distance between feature vectors.

We assume $\sigma_{\mathcal{X}} = \sigma_{\mathcal{Y}} = \alpha \cdot \sigma$, where $\alpha > 0$ and σ is the variance of Euclidean distances between all instances of all bags.

The value of JGS between MI objects cannot exceed one, and if the instances of two MI objects are far from each other, it is close to zero. To obtain a high JGS, it is required that two MI objects have common or similar instances as many as possible.

In contrast to integrating the joint density of potential instances, JGD uses the square differences between two density functions and it is a measure of dissimilarity. The definition is shown as follows.

Definition 3 *(Joint Gaussian Distance). Given two MI objects $\mathcal{X}, \mathcal{Y} \in \mathscr{P}(\mathbb{R}^D)$, JGD sums up the square differences between their Potential Multiple-Instance Density values over the space $\mathbb{R}^D$.*

$$d_{JGD}(\mathcal{X}, \mathcal{Y}) = \left(\int_{\mathbb{R}^D} (f_{\mathcal{X}}(t) - f_{\mathcal{Y}}(t))^2 \, dt \right)^{\frac{1}{2}} \tag{3}$$

$$= \sqrt{d_{JGS}(\mathcal{X}, \mathcal{X}) + d_{JGS}(\mathcal{Y}, \mathcal{Y}) - 2d_{JGS}(\mathcal{X}, \mathcal{Y})}$$

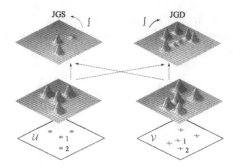

Fig. 2. Demostration of JGS and JGD between MI objects $\mathcal{U}, \mathcal{V}$ in a two-dimensional space.

Figure 2 demonstrates JGS and JGD between two MI objects $\mathcal{U}, \mathcal{V}$, of which the instances are marked by blue dots and green crosses, respectively. Two Potential Multiple-Instance Density functions are generated from the corresponding MI objects. The most similar instances between $\mathcal{U}, \mathcal{V}$ are $u_1 \& v_1$ and $u_2 \& v_2$. As a measure of similarity, generally JGS integrates the similar parts of instances, as shown in the top-left of the figure. On the contrary, JGD mainly integrates the rest part, which includes two instances in $\mathcal{U}$ and three instances in $\mathcal{V}$, as shown in the top-right of the figure.

As mentioned earlier, JGD is a metric for MI objects. Having MI objects represented in this way, we can also extend several similarity measures designed for Gaussian Mixture Models to MIL [29–31], but none of them is a metric.

4   Index

In this part we discuss index techniques for querying MI objects with JGS and JGD. Since JGS is not a metric, we need to design a specialized index structure to ensure the query efficiency and accuracy. As for JGD, we employ VP-tree, a hierarchical structure, directly to speed up both k-NN queries and range queries while guaranteeing the accuracy.

Instance Based Index. Due to the potential unequal numbers of instances in MI objects, traditional index techniques like R-tree cannot be used on MI data. To tackle this problem, on basis of GCI [23], we introduce Instance based Index for MI objects using JGS. Firstly we store all the instances of MI objects into a Gauss-tree [22] which supports efficient queries for Gaussian distributions, thus Instance based Index shares the same insertion and deletion strategies with the Gauss-tree. Then we build an extra structure to locate and store potential candidates.

As shown in Fig. 3, MI objects are decomposed into instances and stored in a Gauss-tree. Given a query object $\mathcal{X}_Q = \{x_j\}_{j=1}^{n_Q}$, where n_Q is the number of

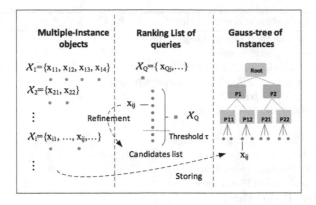

Fig. 3. Instance based Index for querying MI objects with JGS.

instances in $\mathcal{X}_Q$, we start the ranking of instances by JGS between them and $\mathcal{X}_Q$, and get the candidates list of MI objects. As for the query processing, we assume that we always have a pruning threshold τ, below which the corresponding objects of the instances are not of interest. Only for these instances that have higher JGS than τ, their corresponding MI objects will be retrieved to execute the expensive calculation of JGS between MI objects, which we call the refinement. τ can either be defined by the user in range queries, or be the k-th ranked JGS with the query object in k-NN queries. In the latter case we start with $\tau = 0$ and update it whenever we find a greater k-th JGS than τ. For instances that have lower JGS than τ, we can safely exclude the corresponding MI objects if they have not been retrieved yet.

Given an index node $P = [\check{w}_p, \hat{w}_p; \{\check{x_{pi}}, \hat{x_{pi}}\}_1^D; \check{n}_p, \hat{n}_p]$, in the prune stage of instance candidates, we determine whether or not this node contains any instance that has a higher JGS than the threshold τ by its upper bound $\hat{d}_{\mathrm{JGS}}(\mathcal{X}_Q, P)$ shown as follows.

$$
\begin{aligned}
\hat{d_{\mathrm{JGS}}}(\mathcal{X}_Q, P) &= \sum_{x_j \in \mathcal{X}_Q} \hat{d}_{\mathrm{JGS}}(x_j, P) \\
&= \sum_{x_j \in \mathcal{X}_Q} \prod_{1 \le i \le D} \hat{d}_{\mathrm{JGS}}(x_{ji}, x_{pi})
\end{aligned}
\tag{4}
$$

where $\hat{d}_{\mathrm{JGS}}(x_{ji}, x_{pi})$ is the i-th dimensional upper bound of JGS between a query instance x_j and instances x_p stored in a node in the Gauss-tree, and it can be reached when the following conditions are met:

$$
\begin{cases}
w_p = \hat{w}_p \\
x_{pi} = \check{x_{pi}} & \text{if} \quad x_{ji} < \check{x_{pi}} \\
x_{pi} = \hat{x_{pi}} & \text{if} \quad x_{ji} > \hat{x_{pi}} \\
x_{pi} = x_{ji} & \text{if} \quad \check{x_{pi}} \le x_{ji} \le \hat{x_{pi}}
\end{cases}
\tag{5}
$$

Algorithm 1. Instance based Index for the k-NN Query

Data: int k, Node *root*, Query Object $\mathcal{X}_Q$
Result: PriorityQueue *results*

1   PriorityQueue *results* = new PriorityQueue() ;   /* Ascending */
2   PriorityQueue *activePages* = new PriorityQueue() ;   /* Descending */
3   *results*.put(-1, -MAX_REAL);
4   *activePages*.put(*root*, MAX_REAL);
5   $\tau = 0$;
6   **while** *activePages*.isNotEmpty*()* and
      results.getMinJGS*()*<*activePages*.getMaxJGS*()* **do**
7      $P = $ *activePages*.getFirstPage;
8   *activePages*.removeFirstPage();
9   **if** P.isDataPage*()* **then**
10   Entry $E = P$.data;
11   **if** $d_{\widehat{JGS}}(\mathcal{X}_Q, P) > \tau$ **then**
12          $\mathcal{X}_{candidate} = E$.getMIobject;
13   *results*.put($\mathcal{X}_{candidate}$, $d_{JGS}(\mathcal{X}_{candidate}, \mathcal{X}_Q)$);
14   **if** *results*.size> k **then**
15   *results*.removeFirst;
16        $\tau = $ *results*.getMinJGS();
17   **else**
18   *children* = P.getChildren();
19   **while** *children*.hasMoreElements*()* **do**
20   *child* = *children*.getNextElement();
21   *probability* = $d_{\widehat{JGS}}(\mathcal{X}_Q, child)$;
22   *activePages*.put(*child*,*probability*);

The pseudo code in Algorithm 1 shows Instance based Index for k-NN queries. As for range queries, an unknown number of possible candidates for a query object are returned by fixing threshold τ as a given parameter T.

Index for Queries in Metric Spaces. To index data in metric spaces, metric trees exploit the metric property, the triangle inequality, to have more efficient access to data. Because of the metric properties of JGD, various metric trees can be employed to speed up queries for MI objects with JGD. In this paper we use VP-tree to evaluate the performance of JGD.

4.1 Time Complexity

Given two MI objects in a D-dimensional space, both JGS and JGD have the same time complexity as that of Hausdorff distance, $O((m+n)D)$, where m and n are the cardinalities of MI objects.

To apply k-NN search for a query object in a database of N MI objects that have maximally m instances in each object, the time complexity of the linear scan is $O(Nm)$. Storing these N MI objects into Instance based Index, the average query time complexity is $O(log(Nm)) + \alpha O(Nm)$. α varies in $(0, 1]$, and it is related to the distribution of instances and the setting of the Gauss-tree.

5 Experimental Evaluations

In this section, we provide experimental evaluations on both synthetic and real-world data to show the effectiveness and efficiency of two measures and indexes.

All experiments are implemented[1] with Java 1.7, and executed on a regular workstation PC with 3.4 GHz dual core CPU equipped with 32 GB RAM. To keep the consistency of the codes, we use the reciprocal value of JGS as its dissimilarity value. For all experiments, we use the 10-fold cross validation and report the average results over 100 runs.

5.1 Data Sets

Synthetic data[2] is generated from a normal distribution. It varies in the number of MI objects, the number of instances in each object and the dimensionality.

Musk data[3] is a benchmark data for MIL. It has two data sets, Musk 1 and Musk 2, the details of which have been described by T. Dietterich et al. [1]. Fox, Tiger and Elephant data[4] is another benchmark data. It is generated from image data sets after preprocessing and segmentation [3]. Besides these data sets, we also use two other real-world data sets[5], CorelDB data and Weather data. CorelDB data consists of extracted features from images. Each image is smoothed by a Gaussian filter and then it generates a 9×9 grid of pixels of which the 7×7 non-border are chosen as instances. The features of each instance are color differences between a pixel and its neighbours. Weather data is the historical weather data of airports around the world. Each instance of airports is the average statistics in a month. We use the main categories of Köppen climate classification system to label each airport.

5.2 Parameter Setting

To evaluate the influence of parameter α in Eq. 2, we perform 1-NN classifications on Musk data. The classification accuracies when varying α are shown in Fig. 4. The accuracies of both data sets shoot up before α reaches 1, especially for JGS, and level off afterward. Therefore, we choose $\alpha = 1$ for all the following experiments.

---

1 https://drive.google.com/open?id=0B3LRCuPdnX1BMFViblpaS1VKZmM.
2 https://drive.google.com/open?id=0B3LRCuPdnX1BVHFjeWpiLWF3M2M.
3 https://archive.ics.uci.edu/ml/machine-learning-databases/musk/.
4 http://www.miproblems.org/datasets/foxtigerelephant/.
5 https://drive.google.com/open?id=0B3LRCuPdnX1BYXpGUzlxYVdsSDA.

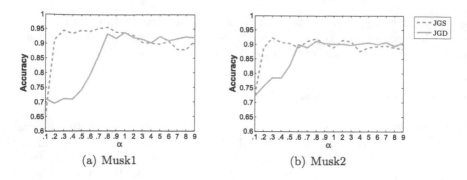

Fig. 4. Classification accuracies on Musk data.

5.3 Effectiveness

In this part, we evaluate the performances of proposed similarity measures on both supervised and unsupervised learning. The parameters of PIM are set to the optimum in the original paper, where the variance of template distribution is 5000 and the number of samples is 1000. For ϕ-quantile distance, ϕ is set to 0.5 as suggested in the original paper.

Classification on Real-World Data Sets. Since we are not interested in tuning the classification accuracy to its optimum, k-NN rather than the other more complex techniques is used to compare the effectiveness of similarity measures here.

The accuracies of classification on seven real-world data sets are shown in Table 1. JGD achieves the best performance on six data sets, and its result is still considerable on Elephant data. The performance of JGS is moderate except on Weather data.

Table 1. Classification results of k-NN on real-world data sets ($k = 10$)

	Musk1	Musk2	Fox	Tiger	Elephant	CorelDB	Weather	Avg.
Hausdorff	.716 ± .134	.707 ± .155	.655 ± .108	.774 ± .094	.829 ± .087	.803 ± .059	.469 ± .026	.707
SMD	.704 ± .143	.712 ± .165	.668 ± .094	.764 ± .099	.810 ± .088	.856 ± .050	.471 ± .026	.712
Chamfer	.716 ± .141	.720 ± .154	.658 ± .111	.796 ± .093	.816 ± .084	.856 ± .050	.470 ± .026	.718
EMD	.729 ± .138	.705 ± .163	.668 ± .104	.809 ± .079	**.835 ± .086**	.873 ± .051	.470 ± .028	.727
Netflow	.729 ± .138	.725 ± .164	.669 ± .103	.811 ⊥ .078	**.835 ± .086**	.873 ± .051	.470 ± .028	.730
minHausd	.729 ± .137	.725 ± .144	.674 ± .104	.783 ± .096	.806 ± .099	.436 ± .063	.299 ± .024	.635
ϕ-quantile	.669 ± .174	.654 ± .157	.638 ± .115	.751 ± .101	.797 ± .091	.793 ± .063	.432 ± .026	.676
PIM	.723 ± .152	.702 ± .169	.667 ± .099	.719 ± .092	.776 ± .095	.871 ± .049	.471 ± .026	.704
BWmean	.711 ± .163	.737 ± .131	.600 ± .103	.703 ± .110	.588 ± .113	**.878 ± .047**	.472 ± .026	.670
JGS	.761 ± .136	.718 ± .149	.656 ± .116	.778 ± .107	.821 ± .081	.837 ± .057	.161 ± .026	.676
JGD	**.871 ± .109**	**.801 ± .142**	**.694 ± .101**	**.813 ± .083**	.808 ± .091	**.878 ± .051**	**.477 ± .030**	**.763**

Table 2. Clustering results of k-medoids

	CorelDB (k=4)		Weather (k=5)	
	Purity	NMI	Purity	NMI
Hausdorff	.691 ± .052	.499 ± .078	.668 ± .060	.368 ± .032
SMD	.805 ± .065	.663 ± .057	.670 ± .049	**.380 ± .030**
Chamfer	.811 ± .070	.667 ± .062	.669 ± .053	.378 ± .035
EMD	.809 ± .072	.716 ± .066	.649 ± .058	.365 ± .035
Netflow	.824 ± .065	.730 ± .052	.651 ± .056	.368 ± .032
minHausdorff	.680 ± .063	.500 ± .066	.537 ± .041	.202 ± .044
ϕ-quantile	.761 ± .056	.683 ± .059	.635 ± .037	.335 ± .020
PIM	.808 ± .068	.671 ± .064	.647 ± .057	.366 ± .030
BWmean	.808 ± .076	.705 ± .079	.529 ± .035	.207 ± .046
JGS	.595 ± .061	.372 ± .093	.613 ± .065	.355 ± .036
JGD	**.825 ± .081**	**.736 ± .072**	**.673 ± .057**	.374 ± .033

Clustering on Real-World Data Sets. We perform clustering experiments to compare the usability of proposed similarity measures for unsupervised data mining. k-medoids is used in this paper because unlike k-means, it works with arbitrary similarity measures. We evaluate clustering results with two widely used criteria, Purity and Normalized Mutual Information (NMI).

Evaluation results on CorelDB data and Weather data that have more than two classes are shown in Table 2. We can see that the performance of JGD is the best or comparable to the best on this task, while JGS achieves a moderate performance.

5.4 Efficiency

In this part, we compare the efficiency of JGS, JGD and the other similarity measures. We start with the time cost[6] of all the similarity calculations when varying the dimensionality and the number of instances of each MI object, and then investigate the performances of five metrics supported by VP-tree, as well as JGS supported by Instance based Index.

Time Complexity. All the similarity measures compared in this paper have a linear relation with the dimensionality and the number of instances in each MI object. Since the curves of SMD and Chamfer almost duplicate that of Hausdorff, their results are not included in Fig. 5.

Due to the inherent complexity of EMD, Netflow and BWmean distance, the influence of dimensionality becomes evident after the dimensionality reaches 512, as shown in Fig. 5(a), where the number of instances is fixed to ten.

---

[6] The time cost in this paper refers to the CPU time.

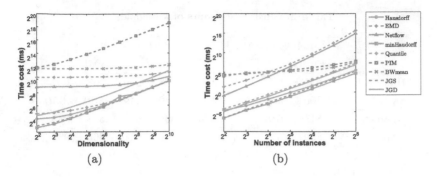

Fig. 5. Time cost of similarity calculations on synthetic data.

Hausdorff distance, SMD, Chamfer, minHausdorff distance and JGS are the most efficient measures. JGD costs slightly more run-time than these relatively simple measures, but it is much efficient than sophisticated measures like PIM. Fixing the data dimensionality to two, the run-time of similarity measures increases linearly with the number of instances in each MI object, and the performance of JGD is almost comparable with that of the most efficient techniques like Hausdorff distance (Fig. 5(b)).

Index. We study the scalability of five metrics with VP-tree and JGS with Instance based Index here. The capacity of nodes in the VP-tree is set to 32, while the minimum and maximum node capacity of the Instance based Index are set to 10 and 50, respectively.

Firstly linear scan queries are applied on synthetic data, and there are ten two-dimensional instances in each MI object. As shown in Fig. 6, the run-time of all six measures increase linearly with the number of objects. JGD and JGS have

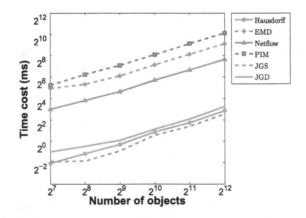

Fig. 6. Time cost of 1-NN queries using linear scan on synthetic data.

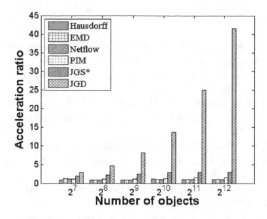

Fig. 7. Acceleration ratio of 1-NN queries using indexes on synthetic data. * indicates that Instance based Index is used, while VP-tree is applied for other measures.

almost the same performance as Hausdorff distance which is the most efficient among all the proposed techniques.

To evaluate the performance of five metrics and JGS when using indexes, we report the ratio of linear scan query time and the index query time on synthetic data. The higher the ratio is, the more the similarity measure benefits from indexes. As shown in Fig. 7, JGD profits the most and its acceleration ratio is much higher than those of the others for this experiment. As for JGS with Instance based Index, it achieves higher speed-up rates than the left four metrics.

6   Conclusion and Future Work

In this paper, we have evaluated JGS and JGD for MIL. They use all the information of MI objects and they are robust to noise. Evaluations on both synthetic and real-world data demonstrate the better performances of them than the other similarity measures, especially for JGD.

To achieve more efficient queries for MI objects, we have introduced Instance based Index using JGS. To the best of our knowledge, Instance based Index is the very first specialized dynamic index structure designed for MI objects. For JGD, it has the ability to employ any metric tree to accelerate queries because of its metric properties. The performance of JGD on VP-tree significant outperforms the other metrics.

For the future work, a specialized index for JGD is a promising perspective to obtain a better performance than existing index structures. Making use of the characteristic of MI objects, the customized index could exploit the potential of efficient queries for MIL.

References

1. Dietterich, T.G., Lathrop, R.H., Lozano-Pérez, T.: Solving the multiple instance problem with axis-parallel rectangles. Artif. Intell. **89**(1–2), 31–71 (1997)
2. Chen, Y., Wang, J.Z.: Image categorization by learning and reasoning with regions. J. Mach. Learn. Res. **5**, 913–939 (2004)
3. Andrews, S., Tsochantaridis, I., Hofmann, T.: Support vector machines for multiple-instance learning. In: Advances in Neural Information Processing Systems 15, pp. 561–568 (2002)
4. Guan, X., Raich, R., Wong, W.: Efficient multi-instance learning for activity recognition from time series data using an auto-regressive hidden markov model. In: ICML, pp. 2330–2339 (2016)
5. Xu, X.: Statistical learning in multiple instance problems. Master's thesis, University of Waikato (2003)
6. Wang, J., Zucker, J.: Solving the multiple-instance problem: a lazy learning approach. In: ICML, pp. 1119–1126 (2000)
7. Zhang, W., Lin, X., Cheema, M.A., Zhang, Y., Wang, W.: Quantile-based KNN over multi-valued objects. In: ICDE, pp. 16–27 (2010)
8. Yianilos, P.N.: Data structures and algorithms for nearest neighbor search in general metric spaces. In: ACM/SIGACT-SIAM SODA, pp. 311–321 (1993)
9. Amores, J.: Multiple instance classification: review, taxonomy and comparative study. Artif. Intell. **201**, 81–105 (2013)
10. Hausdorff, F., Aumann, J.R.: Grundzüge der mengenlehre. Veit (1914)
11. Niiniluoto, I.: Truthlikeness, vol. 185. Springer Science & Business Media, Dordrecht (2012)
12. Belongie, S.J., Malik, J., Puzicha, J.: Shape matching and object recognition using shape contexts. Pattern Anal. Mach. Intell. **24**(4), 509–522 (2002)
13. Rubner, Y., Tomasi, C., Guibas, L.J.: The earth mover's distance as a metric for image retrieval. Int. J. Comput. Vis. **40**(2), 99–121 (2000)
14. Ramon, J., Bruynooghe, M.: A polynomial time computable metric between point sets. Acta Inf. **37**(10), 765–780 (2001)
15. He, X.: Multi-purpose exploratory mining of complex data. Ph.D. dissertation, Ludwig-Maximilians-Universität München (2014)
16. Sørensen, L., Loog, M., Tax, D.M.J., Lee, W., de Bruijne, M., Duin, R.P.W.: Dissimilarity-based multiple instance learning. In: IAPR, pp. 129–138 (2010)
17. Fukui, T., Wada, T.: Commonality preserving multiple instance clustering based on diverse density. In: Jawahar, C.V., Shan, S. (eds.) ACCV 2014. LNCS, vol. 9010, pp. 322–335. Springer, Cham (2015). doi:10.1007/978-3-319-16634-6_24
18. Guillaumin, M., Verbeek, J., Schmid, C.: Multiple instance metric learning from automatically labeled bags of faces. In: Daniilidis, K., Maragos, P., Paragios, N. (eds.) ECCV 2010. LNCS, vol. 6311, pp. 634–647. Springer, Heidelberg (2010). doi:10.1007/978-3-642-15549-9_46
19. Jin, R., Wang, S., Zhou, Z.: Learning a distance metric from multi-instance multi-label data. In: 2009 IEEE Computer Society Conference on Computer Vision and Pattern Recognition (CVPR 2009), Miami, Florida, USA, 20–25 June 2009, pp. 896–902 (2009)
20. Bentley, J.L.: Multidimensional binary search trees used for associative searching. Commun. ACM **18**(9), 509–517 (1975)
21. Guttman, A.: R-trees: a dynamic index structure for spatial searching. In: Proceedings of Annual Meeting, SIGMOD 1984, Boston, Massachusetts, 18–21 June 1984, pp. 47–57 (1984)

22. Böhm, C., Pryakhin, A., Schubert, M.: The gauss-tree: efficient object identification in databases of probabilistic feature vectors. In: ICDE, p. 9 (2006)
23. Zhou, L., Wackersreuther, B., Fiedler, F., Plant, C., Böhm, C.: Gaussian component based index for GMMs. In: ICDM, pp. 1365–1370 (2016)
24. Ciaccia, P., Patella, M., Zezula, P.: M-tree: an efficient access method for similarity search in metric spaces. In: VLDB, pp. 426–435 (1997)
25. Kriegel, H.-P., Pryakhin, A., Schubert, M.: An EM-approach for clustering multi-instance objects. In: Ng, W.-K., Kitsuregawa, M., Li, J., Chang, K. (eds.) PAKDD 2006. LNCS, vol. 3918, pp. 139–148. Springer, Heidelberg (2006). doi:10.1007/11731139_18
26. Wei, X., Wu, J., Zhou, Z.: Scalable multi-instance learning. In: ICDM, pp. 1037–1042 (2014)
27. Vatsavai, R.R.: Gaussian multiple instance learning approach for mapping the slums of the world using very high resolution imagery. In: KDD, pp. 1419–1426 (2013)
28. Zhou, L., Plant, C., Böhm, C.: Joint gaussian based measures for multiple-instance learning. In: ICDE, pp. 203–206 (2017)
29. Sfikas, G., Constantinopoulos, C., Likas, A., Galatsanos, N.P.: An analytic distance metric for gaussian mixture models with application in image retrieval. In: Duch, W., Kacprzyk, J., Oja, E., Zadrożny, S. (eds.) ICANN 2005. LNCS, vol. 3697, pp. 835–840. Springer, Heidelberg (2005). doi:10.1007/11550907_132
30. Jensen, J.H., Ellis, D.P.W., Christensen, M.G., Jensen, S.H.: Evaluation of distance measures between gaussian mixture models of MFCCs. In: ISMIR, pp. 107–108 (2007)
31. Cui, S., Datcu, M.: Comparison of kullback-leibler divergence approximation methods between gaussian mixture models for satellite image retrieval. In: IGARSS, pp. 3719–3722 (2015)

Novel Indexing Strategy and Similarity Measures for Gaussian Mixture Models

Linfei Zhou[1], Wei Ye[1], Bianca Wackersreuther[1], Claudia Plant[2], and Christian Böhm[1(✉)]

[1] Ludwig-Maximilians-Universität München, Munich, Germany
{zhou,ye,wackersb,boehm}@dbs.ifi.lmu.de
[2] University of Vienna, Vienna, Austria
claudia.plant@univie.ac.at

Abstract. Efficient similarity search for data with complex structures is a challenging task in many modern data mining applications, such as image retrieval, speaker recognition and stock market analysis. A common way to model these data objects is using Gaussian Mixture Models which has the ability to approximate arbitrary distributions in a concise way. To facilitate efficient queries, indexes are essential techniques. However, due different numbers of components in Gaussian Mixture Models, existing index methods tend to break down in performance. In this paper we propose a novel technique Normalized Transformation that reorganizes the index structure to account for different numbers of components in Gaussian Mixture Models. In addition, Normalized Transformation enables us to derive a set of similarity measures on the basis of existing ones that have close-form expression. Extensive experiments demonstrate the effectiveness of proposed technique for Gaussian component-based indexing and the performance of the novel similarity measures for clustering and classification.

1 Introduction

Information extraction systems face great challenges in the representation and analysis of the data, especially with the rapid increase in the amount of data. Take player statistics for example, far more than field goal made, rebounds and etc., SportVU utilizes six cameras to track the real-time positions of NBA players and the ball 25 times per second [1]. Comprehensive and sophisticated data generated by SportVU provides a possibility to make the best game strategy or to achieve the most effective team building, but it increases the difficulty of following modeling and analysis as well. Besides, many modern applications like speaker recognition systems [2,3], content-based image and video retrieval [4,5], biometric identification and stock market analysis not only can benefit from the retrieval and analysis of complex data, or the distributions of data, but also are limited by them.

Various statistical models have been proposed in this actively investigated research field. As a general class of Probability Density Functions (PDF),

© Springer International Publishing AG 2017
D. Benslimane et al. (Eds.): DEXA 2017, Part II, LNCS 10439, pp. 158–171, 2017.
DOI: 10.1007/978-3-319-64471-4_14

Gaussian Mixture Models (GMMs) consist of a weighted sum of univariate or multivariate Gaussian distributions, allowing a concise but exact representation of data distributions. Storing complex data as GMMs will dramatically reduce the resource consumption and guarantees the accuracies of retrieval operations.

Besides the data representation, another important aspect is the design of similarity measures that aims at facilitating indexes and further analysis. Matching probability [6] sums up the joint probabilities of two PDFs, and for GMMs it has s closed-form expression which is essential for efficient calculations. What is more, several similarity measures that have closed-form expressions can be reformed into the functions of matching probability [7–9]. Since GMMs might have different numbers of Gaussian components in them, traditional indexes designed for fixed-length vectors cannot be applied directly. For the indexes of distributions, such as U-tree, their performances deteriorate on mixture models [10]. Storing the components, instead of GMMs, into entries, both Gaussian Component based Index (GCI) [11] and Probabilistic Ranking Query (PRQ) [12] provide solutions for efficient range queries and nearest-neighbour queries on GMMs using matching probability. However, the efficiency of the two indexes vary with the distributions of components in GMMs because of the settings of nodes, which encourages us to improve the situation. As we will demonstrate, the main contributions of this paper are:

- We introduce a generalization technique called Normalize Transformation. After Normalize Transformation, indexes based on matching probability can achieve the better performances of queries on GMMs.
- Normalize Transformation enables us to derive a set of new similarity measures from the existing ones. The normalized versions of the similarity measures share the same time complexity of their origins.
- Our experimental evaluation demonstrates the efficiency of filtering in GCI using normalized matching probability and the better performances of normalized similarity measures over their origins.

The rest of this paper is organized as follows: In Sect. 2, we survey the previous work. Section 3 gives the basic definition of GMM and matching probability. Section 4 introduces the motivation of the Normalized Transformation, and demonstrates how it works. Section 5 shows the experimental studies for verifying the efficiency and effectiveness of the proposed similarity measures. Section 6 summarizes the paper.

2 Related Work

This section gives a survey and discussion of similarity measures and indexes for GMMs in previous work.

2.1 Definitions of Similarity

Similarity measures for GMMs can be grouped into two categories, having closed-form expressions for GMMs or not. For measures that have no closed-form expression, Monte Carlo sampling or other approximation approaches are applied, which may be time consuming or imprecise.

Kullback-Leibler (KL) divergence [13] is a common way to measure the distance between two PDFs. It has a closed-form expression for Gaussian distributions, but no such expression for GMMs exists.

To compute the distance between GMMs by KL divergence, several approximation methods have been proposed. For two GMMs, a commonly used approximation for KL divergence between them is Gaussian approximation. It replaces two GMMs with two Gaussian distributions, whose means and covariance matrices depend on those of GMMs. Another popular way is to use the minimum KL divergence of Gaussian components that are included in two GMMs. Moreover, Hershey et al. [14] have proposed the product of Gaussian approximation and the variation approximation, but the former tends to greatly underestimate the KL divergence between GMMs while the latter does not satisfy the positivity property. Besides, Goldberger et al. [15] have proposed the matching based KL divergence (KLm) and the unscented transformation based KL divergence (KLt). KLm works well when the Gaussian elements are far apart, but it cannot handle the overlapping situations which are very common in real-world data sets. KLt solves the overlapping problem based on a non-linear transformation. Cui et al. [16] have compared the six approximation methods for KL divergence with Monte Carlo sampling, where the variation approximation achieves the best result quality, while KLm gives a comparable result with a much faster speed.

Besides the approximation similarity methods for GMMs, several methods with closed-form expression have been proposed. Helén et al. [7] have described a squared Euclidean distance, which integrates the squared differences over the whole feature space. It has a closed-form expression for GMMs. Sfikas et al. [8] have presented a KL divergence based distance C2 for GMMs. Jensen et al. [9] used a normalized L2 distance to measure the similarity of GMMs in mel-frequency cepstral coefficients from songs. Beecks et al. have proposed Signature Quadratic form Distance for modeling image similarity in image databases [17].

2.2   Indexing Techniques

For the indexes of GMMs, there are several techniques available, including universal index structures designed for uncertain data and GMM-specific methods.

U-tree provides a probability threshold retrieval on general multi-dimensional uncertain data [10]. It pre-computes a finite number of Probabilistically Constrained Regions (PCRs) which are possible appearance regions with fixed probabilities, and uses them to prune unqualified objects. Although U-tree works well with single-peak PDFs, its effectiveness deteriorates for mixture models such as GMMs. The reason behind this is that it is difficult for PCRs to represent mixture models, especially when the component numbers increase.

Rougui et al. [18] have designed a bottom-up hierarchical tree and an iterative grouping tree for GMM-modeled speaker retrieval systems. Both approaches provide only two index levels, and are lack of a convenient insertion and deletion strategy. Furthermore, they can not guarantee reliable query results.

Instead of index curves as spatial objects in feature spaces, Probabilistic Ranking Query (PRQ) technique [12] and Gaussian Component based Index [11] search the parameter space of the means and variances of GMMs. However, PRQ can not guarantee the query accuracy since it assumes that all the Gaussian components of candidates have relatively high matching probabilities with query objects, which is not common in general cases. For both indexes, their prune strategies are highly effected by the distributions of Gaussian components.

3   Formal Definitions

In this section, we summarize the formal notations for GMMs. A GMM is a probabilistic model that represents the probability distribution of observations. The definition of the GMM is shown as follows.

Definition 1 *(Gaussian Mixture Model). Let $x \in \mathbb{R}^D$ be a variable in a D-dimensional space, $x = (x_1, x_2, ..., x_D)$. A Gaussian Mixture Model $\mathcal{G}$ is the weighted sum of m Gaussian functions, defined as:*

$$\mathcal{G}(x) = \sum_{1 \leq i \leq m} w_i \cdot \mathcal{N}_i(x) \tag{1}$$

where $\sum_{1 \leq i \leq m} w_i = 1$, $\forall i \in [1, m], w_i \geq 0$, and Gaussian component $\mathcal{N}_i(\mathrm{x})$ is the density of a Gaussian distribution with a covariance matrix Σ_i:

$$\mathcal{N}_i(\mathrm{x}) = \frac{1}{\sqrt{(2\pi)^D |\Sigma_i|}} \exp\left(-\frac{1}{2}(\mathrm{x} - \mu_i)^T \Sigma_i^{-1}(\mathrm{x} - \mu_i) \right)$$

As we can see in Definition 1, a GMM can be represented by a set of m components, and each of them is composed of a mean vector $\mu \in \mathbb{R}^D$ and a covariance matrix $\Sigma \in \mathbb{R}^{D \times D}$. It is worth noting that only for GMMs that have diagonal covariance matrices, matching probability for GMMs has closed-form expressions, so are the other similarity measures[1]. The definition of matching probability is shown as follows.

Definition 2 *(Matching Probability [6]). Let $\mathcal{G}_1$ and $\mathcal{G}_2$ be two GMMs with diagonal covariance matrices, and they have m_1 and m_2 Gaussian components, respectively. Let x be a feature vector in $\mathbb{R}^D$. Matching probability between $\mathcal{G}_1$ and $\mathcal{G}_2$ can be derived as:*

$$mp(\mathcal{G}_1, \mathcal{G}_2) = \int_{\mathbb{R}^D} \mathcal{G}_1(x) \mathcal{G}_2(x) \, dx$$

$$= \sum_{i=1}^{m_1} \sum_{j=1}^{m_2} w_{1,i} w_{2,j} \prod_{l=1}^{D} \frac{e^{-\frac{(\mu_{1,i,l} - \mu_{2,j,l})^2}{2(\sigma_{1,i,l}^2 + \sigma_{2,j,l}^2)}}}{\sqrt{2\pi(\sigma_{1,i,l}^2 + \sigma_{2,j,l}^2)}} \tag{2}$$

---

[1] To the best of our knowledge.

where $\sigma_{1,i,l}$ and $\sigma_{2,j,l}$ are the l-th diagonal elements of $\Sigma_{1,i}$ and $\Sigma_{2,j}$, respectively.

Matching probability between two GMMs cannot exceed one, and if the two GMMs are very disjoint, it is close to zero. To obtain a high matching probability, it is required that two GMM objects have similar shapes, i.e. similar parameters (μ, σ^2, w).

4   Normalized Transformation

In this section, we introduce Normalized Transformation. At first the motivation of Normalized Transformation is given. Secondly the details of this technique and the improvement of nodes in GCI are described. Thirdly we derive a set of novel similarity measures from the previous work using Normalized Transformation.

4.1   Motivation

Because of the potential unequal number of components and the complex structures of mixture models, traditional indexes can not be applied on GMMs directly. To tackle the problem, GCI provides an intuitive solution that stores the Gaussian components in a parameter space and prunes unqualified GMM candidates in a conservative but tight way [11].

Given N GMM-modeled objects, of which the maximum Gaussian components is m, we store the n components into GCI with the minimum number of entries in each node being r. In this case, the time complexity of average queries by GCI is $O\left(log_r\left(N\binom{m}{n}\right)\right) + \alpha O\left(Nm\right)$, where α refers to the percentage of the retrieved GMMs over all the objects. In this expression, the elementary operation of the first part is the minimum bounding rectangle calculation, and it is cheaper to calculate than that of the second part, matching probabilities between GMMs, especially when GMMs have large numbers of components. α varies in $(0, 1]$, and it is related to data distributions and the settings of the index. In the worst case, i.e., all the entries in the index have to be refined, the second part of the time complexity will be equal to that of the linear scan: $O\left(Nm\right)$.

In GCI, each entry stores Gaussian components $g_i = w_i\mathcal{N}(\mu_i, \sigma_i^2)$. For efficient queries, GCI derives the upper bound of matching probability, $\hat{d_{mp}}(\mathcal{G}_q, P)$, between a query object $\mathcal{G}_q$ and a node of entries $P = [\check{w}, \hat{w}, \check{\mu}, \hat{\mu}, \check{\sigma^2}, \hat{\sigma^2}]$. This upper bound is used for filtering unqualified components safely. Obviously, it is reached when $\hat{w}$ is taken. Take Fig. 1 for example, the upper bound of matching probability between a query component and stored components is determined by the weight of the highest component d and the (μ, σ^2) of three left-bottom components, **a**, **b** and **c**, which have the similar (μ, σ^2) with the query component. Under this circumstance, GCI gets a very high value of $\hat{d_{mp}}(\mathcal{G}_q, P)$. However, having very small wights, components **a**, **b** and **c** play tiny influence in the corresponding GMMs. As for the main components of the corresponding GMMs, components **d** and **e** are very disjoint with the query components. Thus the components stored in this node have no strong proof to be refined. The high value of

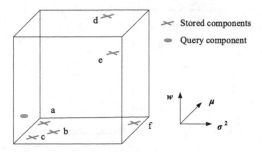

Fig. 1. Demonstration of a node P of GCI for univariate GMMs. Green crosses indicate the stored entries of P, and blue dot indicates one of query component of a query GMM. (Color figure online)

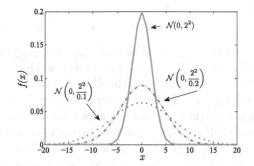

Fig. 2. Normalized Transformation of a Gaussian component in an univariate space. The red solid line indicates an original component with a mean of zero and a standard variation of two. The green dash line and green dot line indicate two normalized components that have a weight of 0.1 and 0.2, respectively. (Color figure online)

$\hat{d_{mp}}(\mathcal{G}_q, P)$, however, will lead to a set of unnecessary and expensive matching probability calculations between the corresponding GMMs and the query GMM.

The conservative strategy guarantees the accuracy of queries, but unnecessary calculations are always very willing to be excluded when possible, i.e., achieving a lower rate of refinement, which leads us to a normalized way to simplify the issue and avoid the situation above.

4.2   Normalized Indexing Strategy

In this paper, we propose Normalized Transformation g'_i for a GMM component $g_i = \mathcal{N}(\mu_i, \sigma_i^2)$ with a weight w_i:

$$g'_i = \mathcal{N}\left(\mu_i, \frac{\sigma_i^2}{w_i}\right) \tag{3}$$

Take a Gaussian component $\mathcal{N}(0, 2^2)$ for example, as demonstrated in Fig. 2, the distributions of two normalized components (green dash line and green dot

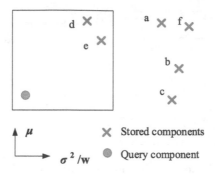

Fig. 3. Demonstration of the normalized node P' of GCI for univariate GMMs. Green crosses indicate the stored components, and blue dot indicates one of query component of a query GMM. (Color figure online)

line) are more flat than the original Gaussian distribution (red solid line). For a Gaussian component in a GMM, the smaller the weight is, the smaller the contribution of this component makes to the GMM. The normalized component keeps the same trend. The transformed variance $\sigma'_i = \sigma_i / \sqrt{w_i}$ becomes greater with the decrease of the weight w_i, making the normalized component more flat.

Storing normalized GMMs in GCI, the demonstration node P in Fig. 1 will be transformed into a rectangle P' in the parameter space of μ and σ^2/w, as shown in Fig. 3. Stored components **a**, **b**, and **c** that have similar μ and σ with the query component but tiny wights now are separated from the original node. In the present scene, the upper bound of matching probability $\hat{d}_{mp}(\mathcal{G}_q, P')$ provides a more objective reference than $\hat{d}_{mp}(\mathcal{G}_q, P)$ to determine whether the stored components need to be refined or not, thus a more tight prune strategy can be achieved.

4.3 Normalized Similarity Measures

Based on Normalized Transformation, we can derive the normalized matching probability of two GMMs from Eq. 2. It is shown as follows.

$$mp'(\mathcal{G}_1, \mathcal{G}_2) = \sum_{i=1}^{m_1} \sum_{j=1}^{m_2} \prod_{l=1}^{D} \frac{e^{-\frac{(\mu_{1,i,l} - \mu_{2,j,l})^2}{2(\sigma_{1,i,l}^2/w_i + \sigma_{2,j,l}^2/w_j)}}}{\sqrt{2\pi(\sigma_{1,i,l}^2/w_i + \sigma_{2,j,l}^2/w_j)}} \tag{4}$$

Since several similarity measures with closed-form expression for GMMs are the functions of matching probability [7–9,19], we can easily extend them into a set of novel similarity measures. These normalized measures share the same time complexities with their origins, and they have closed-form expressions for GMMs as well. The definitions are shown as follows.

$$d_{\mathrm{SE}'}(\mathcal{G}_1, \mathcal{G}_2) = mp'(\mathcal{G}_1, \mathcal{G}_1) + mp'(\mathcal{G}_2, \mathcal{G}_2) - 2mp'(\mathcal{G}_1, \mathcal{G}_2) \tag{5}$$

$$d_{\mathrm{IED}'}(\mathcal{G}_1, \mathcal{G}_2) = \sqrt{mp'(\mathcal{G}_1, \mathcal{G}_1) + mp'(\mathcal{G}_2, \mathcal{G}_2) - 2mp'(\mathcal{G}_1, \mathcal{G}_2)} \tag{6}$$

$$d_{\mathrm{C}2'}(\mathcal{G}_1, \mathcal{G}_2) = - \log \left(\frac{2mp'(\mathcal{G}_1, \mathcal{G}_2)}{mp'(\mathcal{G}_1, \mathcal{G}_1) + mp'(\mathcal{G}_2, \mathcal{G}_2)} \right) \tag{7}$$

$$d_{\mathrm{NL}2'}(\mathcal{G}_1, \mathcal{G}_2) = 2 \left(1 - \frac{mp'(\mathcal{G}_1, \mathcal{G}_2)}{\sqrt{mp'(\mathcal{G}_1, \mathcal{G}_1) \cdot mp'(\mathcal{G}_2, \mathcal{G}_2)}} \right) \tag{8}$$

5   Experimental Evaluation

In this section, we provide experimental evaluations on synthetic and real-world data sets to show the effectiveness of Normalized Transformation for GCI and the effectiveness of normalized similarity measures on both classification and clustering.

All the experiments are implemented with Java 1.7, and executed on a regular workstation PC with 3.4 GHz dual core CPU equipped with 32 GB RAM. For all the experiments, we use the 10-fold cross validation and report the average results over 100 runs.

5.1   Data Sets

Synthetic data and three kinds of real-world data, including activity data, image data and audio data, are used in the experiments. GMMs are estimated from data using iterative Expectation-Maximization (EM) algorithm[2].

The synthetic data sets[3] are generated by randomly choosing mean values between 0 and 100 and standard deviations between 0 and 5 for each Gaussian component. The weights are randomly assigned, and they sum up to one within each GMM. Since there is no intuitive way to assign class labels for GMMs in advance, here we use the synthetic data sets only for the evaluation of indexes.

Activity Recognition (AR) data[4] is collected from 15 participants performing seven activities. The sampling frequency of triaxial accelerometer is 52 Hz. Assuming that participants complete a single activity in three seconds, we regard the 150 continuous measurements of acceleration on three axes as one data object.

Amsterdam Library of Object Images[5] (ALOI) is a collection of images taking under various light conditions and rotation angles [20]. In this paper we use the gray images recording 100 objects from 72 viewpoints. For ALOI data, every image (192×144) is smoothed by a Gaussian filter with a standard deviation of five.

---

[2] Implementation provided by WEKA at http://weka.sourceforge.net/doc.dev/weka/clusterers/EM.html.

[3] https://drive.google.com/open?id=0B3LRCuPdnX1BSTU3UjBCVDJSLWs.

[4] http://archive.ics.uci.edu/ml/machine-learning-databases/00287/.

[5] http://aloi.science.uva.nl/.

Speaker Recognition[6] (SR) consists of 35 h of speech from 180 speakers. We select the speeches from ten speakers to form our audio data set, the Speaker Recognition (SR) data. The names of the ten speakers are as follows: Aaron, Abdul Moiz, Afshad, Afzal, Akahansson, Alexander Drachmann, Afred Strauss, Andy, Anna Karpelevich and Anniepoo. Every wav file is split into ten fragments, transformed into frequency domain by Fast Fourier Transform. The SR data has ten classes (corresponding to ten speakers), and each of them has 100 GMM objects.

5.2   Effectiveness of Queries in GCI

We study the performance of matching probability and normalized matching probability when using GCI to facilitate efficient queries. GMMs are decomposed into Gaussian components that stored into the entries of GCI. The minimum and maximum node capacity of GCI are set to 100 and 500, respectively. Original Gaussian components are stored when using matching probability as the similarity measure, while normalized Gaussian components are stored for normalized matching probability.

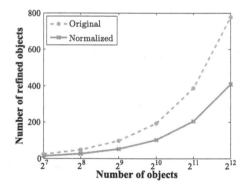

Fig. 4. Number of refined GMMs in GCI when varying the number of stored GMMs in Synthetic data. Each GMM here has ten Gaussian components in a univariate space.

We apply k-Nearest Neighbors (k-NN) queries using both similarity measures when varying the number of GMM objects and report the number of refined objects in Fig. 4. With the increasing number of stored GMM objects, both similarity measures need to refine more and more GMMs, but normalized matching probability significantly reduced the number of expensive calculations between GMMs.

---

[6] http://www.repository.voxforge1.org/downloads/SpeechCorpus/Trunk/Audio/ Main/16kHz_16bit/.

5.3   Effectiveness of Normalized Similarity Measures

Classification. In the evaluation of classification, only k-NN, rather than the other more complex techniques, is used to compare the effectiveness of the similarity measures, since we are not interested in tuning the classification accuracy to its optimum.

We start with experiments on SR data sets when varying k in k-NN, and the classification is applied based on original and normalized similarity measures. Classification accuracies are shown in Fig. 5. From this figure we can see that all four normalized similarity measures outperform their origins, and all the accuracies slightly decrease with the increase of k.

Fixing k as 1, we report the classification results on AR data when varying the number of Gaussian components in estimated GMMs. As shown in Fig. 6, the normalized similarity measures achieve better classification results than the origins in most cases when the number of Gaussian components is high enough. Only at starting points, normalized similarity measures, especially for NL2, have lower accuracies than their origins. GMMs have better representations of data objects with the increase of components number, however, the training time of EM algorithm increases at the same time. To tune the number of Gaussian

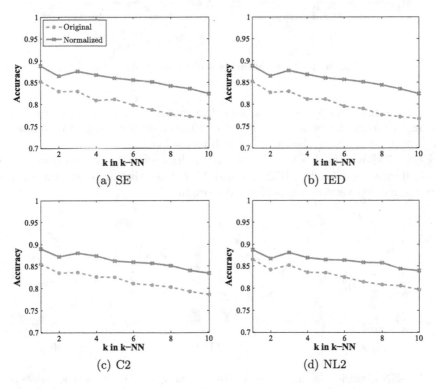

Fig. 5. Classification accuracies on SR data. For each data object, a five-component GMM is estimated.

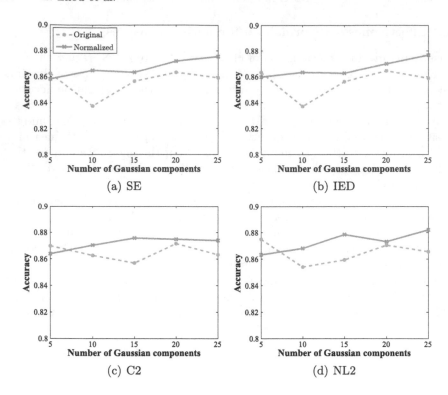

Fig. 6. 1-NN classification accuracies on AR data.

components into the optimum for a given similarity measure, Bayesian Information Criterion can be applied. For the following experiments, we choose the component numbers by the rule of thumb instead.

Figure 7 shows the 1-NN classification results on three read-world data sets. All similarity measure, SE, IED, C2 and NL2, have similar performances, and their normalized versions outperform the origins.

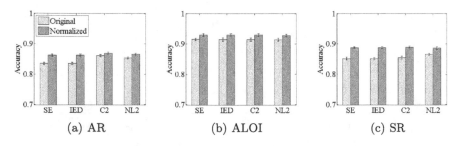

Fig. 7. 1-NN classification accuracies on three real-world data sets. The numbers of Gaussian components in each GMM object for (a), (b) and (c) are ten, five and five, respectively.

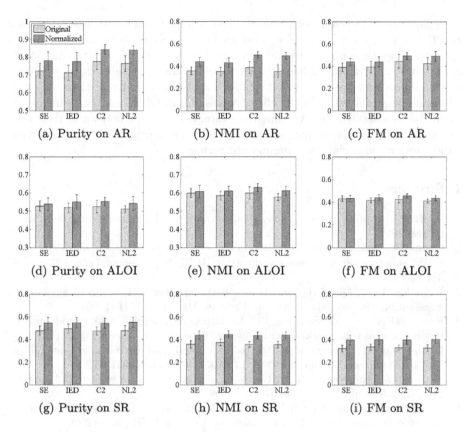

Fig. 8. Evaluations of k-medoids clustering results on three real-world data sets. The components numbers of GMMs are the same as these in Fig. 7. The k for three data are seven, ten and ten, respectively.

Clustering. We perform clustering experiments to compare the usability of the normalized similarity measures for unsupervised data mining. Instead of k-means algorithm, the k-medoids is used since it works with arbitrary similarity measures, making it more suitable here. We evaluate the clustering results using three widely used criteria, Purity, Normalized Mutual Information (NMI) and F1 Measure (FM).

Figure 8 illustrates the evaluation of clustering results when using different similarity measures on three real-world data sets. All three criteria have the same pattern for all the similarity measures on three data sets. The normalized similarity measures have a better performance than their origins.

6   Conclusions

In this paper, we have introduced Normalized Transformation that aims to improve the retrieval performance of index for Gaussian Mixture Models, and it also enable us to derive a set of normalized similarity measures from proposed ones that have closed-form expressions for GMMs. These normalized similarity measures share the same time complexities with their origins. Queries on GMMs using Gaussian Component based Index have illustrated the effectiveness of Normalized Transformation, achieving a much lower refinement rate than the original matching probability. For the effectiveness of normalized similarity measures, we have demonstrated the experimental evaluations on the real-world data sets. The normalized similarity measures outperform their origins on different types of data sets in both classification and clustering.

References

1. STATS description. https://www.stats.com/sportvu-basketball-media/. Accessed 25 Feb 2017
2. Campbell, W.M., Sturim, D.E., Reynolds, D.A.: Support vector machines using GMM supervectors for speaker verification. IEEE Sig. Process. Lett. **13**(5), 308–311 (2006)
3. Reynolds, D.A., Quatieri, T.F., Dunn, R.B.: Speaker verification using adapted gaussian mixture models. Digit. Sig. Process. **10**(1–3), 19–41 (2000)
4. KaewTraKulPong, P., Bowden, R.: An improved adaptive background mixture model for real-time tracking with shadow detection. In: Remagnino, P., Jones, G.A., Paragios, N., Regazzoni, C.S. (eds.) Video-Based Surveillance Systems, pp. 135–144. Springer, Boston (2002). doi:10.1007/978-1-4615-0913-4_11
5. Zivkovic, Z.: Improved adaptive Gaussian mixture model for background subtraction. In: ICPR, pp. 28–31 (2004)
6. Böhm, C., Pryakhin, A., Schubert, M.: The Gauss-tree: efficient object identification in databases of probabilistic feature vectors. In: ICDE, p. 9 (2006)
7. Helén, M.L., Virtanen, T.: Query by example of audio signals using Euclidean distance between Gaussian mixture models. In: ICASSP, vol. 1, pp. 225–228 (2007)
8. Sfikas, G., Constantinopoulos, C., Likas, A., Galatsanos, N.P.: An analytic distance metric for gaussian mixture models with application in image retrieval. In: Duch, W., Kacprzyk, J., Oja, E., Zadrożny, S. (eds.) ICANN 2005. LNCS, vol. 3697, pp. 835–840. Springer, Heidelberg (2005). doi:10.1007/11550907_132
9. Jensen, J.H., Ellis, D.P., Christensen, M.G., Jensen, S.H.: Evaluation of distance measures between Gaussian mixture models of MFCCs. In: ISMIR, pp. 107–108 (2007)
10. Tao, Y., Cheng, R., Xiao, X., Ngai, W.K., Kao, B., Prabhakar, S.: Indexing multidimensional uncertain data with arbitrary probability density functions. In: VLDB, pp. 922–933 (2005)
11. Zhou, L., Wackersreuther, B., Fiedler, F., Plant, C., Böhm, C.: Gaussian component based index for GMMs. In: ICDM, pp. 1365–1370 (2016)
12. Böhm, C., Kunath, P., Pryakhin, A., Schubert, M.: Querying objects modeled by arbitrary probability distributions. In: Papadias, D., Zhang, D., Kollios, G. (eds.) SSTD 2007. LNCS, vol. 4605, pp. 294–311. Springer, Heidelberg (2007). doi:10.1007/978-3-540-73540-3_17
13. Kullback, S.: Information Theory and Statistics. Courier Dover Publications, New York (2012)
14. Hershey, J.R., Olsen, P.A.: Approximating the Kullback Leibler divergence between Gaussian mixture models. In: ICASSP, pp. 317–320 (2007)

15. Goldberger, J., Gordon, S., Greenspan, H.: An efficient image similarity measure based on approximations of KL-divergence between two Gaussian mixtures. In: ICCV, pp. 487–493 (2003)
16. Cui, S., Datcu, M.: Comparison of Kullback-Leibler divergence approximation methods between Gaussian mixture models for satellite image retrieval. In: IGARSS, pp. 3719–3722 (2015)
17. Beecks, C., Ivanescu, A.M., Kirchhoff, S., Seidl, T.: Modeling image similarity by Gaussian mixture models and the signature quadratic form distance. In: ICCV, pp. 1754–1761 (2011)
18. Rougui, J.E., Gelgon, M., Aboutajdine, D., Mouaddib, N., Rziza, M.: Organizing Gaussian mixture models into a tree for scaling up speaker retrieval. Pattern Recogn. Lett. **28**(11), 1314–1319 (2007)
19. Zhou, L., Ye, W., Plant, C., Böhm, C.: Knowledge discovery of complex data using Gaussian mixture models. In: DaWaK (2017)
20. Geusebroek, J.-M., Burghouts, G.J., Smeulders, A.W.: The Amsterdam library of object images. Int. J. Comput. Vis. **61**(1), 103–112 (2005)

A Concurrency Control Protocol that Selects Accessible Replicated Pages to Avoid Latch Collisions for B-Trees in Manycore Environments

Tomohiro Yoshihara[1,2](✉) and Haruo Yokota[2]

[1] Research and Development Group, Hitachi, Ltd.,
Kanagawa 244–0817, Japan
tomohiro.yoshihara.rj@hitachi.com
[2] Department of Computer Science, Tokyo Institute of Technology,
Tokyo 152–8552, Japan
yokota@cs.titech.ac.jp

Abstract. In recent years, microprocessor vendors aiming for dramatic performance improvement have introduced manycore processors with over 100 cores on a single chip. To take advantage of this in database and storage systems, it is necessary for B-trees and their concurrency control to reduce the number of latch collisions and interactions among the cores. Concurrency control methods such as physiological partitioning (PLP), which assigns cores to partitions in a value–range partition, have been studied. These methods perform effectively for nearly static and uniform workloads using multicore processors. However, their performance deteriorates significantly if there is major restructuring of B-trees against skew and for changing workloads. The manycore approach has a high likelihood of causing workload skew, given the lower power of each core, with an accompanying severe degradation in performance. This issue is critical for database and storage systems, which demand consistent high performance even against dynamic workloads. To address this problem, we propose an efficient new concurrency control method suitable for manycore processor platforms, called the selecting accessible replicated pages (SARP) B-tree concurrency control method. SARP achieves a consistent high performance with robustness against workload skew by distributing the workload to many cores on manycore processors, while reducing latch collisions and interactions among the cores. By applying parallel B-trees to shared-everything environments, SARP selects execution cores and access paths that distribute workloads widely to cores with appropriate processor characteristics. Experimental results using a Linux server with an Intel Xeon Phi manycore processor demonstrated that the proposed system could achieve a throughput of 44 times that for PLP in the maximum-skew case and could maintain the throughput at 66% of a throughput for uniform workloads.

1 Introduction

Recently, microprocessor vendors have found it increasingly difficult to make CPUs with higher clock rates than those in previous-generation CPUs. Instead,

© Springer International Publishing AG 2017
D. Benslimane et al. (Eds.): DEXA 2017, Part II, LNCS 10439, pp. 172–189, 2017.
DOI: 10.1007/978-3-319-64471-4_15

improved performance has been achieved by developing multicore CPUs, which have multiple microprocessor "cores" on a single chip. However, increasing the performance of multicore processors is also becoming difficult, because of physical limitations on placing more cores of the standard type on a single chip. Microprocessor vendors are therefore developing single-chip "manycore" CPUs, such as the 72-core Intel Xeon Phi, and are looking to processors with hundreds of cores in the future. This is possible because the performance (frequency and instruction per clock) of each core on a manycore platform is lower than the performance of each core on a multicore platform, but the overall performance of a manycore chip can be higher than that for a multicore platform. The earliest manycore processors could be used only in combination with a separate main processor and could support only small-capacity main memories. They were applied to specific workloads as auxiliary processors to the main processor, and the scope of their application was narrow. Manycore processor vendors are now changing the target applications for manycore processors. The latest manycore processors can perform independently of a main processor and can support large-capacity main memories. They can not only execute applications suited to mainstream multicore processors but also complex major applications such as database systems and storage subsystems.

On the other hand, distributed approaches and distributed software platforms such as Hadoop [3] are also widespread, and share-nothing environments are typically mainstream. In practice, many enterprise applications will be constructed as distributed applications on a plurality of multicore server clusters. Therefore, it is necessary that methods suited to multicore/manycore platforms are extended to distributed environments.

Multicore processors are already widely used, with many concurrency control protocols for shared-everything environments being proposed for such systems [4,6,11–16,18–22]. The main aim of these protocols is to be latch-free in a multicore-processor environment. Although these methods can improve the performance for a single node dramatically, they focus on having a single node in a shared-everything environment, making it difficult for them to be extended to the distributed environments found in the real world.

Physiological partitioning (PLP) [19] and ATraPos [20] achieve their latch-free status by assigning processor cores to each partition in a value–range partition. These methods not only free B-trees from acquiring latches but also consider the allocation of caches in processors. Approaches based on this value–range partition have the potential to extend to distributed environments. Although these methods are effective against nearly uniform and unchanged workloads on multicore processors, they are vulnerable to dynamic skew in workloads. Their performance deteriorates significantly when there is major restructuring of B-trees to avoid skew and when the workload changes. Furthermore, manycore systems have a high likelihood of causing workload skew, given the lower power of each core, with an accompanying severe degradation in performance. This issue is critical for database and storage systems used as in mission-critical online transaction processing, which cannot allow for a decline in performance and demand consistent high performance even with dynamic workloads.

In this paper, we propose a new concurrency control method for manycore environments, called the selecting accessible replicated pages (SARP) B-tree concurrency control method. SARP achieves consistent high performance with robustness against large-scale skew of workloads by distributing the workload to many cores in a manycore system, while reducing latch collisions and interactions among the many cores. SARP is based on a parallel B-tree for shared-everything environments and concurrency control, but in the future, it could be extended to shared-nothing environments that share no memory or disk among processors attract a great deal of attention. It selects a strategy for distributing the workload to many cores by applying a parallel B-tree to the shared-everything environment, using a distribution that is appropriate to the characteristics of each processor. This treats effectively any large skews in the workload and minimizes the redirection communication among the cores. SARP also introduces three techniques for reducing the more frequent latch collisions from large-scale workload skew when using manycore processors. Experimental results using a Linux server with an Intel Xeon Phi manycore processor demonstrated that the proposed method could achieve a throughput of 44 times that for PLP in the maximum-skew case and could maintain the throughput at 66% of a throughput for uniform workloads.

The remainder of the paper is organized as follows. First, we review related work in Sect. 2. Section 3 outlines some prior technologies, namely the parallel B-tree, the Fat-Btree, and LCFB. The proposed method in this paper is based on these technologies. In Sect. 4, we describe our proposed new concurrency control method for B-tree structures on manycore platforms. Our experimental results are reported in Sect. 5. The final section presents the conclusions of this paper.

2   Related Work

Adopting a latch-free approach reduces cache pollution that processor cache for each core purged by each core in a multicore processor environment. The Foster B-tree [6] combines the advantages of B-link trees, symmetric fence keys, and write-optimized B-trees [5]. OLFIT [4], Mass-tree [18], and Silo [22] use the technique of checking the version of pages before and after reading pages instead of latching for page reads. The Master-tree [13] was proposed for a rich NVRAM environment, as a combination of the Foster B-tree and Silo. Other proposals [11,14,16] use read-copy-updates and compare-and-swap techniques. Intel Transactional Synchronization Extensions [12,15] have also been used. PALM [21] uses a bulk synchronous parallel technique.

PLP [19] achieves latch-free status by assigning processor cores to each partition in a value–range partition. It uses a multiroot B-tree for the value–range partition. Each core in the partition can avoid latches because at most one core accesses each page in each partition. ATraPos [20], which is based on PLP, is a storage manager design that is aware of the nonuniform access latencies of multisocket systems. To reduce the latches for system states in PLP, ATraPos divides the system state into states for each CPU socket. ATraPos also supports dynamic repartitioning for a multiroot B-tree based on the usage of CPU cores.

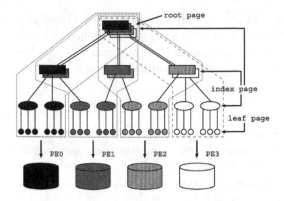

Fig. 1. Fat-Btree.

Table 1. Latch matrix.

Mode	IS	IX	S	SIX	X
IS	✓	✓	✓	✓	
IX	✓	✓			
S	✓			✓	
SIX	✓				
X					

PLP and ATraPos not only free the B-tree from acquiring latches but also consider the allocation of caches in processors, and these methods are effective traditional methods for multicore/manycore processors for nearly unchanging and uniform workloads. Moreover, these approaches, based on multi B-trees and the value–range partition, can potentially extend to distributed environments. However, they are vulnerable to workload skew, and it would be difficult for them to perform well with over 100 cores in a multicore system. Both PLP and ATraPos need to largely reconstruct the B-tree structure to rebalance the skewed workload for each partition. Reconstructing the B-tree structure in either PLP or ATraPos will severely affect the performance. For a manycore platform, the executing capacity of each core is low. Therefore, the impact of reconstruction for manycore processors will be even larger than for multicore processors, and this issue will become critical.

3   Background

3.1   Fat-Btree

A Fat-Btree [23] is a form of parallel B-tree in which the leaf pages of the B^+-tree are distributed among the processing elements (PEs). Each PE has a subtree of the whole B-tree containing the root node, and intermediate index nodes between the root node and leaf nodes allocated to that PE. Fat-Btrees have the advantage of being parallel B-trees (which hashing does not have) and they are naturally height balanced. Figure 1 shows an example of a Fat-Btree using four PEs.

Although the number of copies of an index node increases with the node's proximity to the root node of the Fat-Btree, the update frequency of these nodes is relatively low. On the other hand, leaf nodes have a relatively high update frequency, but are not duplicated. Consequently, nodes with a higher update frequency have a lower synchronization overhead. Moreover, for Fat-Btrees, index pages are only required for locating the leaf pages stored in each PE. Therefore, Fat-Btrees will have a high cache hit rate if the index pages are cached in each PE. Because of this high cache hit rate, update and search processes can be executed quickly, compared to those for conventional parallel B-tree structures.

3.2 SMOs

Structure-modification operations (SMOs) involve page-split and page-merge operations. Page splits are performed when pages become full, whereas page merges are performed when pages become empty. B-trees in real database systems usually perform only page merges when pages become empty. This follows nodes not being required to fill beyond being half full because this was not found to decrease occupancy significantly for practical workloads [10]. Because the consistency of the B-tree must be guaranteed when SMOs occur, concurrency control for the B-tree is necessary. Concurrency control is also very important because it largely influences the system throughput when SMOs occur.

3.3 Latches

Some kind of concurrency control method for the B-tree is necessary to guarantee consistency. Instead of using locks, fast and simple latches are usually used for concurrency control during a traversal of the index nodes in the B-tree [7]. A latch is a form of semaphore and the latch manager does not have a deadlock-detection mechanism. Therefore, the concurrency control for a B-tree node should be deadlock free.

In this paper, a latch is assumed to have five modes: IS, IX, S, SIX, and X, as shown in Table 1 [7]. The symbol "√" means that the two modes are compatible, i.e., two or more transactions can hold a latch at the same time.

Because parallel B-tree structures, including the Fat-Btree, have duplicated nodes, a special protocol for the distributed latch manager is required to satisfy latch semantics. Requested IS and IX mode latches can be processed only on a local PE, whereas the other modes must be granted on all the PEs storing the duplicated nodes to be latched. That is, the IS and IX modes have much smaller synchronization costs than the S, SIX, and X modes, which require communication between the PEs. The S, SIX, and X mode latches on remote copies are acquired by using their pointers. In addition, such latches must be set in linear order to avoid deadlock, such as ordering by logical PE number. This means that the synchronization cost will grow in proportion to the number of PEs storing a copy of the node to be latched.

3.4 LCFB

LCFB [24] is a concurrency control protocol that improves the performance of parallel B-trees. It requires no latch-coupling in optimistic processes and reduces the amount of communication between PEs during a B-tree traversal. To detect access-path errors in the LCFB protocol caused by the removal of latch-coupling, we assign boundary values to each index page. The access-path errors show that it mistakes to reach a leaf page not having target key and value. Because a page split may cause page deletion in a Fat-Btree, we also propose an effective method for handling page deletions without latch-coupling.

In traditional protocols with latch-coupling, a traversal reaches a leaf using latch-coupling with IS or IX latches. For the transfer of access requests to another PE (called the redirection), a message is sent to a destination PE to acquire a latch on the child, to the source PE to release the latch on the parent, and to the destination PE to process the child in the destination PE. The redirection requires three sequential messages per transfer. If a traversal reaches a leaf without latch-coupling, as in LCFB, only one message per transfer is required. LCFB improves the response time by reducing the need for network communication. Because the cost of network communication is large during traversal, LCFB can obtain some dramatic effects from relatively small changes.

Searches of LCFB use latch-coupling with IS latches on index nodes and use S latches on leaf nodes. Inserts in LCFB basically use latch-coupling with IX latches on index nodes and use X latches on leaf nodes. When structure-modification operations (SMOs) occur, inserts in LCFB also use X latches on the minimum-necessary number of index nodes. The LCFB is defined more precisely in [24].

4 Proposed Method

We propose SARP, a new concurrency control method that improves the performance of B-trees in manycore environments. SARP applies the parallel B-tree for a shared-nothing environment to shared-everything environments. It assigns each PE and partition of a parallel B-tree to a logical core on a manycore/multicore platform, as described in Sect. 4.1. SARP's basic execution of search and insert operations are based on the LCFB and the Fat-Btree, using the three proposed techniques described in Sect. 4.6. On B-tree traversal, each core preferentially accesses pages on the assigned partition, selecting an accessible and optimal page. It then either accesses the directory for the selected page or redirects to another core, as described in Sect. 4.2. It executes this selection on each B-tree node until it reaches a target leaf page. SARP introduces three techniques for reducing the frequent latch collisions caused by the large workload skew on manycore systems, as described in Sects. 4.3–4.5.

4.1 Assigning Parallel B-Tree Partitions to a Logical Core

Figure 2 is an overview of the proposed parallel B-tree system used for the experiments described in Sect. 5. All logical cores have one client module each and one PE. To bind each module to its specified core, we use the ability for an operating system (OS) to bind one or more processes to one or more processors: a technique called CPU affinity [17]. All PEs share the parallel B-tree and all B-tree

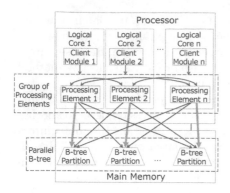

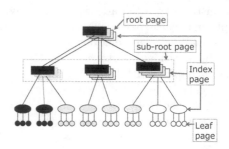

Fig. 2. System overview.   **Fig. 3.** Replicating the subroot index pages.

Partitions. SARP assigns each partition to a particular PE, called the owner PE in this paper. The PEs in SARP access their own partition on a priority basis but can access other partitions if needed. Traditional PLP or LCFB also assigns each partition on the Fat-Btree to a particular PE, but a PE can access only its own partition. To access data in other partitions, redirection to another PE is needed. SARP not only distributes workloads to many PEs on manycore processors, but also reduces the cost of redirection communications, depending on the situation. In our experimental system, client modules send requests to PEs, which execute operations on the parallel B-tree. We are dealing with various applications such as database management systems, storage systems, and their convergent systems. For such applications, the B-tree will be partitioned into modules with processors, just like our experimental system.

One of the benefits of CPU affinity is its ability to optimize cache performance. Scheduling a software process to execute on a specified processor core can result in efficient use of that processor by reusing preloaded resources such as the data in the cache levels. Many OSs, from Windows Vista to Linux, already provide a system call to set the CPU affinity for a process. By invoking this affinity, processes can dedicate resources such as processor execution time and processor caches to a PE's own resources in shared-nothing environments.

4.2   Selecting the Next Page

To scale the access throughput to a narrow range of partitions by distributing workloads, SARP selects PEs suited to accessing each page using the characteristics of the processor architectures. In traditional methods and shared-nothing environments, the owner PE can process its own page but has to send redirections to the owner PE to process other pages. On the other hand, SARP's shared-everything environment allows non-owner PEs to process any page as necessary, with the PE processing a parent page trying to continue to process the next page.

The important characteristic of a processor architecture is the relationship between the bus bandwidth among the processor cores and the processing power of each core when sending or receiving redirections. If the bandwidth is high, the cost of reading other PE's pages via the bus is low. Therefore, workloads should tend to be distributed across the available PEs. If the processing power is low, the cost of sending and receiving redirection is high. Therefore, the PE should continue processing without redirection.

A manycore processor has the characteristic that the bandwidth is high and the processing power is low. Therefore, the PE first receiving a request should continue without redirection. In this way, SARP can distribute workloads to all cores on the manycore processor. A multisocket multicore processor has the characteristics that the processing power is high, and the bandwidth is high within the processor sockets but is low between sockets. Therefore, if the PE first receiving a request can go to the next page within its socket, then it should continue without redirection. If not, it should send a redirection to the PE associated with the next page. With this strategy, SARP can distribute workloads mainly within each socket on a multisocket multicore processor.

4.3 Redirection by Detecting a Latch Collision (Technique A)

To avoid latched pages and to traverse to the leaf nodes via access paths without latch collisions, SARP employs a technique that sends redirections to prior owner PEs when it detects latch collisions (Technique A). The prior owner PE shows the one PE that has a right to update each page in a SMO. Consider a PE trying to acquire a latch in a traditional method. If another PE is engaged in acquiring another latch or if the new latch and other existing latches are not compatible, the PE will wait for these conditions to change. In this situation, SARP will send redirections to prior owner PEs. This will reduce the number of PEs that concurrently acquire latches on the same page and will reduce the likelihood of latch collisions. By combining this with the technique described in Sect. 4.5, it will be redirecting to latch-free pages.

4.4 Replicating the Subroot Index Pages (Technique B)

To widen the access paths without latch collisions, SARP adopts a technique that replicates index pages at the next level of the root (subroot index pages) for all PEs (Technique B), as shown in Fig. 3. Latch collision at the root is infrequent, except during SMOs, because a number of parallel B-trees, including Fat-Btree, have root indices for all partitions. On the other hand, latch collision in subroots can be very frequent, depending on the condition of the B-tree structure. In the common worst case, most requests could be concentrated on a single subroot page following workload skew. To address this issue, SARP keeps copies of all subroot index pages when subroots are created by SMOs on the roots or the subroots, whereas unnecessary pages are deleted in traditional parallel B-tree methods.

4.5   Being Latch-Free in the Assigned Index Pages (Technique C)

To create access paths that are partially latch-free during traversal, SARP allows only the owner PEs assigned to the partition to perform SMOs on that partition and to traverse partially while latch-free (Technique C). SARP considers whether the PE owns the page when accessing a page. If the PE does not own the page, it processes it in the conventional way that the PE acquires the IS or IX latch like conventional LCFB on the pages. If the PE owns the page, it checks the mode of the latch on that page. If the mode is X, it processes it in the conventional way. If the mode is not X, it locates entries in the page without the latch. SARP can perform this technique because although SARP allows non-owners to access the pages, it allows only the owner to modify the page. If an X latch is not acquired on the page the owner PE is accessing, the SMO does not occur on that page at that time, waiting until the owner PE finishes or interrupts the current traversal. Although Techniques A and B will be mainly used by the PE assigned to each partition, the three techniques together have a synergistic effect.

4.6   Search and Insert

In this section, we describe the operations for search and insert using the proposed techniques. Traversal for searches and inserts by SARP is based on LCFB, but SARP uses additional considerations when accessing each page and selecting the next page. The following steps are performed for each index page until the traversal reaches the leaf nodes.

(1) When a PE tries to access a page, check if the PE owns the page.
(2) If the PE is the owner, check the mode of the latch. Otherwise, go to (4).
(3) Following the check in (2), if the mode of the latch is not X, go to (5).
(4) Try to acquire a latch in the usual way. If the PE detects a latch collision, send a redirection to an owner. The owner then continues from (1).
(5) Release the latch on the parent page if needed.
(6) Read the page and locate the next pages.
(7) Check if the next pages include pages that do not require redirection. If it does, send a redirection to an owner.
(8) Go back to (1) with the next page.

The behavior of SARP on a leaf page is identical to that of LCFB, and the SMOs in SARP are identical to those of LCFB except for the replicating of subroots.

5   Experiments

To investigate whether the proposed methods are effective, we implemented PLP and the proposed method on Linux servers. Moreover, to test the effectiveness of the three techniques for avoiding latch collisions proposed in Sect. 4, we prepared five configurations comprising various combinations of these techniques.

Table 2. Combinations of proposed techniques evaluated in experiments.

	SARP	SARP-R	SARP-RR	SARP-RRL
Technique A		✓	✓	✓
Technique B			✓	✓
Technique C				✓

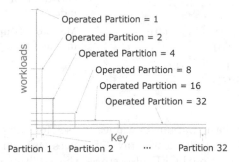

Fig. 4. Example of workloads in the experiments.

Table 2 shows the combinations of proposed techniques evaluated in the experiments. Technique A was described in Sect. 4.3, Technique B was described in Sect. 4.4, and Technique C was described in Sect. 4.5. The PLP used a multi-rooted B+Tree [19], and the proposed method used a Fat-Btree and the extension of a Fat-Btree using Technique B. Because we were interested in the performance at the intermediate point at which workloads change, we measured the performance of a PLP that does not reconstruct the B-tree because of workload skew in this experiment. If there was reconstruction in the PLP or the ATraPos functions, the performance during reconstruction would be dramatically lower than the performance in this experiment. Figure 4 shows an example of workloads in the experiments in Sect. 5.5. The horizontal and vertical axes are the number of workloads and the key address of B-tree respectively. If operated partition is 32, the workload are sent to all 32 partitions and is uniform in all partition. If operated partition is 1, the workloads are sent to only partition 1 and is uniform in partition 1.

We evaluated the performance of our proposed method for the three different platforms shown in Table 3. The experiments reported in this paper were organized as follows. First, as described in Sects. 5.1–5.3, we evaluated each method under varying conditions using a previous-generation Xeon Phi 5110P, to check the effectiveness of each method under varying conditions on a manycore platform. As described in Sect. 5.4, we evaluated each method using the latest Xeon Phi 7210, which showed that the proposed methods are also efficient on the latest manycore platforms. As described in Sect. 5.5, we evaluated each method on a mainstream multicore platform Xeon E5-4650, showing that the proposed methods are effective for mainstream servers.

Table 3. Experimental environments.

Processor	Intel Xeon Phi 5110P	Intel Xeon Phi 7210	Intel Xeon E5-4650
Sockets	1	1	4
Cores/Socket	60	64	8
Frequency	1.053 GHz	1.3 GHz	2.7 GHz
Memory	8 GB	192 GB	64 GB

Table 4. Parameters used for the experiments.

Page size:	4 KByte
Tuple size:	8 Byte
Max no. of entries in an index node:	240
Max no. of tuples in a leaf node:	240

Table 4 gives the basic parameter values we set for the experiments. These values were chosen to identify clearly any differences between the protocols. The proposed method based on LCFB also stores the additional upper and lower boundary values in a node. Therefore, the traditional methods and the proposed methods should have different fanout. Because the difference is very small, we ignored it in this paper.

5.1   Comparison with Different Numbers of Operational Partitions

120 client modules sent requests to the PEs containing a B-tree with 20 M tuples, for 10 s. The access frequencies were uniform for the operational B-tree partitions. The number of PEs was fixed at 120 on 60 physical cores and 120 logical cores, and the update ratio was fixed at 25%. Figure 5 shows the throughput for the five concurrency control methods as the number of B-tree partitions operated on by client modules varies from 1 to 120. Figure 6 shows the execution time for the five concurrency controls when the number of operational partitions was one.

When the number of operational partitions is reduced, the PLP performance decreases sharply. This is because most of the manycore processor cores are free and total execution times for the manycore processor are very low, as shown in Fig. 6. In contrast to PLP, the proposed methods can provide reasonable throughput even if the number of operational partitions is low. Reductions in throughput for the proposed methods are much less than those for PLP even when the workload skew occurs. This is because the reduction of latch collisions by the three proposed techniques reduces not only the execution times for acquiring latches but also the waiting times for other requests to process latches, as shown in Fig. 6. The distribution of workloads in SARP reduces idle times. SARP-R's Technique A substantially reduces the execution time for acquiring latches and the idle time caused by waiting for other requests to process latches. SARP-RR's Technique B also substantially reduces this idle time. SARP-RRL's Technique C reduces the execution times for acquiring latches and the idle time.

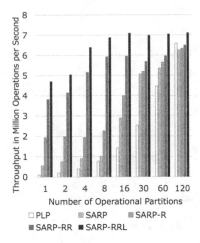

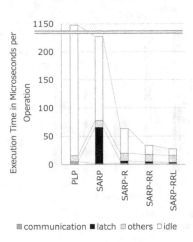

Fig. 5. Comparison with different numbers of operational partitions on the Xeon Phi 5110P.

Fig. 6. Execution time when one partition was operated by the Xeon Phi 5110P.

In addition, when the number of operational partitions is 120, SARP-RRL improves on PLP even under conditions for which PLP is most effective. This is because the effects of the combination of Techniques A, B, and C and the reduction in communications exceeds the effects of latch-free operation by PLP.

SARP-RRL achieved a throughput of 44 times that for PLP when the number of operational partitions was one, and maintained the throughput at 66% of a throughput for uniform workloads.

5.2 Comparison with Changing Update Ratio

120 client modules sent requests to the PEs containing a B-tree with 20 M tuples, for 10 s. The access frequencies were uniform among all B-tree partitions, and the number of PEs was fixed at 120 on 60 physical cores and 120 logical cores. Figure 7 shows the throughput for the five concurrency control methods as the update ratio changes from 0% to 100%.

When the update ratio was 0%, the throughput for all proposed methods is higher than those for PLP. Unlike PLP, the effect of reduced communications among the PEs exceeds the increase in the cost of latching for traversals in the proposed method. On the other hand, when the update ratio is high, only SARP-RRL offers a slight improvement over PLP. This is because PLP is latch-free for SMOs and the proposed system has to acquire latches during SMOs. The increase in cost for latches for traversals and SMOs in most versions of the proposed methods exceeds the effect of reduced communications among PEs. However, because SARP-RRL can reduce latch acquisition via Technique C, the effect of reducing communications among PEs can then exceed the cost of latches for traversals and SMOs.

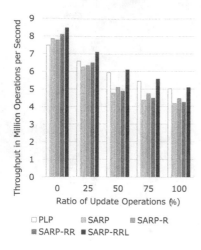

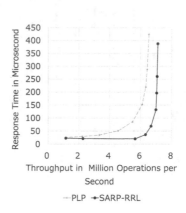

Fig. 7. Comparison of concurrency control protocols with changing update ratio.

Fig. 8. Comparison of response time.

Table 5. Workload parameter values.

Concurrent workloads	Delay [microsecond]
2880	0
1920	0
1440	0
960	0
480	0
240	0
120	0
120	50
120	100

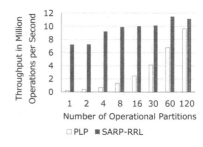

Fig. 9. Comparison with different numbers of operational partitions on the Xeon Phi 7210.

5.3   Comparison of Response Times

120 client modules sent requests to the PEs containing a B-tree with 20 M tuples, for 10 s. The access frequencies were uniform among all B-tree partitions, the number of PEs was fixed at 120 on 60 physical cores and 120 logical cores, and the update ratio was fixed at 25%. Figure 8 shows the response time for client modules corresponding to the throughput for PLP and SARP-RRL as the workloads change in accordance with Table 5. The delay shows the delay time between received time of a previous operation and sending time of a next operation.

SARP-RRL offers an improvement in response time when compared to PLP. This is because SARP-RRL requires less communication among the PEs.

SARP-RRL redirects requests during traversal of the B-tree only when latch collisions occur on manycore processors. It does not have to redirect requests to other PEs for uniform workloads. SARP-RRL's response time is 20% of that of PLP for the case of clients requesting some 6,000,000 operations per second and SARP-RRL is 500% better than PLP on the benchmark of response time in high workloads.

5.4   Comparison of Different-Generation Manycore CPUs

120 client modules sent requests to the PEs containing a B-tree with 20 M tuples, for 10 s. The access frequencies were uniform for the operational B-tree partitions, the number of PEs was fixed at 120 on 60 physical cores and 120 logical cores, and the update ratio was fixed at 25%. Figure 9 shows the throughput for the two concurrency control methods on the Xeon Phi 7210 as the numbers of B-tree partitions operated on by client modules changes from 1 to 120. Figure 10(a) shows the execution times divided into processor behaviors for the Xeon Phi 7210 and Fig. 10(b) shows the execution times divided into processor behaviors for the Xeon Phi 5110P, as described in Sect. 5.1, when the number of B-tree partitions operated on by client modules is one. In Fig. 10, dark gray bar charts show core execution times, light gray bar charts show data cache miss times, black bar charts show instruction cache miss times, and white bar charts show TLB miss times. Figure 10 is based on the latencies given in [8,9] and the performance counters given in [8,9].

Similarly to the results for the manycore processor described in Sect. 5.1, the proposed methods using the latest manycore processor substantially improve on the throughput achieved by PLP when the workload skew is large. This is because

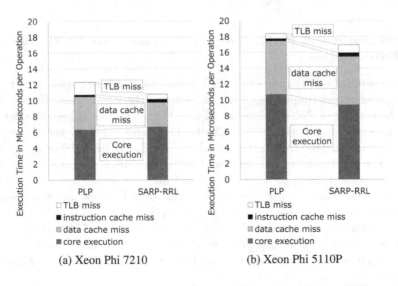

(a) Xeon Phi 7210   (b) Xeon Phi 5110P

Fig. 10. Comparison using different-generation manycore CPUs.

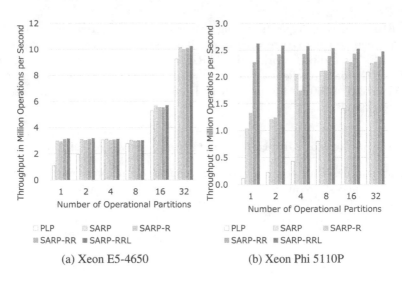

(a) Xeon E5-4650   (b) Xeon Phi 5110P

Fig. 11. Comparison of throughput for multicore and manycore platforms.

the characteristics of the latest manycore processor are mostly unchanged. This can be inferred from the fact that the rates for each factor of the execution times when divided into processor behaviors are similar, as shown in Fig. 10. The effects of the clocking up of each core and the improvement in core mechanisms improve the performance of each factor almost evenly. Therefore, this result indicates that the current design of manycore processors is ongoing, and the proposed methods will be effective not only for current manycore processors but also for manycore processors in the near future.

5.5   Comparison of Performance in Manycore and Multicore Environments

32 client modules sent requests to the PEs containing a B-tree with 20 M tuples, for 10 s. The access frequencies were uniform for the operational B-tree partitions, the number of PEs was fixed at 32 on 32 physical and 32 logical cores, and the update ratio was fixed at 25%. Figures 11(a) and (b) show the throughput for the five concurrency control methods on the Xeon E5-4650 and the Xeon Phi 5110P, respectively, as the number of B-tree partitions operated on by client modules changes from 1 to 32. Figures 12(a) and (b) show the execution time for the five concurrency control methods on the Xeon E5-4650 and the Xeon Phi 5110P, respectively. Figures 13(a) and (b) shows the execution times divided into processor behaviors on the Xeon Phi 7210 and the Xeon Phi 5110P, respectively, when the number of B-tree partitions operated on by client modules is one. Figure 13 is based on the latencies given in [1,8] and the performance counters given in [2,8].

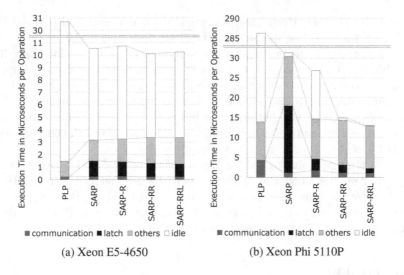

(a) Xeon E5-4650   (b) Xeon Phi 5110P

Fig. 12. Comparison of processing details for multicore and manycore platforms.

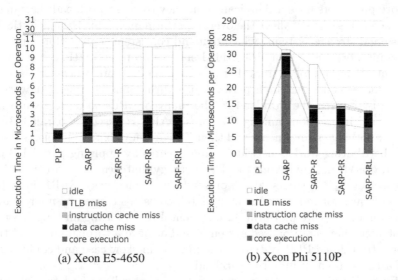

(a) Xeon E5-4650   (b) Xeon Phi 5110P

Fig. 13. Comparison based on performance counters for multicore and manycore platforms.

The improvements achieved by the proposed method for the Xeon-based platform are smaller than those for the Xeon Phi-based platform, even when the number of B-tree partitions operated on by client modules is one. There are two reasons for this. First, the proposed system distributes requests only to PEs within the same socket of the Xeon processor because the throughput deteriorates when requests are distributed between Xeon sockets. Second, one PE on the socket can monopolize all the capacity of the last-level cache shared within the Xeon socket. Even when the workload is concentrated to one PE on PLP, this PE monopolizes all the capacity of the last-level cache.

Moreover, the three techniques for the reduction of latch collisions have little effect. This is mainly because the time to access data in the main memory is dominant for the Xeon, as shown in the comparison of Figs. 13(a) and (b). The ratio of data miss time to usage time in Xeon is much larger than that in Xeon phi. It is also because the performance impact of latch collisions among the eight PEs per socket on the Xeon is insignificant, as shown in the comparison of Figs. 12(a) and (b).

There is a single point for which the throughput of SARP-R exceeds that of SARP-RR on the Xeon Phi. This is when the number of B-tree partitions operated on by client modules is four, as shown in Fig. 11(b). It occurs because the impact of reduced usage of processor by redirects to prior PEs exceeds the performance impact of latch collisions on the total of 32 PEs.

6   Conclusion

Manycore processors have a significant tendency to cause workload skew in each core, which can seriously affect the overall performance. This issue is critical for database and storage systems, which demand consistent high performance even for dynamic workloads. To address this issue, we propose SARP, a new concurrency control method for B-trees that is suited to manycore processor platforms. SARP achieves consistent high performance with robustness against workload skew by distributing the workload to the manycore PEs so as to reduce latch collisions and interactions among the cores. SARP is based on parallel B-trees for shared-nothing environments and their concurrency control. It selects a strategy for distributing workloads widely among cores by applying parallel B-trees to shared-everything environments. The strategy matches the characteristics of each processor, and it treats large workload skews effectively. SARP introduces three techniques for reducing the many latch collisions caused by many parallel contexts on a manycore platform. Experimental results using an Intel Xeon Phi demonstrated that the proposed system could achieve a throughput of 44 times that for a traditional PLP system in the maximum-skew case and could sustain 66% of a throughput in uniform workloads.

In future studies, we plan to extend the proposed method to distributed environments on plural manycore or multicore servers. Moreover, we would like to develop the method to consider factors other than hardware configuration when selecting a PE for execution.

References

1. Intel 64 and ia-32 architectures optimization reference manual. http://www.intel.com/content/dam/doc/manual/64-ia-32-architectures-optimization-manual.pdf
2. Intel 64 and ia-32 architectures software developers manual. http://download.intel.com/design/processor/manuals/253668.pdf

3. Apache: Hadoop. http://hadoop.apache.org/
4. Cha, S.K., Hwang, S., Kim, K., Kwon, K.: Cache-conscious concurrency control of main-memory indexes on shared-memory multiprocessor systems. In: Proceedings of VLDB 2001, pp. 181–190 (2001)
5. Graefe, G.: Write-optimized B-trees. In: Proceedings of VLDB 2004, pp. 672–683 (2004)
6. Graefe, G., Kimura, H., Kuno, H.A.: Foster b-trees. ACM Trans. Database Syst. **37**(3), 17 (2012)
7. Gray, J., Reuter, A.: Transaction Processing: Concepts and Techniques. Morgan Kaufmann, San Mateo (1992)
8. Jeffers, J., Reinders, J.: Intel Xeon Phi Coprocessor High Performance Programming, 1st edn. Morgan Kaufmann, San Francisco (2013)
9. Jeffers, J., Reinders, J., Sodani, A.: Intel Xeon Phi Processor High Performance Programming, 2nd edn. Morgan Kaufmann, San Francisco (2016)
10. Johnson, T., Shasha, D.: Utilization of B-trees with inserts, deletes and modifies. In: Proceedings of PODS 1989, pp. 235–246 (1989)
11. Jung, H., Han, H., Fekete, A.D., Heiser, G., Yeom, H.Y.: A scalable lock manager for multicores. In: Proceedings of ACM SIGMOD 2013, pp. 73–84 (2013)
12. Karnagel, T., Dementiev, R., Rajwar, R., Lai, K., Legler, T., Schlegel, B., Lehner, W.: Improving in-memory database index performance with intel® transactional synchronization extensions. In: Proceedings of HPCA 2014, pp. 476–487 (2014)
13. Kimura, H.: FOEDUS: OLTP engine for a thousand cores and NVRAM. In: Proceedings of ACM SIGMOD 2015, pp. 691–706 (2015)
14. Kissinger, T., Schlegel, B., Habich, D., Lehner, W.: KISS-Tree: smart latch-free in-memory indexing on modern architectures. In: Proceedings of DaMoN 2012, pp. 16–23 (2012)
15. Leis, V., Kemper, A., Neumann, T.: Exploiting hardware transactional memory in main-memory databases. In: Proceedings of ICDE 2014, pp. 580–591 (2014)
16. Levandoski, J.J., Lomet, D.B., Sengupta, S.: The bw-tree: a b-tree for new hardware platforms. In: Proceedings of ICDE 2013, pp. 302–313 (2013)
17. Love, R.: Linux J. 2003(111), 8 (2003)
18. Mao, Y., Kohler, E., Morris, R.T.: Cache craftiness for fast multicore key-value storage. In: Proceedings of EuroSys 2012, pp. 183–196 (2012)
19. Pandis, I., Tözün, P., Johnson, R., Ailamaki, A.: PLP: page latch-free shared-everything OLTP. PVLDB **4**(10), 610–621 (2011)
20. Porobic, D., Liarou, E., Tözün, P., Ailamaki, A.: Atrapos: adaptive transaction processing on hardware islands. In: Proceedings of ICDE 2014, pp. 688–699 (2014)
21. Sewall, J., Chhugani, J., Kim, C., Satish, N., Dubey, P.: PALM: parallel architecture-friendly latch-free modifications to B+ trees on many-core processors. PVLDB **4**(11), 795–806 (2011)
22. Tu, S., Zheng, W., Kohler, E., Liskov, B., Madden, S.: Speedy transactions in multicore in-memory databases. In: Proceedings of ACM SIGOPS 2013, pp. 18–32 (2013)
23. Yokota, H., Kanemasa, Y., Miyazaki, J.: Fat-Btree: an update-conscious parallel directory structure. In: Proceedings of ICDE 1999, pp. 448–457 (1999)
24. Yoshihara, T., Dai, K., Yokota, H.: A concurrency control protocol for parallel b-tree structures without latch-coupling for explosively growing digital content. In: Proceedings of EDBT 2008, pp. 133–144 (2008)

Non-order-preserving Index for Encrypted Database Management System

Wai-Kit Wong, Kwok-Wai Wong$^{(\boxtimes)}$, Ho-Yin Yue, and David W. Cheung

Hang Seng Management College, University of Hong Kong, Hong Kong, China
{wongwk,stephenwong,willyyue}@hsmc.edu.hk, dcheung@cs.hku.hk

Abstract. Data confidentiality is concerned in Database-as-a-Service (DBaaS) model. Encrypted database management system (EDBMS) addresses this concern by the data owner (DO) encrypting its private data before storing them in the database hosted by a third party service provider (SP). Indexing at SP over encrypted data is not straightforward. Most existing indexing methods are either *order-preserving*, or requiring DO to involve in query computation. Order-preserving index is vulnerable to *inference analysis*. Having DO to compute query beats the purpose of DBaaS model which is to delegate the database works of DO to SP. We developed a non-order-preserving indexing method that does not require DO's involvement in query processing at SP. Our empirical study shows that our indexing method can reduce selection processing cost by an order of magnitude compared to the case without the index.

1 Introduction

Data confidentiality is concerned in database-as-a-service (DBaaS) model. When a data owner (DO) uses the database service hosted by a third party service provider (SP), there is security risk that SP is compromised by an attacker and steals DO's data. In encrypted database management system (EDBMS), DO encrypts its data before storing them at SP. Data confidentiality is addressed but it raises a challenge for SP to compute query over encrypted data. Many works have been done to facilitate query processing in EDBMS. In general, the developed methods are more expensive than query processing over plain data. Some works also develop order-preserving indexing mechanism so as to improve query processing speed. However, order-preserving schemes have been shown to be highly vulnerable to inference analysis [12,16]. Thus, a *non-order preserving* indexing method is desired.

1.1 Related Work

Query processing in EDBMS. There are numerous approaches developed to facilitate query processing in EDBMS. In TrustedDB [4] and Cipherbase [2,3], a trusted machine (TM) is deployed at SP (e.g., TrustedDB uses IBM PCI Cryptographic Coprocessor and Cipherbase uses FPGA). TM is assumed not to be compromised by attacker. DO sends the decryption key to TM while encrypted

© Springer International Publishing AG 2017
D. Benslimane et al. (Eds.): DEXA 2017, Part II, LNCS 10439, pp. 190–198, 2017.
DOI: 10.1007/978-3-319-64471-4_16

data is stored at SP. Any query processing related to encrypted data is then done by TM.

In SDB [10,22] and Sharemind [5], secret sharing is used to 'encrypt' data. Each data item is split into several shares. Some shares are held by DO and the others are held by SP[1]. Without collecting DO's shares, SP cannot recover any plain value. To process a selection for an encrypted tuple, SDB and Sharemind each developed different secure multi-party computation (MPC) [8] protocols, which involve message exchange between parties and local computation at each party.

The above systems do not reveal the order of tuples to SP. Our indexing method can be applied in these systems.

In CryptDB [18]/MONOMI [21], order preserving encryption (OPE) is used to support selection processing. OPE possesses a property that $x > y$ if $E(x) > E(y)$ for any x, y. OPE allows SP to observe the plain comparison result and so SP can easily process selection query. Since CryptDB/MONOMI leak ordering information to SP already, our non-order-preserving index does not apply to them.

Indexing in EDBMS. Some EDBMS has developed/used some indexing methods. In Cipherbase [2,3], an encrypted B+-tree (EBT) is used for indexing. The encrypted tuples are organized in a B+-tree structure while all the keys in EBT are encrypted. This allows SP to observe the total order of plain values of tuples from EBT. Such *order-preserving* property is vulnerable to inference analysis [12,16]. We will discuss more about inference analysis in next section.

In CryptDB [18], since data is encrypted using OPE [13,17], indexing can be done directly on encrypted data. Similar to EBT, the order-preserving property leaves EDBMS at high security risk against inference analysis.

There are other indexing methods. SDB [10,22] uses domain partitioning method [9,11] to build an index but it leaks significant information about plain data to SP. Client-centric approach [6,7,14,19] allows DO to traverse the encrypted index stored at SP securely but DO has a significant amount of workload, which is not suitable in the cloud environment The above methods are considered not suitable in our problem.

Inference analysis An attacker is assumed to obtain an approximate distribution of plain data as background knowledge, e.g., from information published from other sources. The attacker compares the ordered encrypted data with the known distribution of plain data and outputs a guess of the plain value of each encrypted value. From the results of [12,16] and our experiment in Sect. 3.1, the attacker can recover plain data accurately from the ordering information. EDBMS is not secure if it leaks order information.

---

[1] Sharemind uses multiple service providers instead of one serive provider in SDB.

Privacy preserving applications Our indexing method is similar to data perturbation techniques used in privacy preserving applications [1,15,20], where DO generates a 'safe' dataset from the original plain dataset and lets others view the 'safe' dataset freely. They differ in the ways to perturb the data, e.g., a plain value 3 may be generalized to a range '1–10' [15,20] or become 5 after adding noise [1]. Existing data perturbation methods are not suitable in our case since these methods have not considered what information SP can observe in query processing.

1.2 Scope of Our Work

We aim to develop an indexing method to facilitate selection processing in EDBMS. The selection criteria φ is in the form of a propositional formula of comparison predicates. Each comparison predicate is in the form of x op y where x, y can be (numeric) attributes in the table or user-defined query parameters and op is one of the following: $>, \geq, =, <, \leq$. Since EDBMS can process the comparison predicates one by one, our discussion in the rest of paper will focus on the case with one single comparison predicate.

Our security goal is that SP does not observe more information about plain data from the index than the case no indexing is implemented.

2 Our Proposed Indexing Method - SanTable

2.1 What SP Can See in Existing EDBMS

First, we need to know what SP can see in existing EDBMS and make sure our indexing method does not leak more information than that. EDBMS like Cipherbase does not reveal the total order of plain data to SP. It only reveals to SP *whether an encrypted tuple satisfies a comparison predicate*. Then, SP can make use of such knowledge to derive some information about plain data.

Consider a comparison predicate with the form of 'X $> c$' where X is an attribute and c is a query parameter determined by DO. (SP does not know the value of c.) The predicate divides the tuples in the database into two groups. One group G_l contains all tuples with plain value on X greater than c and the other group G_s contains all other tuples small than or equal to c. On the other hand, SP does not know which of any two tuples in G_l is larger. This form a *partial order knowledge* between two groups of encrypted tuples. Similarly, SP can observe similar partial order knowledge from other comparison predicates with different query parameters. SP can combine the partial order knowledge observed together.

Suppose there are n predicates in the form of 'X op c_i' where op can be $<, >, \geq$ or $\leq$ and c_i is the query parameter of i-th predicate. Each predicate divides the encrypted tuples into two group G_{li} and G_{si} where G_{li} (G_{si} resp.) represents the group with larger values (smaller values resp.). Without loss of generality, we assume $c_1 < c_2 < ... < c_n$. This divides the data domain into

$n + 1$ sub-domains. Tuples within the same sub-domain have the same outputs on all n predicates. For example, if an encrypted tuple $\bar{t}$ has a value x on X s.t. $c_1 < x < c_2$, $\bar{t} \in G_{l1}$ and $\bar{t} \in G_{si}$ for $i > 1$. Observe that $\bar{t}$ belongs to the group with larger values G_{li} once out of n predicates. If $x < c_1$, $\bar{t}$ belongs to G_{si} for all predicates. Let P_k be the group of encrypted tuples that belongs to G_{li} k times out of n predicates. We can order the groups as: $P_0 < P_1 < ... < P_n$. ($P_i < P_j$ denotes that all tuples in P_i must be smaller on X than any tuple in P_j.)

In summary, although SP cannot see the plain values of data, SP can divide the encrypted tuples into $n + 1$ ordered group with n comparison predicates. Although SP does not obtain the total order, the partial order information may allow an attacker to recover plain data using inference analysis. We will empirically evaluate such security threat in Sect. 3.1 and show that the attacker cannot recover as accurate as the case with total order information.

2.2   Building SanTable Using Inference Analysis

Following the observation in Sect. 2.1 that SP derives a partial order of data from queries, SP can further applies inference analysis on this partial order to estimate the plain values of data. In our case, DO simulates the above scenario and obtains the estimated values of its plain data. Next, DO *sanitizes* the estimated data with a system parameter we call maximum error ϵ. ϵ is not revealed to SP. We will talk about how to select ϵ in Sect. 2.6. The purpose of ϵ is to bound the error between estimated data and original plain data and thus facilitate query computation over estimated data.

In the above inference analysis simulation, DO divides the tuples into (partially ordered) groups. When there is a data item in a group that the error may exceed ϵ, all tuples in this group are labeled as *'outliers'*. For all outliers, DO simply tells SP that their estimated values are 'null'. For other tuples, DO sends the estimated values to SP. The resulted information given to SP is referred as sanitized data, which constitutes our index - Sanitized table (SanTable).

Since SanTable is stored in plain at SP, SP can use standard indexing mechanism, e.g., B+-tree, to support efficient search on SanTable.

2.3   Use of SanTable in Query Processing

SanTable contains estimated values of encrypted data with bounded error. Given a query predicate p, SanTable allows SP to divide the encrypted tuples into 3 groups: (1) Winner: encrypted tuples that must satisfy p; (2) Loser: encrypted tuples that must dissatisfy p; and (3) Not sure: encrypted tuples that are not sure whether they satisfy p. Only the group "Not sure" is required to be processed by the EDBMS (using more expensive cryptographic methods) to confirm whether the encrypted tuples are true answers. Since there are much fewer tuples to be processed by EDBMS and processing over SanTable (which contains estimated data in plain) is very efficient, the overall query processing cost is reduced. We will talk in the next section about how a query predicate can be translated for the above processing.

2.4   Selection Predicate Translation for SanTable

With the sanitized value $\widehat{x}$ and maximum error ϵ, the plain value of x lies between $\widehat{x} - \epsilon$ and $\widehat{x} + \epsilon$. Given a comparison predicate p, we can easily translate the selection predicate. However, if DO simply issues this translated predicate to SP, SP can easily recover the query parameter c. In our case, we also need to hide the query parameters from SP.

Before we present our protection method, we discuss what SP can observe about the query parameter c in existing EDBMS without using SanTable. Recall that SP can divide encrypted tuples into n (partially ordered) groups: P_1, P_2, ..., P_n. The (estimated) plain values of boundaries of each group P_i can be observed from the known data distribution. We use $P_i[low]$ and $P_i[high]$ to denote the values of boundaries of P_i. SP can observe from the query result that the query parameter c falls in the boundary of some group. For example, SP sees that some encrypted tuples in a group P_k satisfy the predicate but other tuples in P_k do not, SP can derive that $P_k[low] \leq c \leq P_k[high]$.

Note that the above information is derived from existing EDBMS without any index. Our method should not let SP further refine the query parameter c to be more accurate than the above uncertain range. Thus, our query predicate translation is done as shown in Table 1.

Table 1. Selection predicate translation for SanTable. c_l and c_u are the estimated range of c in DO's simulated analysis of c.

p	p_w	p_{ns}
$X = c$	N.A	$c_l - \epsilon \leq \widehat{X} \leq c_u + \epsilon$ or $\widehat{X}$ is null
$X > c$	$\widehat{X} > c_l + \epsilon$	$c_l - \epsilon < \widehat{X} \leq c_u + \epsilon$ or $\widehat{X}$ is null
$X \geq c$	$\widehat{X} \geq c_l + \epsilon$	$c_l - \epsilon \leq \widehat{X} < c_u + \epsilon$ or $\widehat{X}$ is null
$X < c$	$\widehat{X} < c_u - \epsilon$	$c_l - \epsilon \leq \widehat{X} < c_u + \epsilon$ or $\widehat{X}$ is null
$X \leq c$	$\widehat{X} \leq c_u - \epsilon$	$c_l - \epsilon < \widehat{X} \leq c_u + \epsilon$ or $\widehat{X}$ is null

2.5   Security Proof Sketch

Our scheme provides two additional information to SP (who may be compromised by attacker): (i) sanitized data SanTable; and (ii) translated queries. For item (i), even if SanTable is not computed by DO, the attacker can also compute estimated data similar to SanTable on its own using inference analysis. For item (ii), since the queries are translated independent to data, data confidentiality is not affected.

On the other hand, query parameters are protected in our method. Note that SP can derive in existing EDBMS that the query parameter c lies in some range, $c_l < c < c_u$. In our modified translation, any query parameter c and c' such that $c_l < c < c' < c_u$ would result in the same translated predicate. So, SP cannot distinguish whether the query parameter is c or c'. This implies SP cannot refine the query parameter except the already known information observed without SanTable.

2.6   Selecting System Parameter: Maximum Error ϵ

ϵ affects number of outliers and probability of a group satisfying the predicate for "Not Sure". With a larger ϵ, there are fewer outliers but more groups will fall into "Not Sure". A good ϵ balances the two factors. A good way to select ϵ is to optimize the query workload in the future.

For example, if query workload is uniformly distributed, expected number of encrypted tuples in NotSure can be computed as

$$\sum_{P_i \notin Outlier} \{\sum_{j=1}^{n} \frac{P_j[high] - P_j[low]}{|D|} \Pr(P_i \text{satisfies query})|P_i|\} + \sum_{P_i \in Outlier} |P_i|$$

where $|D|$ is the data domain size and P_i denotes the partial ordered groups of tuples derived from query result (see Sect. 2.1).

$\Pr(P_i \text{satisfies query})$ is either 1 or 0 depending on whether the query parameter falls into the boundary of the P_i. Optimal value of ϵ can be found by a simple mathematical optimization problem to minimize the above value.

3   Empirical Studies

3.1   Security Evaluation

We evaluate how secure SanTable is against inference analysis. For comparisons, we implemented two methods. One is order-preserving scheme, denoted as OPE. The attacker knows the order of data. The other one is an EDBMS without any index, denoted as "EDBMS". EDBMS represents the best security level any index can achieve. To build SanTable, DO randomly generates 250 queries. The attacker is assumed to receive 250 queries in EDBMS.

We simulated an attack using inference analysis on two real datasets: Hospital Inpatient Discharges 2013[2] and Twitter[3]. (We attacked on the attribute "Total Charges" of Hospital Inpatient Discharges 2013 and "Number of Followers" of Twitter.) For Hospital Inpatient Discharges 2013, we used Hospital Inpatient Discharges 2012 to generate an approximate distribution of data as background knowledge of attacker. For Twitter, we used the exact distribution of the dataset as attacker's background knowledge to simulate the case where the attacker has acquired a highly accurate background knowledge.

We measure the attack accuracy (how many data items are accurately recovered). Following [12], we say a data item x is recovered when the recovered value $\hat{x}$ is within a very small error tolerance α, i.e., $\frac{|x-\hat{x}|}{x} \leq \alpha$. Figure 1 shows the attack accuracy of the inference analysis varying α from 0.1% to 2.5%.

The results show revealing the order information to SP leads to a significant security threat. Order-preserving indices are not recommended. EDBMS and SanTable demonstrate good resistance against inference analysis attack. SanTable achieves our security goal, which is to maintain the same security level as EDBMS without any indexing.

---

[2] http://www.health.ny.gov/statistics/sparcs/datadic.htm.

[3] https://snap.stanford.edu/data/.

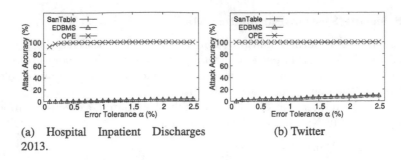

(a) Hospital Inpatient Discharges 2013.

(b) Twitter

Fig. 1. Attack accuracy using inference analysis varying error tolerance

3.2   Efficiency Evaluation

We evaluate the performance of query processing using SanTable. Cipherbase (without indexing) is implemented as a baseline competitor, denoted as 'Baseline'. We did not implement other competitors as they do not provide the same security guarantee and/or require DO's involvement in query processing. To build SanTable, DO randomly generates 250 queries.

We used TPC-H[4] datasets and queries. TPC-H is a decision support benchmark with 22 queries. The size of database varies from size order 1 (about 1 GB) to 10 (about 10 GB). We extracted 8 selections from the workload as shown in Table 2.

Table 2. Selections (denoted S_i) in TPC-H

	Table	Attribute	Type (Relational Algebra)
S_1	Lineitem	shipdate	$\sigma_{shipdate<c}$ or $\sigma_{shipdate>c}$
S_2	Lineitem	shipdate	$\sigma_{c_i<shipdate<c_j}$
S_3	Lineitem	receiptdate	$\sigma_{c_i<receiptdate<c_j}$
S_4	Lineitem	quantity	$\sigma_{quantity<c}$ or $\sigma_{quantity>c}$
S_5	Lineitem	quantity	$\sigma_{c_i<quantity<c_j}$
S_6	Lineitem	discount	$\sigma_{c_i<discount<c_j}$
S_7	Order	orderdate	$\sigma_{orderdate<c}$ or $\sigma_{orderdate>c}$
S_8	Order	orderdate	$\sigma_{c_i<orderdate<c_j}$

We measured the average processing time of different methods out of 500 random queries. Figure 2 shows the experiment result.

We make the following observations from the experiment results:

– SanTable has an average improvement of at least an order of magnitude over Baseline, except S_6. SanTable is very effective in general.

---

[4] http://www.tpc.org/tpch/.

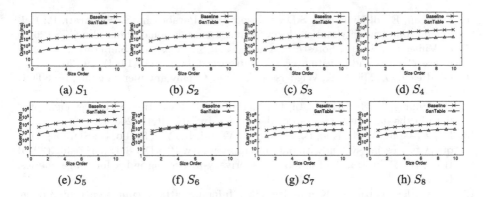

Fig. 2. Average processing time of different selection S_i of different methods

- For S_6, there are only 19 distinct values in this attribute. A small uncertain range can easily cover a large number of tuples. Yet, SanTable provides a small improvement over Baseline.
- SanTable scales well with increasing size of the database.

4   Conclusions and Future Work

We developed SanTable, an indexing method for EDBMS that can effectively save query cost. SanTable is designed so that it does not leak information about plain data beside what SP can already see in existing EDBMS. In the future, we will study how to make use of SanTable to optimize processing cost of more complex queries.

Acknowledgements. The paper is supported by FDS grant (UGC/FDS14/E05/14) and GRF Grant 17201414 from Hong Kong Research Grant Council.

References

1. Agrawal, R., Srikant, R.: Privacy-preserving data mining. In: SIGMOD (2000)
2. Arasu, A., Blanas, S., Eguro, K., Joglekar, M., Kaushik, R., Kossmann, D., Ramamurthy, R., Upadhyaya, P., Venkatesan, R.: Secure database-as-a-service with cipherbase. In: SIGMOD (2013)
3. Arasu, A., Eguro, K., Joglekar, M., Kaushik, R., Kossmann, D., Ramamurthy, R.: Transaction processing on confidential data using cipherbase. In: ICDE (2015)
4. Bajaj, S., Sion, R.: Trusteddb: a trusted hardware based database with privacy and data confidentiality. In: SIGMOD (2011)
5. Bogdanov, D., Jagomägis, R., Laur, S.: A universal toolkit for cryptographically secure privacy-preserving data mining. In: Chau, M., Wang, G.A., Yue, W.T., Chen, H. (eds.) PAISI 2012. LNCS, vol. 7299, pp. 112–126. Springer, Heidelberg (2012). doi:10.1007/978-3-642-30428-6_9

6. Damiani, E., di Vimercati, S.D.C., Jajodia, S., Paraboschi, S., Samarati, P.: Balancing confidentiality and efficiency in untrusted relational dbmss. In: CCS (2003)
7. di Vimercati, S.D.C., Foresti, S., Paraboschi, S., Pelosi, G., Samarati, P.: Shuffle index: efficient and private access to outsourced data. TOS **11**(4), 19 (2015)
8. Goldreich, O., Micali, S., Wigderson, A.: How to play any mental game. In: STOC (1987)
9. Hacigümüs, H., Iyer, B.R., Li, C., Mehrotra, S.: Executing SQL over encrypted data in the database-service-provider model. In: SIGMOD (2002)
10. He, Z., Wong, W.K., Kao, B., Cheung, D.W., Li, R., Yiu, S., Lo, E.: SDB: a secure query processing system with data interoperability. PVLDB **8**(12), 457–462 (2015)
11. Hore, B., Mehrotra, S., Tsudik, G.: A privacy-preserving index for range queries. In: VLDB (2004)
12. Islam, M.S., Kuzu, M., Kantarcioglu, M.: Inference attack against encrypted range queries on outsourced databases. In: CODASPY (2014)
13. Kadhem, H., Amagasa, T., Kitagawa, H.: A secure and efficient order preserving encryption scheme for relational databases. In: KMIS (2010)
14. Li, R., Liu, A.X., Wang, A.L., Bruhadeshwar, B.: Fast range query processing with strong privacy protection for cloud computing. PVLDB **7**(14), 1953–1964 (2014)
15. Machanavajjhala, A., Kifer, D., Gehrke, J., Venkitasubramaniam, M.: L-diversity: privacy beyond k-anonymity. TKDD **1**(1), 3 (2007)
16. Naveed, M., Kamara, S., Wright, C.V.: Inference attacks on property-preserving encrypted databases. In: SIGSAC (2015)
17. Popa, R.A., Li, F.H., Zeldovich, N.: An ideal-security protocol for order-preserving encoding. In: SP (2013)
18. Popa, R.A., Redfield, C.M.S., Zeldovich, N., Balakrishnan, H.: Cryptdb: protecting confidentiality with encrypted query processing. In: SOSP (2011)
19. Shmueli, E., Waisenberg, R., Elovici, Y., Gudes, E.: Designing secure indexes for encrypted databases. In: Jajodia, S., Wijesekera, D. (eds.) DBSec 2005. LNCS, vol. 3654, pp. 54–68. Springer, Heidelberg (2005). doi:10.1007/11535706_5
20. Sweeney, L.: k-anonymity: a model for protecting privacy. IJUFKS **10**(5), 557–570 (2002)
21. Tu, S., Kaashoek, M.F., Madden, S., Zeldovich, N.: Processing analytical queries over encrypted data. PVLDB **6**(5), 289–300 (2013)
22. Wong, W.K., Kao, B., Cheung, D.W., Li, R., Yiu, S.: Secure query processing with data interoperability in a cloud database environment. In: SIGMOD (2014)

Data Warehouse and Data Stream Warehouse

A Variety-Sensitive ETL Processes

Nabila Berkani[1(✉)] and Ladjel Bellatreche[2]

[1] École nationale Supérieure d'Informatique (ESI), Algiers, Algeria
n_berkani@esi.dz
[2] LIAS/ISAE-ENSMA – Poitiers University, Poitiers, France
bellatreche@ensma.fr

Abstract. Nowadays, small, medium and large companies need advanced data integration techniques supported by tools to analyse data in order to deliver real-time alerts and trigger automated actions, etc. In the context of rapidly technology changing, these techniques have to consider two main issues: **(a)** the variety of the huge amount of data sources (ex. traditional, semantic, and graph databases) and **(b)** the variety of storage platforms, where a data integration system may have several stores, where one hosts a particular type. These issues directly impact the efficiency and the deployment flexibility of ETL (Extract, Transform, Load). In this paper, we consider these issues. Firstly, thanks to Model Driven Engineering, we make generic different types of data sources. This genericity allows overloading the ETL operators. To show the benefit of this genericity, several examples of instantiation are described covering relational, semantic and graph databases. Secondly, a Web-service-driven approach for orchestrating the ETL flows is given. Thirdly, we present a fusion procedure that merges the set of heterogeneous instances and deployed according their favorite stores. Finally, our finding is validated through a proof of concept tool using the LUBM benchmark and YAGO $\mathcal{KB}$ and deployed in Oracle RDF Semantic Graph 12c.

1   Introduction

The past few decades have witnessed a spectacular explosion in the quantity of data sources available in various types and formats. This situation pushes small, medium and large companies to exploit this mine of data in order to achieve a high decision making in science, society, health, etc. This usually passes through the data integration process [11]. Plenty of commercial and open sources data integration solutions and tools exist in the market. When source data are extracted and materialized in an integration system such as a data warehouse, more specific techniques and tools implementing ETL (Extract, Transform, Load) are widely used [20]. Oracle Warehouse Builder, SAP Data Service, Talent Studio for Data Integration, IBM Infosphere Warehouse Edition, etc. are examples of these tools. The maturity of the ETL motivates researchers to make generic its whole workflow [5,26]. An ETL algebra composed of 10 generic operators (*Retrieve, Extract, Convert, Filter, Merge, Join, Union, Aggregate, Delete and Store*) has been proposed [20]. The signature of each operator is personalized

© Springer International Publishing AG 2017
D. Benslimane et al. (Eds.): DEXA 2017, Part II, LNCS 10439, pp. 201–216, 2017.
DOI: 10.1007/978-3-319-64471-4_17

according the type of data sources and the target data warehouse ($\mathcal{TDW}$). By examining deeply the ETL techniques, we figure out that they mainly concentrate on the *traditional types of data* such as relational databases [24] – which has reigned several decades – and recently semantic databases ($\mathcal{SDB}$) [3,14]. In the context of rapidly technology changing, several new types of databases appear (e.g., Graph databases, NoSQL, Time Series, Knowledge bases ($\mathcal{KB}$) such as Yago [21]) and consequently they became candidate for the data integration. This phenomenon is called by the *Variety of sources* [7].

The variety does not only impact data sources, but also the storage of the $\mathcal{TDW}$, where multi-stores are well-adapted to achieve high performance of data accesses. More concretely, we passed from $(n-1)$ scenario, where n heterogeneous sources are integrated into $\mathcal{TDW}$ deployed on one store to $(n-m)$ scenario, where the $\mathcal{TDW}$ may be deployed in several stores, where each one may store a specific type of data [10]. To deal with the variety of sources, we propose the usage of Meta-Driven Engineering (MDE) and then overload the ETL operators to deal with specific type of source. Inspired from the Meta-Object Facility (MOF)[1], we make generic different sources in order to deal with their variety. The MOF describes a generic framework in which the abstract syntax of modelling languages can be defined [15]. It has a hierarchical structure composed of four layers of meta-data corresponding to the different levels of abstraction: the instance layer (M0), the model layer (M1), the meta-model layer (M2) and the meta-meta-model layer (M3). Each layer defines a level to ensure the consistency and the correctness of the instance model syntax and semantics at each level of abstraction.

This generic model can be easily exploited by ETL operators, where each one is overload. An operator overloading (as in C++ language) allows a programmer to define the behaviour of an operator applied to objects of a certain class the same way methods are defined [6]. In our context, each ETL operator will be overload to deal with the diversity of each type of sources.

To offer the designers the possibility to deploy their $\mathcal{TDW}$ on a multi-stores, we exploit the *Store operator* of ETL. It can be associated with a Service Web that orchestrates the ETL flows and distributes the data over the stores according their storage formats.

In this paper, we detail our generic model using MDE. We give examples of its instantiation from three types of data sources: relational, semantic and graph databases. The ETL operators are then overloaded for these types. Thanks to the *Store* operator, the multi-store deployment is guaranteed. Our proposal is implemented and experimented.

The rest of this paper is organized as follows. We give an overview on the evolution of the ETL in Sect. 2. In Sect. 3, we give a formalization of three main classes of databases (relational, semantic and graph) and a motivating example. In Sect. 4, a generalization of ETL elements are given by the means of MDE techniques and the process to overload the ETL operators. In Sect. 5, the deployment methodology of a data warehouse on multi-store system is developed.

---

[1] http://www.omg.org/mof/.

A case study is proposed and various experiments are presented. Section 7 concludes the paper.

2   Related Work

In this section, we give an overview of the most important studies on the ETL and the efforts to making them variety-sensitive. The first studies on ETL dealt with sources considering their physical implementations such as: (i) their deployment platforms (centralized, parallel, etc.) and (ii) their storage models (e.g. tables, files). In [23], a set of algorithms was proposed to optimize the physical ETL design. Simitsis et al. [18] have proposed algorithms for optimizing the efficiency and the performance of ETL process. Other non-functional requirements such as freshness, recoverability, and reliability have also been considered [19]. The work of [13] proposes an automated data generation algorithm assuming the existing physical models for ETL to deal with the problem of data growing. In order to hide the physical implementations, several research efforts have been proposed. The first category of these studies attempts to consider the logical level of data sources. In this perspective, [27] proposed an ETL workflow modelled as a graph, where its nodes represent activities, record-sets, attributes, and its edges describe the relationships between nodes that define ETL transformations. In [25], a formal ETL logical model is given using L_{DL} [17] as a formal language for expressing the operational semantics of ETL activities. The second category of these studies considered the conceptual level of sources. Approaches based on ad-hoc formalisms [26], on standard languages using UML [22], model driven architecture (MDA) [8], BPMN [1,28] and mapping modelling [4,12] have been proposed. The third category use ontologies as external resources to facilitate and automate the conceptual design of ETL process. [20] automated the ETL process by constructing an OWL ontology linking schemes of semi-structured and structured (relational) sources to a target data warehouse ($\mathcal{DW}$) schema. Other studies like [14] consider data source provided by the semantic Web and annotated by OWL ontologies. However, the ETL process in this work is dependent on the storage model used for instances which is the triples.

Based on this brief overview, we figure out the effort deployed by the research community in generalizing the ETL processes by going from the physical level to the semantic level of the sources. In [27], a generic model of ETL activities that plays the role of a pivot model has been proposed, but without MDE techniques. [20] has defined an ETL algebra with 10 generic operators. The main drawbacks of these approaches are: they deal with traditional types of sources (relational and XML schemes) and they make an implicit assumption that the data warehouse is deployed on one system usually relational.

3   Background and a Motivating Example

In this section, we give an overview on the most important types of databases: relational, semantic and graph databases adopted by a large number of sources.

Then, a motivating example is considered to illustrate the basic ideas behind our proposal.

3.1   Formalization of Databases

A relational database is defined by set of tables, attributes, instances and constraints.

A semantic database (SDB) is formally defined as follows [3]: $<OM, I, Pop, SL_{OM}, SL_I>$, where:

- OM: $<C, R, Ref, formalism>$ is the ontology model of the SDB; where C and R denote respectively concepts and roles of the model; Ref is a function defining terminological axioms of a DL TBOX (Terminological Box) [2], (e.g., Ref(Student) $\rightarrow$ (Person $\cap$ $\forall$ takesCourse(Person, Course))) and $Formalism$ is the formalism followed by the global ontology model like RDF, OWL, etc.);
- I: presents the instances (the ABox) of the SDB;
- Pop: $C \rightarrow 2^I$ is a function that relates each concept to its instances;
- SL_{OM}: is the Storage Layout of the ontology model (vertical, binary or horizontal) [9]; and
- SL_I: is the Storage Layout of the instances I.

A graph database usually used to represent knowledge bases through a graph G whose nodes (V), edges (E) and labels (L_v, L_e) represent respectively classes, instances and data properties, object properties and **DL** constructors. Neo4J[2] is an example of a storage system of graph databases.

3.2   Motivating Example

To explicit the basic ideas behind our proposal, let us consider a scenario, where a governmental organisation wants constructing a data warehouse to analyse the performance of students in universities. To do so, this organisation considers four data sources with a high variety. The particularity of these sources is that they are derived from the benchmark related to the universities (LUMB[3]) and the Yago[4] knowledge base. The details of these sources are given below: S_1 is a MySQL relational databases with the following schema composed of tables and attributes: *Student(name), Course(title), University*, S_2 is a Berkeley XML DB with a schema composed of elements and attributes: *GraduateStudent(name), GraduateCourse(title), University*, S_3 is an Oracle RDF SDB composed of classes, properties: *Student(name), Publication, University*, and S_4 is a Neo4j Graph DB with nodes, edges: *Person, Student(name), Publication, PublicationAuthor, University*.

The obtained warehouse has two stores Semantic Oracle and Mongodb. In this context, the different ETL operators have to be overloaded to deal with

---

[2] https://neo4j.com/product/.
[3] http://swat.cse.lehigh.edu/projects/lubm/.
[4] www.yago-knowledge.org/.

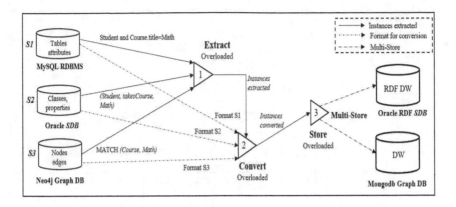

Fig. 1. An example of ETL operator overloading

this variety. Figure 1 describes the whole architecture of the ETL process, where *Extract* and *Convert* operators are overloaded. As we see, they have the same name, but different signatures. Based on the format of each store, the Store operator is also overloaded.

4   Generalisation of ETL Elements

Before discussing our proposal in overloading ETL operators, we first formalize the ETL process and its operators. An ETL process is defined as 5-tuples as follows: $<InputSet, OutputSet, Operator, Function, ETLResul>$, where:

InputSet: represents a finite set of input elements describing data sources. Each source has its own format and storage layout. To make generic the representation of data sources, we propose to generalize them using MOF initiatives. The obtained meta-model is composed of conceptual entities and their attributes. In addition, links between entities are also represented via associations. We also represent several semantically restrictions, such as primary and foreign keys. Figure 2, part (a), illustrates the fragment of our meta-model. Table 1 is an instantiation of relational, semantic and graph databases sources.

OutputSet: is a finite set of intermediate or target elements. The output of the ETL process can be either the intermediate output (sub process) or the final output (ETL process). The final output corresponds to the target data stores, where the schema of each store can be seen as an instance of our meta-model (part (a) of Fig. 2).

Operator: is a set of operators commonly encountered during the ETL process in [20]. By analysing these operators, we propose to decompose them into four categories: (1) loading class, (2) branching class, (3) merging class and (4) activity class.

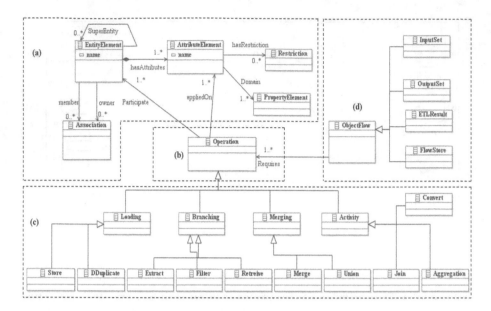

Fig. 2. Excerpt of ETL meta models

Table 1. Sample of InputSet and OutputSet databases.

Elements	Databases		
	Relational	Ontological	Graph
Entity	Table	Class	Node
Association	Table	Object_property	Edge
Attribute	Column	Data_property	Node
Property	Domain of values	Data type	Domain of values
Restriction	Primary key	SameAs	Node

- **Branching Class:** delivers multiple output-sets which can be further classified in Filter operations based on conditions or Extract and Retrieve operations that handle with the appropriate portion of selected data.
- **Merging Class:** fuses multiple data incoming from data sources. We identify two possible operations: (i) Merge operation applied when the data belong to attributes related to entities of the same source; (ii) Union operation applied when data belong to entities incoming from different data sources
- **Activity Class:** represents points in the process where work is performed. It corresponds to all operations of join conversion and aggregation. Join operations is applied when data belong to different entities. Conversion operation is applied on data having different format in order to unify it and adapt it to the target data stores. The aggregation operation is done depending on the schema of the target data stores applying needed functions (count, sum, avg, max, min).

- **Loading Class:** represents the point of data quality by the detection of duplicated data and cleaning them before their loading in the target data store.

Based on this, we propose a meta models of these operations (part (c) of Fig. 2). The generic formalization of each operator is given by:

- $Retrieve(S, E, A, R)$: retrieves data D of attributes A related to entities E from Source S;
- $Extract(S, E, A, R, CS)$: enables the data D extraction of A related to entities E from source S satisfying constraint CS;
- $Merge(S, E_1, E_2, A_1, A_2, D_1, D_2)$: merges data D_1 and D_2 belonging to the source S;
- $Union(S_1, S_2, A_1, A_2, D_1, D_2)$: unifies data D_1 and D_2 belonging to different sources S_1 and S_2 respectively;
- $Filter(S, E, A, D, CS)$: filters incoming data D, allowing only values satisfying constraints CS;
- $Join(S, E_1, E_2, A_1, A_2, D_1, D_2)$: joins data D_1 and D_2 having common attributes A_1 and A_2;
- $Convert(S, E, A, D, F_S, F_T)$: converts incoming data D from the format F_S of source S to the format of the target data store F_T;
- $Aggregate(S, E, A, D, F)$: aggregates incoming data D applying the aggregation function F (count, sum, avg, max) defined in the target data-store.
- $DD(D)$: detects and deletes duplicate values on the incoming data D;
- $Store(T, E, A, D)$: loads data D of attributes A related to entities E in the target data store T.

Function: is a function over a subset of Input-Set applied in order to generate data satisfying restrictions defined by any ETL operator.

ETLResult: is a set of output elements representing the flow.

4.1   Overloading Operators

In this section, we show the mechanism to overload ETL operators by considering semantic and graph databases.

In the case of a semantic database, the signature of overload operators is as follows:

- $Retrieve(S_i, C, I)$: retrieves instances I related to classes C from Source S_i;
- $Extract(S_i, C, I, CS)$: extracts instances I related to classes C from source S satisfying constraint CS;
- $Merge(S_i, C_1, C_2, I_1, I_2)$: merges instances I_1 and I_2 related to the classes C_1 and C_2 respectively and belonging to the same source S_i;
- $Union(S_1, S_2, C_1, C_2, I_1, I_2)$: unifies instances I_1 and I_2 related to C_1 and C_2 respectively and belonging to different sources S_1 and S_2 respectively;
- $Filter(S_i, C, I, CS)$: filters incoming instances I related to C, allowing only the values satisfying constraints CS;
- $Join(S_i, C_1, C_2, I_1, I_2)$: joins instances I_1 and I_2 related to C_1 and C_2 respectively and having common object properties;
- $Convert(S_i, C, I, F_S, F_{TDW})$: converts incoming instances I from the format F_S of source S_i to the format of the target $\mathcal{TDW}$ (F_{TDW});
- $Aggregate(S, C, I, F)$: aggregates incoming instances I related to C applying the aggregation function F (count, sum, avg, max) defined in the $\mathcal{TDW}$.
- $DD(I)$: detects and deletes duplicate values on the incoming instances I;
- $Store(\mathcal{TDW}, I)$: loads instances I in target $\mathcal{TDW}$.

In the case of a graph database, the signature of overload operators is as follows:

- $Retrieve(G, V_j, L_j)$: retrieves a node V_j having an edge labeled by L_j of G.
- $Extract(G, V_j, CS)$: extracts, from G, the node V_j satisfying CS.
- $Convert(G, G_T, V_i, V_T)$: converts the format of the node V_i to the format of the target node V_T. The conversion operation is applied at instance level.
- $Filter(G, V_i, CS)$: applied on V_i node, allowing only instances satisfying CS.
- $Merge(G, V_i, V_j)$: merge instances denoted by nodes V_i, V_j in same graph G.
- $Union(G, G_T, V_i, V_j, E_j)$: links nodes that belongs to different sources. It adds in the target graph G_T, both nodes V_i and V_j and link them by an edge E_j.
- $Join(G, G_T, V_i, V_j, E_j)$: joins instances whose corresponding nodes are $V_i \in G$ and $V_j \in G_T$. They are linked by an *object property* defined by an edge E_j.
- $Store(G_T, V_j)$: loads instances denoted by nodes V_j to the target graph G_T.
- $DD(G_T, CS)$: sorts the graph G_T based on CS and detects duplication.
- $Aggregate(G_T, V_j, Op)$: aggregates instances represented by the nodes V_j.

Some primitives need to be added to manage the ETL operations required to build the ETLgraph such as:

- $AddNode(G_T, V_j, E_j, L_j)$: adds node V_j, edge E_j, label L_j required to G_T.
- $UpdateNode(G_T, V_j, E_j, L_j)$: updates node V_j, edge E_j, label L_j in G_T.
- $RenameNode(G_T, V_j, E_j, L_j)$: renames node V_j, edge E_j, label L_j in G_T.
- $DeleteNode(G_T, V_j, E_j, L_j)$: deletes node V_j, edge E_j, label L_j from G_T.
- $SortGraph(G_T, V_j, CS)$: sorts nodes of G_T based on some criteria CS to improve search performance.

Our goal is to facilitate, manage and optimize the design of the ETL process during the initial design, deployment phase and during the continuous evolution of the $\mathcal{TDW}$. For that, we enrich the existing ETL operators with *split*, *context* and *Link* operators elevating the clean-up and deployment of ETL process at the conceptual level.

- $Split(G, G_i, G_j, CS)$: splits G into two sub-graphs G_i and G_j based on CS.
- $Link(G_T, V_i, V_j, CS)$: links two nodes V_i and V_j using the rule CS.
- $Context(G, G_T, CS)$: extracts from the graph G a sub-graph G_T that satisfies the context defined by restrictions CS using axioms.

5   Deployment on a Multistores System

In this section, we propose a methodology to satisfy the $n - m$ scenario discussed in the Introduction. To do so, we have to consider three issues: consolidation of schemas, fusion of instances, and deployment.

5.1   Consolidation ETL Algorithm

Algorithm 1 describes in details the overloading of ETL operators in the context of semantic and graph data sources. It is based on mappings defined between data sources schemes and global schema. We used mappings described in [3].

5.2   Fusion Procedure

In this section, we propose a fusion method to merge different input data sources representations based on the target model chosen by the designer. Our solution is based on the *Graph Property* model presented above [16]. The property graph is common because modellers can express other types of models or graphs by

Algorithm 1. Overloading ETL Process Algorithm

Input: $\mathcal{IO}$ or Contextual $\mathcal{KB}$, S_i: Local sources $\mathcal{SDB}$
Output: $\mathcal{TDW}$ (schema + instances)

1: $V_{Si} := \emptyset$; $\text{ETL}_G := \text{Graph}(\text{Tbox}(kb))$; $V_{kb} := \text{GetNodes}(kb)$;
2: **if** Input is $\mathcal{IO}$ **then**
3:   $Input_{cond} := (C : \text{Class of ontology IO})$;
4: **else if** Input is $\mathcal{KB}$ **then**
5:   $Input_{cond} := (V_i \in V_{kb} \wedge (V_i \text{ isClass}))$;
6: **end if**
7: **for** each $Input_{cond}$ **do**
8:   **for** Each S_i **do**
9:     **if** *Equivalent or complete mappings* $(N_{S_i}, N_{KB}) \vee (C_{S_i}, C_{IO})$ **then**
10:       **if** Input is $\mathcal{IO}$ **then**
11:         $C := \text{IdentifyClass}(C_{TDW}, C_i)$;
12:       **else if** Input is $\mathcal{KB}$ **then**
13:         $V_i := \text{IdentifyNode}(\text{ETL}_G, V_i)$;
14:         $E_i := \text{IdentifyEdge}(\text{ETL}_G, E_i)$;
15:       **end if**
16:     **else if** *sound or overlap mappings* $(N_{S_i}, N_{KB}) \vee (C_{S_i}, C_{IO})$ **then**
17:       **if** Input is $\mathcal{IO}$ **then**
18:         $\text{Const} := \text{ExtractConstraint}(C_{TDW}, C_i)$;
19:       **else if** Input is $\mathcal{KB}$ **then**
20:         $\text{Const} := \text{ExtractNeighbor}(\text{ETL}_{Graph}, V_i)$;
21:       **end if**
22:     **end if**
23:     **if** (Input is $\mathcal{IO}$) **then**
24:       **if** (Const isDataTypeProperty) **then**
25:         $I := \text{Convert}(C_j, I, \text{const})$;
26:         $I := \text{Filter}(C_j, I, \text{Const})$;
27:       **else if** (Const isObjectProperty) **then**
28:         $I := \text{Join}(C_j, C_i, I, \text{Const})$;
29:       **else if** (Const isAxiom) **then**
30:         $I := \text{Aggregate}(C_j, C_i, I, \text{Const})$;
31:       **end if**
32:       $I := \text{MERGE}(C_{S_i}, I)$; $I := \text{UNION}(C_{S_i}, C_{IO}, I)$;
33:       $\text{STORE}(IO, C_i, DD(I))$;
34:     **else if** (Input is $\mathcal{KB}$) **then**
35:       **if** (Const isDataTypeProperty) **then**
36:         $I := \text{Convert}(V_j, I, \text{const})$;
37:         $I := \text{Filter}(V_j, I, \text{Const})$;
38:       **else if** (Const isObjectProperty) **then**
39:         $I := \text{Join}(V_j, V_i, I, \text{Const})$;
40:       **else if** (Const isAxiom) **then**
41:         $I := \text{Aggregate}(V_j, V_i, I, \text{Const})$;
42:       **end if**
43:       $I := \text{MERGE}(N_{Si}, I)$; $I := \text{UNION}(N_{Si}, N_{kb}, I)$;
44:       **for** Each I **do**
45:         $\text{ETL}_G := \text{addEdge}(\text{ETL}_G, N_i, \text{edge}, I)$;
46:         $\text{ETL}_G := \text{addNode}(\text{ETL}_G, N_i, \text{edge}, I)$;
47:       **end for**
48:       $\text{ETL}_G := \text{Filter}(\text{ETL}_G, DD(N_i), \text{Null-values})$;
49:       $\text{STORE}(\text{ETL}_G)$;
50:     **end if**
51:   **end for**
52: **end for**

adding or abandoning particular elements. To do so, we propose to use the primitives proposed previously. They enable designers to adds, deletes and renames graph elements in order to manage the ETL flow generated and adapt it to the target storage layout chosen. An example of *addnode* primitive is done as follows:

```
- Sparql query language :
construct ?V
where {GRAPH  :?G {?V rdf:type name-space:Class}}
- Using Cypher query language for Neo4j graph database:
MERGE (<node-name>:<label-name>
{<Property1-name>:<Pro<rty1-Value> ..... <Propertyn-name>:<Propertyn-Value>}
```

On the basis of items presented previously, we have identified three particular cases:

Deployment of $\mathcal{KB}$ on $\mathcal{SDB}$: the RDF graph allowing the representation of $\mathcal{KB}$ deployed on $\mathcal{SDB}$ can be obtained by restricting labels of the nodes and edges to Uniform Resource Identifiers (URIs) and not allowing node/edge attributes;

Deployment of $\mathcal{KB}$ on graph database: using graph property having directed, labelled, attributed nodes and edges will allow a deployment of $\mathcal{KB}$ on a graph system;

Deployment of traditional data on graph: starting from a property graph, we generate a standard semantic graph by discarding the nodes/edges attributes. Having a semantic graph, we consider the nodes as attributes/data of traditional data, labels nodes are either *attributes or data*, edges as relationships between data and attributes of traditional data, labels edges can be either *has_data or has_attributes*.

5.3   Deployment of ETL Process

Storage deployment models can follow different representations according to specific requirements. A $\mathcal{TDW}$ can be deployed using horizontal, vertical, hybrid models, NoSQL, etc. [9]. In our case, we choose to deploy the $\mathcal{TDW}$ into vertical representation using Oracle DBMS which offers a storage model to represent instances and graphs using Oracle RDF Semantic Graph. We translated the $\mathcal{TDW}$ schema into vertical relational model, then generated an N-Triple file, load it into a staging table using Oracle's SQL*Loader utility. We applied the ETL Algorithm to populate the target schema.

6   Experimental Study

In order to illustrate the feasibility of our approach, we use our motivating example (cf. Section 3). We choose Oracle semantic database system to implement the sources and the warehouse. Oracle 12c release 2 delivers *RDF Semantic Graph features* as part of Oracle Spatial and Graph. With native support for RDF and OWL standards for representing semantic data, with SPARQL for query language. Oracle has defined two subclasses of DLs: OWLSIF and a richer fragment OWLPrime. Note that *OWLPrime* limits the expressive power of DL formalism in order to ensure decidable query answering. The proposed Algorithm 1 was implemented using the overload of ETL operators in order to integrates the created sources into the DW taking in account their heterogeneity. Note that generic ETL operators defined in the previous section are expressed on the conceptual level. Therefore, each operation has to be translated according the logical level of the target DBMS (Oracle). Oracle offers two ways for querying semantic data: SQL and SPARQL. We choose SPARQL to express this translation. Here an example of $\mathcal{KB}$ aggregation ETL operator translation to SPARQL:

```
PREFIX yago: http://yago-knowledge.org/resource/yago.owl#
AGGREGATE: Aggregates incoming record-set.
Select (Count(?Instance) AS ?count) Where {
GRAPH  :?G {?Instance rdf:type yago:Class}}
Group By ?Instance.
```

The proposed tool is implemented in Java language and uses JENA API to access ontologies and a $\mathcal{KB}$. Each generic ETL operator is implemented as a Web Service Restful using Java overload polymorphism implementation. The restful web service is implemented is such way to consider the overload resolution. Each ETL operator is overloaded by determining the most appropriate definition to use. It compares the argument type used to call the appropriate service restfull with the parameter types specified in the definitions. This will allow managing the different representations of input data (instances and graph). The proposed ETL algorithm consists then in orchestrating the Web services.

Each Web service that accesses the persistent storage is implemented using Data Access Object (DAO) Design patterns[5]. DAO implements the access mechanism required to handle the different input representations. The DAO solution abstracts and encapsulates all access to persistent storage, and hides all implementation details from business components and interface clients. The DAO pattern provides flexible and transparent accesses to different storage layout. In order to obtain a generic implementation of the ETL process, we implemented our solution following service oriented architecture (SOA). SOA offers the loose coupling of the web services defined bellow, and interaction among them. The application implements an orchestration of web services in early binding. Indeed, each web service is implemented in such way that parameters and variables are detected and checked at compile time. Figure 3 describes the whole architecture of the ETL and MultiStore Services.

A demonstration video summarizing the different services offered by our proposal is available at: https://youtu.be/zbtl1qMvPOU.

6.1 Evaluation Study

In this section, we present the performance of our approach through a set of experiments considering an Ontology and large $\mathcal{KB}$. Four criteria are used to

---

[5] http://www.oracle.com/technetwork/java/dataaccessobject-138824.html.

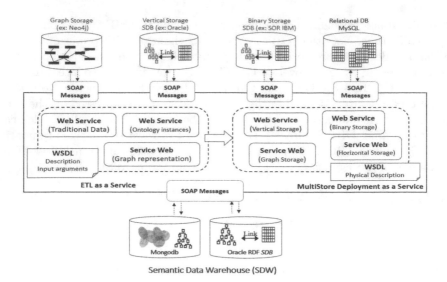

Fig. 3. A general architecture of the ETL and MultiStore Services.

evaluate our proposal: (i) complexity of the proposed ETL algorithm, (ii) evaluation time per ETL operators before and after overloading, (iii) scalability of the ETL process, (iv) inference performance.

Environment of our experiments. Our experiments are based on LUBM ontology and YAGO $\mathcal{KB}$ (version 3.0.2). The architecture of the YAGO system is based on themes. Each theme is a set of facts. A fact is the equivalent of an RDF triple (s,p,o). YAGO has defined the context relation between individuals [21], which we used to extract the set of themes related to our context study which is university domain. The resulting contextual YAGO $\mathcal{KB}$ contains around $5, 9 \times 10^6$ triples. Note that five (5) sets of triples were generated using LUBM benchmark and Yago knowledge base.

(1) *Deployment of Data Sources and $\mathcal{TDW}$*: We have created five Oracle *SDBs* using generated data-sets and deployed the $\mathcal{TDW}$ schema using Oracle *SDB*. We chose N-Triple format (.nt) to load instances using Oracle SQL*Loader.
(2) *Oracle Database Tuning:* $\mathcal{TDW}$ schema was optimized using Btree indexing triples and sparql query hints. Some PL/SQL APIs are also invoked after each load of significant amount of data. The API SEM_PERF.GATHER_STATS Collects stats for sources models and SEM_APIS.ANALYZE_MODEL for $\mathcal{TDW}$ model in the semantic network graph. The memory SGA and PGA are also increased to 2GB.
(3) *Inference Engine:* Oracle has incorporated a reasoner engine defined based on *TrOWL* and *Pellet* reasoners. Oracle provides full support for native inference in the database for RDFS, RDFS++, OWLPRIME, OWL2RL, etc. It uses forward chaining to do the inference. It compiles entailment rules

directly to SQL and uses Oracle's native cost-based SQL optimizer to choose an efficient execution plan for each rule. The following is an example of user defined rules applied, they are saved as records in tables. $Rule_1$: co-author rule: $authorOf(?A1, ?P) \wedge authorOf(?A2, ?P) \rightarrow coAuthor(?A1, ?A2)$.

(4) *Hardware:* Our evaluations were performed on a laptop computer (HP Elite-Book 840 G2) with an Intel(R) CoreTM i5-5200U CPU 2.20 GHZ and 8 GB of RAM and a 500 GB hard disk. We use Windows10 64bits. Cytoscape[6] is used for visualization.

Obtained results. We evaluate our proposal based on the following criteria:

Criterion 1: ETL Algorithm Complexity. The algorithm is implemented based on semantic ontologies (classes and properties) and graph theory, where nodes represents concepts and instances, edges for roles and labels for definitions. We examine the number of iterations of our algorithm to generate ETL process as flow or graph. In this case, we are interesting on the time complexity. The algorithm is based on concepts searches (Tbox for intentional mappings i.e. mappings only between classes and properties and not between instances). The time complexity is $O(n)$, where n represents the number of involved classes or nodes. Figure 4a shows the number of iterations by classes. It indicates a polynomial time. This finding shows the feasibility and efficiency of our approach.

Criterion 2: Evaluation Time Per ETL Operator Before and After Overloading. We run the ETL Algorithm for both scenarios (without overload for ontology and $\mathcal{KB}$, and with overload for both) to populate the target schema of semantic $\mathcal{TDW}$. We measure the time spent to run each ETL operator. Figure 4b shows the results obtained. Our approach improves the performance time spent by overloaded ETL operator in an 18%. This is due to one call of the functions related to ETL operators done by the compiler, instead of multiple calls in a case without overload.

Criterion 3: Scalability of the Proposed Solution. The ETL Algorithm populates the target schema of semantic $\mathcal{TDW}$ using an overload of ETL operators.

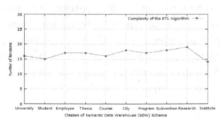

(a) Complexity of the proposed ETL algorithm

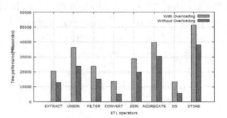

(b) Evaluation time per ETL operator before and after overloading.

Fig. 4. Complexity and evaluation time of the ETL process

---

[6] http://www.cytoscape.org/.

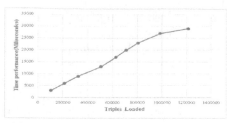
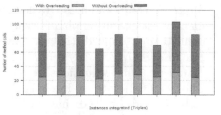

(a) Scalability of the proposed solution.   (b) Number of method calls

Fig. 5. Scalability and number of calls

We measure the time spent to integrate data sources having different sizes. Note that time spent to load all instances is equal 3, 2 min. Figure 5a illustrates the results obtained where for each triple size loaded using overload approach, corresponding time performance is shown in milliseconds. The result remains reasonable w.r.t. the size of the stored instances. This is proof the scalability of our approach.

Criterion 4: Number of Method Calls. We consider a set of SDB participating in the TDW. We run the ETL Algorithm from two different perspectives: first taking in account the overload of ETL process, second without considering it. Figure 5b shows the number of methods calls with and without overloading of ETL operators for each SDB integrated. It clearly demonstrates that number of invocation method without the overload is much higher comparing to the number of method calls using the overload of ETL operators.

7   Conclusion

In this paper, we deal with the variety of data sources and diversity of deployment platforms when constructing a data warehouse. Thanks a Model Driven Engineering techniques, we make generic all elements of the ETL processes. Meta models are proposed for each element. This genericity contributes in overloading all ETL operators in order to reduce their development costs (prototyping) and consequently their performance. Examples of instantiation of three major classes of databases (relational, semantic and graph) are given. Our efforts of genericity facilitates the multi-store deployment. Finally, an evaluation of our proposal to study the effect of overloading on the performance of different operators is also given. A tool available at Youtube is also given. Currently, we are working the scalability of our proposal using considering a large set of dynamic data sources.

References

1. Akkaoui, Z., Mazón, J.-N., Vaisman, A., Zimányi, E.: BPMN-based conceptual modeling of ETL processes. In: Cuzzocrea, A., Dayal, U. (eds.) DaWaK 2012. LNCS, vol. 7448, pp. 1–14. Springer, Heidelberg (2012). doi:10.1007/978-3-642-32584-7_1
2. Baader, F., Calvanese, D., McGuinness, D.L., Nardi, D., Patel-Schneider, P.F. (eds.): The Description Logic Handbook: Theory, Implementation, and Applications. Cambridge University Press, Cambridge (2003)
3. Berkani, N., Bellatreche, L., Khouri, S.: Towards a conceptualization of ETL and physical storage of semantic data warehouses as a service. Cluster Comput. 16(4), 915–931 (2013)
4. Calvanese, D., Lenzerini, M., Nardi, D.: Description logics for conceptual data modeling. In: Chomicki, J., Saake, G. (eds.) Logics for Databases and Information Systems, pp. 229–263. Springer, Boston (1998). doi:10.1007/978-1-4615-5643-5_8
5. Casati, F., Castellanos, M., Dayal, U., Salazar, N.: A generic solution for warehousing business process data. In: VLDB, pp. 1128–1137 (2007)
6. Craig, I.: The Interpretation of Object-Oriented Programming Languages. Springer, London (2002). doi:10.1007/978-1-4471-0199-4
7. Dong, X.L., Srivastava, D.: Big data integration. PVLDB 6(11), 118 (2013)
8. Mazón, J.-N., Trujillo, J.: An MDA approach for the development of data warehouses. In: JISBD, p. 208 (2009)
9. Jean, S., Bellatreche, L., Ordonez, C., Fokou, G., Baron, M.: OntoDBench: interactively benchmarking ontology storage in a database. In: Ng, W., Storey, V.C., Trujillo, J.C. (eds.) ER 2013. LNCS, vol. 8217, pp. 499–503. Springer, Heidelberg (2013). doi:10.1007/978-3-642-41924-9_44
10. Kolev, B., Valduriez, P., Bondiombouy, C., Jiménez-Peris, R., Pau, R., Pereira, J.: CloudMdsQL: querying heterogeneous cloud data stores with a common language. Distrib. Parallel Databases 34(4), 463–503 (2016)
11. Lenzerini, M.: Data integration: a theoretical perspective. In: ACM SIGACT-SIGMOD-SIGART Symposium on Principles of Database Systems, pp. 233–246 (2002)
12. Luján-Mora, S., Vassiliadis, P., Trujillo, J.: Data mapping diagrams for data warehouse design with UML. In: Atzeni, P., Chu, W., Lu, H., Zhou, S., Ling, T.-W. (eds.) ER 2004. LNCS, vol. 3288, pp. 191–204. Springer, Heidelberg (2004). doi:10.1007/978-3-540-30464-7_16
13. Nakuçi, E., Theodorou, V., Jovanovic, P., Abelló, A.: Bijoux: data generator for evaluating ETL process quality. In: ACM DOLAP, pp. 23–32 (2014)
14. Nebot, V., Berlanga, R.: Building data warehouses with semantic web data. Decis. Support Syst. 52(4), 853–868 (2012)
15. Raventós, R., Olivé, A.: An object-oriented operation-based approach to translation between MOF metaschemas. Data Knowl. Eng. 67(3), 444–462 (2008)
16. Rodriguez, M.A., Neubauer, P.: Constructions from dots and lines. CoRR, abs/1006.2361 (2010)
17. Shmueli, O., Tsur, S.: Logical diagnosis of LDL programs. New Gener. Comput. 9(3/4), 277–304 (1991)
18. Simitsis, A., Vassiliadis, P., Sellis, T.-K.: Optimizing ETL processes in data warehouses. In: ICDE, pp. 564–575 (2005)
19. Simitsis, A., Wilkinson, K., Dayal, U., Castellanos, M.: Optimizing ETL workflows for fault-tolerance. In: ICDE, pp. 385–396 (2010)

20. Skoutas, D., Simitsis, A.: Ontology-based conceptual design of ETL processes for both structured and semi-structured data. Int. J. Semant. Web Inf. Syst. **3**(4), 1–24 (2007)

21. Suchanek, F.M., Kasneci, G., Weikum, G.: Yago: a core of semantic knowledge. In: WWW, pp. 697–706 (2007)

22. Trujillo, J., Luján-Mora, S.: A UML based approach for modeling ETL processes in data warehouses. In: Song, I.-Y., Liddle, S.W., Ling, T.-W., Scheuermann, P. (eds.) ER 2003. LNCS, vol. 2813, pp. 307–320. Springer, Heidelberg (2003). doi:10. 1007/978-3-540-39648-2_25

23. Tziovara, P., Vassiliadis, P., Simitsis, A.: Deciding the physical implementation of ETL workflows. In: DOLAP, pp. 49–56 (2007)

24. Vassiliadis, P.: A survey of extract-transform-load technology. IJDWM **5**(3), 1–27 (2009)

25. Vassiliadis, P., Simitsis, A., Georgantas, P., Terrovitis, M., Skiadopoulos, S.: A generic and customizable framework for the design of etl scenarios. Inf. Syst. **30**(7), 492–525 (2005)

26. Vassiliadis, P., Simitsis, A., Skiadopoulos, S.: Conceptual modeling for ETL processes. In: DOLAP, pp. 14–21 (2002)

27. Vassiliadis, P., Simitsis, A., Skiadopoulos, S.: Modeling ETL activities as graphs. In: DMDW, pp. 52–61 (2002)

28. Wilkinson, K., Simitsis, A., Castellanos, M., Dayal, U.: Leveraging business process models for ETL design. In: Parsons, J., Saeki, M., Shoval, P., Woo, C., Wand, Y. (eds.) ER 2010. LNCS, vol. 6412, pp. 15–30. Springer, Heidelberg (2010). doi:10. 1007/978-3-642-16373-9_2

Integrating the R Language Runtime System with a Data Stream Warehouse

Carlos Ordonez[(✉)], Theodore Johnson, Simon Urbanek,
Vladislav Shkapenyuk, and Divesh Srivastava

AT&T Labs - Research, 33 Thomas St, New York, NY 10007, USA
carlos.ordonez@acm.org

Abstract. Computing mathematical functions or machine learning models on data streams is difficult: a popular approach is to use the R language. Unfortunately, R has important limitations: a dynamic runtime system incompatible with a DBMS, limited by available RAM and no data management capabilities. On the other hand, SQL is well established to write queries and manage data, but it is inadequate to perform mathematical computations. With that motivation in mind, we present a system that enables analysis in R on a time window, where the DBMS continuously inserts new records and propagates updates to materialized views. We explain the low-level integration enabling fast data transfer in RAM between the DBMS query process and the R runtime. Our system enables analytic calls in both directions: (1) R calling SQL to evaluate streaming queries; transferring output streaming tables and analyzing them with R operators and functions in the R runtime, (2) SQL calling R, to exploit R mathematical operators and mathematical models, computed in a streaming fashion inside the DBMS. We discuss analytic examples, illustrating analytic calls in both directions. We experimentally show our system achieves streaming speed to transfer data.

1 Introduction

Big data analytics is notoriously difficult when there exist multiple streams and there is a need to perform advanced statistical analysis on them, beyond SQL queries. In our case, we focus on enabling statistical analytics on a data stream warehouse (DSW) [8], where the goal is to analyze multiple streams of networking data (logs). Big data analytics research is active. From a systems perspective, parallel database systems and Apache Hadoop world (HDFS, MapReduce, Spark, and so on) are currently the main competing technologies [13], to efficiently analyze big data, both based on automatic data parallelism on a shared-nothing architecture. It is well established a parallel DBMS [4] is much faster than MapReduce for analytical computations on the same hardware configuration [14]. Spark [16] has become a faster main-memory alternative substituting

---

Research work conducted while the first author was a visiting research scientist with AT&T. Carlos Ordonez current affiliation: University of Houston, USA.

D. Benslimane et al. (Eds.): DEXA 2017, Part II, LNCS 10439, pp. 217–231, 2017.
DOI: 10.1007/978-3-319-64471-4_18

MapReduce, but still trailing parallel DBMSs. Our protoype follows the DBMS approach. Database systems research has introduced specialized systems based on rows, columns, arrays and streams to analyze big data. Moreover, UML has been extended with data mining [17]. However, the integration of mathematical languages, like R, with SQL, remains a challenge [6]. Currently, R is the most popular open-source platform to perform statistical analysis due to its powerful and intuitive functional language, extensive statistical library, and interpreted runtime. Unfortunately, as noted in the literature, even though there exists progress to scale R to large data sets and perform parallel processing, such as IBM Ricardo (R+MapReduce) [3], SciDB-R (R+SciDB) [15], HP Distributed R suite (R+Vertica), R remains inadequate and slow to analyze streams.

1.1   Motivation: SQL and R Languages

From a language perspective, SQL remains the standard query language for database systems, but it is difficult to predict which language will be the standard for statistical analytics: we bet on R. With that motivation in mind, we present STAR (STream Analytics in R), a system to analyze stream data integrating the R and SQL languages. Unlike other R tools, STAR can directly process streaming tables, truly performing "in-database" analytics. STAR enhances R from several angles: it eliminates main memory limitations from R, it can perform data preprocessing with SQL queries, leaving mathematically complex computations as a job for R. STAR enables analytics in both directions closing the loop: (a) R programs can call SQL queries. (b) SQL queries can call R functions. In short, on one hand, users can exploit DBMS functionality (query processing, security, concurrency control and fault tolerance) and on the other hand, call R as needed for mathematical processing.

1.2   Related Work: Other Analytic Systems

The problem of integrating machine learning algorithms with a DBMS has received moderate attention [6,10,12]. In general, such problem is considered difficult due to a relational DBMS architecture [11], the matrix-oriented nature of statistical techniques, lack of access to the DBMS source code, and the comprehensive set of techniques already available in mathematical languages like Matlab, statistical languages like R, and numerical libraries like LAPACK. The importance of pushing statistical and data mining computations into a DBMS is recognized in [1]; this work emphasizes exporting large tables outside the DBMS is a bottleneck and it identifies SQL queries and HDFS/MapReduce (now substituted by Spark) as two complementary mechanisms to analyze large data sets. As a consequence, in modern IT environments, users generally export data sets to a statistical tool or a parallel system, iteratively build several models outside the DBMS, and finally deploy the best model back into the DBMS. Exploratory cube analysis does most of query processing inside the DBMS, but the computation of statistical models (e.g. regression, PCA, decision trees, Bayesian classifiers, neural nets), graph analytics (e.g. shortest path, connectivity, clique detection) and pattern discovery (e.g. itemsets, association rules) is more commonly performed outside a DBMS, despite the fact that DBMSs indeed offer some data mining algorithms and powerful query capabilities. A promising direction bridging DBMS and Big Data technology is to integrate query processing and MapReduce for exploratory cube queries [7].

2   Overview of Interconnected Systems

2.1   Database System for Static Compiled SQL Queries

We first discuss the main features of the data stream warehousing (DSW) system TidalRace [8], contrasting them with features from previous systems. We start with an overview of data stream systems built at AT&T. GigaScope Tool [2] was a first system that could efficiently evaluate a constrained form of SQL on packet-level data streams with recent data as they were flowing in a network interface card. As main limitations, it was not capable of continuously storing streaming data, it could not take advantage of a parallel file system in a cluster of computers and it could not correlate (analytical querying) recent data with historical data. As time went by it became necessary to store summary historical data from streams (orders of magnitude smaller than packet-level data, but still orders of magnitude larger than transactional data) and support standard SQL including arbitrary joins (natural, outer, time band) and all kinds of aggregations (distributive, holistic). Thus the DataDepot DSW [5] was born, with novel storage on top a POSIX parallel file system, supporting standard SQL and incorporating UDFs [9]. More recently, the big data wave brought new requirements and new technology: higher stream volume (more streams with more data), intermittent streams (with traffic spikes), more efficient C++ code for queries, HDFS (instead of a POSIX file system), eventual consistency and advanced analytics beyond SQL queries. These requirements gave birth to TidalRace [8]. TidalRace [8] represents a next-generation data warehousing system specifically engineered for data management of high volume streams, building on long-term experience from the previous systems.

Storage: TidalRace uses a parallel file system (currently HDFS), where time partitions (a small time interval) are the main storage I/O unit for data streams, being stored as large blocks across nodes in the parallel cluster. The storage layout is hybrid: a row store for recent data (to insert stream records and maintain some materialized views), and a column store for large historical tables with recent and old data (to evaluate complex queries). The system provides a DDL with time-oriented extensions. Atomic data types include integers, floats, date/time, POSIX timestamps and strings. Vectors and matrices are supported internally within UDFs in C++ and special SQL access functions. The DBMS supports time-varying schemas, where columns are added or deleted from an existing table over time. This advanced feature is fundamental to keep the system running without interruption concurrently processing insertions, queries and propagating updates to materialized views.

Language: The DBMS provides standard SQL enhanced with time-oriented extensions to query streaming tables. Its SQL offers both distributive aggregations (e.g., sum() and count()) and holistic aggregations (e.g., rank, median, OLAP functions). User-defined functions (UDFs) are available as well: scalar and user-defined aggregates (especially useful for analytics), programmable in the C language. Query processing is based on compiling SQL queries into C code, instead of producing a query plan in an internal representation, which allows most optimizations at compile time. Materialized views, based on SQL queries combining filters, joins and aggregations, are a fundamental feature.

Processing: The database is refreshed by time partition, being capable of managing out-of-order arrival of record batches, intermittent streams and streams with varying speed (e.g. traffic spikes). That is, the system is robust to ingest many diverse streams traveling in a large network. The system uses MVCC (lock-free), which provides read isolation for queries when they are processed concurrently with insertions. The system provides ACID guarantees for base tables (historical tables) and database metadata (schema info, time partition tracking), and eventual consistency for views (derived tables). Therefore, queries, including those used in views, read the most up-to-date version, which is sufficient to compute queries with joins and aggregations on a time window. Query processing is multi-threaded, where threads are spawned at evaluation time by the query executable program. A key feature are materialized views, which are periodically updated when inserting records. Materialized views are computed with SQL queries combining selection filters on time partitions, time band joins and aggregations. We emphasize that every query and view should have a time range, where such time range generally selects the most recent data. The DBMS operates with a minimal time lag between data stream loading and querying (1–5 minutes) and efficiently propagates insertion of new records and removes old records to update materialized views.

In Fig. 1 we show analytic applications at AT&T, where base tables are periodically and asynchronously appended by time partition, ingesting different streams, and derived tables are materialized views of compiled SQL queries. Derived tables are periodically updated with either incremental computation when feasible (i.e. joins, distributive aggregations, approximate histograms) or total recomputation of complex mathematical models during low stream traffic or low usage periods. In general, STAR pulls data from streams and pushes new records to derived tables to compute aggregations, descriptive statistics, histograms and machine learning models using the R language.

2.2   R Dynamic Runtime for Interpreted Functional Programs

Storage: R provides atomic data types and data structures like most programming languages. Atomic data types have a 1-1 correspondence to data types in the C language and include integer, real, string and timestamp, where only strings have different storage from C. For data structures R provides vectors, matrices, data frames, and lists. Vectors represent a collection of elements of

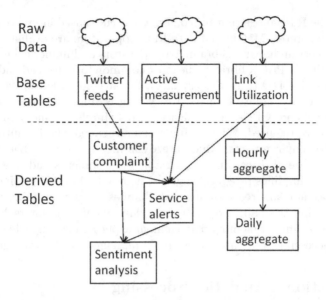

Fig. 1. TidalRace data stream warehousing system end-user applications.

the same atomic data type, especially real and integer. Vectors are stored as one contiguous block dynamically allocated. Matrices are 2-dimensional array of real numbers, also stored as one block in column major order dynamically allocated. Data frames are a list of columns of diverse data types, where each column is a C array dynamically allocated. Lists are the most general and can have elements of diverse data types, including atomic data types and even nested data structures. Therefore, it is feasible to create lists mixing lists, matrices and vectors. Vectors and matrices are easy to manipulate either as a single block or cell by cell. Data frames and lists are the most difficult to manage, easy to manipulate in R itself, but more difficult in the C language. Lists, in their most general form, are the most difficult to manipulate and transfer to the database system.

Language: By its original specification based on S, R is fundamentally functional, but it also incorporates imperative programming statements (i.e. if-then-else and loops). The data type of a variable is totally dynamic because it can change any time with a new value assignment. R also incorporates object-oriented features that enable the creation of new data types, libraries and reusable functions. To manipulate vectors, matrices and data frames R provides simple, yet powerful, subscript operators to access rows, columns and individual cells. Finally, function arguments are named which allows passing argument values in different order and providing default values. Based on our experience, the two most difficult integration aspects are correctly building data structures and setting up all function parameters before calling an R function. From a performance perspective, the most difficult programming aspect is following a functional style over an imperative style (i.e. C++) or declarative style (i.e. SQL), which requires a different algorithmic approach.

Processing: The R language runtime combines a script-based interactive shell and a dynamic interpreter. Matrix and data frame operators are evaluated in the C language and certain linear algebra matrix operators and numerical methods are evaluated by the LAPACK library. The most common form to read and write files is via plain text file I/O (e.g. with csv files), but it is feasible to perform binary I/O with pre-defined binary formats. Processing is single-threaded, which does not exploit multicore CPUs, but which requires simpler memory management since no locks are required. When R functions are called, the R runtime creates nested variable environments, which are dynamically scoped. The R garbage collector takes care of discarding old variable environments and releasing main memory, which simplifies programming, but it may be inefficient. Since the core R runtime has not suffered fundamental changes since it was born, its main memory footprint is small, especially nowadays when computers have ample RAM. The most difficult memory management aspect is tracking when a variable is no longer accessible, but this is reasonably managed by R's garbage collector.

3   Bidirectional Analytic Processing

3.1   STAR Architecture

Our system provides a fully bidirectional programming API: an R program can call (evaluate) any SQL query and its results are seamlessly and efficiently transferred into R. Alternatively, SQL can call any R function via UDFs. Both complementary interfaces are explained below. Our integrated system architecture is shown in Fig. 2.

STAR makes two strong, but reasonable, assumptions to process stream data: (1) the result table from an SQL query with a time range generally fits in RAM. (2) for those result tables from SQL queries that cannot fit in RAM they generally represent materialized views (i.e. pre-processed data streams), which SQL can handle.

The key issues to integrate R with a DBMS are understanding main memory management, layout of vectors and matrices in RAM, building data frames as a set of columns, setting up R function calls, access serialization and properly configuring the operating system environment. Main memory management is significantly different in both systems. R has a garbage collector and the runtime is single threaded. R can address main memory with 64 bits, but integers for subscripts to access data structures are internally 32 bits. On the other hand, the C++ in the DBMS uses a flat 64 bit memory space also with a single thread per compiled query, but no garbage collector. Therefore, each system works as a separate OS process with its own memory space. In addition, since both systems internally have different data structure formats it is necessary to transfer and cast atomic values between them. A fundamental difference with other systems, integrating R and a parallel data system, is that building data structures and transferring them is done only in main memory, copying atomic values as byte sequences in most cases, moving memory blocks from one system to the other and

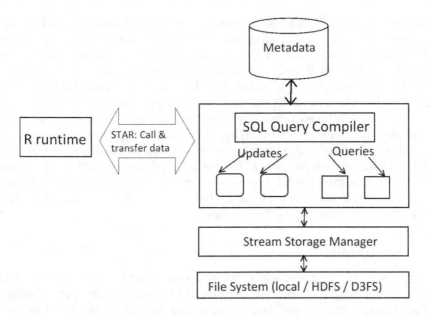

Fig. 2. STAR architecture: R ⇔ DBMS.

avoiding creating files. In other words, we developed a direct binding between R and DBMS runtimes, bypassing slow network communication protocols.

3.2 Mapping Data Types and Assembling Data Structures

STAR exchanges data between R and the DBMS with a careful mapping between atomic values. Data structures like vectors, matrices, data frames and tables are built from atomic values. The atomic mapping is defined as follows: A real in R and the DBMS are both internally a C double in their interpreter and SQL compiler, respectively. Integers are more complicated because in C they are tied to the CPU register size, but R has an older code base; 32 bit integers are directly mapped between both systems, but C++ 64 bit integers are problematic because an integer in R internally still uses a 32 bit integer. Therefore, a 64 bit integer in C++ must be mapped to a real in R (a double in C), which avoids potential overflows. Conversely, when transferring from R to the DSW integers are mapped to 32 bit integers. Strings in the DSW are managed storing their length to avoid scanning for the C null terminator, whereas in R strings are vectors of characters. That is, neither system uses standard C or C++ strings. Therefore, strings require more involved manipulation, but since their length is pre-computed and stored they can be easily moved as byte sequences. Finally, POSIX timestamps are integers on the DBMS side, but real (C double) on the R side, which requires type casting going into either system.

Data structures include vectors, matrices and data frames on the R side and tables (including materialized views) in SQL. To achieve maximum efficiency, transferring is always done moving atomic values between systems as byte sequences: string parsing is avoided. We make sure a data frame only contains atomic values (i.e. complying with 1NF), thereby enabling transferring data into an SQL table. Lists in R violate a database first normal form. Therefore, they cannot be transferred into the DBMS, but they can be pre-processed converting them into a collection of data frames. Transferring in the opposite direction, an SQL table is straightforward to convert into an R data frame since the latter is a more general data structure. Converting an SQL table into an R matrix requires considering a sparse or dense matrix storage and how subscripts are represented in SQL. Finally, vectors/matrices in C++ are a mechanism to efficiently transfer and serialize data from the UDF to vectors/matrices in R (which have different storage and require memory protection), but not to perform statistical analysis. That is, they are transient data structures.

3.3   Data Transfer

We start by considering storage layout and processing in both systems. From the DBMS side we aim to process streams which come in row form by default. On the R side, there are two main data structures: data frames and matrices (vectors being a particular case). Notice that in general, matrices come as output from a machine learning algorithm, transforming the input data set into matrix form if needed. Data frames are organized in RAM by column, which means that a single column is stored in contiguous space. The second consideration is size: we assume tables coming from the DBMS cannot fit in main memory. The row to column conversion and limited main memory leads to block-based algorithms, which transform SQL table rows into column-oriented chunks (blocks) of a data frame. Given the bidirectional transfer, one algorithm transfers data from the DBMS to the R runtime and the second one from the R runtime to the DBMS. In order to handle such bidirectional data transfer efficiently we developed a packed (binary, space-efficient) record binary file format, which allows mixing integers, reals, date/time and variable length strings into variable length records. Data values are transferred as transferred back and forth as long byte sequences. That is, there is no parsing to maximize throughput. This packed record binary file is the basic mechanism to transfer data in both directions. Our fastest data transfer algorithms are programmed in the C language and they are exposed as R functions, to provide intuitive extensibility and interoperability (R), and maximum processing speed (C).

We consider two alternatives to program data transfer between both systems: (1) programmed in the R language; (2) programmed in the C language. In alternative (1) data mapping and building data structures is done entirely with existing R data structures, operators and functions. That is, nothing is programmed in the C language and therefore it is elegant and intuitive for analysts to customize. The caveat if alternative (1) is low speed. On the other hand, in alternative (2) we develop new R functions whose source code is programmed in the C language. That is, these functions directly access R internal data structures, especially data frames (somewhat similar to a table) and matrices (multidimensional array). Such data structures are converted to tabular form, as explained below. In alternative (2) function calls are intutitive to the analyst, but the source code is difficult to extend and customize. We emphasize that both alternatives execute within the R runtime in the R language native data structures, with dynamic memory.

Each alternative has different application scenarios. Alternative (1) (R language) is necessary to export machine learning models, which are composed of vectors, matrices and associated statistical metrics. That is, machine learning models cannot be directly converted to relational tables in SQL. Instead, the DBMS can store them as binary objects that can be accessed calling external C functions (i.e. cannot be called in SQL). Notice machine learning models are small: they take little space in main memory. In addition, the R language can provide a flexible way to transform small data sets exploiting R mathematical operators and functions. On the other hand, our system offers R data transfer functions programmed in the C language (alternative (2)). Evidently, the C language is more efficient, especially for large data sets and streams. However, C data transfer functions are tailored to data frames. Therefore, they cannot compute models with R language statements and therefore such C functions are not adequate to transfer models from R to SQL. In summary, data sets are transferred in both directions between the R runtime and the DBMS but machine learning models are transferred only in one direction (from R to the DBMS).

To round up this section, we provide a brief complexity analysis. For a data set having p attributes and n rows the time complexity of data transfer algorithms is linear $O(pn)$ because each table row (SQL) or data frame row (R) is accessed once and each value is touched once as well. In the specific case that the data set is a matrix with d dimensions (all p attributes are real numbers) and n points time complexity is $O(dn)$. Space complexity is the same, but we emphasize the DBMS and R run in different processes: they do not share memory. Therefore, data transfer is required and space consumption doubles. Finally, models consist of a collection of matrices whose size is generally $O(d)$ or $O(d^2)$. That is, they are much smaller than $O(dn)$.

3.4   Calls in Both Directions

We proceed to explain processing in more technical depth.

R calling SQL (evaluate SQL query)

Since R has a flexible script-based runtime and SQL queries in the DBMS are compiled to an executable C program it is not necessary to develop specialized C code to call an SQL query: the SQL query is simply called with a system command call. Transferring data from the evaluated SQL query to R is achieved via the packed record binary format that is converted to data frame format (column-oriented) and then incrementally transferred to a data frame in RAM (via Unix pipes). This format resembles a big network packet, with a header specifying fields and their sizes, followed by a payload with the sequence of packed records.

Notice that since in the DBMS strings have highly variable length then records also have variable length. Therefore, conversion and transfer record by record is mandatory (instead of block by block), but it is efficiently done in RAM, always moving byte sequences. When the output SQL table does not fit in RAM the data set can be processed in a block-by-block fashion in R; the drawback is that most existing R functions assume the entire data set is used as input and therefore they must be reprogrammed. When the algorithm behind the R function is incremental the respective R function is called on each block. This is a common case for transforming columns of a data set with mathematical functions, computing models with gradient descent or when partial data summaries can be merged at the end. Otherwise, when there is no incremental algorithm multiple models must be compared or merged, which is more difficult to do. A data frame containing only real numbers can then be converted to matrix. That is, it is feasible to call most R functions using a data frame or a matrix as input. Further math processing happens in R and in general R mathematical results (models, a set of matrices, diagnostic statistics) are locally stored in R. However, if the output of an R program is a data frame, preferably with a timestamp attribute, it can be converted to our packed binary format and then loaded back into the DBMS.

We list the main programming steps for the R analyst: (1) execute SQL query (which must have a time range); (2) transfer the SQL result table to one R data frame with a simple R variable assignment; (3) call R function on either: (a) entire data frame (once when result table fits in RAM, common case); (b) with a block-based algorithm (iteratively, less common case). Complex R statistical results cannot imported back into the DBMS, due to R being a more general language and its functional computation model. The main reasons behind this limitation are: models are composed of matrices, vectors and associated metrics, not flat tables like SQL; the need to incorporate time ranges on every result so that they become streams as well. However, a data frame containing atomic values can be be converted to an SQL table, but this scenario makes more sense to be managed by the DBMS, as explained below.

SQL calling R (evaluate R expression or call function)

SQL is neither a flexible nor an efficient language to manipulate data structures in main memory, but it offers UDFs programmable in C++, which can be easily called in a SELECT statement. On the other hand, the most flexible mechanism to call R to perform low-level manipulation of data is to embed R code inside C (or C++) code. Since UDFs are C++ code fragments plugged into the DBMS that isolate the programmer from the internals of physical database operators we use them as the main programming mechanism to call R, bypassing files and communication protocols. Specifically, calling R from the UDF C++ code is achieved by building temporary C++ vectors and then converting the set of C++ vectors into an R matrix. Notice we do not convert SQL records to data frame format in R because we assume the R function to call takes a matrix as input, the most useful case in practice. We should mention that directly moving data from an SQL table to an R data frame in embedded R code is significantly more involved to program, but not faster than matrices. R results can be further processed in C++ inside the DBMS and potentially be imported back into a table. Only R results that are a data frame can be transferred back into some SQL table. In general, there exist materialized views which have a dependence on this temporary table. From a query processing perspective when the R result is a data frame the DSW can treat R functions as table user-defined operators, where the size of the result can be known or bounded in advance.

We summarize the main programming steps in the UDF C++ for the SQL developer: (1) Include R header files; (2) load our R library; (3) setup parameters for R function; (4) build matrix in the UDF aggregation phase (row by row, but in RAM); (5) call R function in UDF final phase, (6) write R function results back into the DBMS either as: vector or matrix (accessible via special SQL functions) or table (accessible via SQL queries) when the result is a data frame.

3.5   Examples

We discuss typical analytic examples on network data. These examples illustrate two different needs: (1) an analyst, with basic SQL knowledge, just wants to extract some relevant data from the DBMS to perform compute some machine learning model. (2) a BI person, with advanced SQL and data cubes knowledge, but basic statistical background, wants to compute some mathematical transformation on the data set that is cumbersome or difficult to do in SQL.

R calling SQL: The analyst writes several SQL queries in a script to build a data set extracted by selecting records with a time window and then aggregating columns to create variables for statistical analysis. Then this data set is ideal to be analyzed by R to get descriptive statistics like the mean or standard deviation for numeric variables and histograms for numeric or discrete variables. After the data set is well understood the analyst can exploit R to compute a predictive model such as linear regression (to predict a numeric variable) or classification (to predict a discrete variable). These tasks boil down to writing an SQL script, starting the R environment, sending the SQL script to the DBMS for evaluation, transferring the final SQL table into a data frame and then analyzing the R data frame as needed. Notice that the output of these calls cannot be sent to the DBMS since they are collection of diverse vectors, matrices, arrays and associated statistics and diagnostic metrics.

SQL calling R: In this case there is an experienced SQL user who needs to call R to exploit some complicated mathematical function. A first example is getting the covariance or correlation matrix of all variables in the data set. Many insights are derived from these matrices. Moreover, these matrices are used as input to multidimensional models like PCA. Assume the user builds the data set with SQL queries as explained above, but the user wants to store the correlation matrix in the DBMS. To accomplish this goal, the user simply needs to create a "wrapper" aggregate UDF that incrementally builds a matrix, row by row. The aggregation phase reads each record and converts it to a vector in RAM.

After the matrix is built R is called in the final phase of the UDF. At the end, the correlation matrix is locally stored in the DBMS to be consumed by a C++ program using our vector/matrix library. A second example is analyzing a stream as a time series and smoothing the time series in order to visualize it or analyze it. In this case the user does want to store the smoothed time series back into the DBMS. In more detail, the user wants to call R to solve the Fast Fourier Transform to find the harmonic decomposition of the time series and identify its period. Once the period is known values are averaged with a moving time window. The net result is a time series that is easier to interpret because it has less noise and a periodic pattern has been identified. Assuming the input table has a timestamp and some numeric value an aggregate UDF builds a data frame in the aggregation phase and then it calls R in the final phase to get an output data frame. This "cleaned" time series can be transferred back into the DBMS as a streaming table.

4 Experimental Evaluation

We did not conduct a detailed benchmark of STAR at AT&T due to two main reasons: (1) STAR can work with any DBMS supporting SQL and materialized views. (2) TidalRace, our current DBMS, works with confidential data whose characteristics cannot be disclosed. Instead, we focus on understanding STAR's ability to analyze high-volume streams with low-end hardware (i.e. under pessimistic conditions).

4.1 Hardware and Software

Our benchmark experiments were conducted on a rack server with a Quad-core CPU running at 2.153 GHz (i.e. 4 cores), 4 GB RAM and 1 TB disk. As explained above, STAR was programmed in the R and C languages, providing an API to transfer data and make function/query calls in both directions.

In order to test correctness of results we performed full bi-directional data transfers with large data sets (tens of attributes, millions of records): (1) exporting an SQL table to an R data frame and then exporting the R data frame back to the DBMS to another SQL table. (2) transferring a data frame to an SQL table and then exporting such SQL table back to R as another data frame. We did not test correctness of complex SQL queries or arbitrary R scripts computing models because we do not alter the results returned by each system. Since these tests are basically a Y/N check mark they are omitted.

Our STAR system works with any DBMS supporting SQL. The time to evaluate SQL queries will vary widely depending on the specific query, DBMS storage (e.g. row or column based), indexing data structures for sliding time windows and size of result table. On the other hand, the time to import data into a DBMS will vary widely depending on parallel processing, storage layout and file format. On the other hand, R functions to compute models (e.g. K-means, linear regression, PCA) take a few seconds working on our data set.

Table 1. Comparing languages: transfer time and throughput (10 mixed type attributes, time in secs).

n	R	C		
	bin	csv	bin	recs/sec
100 k	60	1.1	0.011	9.1 M
1 M	604	9.5	0.096	10.0 M
10 M	na	98.1	0.968	9.7 M

We emphasize our packed record binary file allows processing as efficient as possible. It is understood R does not scale well to large n. Therefore, we focus on measuring time after the SQL query result is ready or before the packed binary file is imported into the DBMS.

4.2 Benchmarking Data Transfer

In general, network data streams have few columns (2-10), resembling normalized tables. Most stream data sets have at least one time attribute (typically a timestamp) and some measurement (count or real number). Additional attributes include geographical location, network connection information (source and/or destination), and device information (e.g. MAC, firmware version). Based on this motivation, we use a data set with 10 attributes, including a timestamp, two variable length strings and seven measurements selected from the KDD network intrusion data set obtained from the KDD Cup web site. Each record is about 60 bytes which is wide enough to trigger heavy I/O. Our goal is to compare speed and measure maximum throughput. Table 1 compares speed to transfer data from the DBMS to R, with data transfer programmed in R and data transfer programmed in tuned C code. We stopped execution at one hour. The R language is more than three orders of magnitude (1000 times) slower than C to process the packed binary file. The R language provides built-in routines to read CSV text files, programmed in the C language. In this case our packed binary file is two orders of magnitude faster (100 times). The last column in Table 1 highlights our system is capable of transferring 10 M records/second between the DBMS and R, surpassing DBMS query processing speed and R mathematical speed in most cases (i.e. our system is not a bottleneck to process a large data set despite the fact it is strictly sequential and it reads/writes to secondary storage). We emphasize that any reasonably complex SQL query (mixing joins and aggregations) or R program working on a large data set with 10 M records is likely to take a few seconds or minutes, even with parallel processing. Exporting a model takes only a fraction of one second (e.g. 0.1, 0.5 s) for data sets with up to hundreds of dimensions after the model is computed (refer to Sect. 3.3 for an explanation on data set and matrix sizes). Therefore, we omit time measurements to evaluate specific SQL queries or to export a machine learning model (e.g. K-means clustering, PCA, linear regression, classification).

5   Conclusions

We presented a "low level connector" system to efficiently transfer data and enable calls between SQL and the R languages in both directions, thereby removing R main memory limitations, allowing SQL to perform mathematical computations, achieving streaming speed and improving interoperability between both systems. Our system defends the idea of combining sequential processing in R with a streaming computation model, where the stream is either: a query result table coming from the DBMS or a transformed data set coming from R imported back into the DBMS. We introduced a packed binary file which allows efficient data transfer in both directions for data sets having fixed (integers, reals, date, timestamp) and variable length columns (strings). We provide functions that allow calling R functions from SQL and calling SQL queries from R. In addition, we provide external functions to transfer machine learning models from R to the DBMS, as objects. Benchmark experiments show data transfer functions programmed in C are orders of magnitude faster than functions programmed in R. However, such C functions can only convert data frames into SQL tables and vice-versa. That is, they cannot convert mathematical models, consisting of matrices, vectors and statistics, into SQL tables. On the other hand, functions programmed in the R language are slow to transfer data sets, but efficient and intuitive to export models from R to the DBMS. Despite its limitations, we believe R will remain as a major alternative to perform statistical and even more general mathematical processing on large data sets and streams. Our prototype is a step in that direction.

Our bidirectional transfer/call approach offers many research opportunities. Scaling R beyond RAM limits and exploiting parallel processing remain important research issues. Specifically, we want to develop incremental machine learning algorithms for large SQL tables that can call R mathematical operators and functions. Parallel processing in R challenging since its architecture is single-threaded, but matrix operators and numerical methods are highly parallel. At a more fundamental level we will keep studying how to transform SQL tables into matrices and vice-versa and how to exploit SQL queries on transformed data frames and matrices.

References

1. Cohen, J., Dolan, B., Dunlap, M., Hellerstein, J., Welton, C.: MAD skills: new analysis practices for big data. In: Proceedings of VLDB Conference, pp. 1481–1492 (2009)
2. Cranor, C., Johnson, T., Spataschek, O., Shkapenyuk, V.: Gigascope: a stream database for network applications. In: Proceedings of ACM SIGMOD (2003)
3. Das, S., Sismanis, Y., Beyer, K.S., Gemulla, R., Haas, P.J., McPherson, J.: RICARDO: integrating R and hadoop. In: Proceedings of ACM SIGMOD Conference, pp. 987–998 (2010)

4. Ghazal, A., Crolotte, A., Bhashyam, R.: Outer join elimination in the teradata RDBMS. In: Galindo, F., Takizawa, M., Traunmüller, R. (eds.) DEXA 2004. LNCS, vol. 3180, pp. 730–740. Springer, Heidelberg (2004). doi:10.1007/978-3-540-30075-5_70

5. Golab, L., Johnson, T., Seidel, J.S., Shkapenyuk, V.: Stream warehousing with DataDepot. In: Proceedings of ACM SIGMOD, pp. 847–854 (2009)

6. Hellerstein, J., Re, C., Schoppmann, F., Wang, D.Z., Fratkin, E., Gorajek, A., Ng, K.S., Welton, C.: The MADlib analytics library or MAD skills, the SQL. Proc. VLDB 5(12), 1700–1711 (2012)

7. Jemal, D., Faiz, R., Boukorca, A., Bellatreche, L.: MapReduce-DBMS: an integration model for big data management and optimization. In: Chen, Q., Hameurlain, A., Toumani, F., Wagner, R., Decker, H. (eds.) DEXA 2015. LNCS, vol. 9262, pp. 430–439. Springer, Cham (2015). doi:10.1007/978-3-319-22852-5_36

8. Johnson, T., Shkapenyuk, V.: Data stream warehousing in Tidalrace. In: CIDR (2015)

9. Ordonez, C.: Building statistical models and scoring with UDFs. In: Proceedings of ACM SIGMOD Conference, pp. 1005–1016. ACM Press, New York (2007)

10. Ordonez, C.: Statistical model computation with UDFs. IEEE Trans. Knowl. Data Eng. (TKDE) 22(12), 1752–1765 (2010)

11. Ordonez, C.: Can we analyze big data inside a DBMS?. In: Proceedings of ACM DOLAP Workshop (2013)

12. Ordonez, C., García-García, J.: Vector and matrix operations programmed with UDFs in a relational DBMS. In: Proceedings of ACM CIKM Conference, pp. 503–512 (2006)

13. Ordonez, C., Song, I.Y.: Relational versus non-relational database systems for data warehousing. In: Proceedings of ACM DOLAP Workshop (2010)

14. Stonebraker, M., Abadi, D., DeWitt, D.J., Madden, S., Paulson, E., Pavlo, A., Rasin, A.: MapReduce and parallel DBMSs: friends or foes? Commun. ACM 53(1), 64–71 (2010)

15. Stonebraker, M., Brown, P., Zhang, D., Becla, J.: SciDB: a database management system for applications with complex analytics. Comput. Sci. Eng. 15(3), 54–62 (2013)

16. Zaharia, M., Chowdhury, M., Franklin, M.J., Shenker, S., Stoica, I.: Spark: cluster computing with working sets. In: HotCloud USENIX Workshop (2010)

17. Zubcoff, J.J., Trujillo, J.: Extending the UML for designing association rule mining models for data warehouses. In: Tjoa, A.M., Trujillo, J. (eds.) DaWaK 2005. LNCS, vol. 3589, pp. 11–21. Springer, Heidelberg (2005). doi:10.1007/11546849_2

Cleaning Out Web Spam by Entropy-Based Cascade Outlier Detection

Sha Wei and Yan Zhu$^{(\boxtimes)}$

School of Information Science and Technology,
Southwest Jiaotong University, Chengdu 610031, China
gxtdws@163.com, yzhu@swjtu.edu.cn

Abstract. Web spam refers to those Web pages where tricks are played to mislead search engines to increase their rank than they really deserved. It causes huge damages on e-commerce and Web users, and threats the Web security. Combating Web spam is an urgent task. In this paper, Web quality and semantic measurements are integrated with the content and link features to construct a more representative characteristic set. A cascade detection mechanism based on entropy-based outlier mining (EOM) algorithm is proposed. The mechanism consists of three stages with different feature groups. The experiments on WEBSPAM-UK2007 show that the quality and semantic features can effectively improve the detection, and the EOM algorithm outperforms many classic classification algorithms under the circumstance of data unbalanced. The cascade detection mechanism can clean out more spam.

Keywords: Web spam cascade detection · Outlier mining · Web quality features · Web semantic features

1 Introduction

Along with the wide usage of Web search engines, Web spammers play tricks to trap search engines into ranking their pages higher than they really deserved, such Web pages are known as Web spam. Web spammers not only gain benefits by cheating Web users and search engines, but also encourage the dissemination of malicious software, malign content and the attack of phishing [1]. A great deal of legal reliable Web applications including e-government and e-business are the victim of Web spam.

Web spam can be divided into four categories: Content-based Spam, Link-based Spam, Cloak/Redirection and Click Spam. Currently, content and link tricks are integrated together for a comprehensive spamming, so researches on ascertaining Web spam by using content and link features are largely conducted [2, 3]. In addition, spam and Web quality have tight relationship [4]. Most of spam pages are of poor quality and have short life cycle, because the spammers want to reap profits quickly and have to dodge various anti-spam techniques, so that they have no intention of organizing and maintaining their pages carefully. Therefore Web spam is viewed as one of Web quality abnormalities.

D. Benslimane et al. (Eds.): DEXA 2017, Part II, LNCS 10439, pp. 232–246, 2017.
DOI: 10.1007/978-3-319-64471-4_19

Another observation is that spam pages show different characteristics in semantics with normal pages. The harmfulness degree of spam pages is mostly demonstrated by terms [5]. Content spam pages tend to be topic-centric, as spammers design their pages relevant to the search queries of some hot topics [6]. Semantic features can be extracted from two sub-dimensions consisting of harmfulness degree and topic characteristics.

In this paper, Web quality and semantic metrics are integrated with content and link features to construct a more representative feature set, which aims at improving the performance of Web spam detection.

Web spam detection is a typical issue of binary classification of data mining and classification algorithms are commonly used [6–9]. However, the detection performance is easily affected by unbalanced data, because spammed pages are very few in contrast with normal pages in the real Web world. Data balancing is a general way for improving the detection quality, e.g., [8, 9], but the results are somewhat not ideal. Another interesting way is outlier mining, which is naturally feasible to discover the minority (outlier) in an unbalanced situation. However, the unsupervised or semi-supervised outlier detection approaches usually are not as powerful as supervised classification approaches.

To deal with the issues above, we propose an Entropy-based Outlier Mining (EOM) algorithm and a cascade detection mechanism to improve the detection performance.

The major contributions of this paper are as follows.

1. EOM algorithm takes full advantages of the outlier mining on unbalanced data. The results of contrast experiments show that EOM outperforms many classification algorithms when data is unbalanced.
2. Quality metrics are extracted in terms of Web sources, content and usage-specific dimensions. Semantic metrics are quantified based on harmfulness degree and topical characteristics. These metrics are integrated with common used features (content, link) together to capture more facts of Web pages and to enhance the discrimination. The experiments show that the new feature set can bring better performance than content and link features only.
3. A cascade detection mechanism is designed based on the EOM algorithm. The detection is divided into three stages: content, link and semantic characteristics detection. The experimental results show the proposed cascade detection framework can improve the performance of detection effectively.

The remainder of the paper is organized as follows. The related works are briefly introduced and compared with our work in Sect. 2. Section 3 addresses EOM algorithm. Section 4 discusses the cascade detection framework and the details of feature quantification. The experiments are introduced in Sect. 5. Finally, the pros and cons of our approach are concluded.

2 Related Works

All researches are investigating two key factors, feature and detection algorithm, for combating spam thoroughly.

As to feature selection, Authors of [2] identified the spam sites and link farm by collecting the link information posted on the SEO forum. By using page content features and link features, Wei et al. achieved good experimental results based on the co-training model [3]. As the spamming has become tricky, new characteristics besides content and link features should be captured. Integrating the Web quality and semantic metrics in our mechanism obviously increases the coverage and representativeness of features. Wang et al. studied the Web content credibility [4], where they considered the spam detection as reliability ranking issue. They introduced Web page information quality as credibility features and designed a machine learning algorithm based on content credibility for detecting Web spam. We also take the Web quality into account in this paper. Different from [3], quality features are extracted from three Web dimensions (information source quality, content quality and usage-specific quality) and are integrated with conventional features together to form a comprehensive feature set.

Authors of [6, 7] proposed topical measures based on semantics analysis with LDA model. The former extracted topical features from whole pages, for example, topic similarity, according to the fact that spam pages tend to be topic-centric. While the latter exploited the information contained within topical variation over sentences and calculated the sentence-level topic assignment and vector to extract features. In our paper, not only topical characteristics are extracted, but a harmfulness degree vocabulary for measuring the perniciousness of page content is built as well. And semantic metrics are quantified based on two sub-dimensions, harmfulness degree and topical characteristics.

As to detection algorithms, decision tree, support vector machine (SVM), etc. are widely used. In addition, ensemble methods can improve the performance of classification algorithms.

However, the performance of classification algorithms is easily affected by unbalanced data. An unbalanced data set can lead to good precision on the majority but very poor precision on the minority. Balancing data before classifying is necessary. Authors of [8] balanced data by using SMOTE method. Authors of [9] adjusted the adaptive function of genetic algorithm to make the method capable to unbalanced data.

Outlier mining algorithms are often used for fraud or anomaly detection, such as credit card fraud, but rarely used for Web spam detection. The unsupervised outlier detection approaches can be commonly classified into statistics-based, density-based, distance-based, and clustering-based. Computing the ranking of nearest-neighbors is used for the detecting outliers, for example, [10] introduced a density-based approach, where the density was computed by the ranks of forward Nearest Neighbor (kNN) and Reverse Nearest Neighbor (RNN). A rank difference based outlier score of an observed data was computed on the basis of its density. This data was detected as outlier if its score was sorted on the top. A semi-supervised outlier mining algorithm is discussed in [11]. The key points of their methods are to rank samples based on their similarity with outliers and to obtain two ranked lists by exchange samples based on minimum entropy. One of lists contains most of outliers.

In this paper, we view spammed Web pages as outliers and propose EOM method for detecting web spam, which is stimulated by the idea of [11].

3   Entropy-based Outlier Mining (EOM)

The Entropy-based Outlier Mining (EOM) approach consists of two steps, Discretizing data, grouping samples to different sets (spam and nonspam). These tasks are both accomplished based on entropy.

3.1   Entropy

Entropy introduced by C.E. Shannon is to define the uncertainty of information. It represents the disorder level of random variables. The higher the entropy is, the more miscellaneous information is.

$x = \{x_1, \ldots, x_k\}$ denotes that a random variable x has k values, $p(x)$ is the probability of x. The entropy $E(x)$ is computed by Eq. (1):

$$E(x) = -\sum_{i=1}^{k} p(x_i) lg\, p(x_i) \tag{1}$$

Given a data set X including n samples with m attributes, which is shown as the matrix below, where v_{ij} denotes the value of ith sample in terms of jth attribute.

$$X = \begin{bmatrix} \vec{v_1} & \cdots & \vec{v_m} \end{bmatrix} = \begin{bmatrix} v_{11} & \cdots & v_{1m} \\ v_{21} & \cdots & v_{2m} \\ \vdots & \ddots & \vdots \\ v_{n1} & \cdots & v_{nm} \end{bmatrix}$$

If m attributes are independent, then $E(\vec{x})$ is as in Eq. (2).

$$\begin{aligned} E(X) &= E(\vec{v_1}) + \ldots + E(\vec{v_m}) \\ &= -[\sum_{i \in \{1,\ldots,n\}} p(v_{i1}) lg\, p(v_{i1}) + \ldots + \sum_{i \in \{1,\ldots,n\}} p(v_{im}) lg\, p(v_{im})] \end{aligned} \tag{2}$$

3.2 Entropy-Based Discretization

The key idea of the EOM approach is all normal Web pages are more or less alike, while spam pages are great different from normal ones. Grouping normal pages and spam pages by exchanging samples based on minimum entropy will result in two sets, the members of one set are mostly normal, while the members of another are spam. The continuous attributes are discretized for the better comparison of similarity. The entropy-based discretization [12] is adopted, which chooses the optimal partition based on minimal entropy and is more effective comparing with unsupervised discretization. In the discretization, the data set is divided into training set and test set. The training set is used to determine the best cut point of each attribute, and the test one is discretized according to the best cut points.

3.3 Sample Exchanging Based on Minimal Entropy

Exchanging Sample based on minimal entropy is to find an outlier set O containing k outliers from test data set D (discretized samples). The objective function of the method is defined as $min\,E(D - O)$, i.e., the entropy of the set D – O is minimal, which demonstrates that the samples are grouped into two sets O and N, N = D – O, and the members of both O and N are as pure as possible. The more similar the samples at the same set are, the more minimal the entropy value is. For test data set D, the exchanging process is as follows.

1. The set O (spam) and N (nonspam, D – O) are initialized by putting k samples (k is far more less than |N|) into O randomly and the rest into N. Then the entropy of N is computed. Because the number of spam samples is much less than that of normal samples as it is in reality, k is then much less than |N|, the majority of the samples in N is normal.
2. A sample s is selected from N randomly. Calculating entropy of N when s is supposed to exchange with one sample from O. The entropy of N is computed one by one when s is exchanged with each different sample of O until all samples of O are checked. Finally, s finds an exchange partner in O, which makes the entropy of N is minimal. The corresponding samples between O and N are really swapped. The value of $E(D - O)$ are renewed with the minimum.
3. Repeat 2 until there is no samples in N unprocessed. If no exchange and no unprocessed samples, the algorithm is terminated.
4. Now, N contains most of similar normal samples, while data in O is dissimilar to N. Therefore, O is the set of spam samples and N is the set of the normal. Besides, k is expected to be close to the real number of spam data to ensure the detected performance.

Algorithm: Exchanging based on Minimum Entropy
Input: k: the number of outliers; D: the test set; |D|: the number of test set; d_i: the ith sample in D.
Output: Sample set O and N
BEGIN
(1) begin for $i \in [1, |D|]$ do
   num_outlier++;
   if *num_outlier* $\leq k$ then
      place $(d_i) := O;$ //The first k samples are put into O.
    else place $(d_i) := N;$ //The rest are put into N.
   end if
  end for
  $Emin := E_0(N);$ //computing the entropy in the initial state.
  flag: = true;
(2) begin while *flag* == true do
   begin for $i \in [1, |N|]$ do   // |N| denotes the number of N
    begin for $j \in [1, |O|]$ do //|O| denotes the number of O
// if the samples are exchanged, then the entropy of the new N is computed.
       if *exchange* (d_i, d_j) then *calculate* $Eij(N)$; end if
      if $Eij(N) < Emin$ then $Emin := Eij(N)$; $J\_min := j$;
      end if
    end for //swapping the two samples with the minimal entropy.
    Based on Emin, $swap(d_i, d_{J\_min})$;
   end for
   if no swapping then *flag: = false;* end if
  end while
END

4   Cascade Detection Mechanism with EOM

4.1   The Detection Framework

A cascade detection mechanism based on EOM is proposed in this section, which contains two parts: feature extraction and EOM cascade detection (see Fig. 1.)

In Fig. 1, four feature sets are prepared in the feature extraction module. Detection is executed in three cascaded stages, content feature-based, link feature-based, and semantic feature-based. Web content and usage quality integrated with content features are used in the first stage, while Web source quality with link features are in the second stage, semantic features are in the third stage, because such a feature division helps the

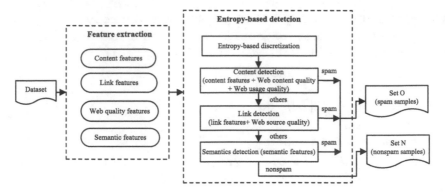

Fig. 1. The framework of spam cascade detection

EOM cascade algorithm to filter different spam pages more accurately. The spam and nonspam pages sets (O and N) determined by three cascade stages are finally output.

4.2   Feature Extraction

4.2.1   Content and Link Features

The content and link features used in our work are provided by dataset WEBSPAM-UK2007. For example, number of words in the page, number of words in the title, PageRank, TrustRank, etc.

4.2.2   Web Quality Features

Most of spam pages are of poor quality. For example, normal pages usually use more feasible expressions (e.g. html for text) while spam pages may use more pictures representing text for avoiding detection. We have crawled all valid homepages (HTML) of WEBSPAM-UK2007 from Internet and analyzed their quality in terms of

Table 1. Some examples of Web quality features

Quality feature	Category
TLDR	Web source quality
TDR	
WATR	
Anti-TrustRank	
Ratio of nouns and of verbs in each sentence	Web content quality
Average length of each sentence	
Ratio of sentences number and the text length of the whole page	
Ratio of strange punctuation	
Number of text/html in the page	Web usage quality
Number of text/java script in the page	
Number of image/jpeg in the page	
Number of application/zip in the page	

several dimensions. 21 quality features are automatically extracted and quantized, which are based on WebQM model [13] proposed by one of authors of this paper. The WebQM model maps the Web quality features to three dimensions, Web source quality, Web content quality and Web information usage quality, which feasibility and effectiveness has been validated. Some of our quality features are given in Table 1, where TLDR, TDR, WART and Anti-TrustRank are 4 measurements of distrust ranking and are calculated using methods proposed in [14–16]

4.2.3   Web Semantic Features

Recently, some methods of Web information security have been discussed. Pages are divided into many categories, each category has a computed reputation and a risk description in [17], where different security features, e.g., URL, domain behavior, etc., are automatically extracted to determine a reputation score. In [18], the information security incidents are graded into four degrees based on three factors, i.e., importance of information system, system damage, and social influence. Lee et al. built an artificial neural network classifier to combat Web pornography content [5]. They published a list of 55 indicative terms including 42 sexually explicit terms and 13 legal terms. The occurrence frequencies of the indicative terms in one Web page are as features. Their classifier was proved effective on filtering pornographic Web pages.

The harmfulness term blacklists are also discussed by Google, WordPress, Forti-guard, etc., because the degree of harmful content are mostly embodied on terms. Following this line of thinking, a Web content category with harmfulness degrees (shown in Table 2) is constructed and a harming term vocabulary is built, where each term belongs to one of categories (some examples in Table 3).

Table 2. Lexical category with harmfulness degree

Extremely harmful	Harmful	Slightly harmful
Pornography	Gambling	Ads
Child Abuse	Plagiarism	Brokerage and Trading
Weapons	Discrimination	
Drug Abuse	Hacking	
Racism		
Violence		
Extremist Groups		

Table 3. Harmfulness term vocabulary (partly)

Pornography	Child abuse	Gambling	Trading	Ads
Escorts	Child pornography	Betting	For sale	Insurance
Sex	Kid pornography	Casino	Credit card	Lose weight
Adult	Abuse	Craps	Payment	Viagra
Swingers	Strike	Keno	Rental	Supplements
Arse	Child molestation	Poker	Online shop	Advertise

Table 4. Semantic features of Web content

	Semantic features
Harmfulness degree features	Number of extremely harmfulness words
	Number of harmfulness words
	Number of slightly harmfulness words
Topic features	The maximum of topic distribution
	The minimum of topic distribution
	The variance of topic distribution

In addition, studies show content spam pages tend to be topic-centric, as spammers design their pages for accurate response to the hot search queries with similar topics [6]. In contrast, the topic distribution of normal pages is relatively wide-ranging and balanced. Therefore, we used LDA model to get the topic distribution and then extracted three topical features as shown in Table 4. LDA model construction is based on [14]. The topic distribution of each document d is denoted as $T(d) = \{t_1, \ldots, t_m\}$, t_i is a term in d. The formula of variance is shown in Eq. (3).

$$TopicVar(d) = \frac{\sum_{i=1}^{m}(t_i - u)^2}{m} \tag{3}$$

Where $u = \sum_{i=1}^{m} t_i/m = 1/m$.

4.3   Cascade Detection Based on EOM

As mentioned in Sect. 4.1 and shown in Fig. 1, the spam page (outlier) detection consists of three stages by using different features combinations, and O_1, O_2, O_3 are outliers found by three stages, separately. In the third stage a multi-category detection is applied, where pages are recognized as one of four types from extremely harmfulness to innocent. This detection can discover those spammed pages slipping through the first two nets. Three outlier sets are merged to the final set O while normal pages are in N.

For example, there are 7 test samples as in Fig. 2, where C, L and S denote respectively the features used in the content detection, link detection and semantic detection.

```
ID    label     C     L     S     harmful degree
1     spam      C1    L1    S1    high
2     nonspam   C2    L2    S1    no
3     spam      C3    L3    S3    mid
4     spam      C4    L4    S4    low
5     nonspam   C5    L5    S5    no
6     nonspam   C6    L6    S6    no
7     nonspam   C7    L7    S7    no
```

Fig. 2. An example of test sample feature vectors

According to the mechanism in Fig. 1, feature vectors as (ID, label, C) of 7 samples are input into the first stage, i.e., content detection. If the sample of ID = 1 is detected, the spam set O_1 contains $(1, spam, C1, L1, S1, high)$. The rest samples are filtered similarly in the 2nd and then in the 3rd stage with link and semantic features. By merging 3 spam sample sub-sets, $O(1; \ldots 3; \ldots; 4; \ldots;)$ and non-spam sample set $N(2, \ldots; 5, \ldots; 6, \ldots; 7, \ldots;)$ are obtained.

5   Evaluation of the Proposed Approach and Mechanism

5.1   Dataset and Evaluation Measures

The original dataset of our experiment is WEBSPAM-UK2007 which contains 105,896,555 pages from 114529 hosts in the UK domain. 321 hosts among them were labeled as spam and 5476 hosts of them as non-spam, the rest are unknown.

In this paper, spam pages are viewed as positive samples, while normal pages as negative. Precision, recall, F1-measure, accuracy and AUC are used for evaluation. Because AUC is computed based on a rank of pages, entropy-based ranking method is introduced for AUC. The samples in O and N are ranked based on entropy denoted in Eq. (4). The method can be understood as finding similarity between a test sample d and all samples of one class (spam or non-spam). The smaller the entropy is, the bigger the similarity.

O is detected by using the same approach at Sect. 3. The entropy between d_i of O and positive samples in training set is computed as $E_+(d_i)$. The entropy between d_i of O and negatives (nonspam) samples in training set is calculated as $E_-(d_i)$. d_i is more similar with spam samples when $E_+(d_i)$ is smaller or $E_-(d_i)$ is bigger. All samples from O are ranked to build two sequences, one descending based on $E_-(d_i)$, another ascending based on $E_+(d_i)$. The position of d_i in O' is determined based on the average of the serial numbers of two sequences and the samples of O' is ascending. Samples in N are ranked at the same way. Finally, sequence of O' is put in front of N' to produce the final sequence of all test samples.

$$E(d_i) = -\sum_{j=1}^{|p|} p(p_j|d_i) lg\, p(p_j|d_i) \tag{4}$$

$$p(p_j|d_i) = \frac{distance(d_i, p_j)}{\sum_{j=1}^{|p|} distance(d_i, p_j)} \tag{5}$$

Where, d_i is a sample in test set. p_j denotes a sample from one of the classes (spam or nonspam) in training set, $|p|$ is the number of samples from the same class. $distance(d_i, p_j)$ is the similarity between d_i and p_j, which is measured by Euclidean distance.

A sample d with m features $a_1, a_2, \ldots, a_m$. The Eq. (6) is used to normalize features in the experiments.

$$a_{i\_new} = \frac{a_i - min(a_i)}{max(a_i) - min(a_i) + 1} \qquad i = \{1, \ldots, m\} \qquad (6)$$

Sample distribution is shown in Table 5. Three comparison experiments (shown in Table 6) are designed to evaluate the performance of our EOM cascade mechanism.

Table 5. Sample distribution used in the EOM algorithm experiments

	Spam (positive samples)	Nonspam (negative samples)	Usage
Training set	152	2724	Discretization
Validation set	56	917	Determing the sample numbers of O_i
Test set	113	1835	Detection based on EOM

5.2 Experiment Results and Analysis

5.2.1 Experiment 1

The sample number of O is determined by k in EOM algorithm (ref. Sect. 3.3). The optimal k is obtained by carrying out the detection using the validation set (ref. Table 5), which is built with 56 spam samples and 917 nonspam samples. The optimal k_1 is found when obtaining the highest F1-measure in the 1st stage, so do k_2 and k_3 in the 2nd and 3rd stages.

The 3-stage cascade detection is implemented. If the real number of spam samples at the first stage is 56, the optimal k_1 with the highest F1-measure is 60. 23 spam data is detected with this k_1. Then rest 33 spam samples are input into the 2nd stage, the optimal k_2 will be determined as 50. Then the rest 27 spam samples are input into 3rd stage, the optimal k_3 is 30. The performance is good when k is close to the real spam value, e.g. k_1 is 60 while the number of spam samples is 56. Because the number of positive or negative samples in validation set to those in test set is about 1:2, multiplying k_1, k_2 and k_3 by 2 can obtain k_i for the regular detection based on the test set, they are 120,100 and 60.

Such a method for finding optimal k can be applied in real applications. Supposing a test set contains 1 million samples, we can choose a small part of the data as the validation set, for example, 1000 samples. The distribution of positive and negative samples in the validation set can be viewed similar to the actual distribution of the test set, when the sampling is done enough times. Labeling the 1000 samples and conducting the spam detection for determining k as mentioned above, the optimal k values for 3-stage detection of 1 million samples are fixed just based on the small validation samples and the proportion (1:1000) of two sets (Table 7).

AUC of this paper outperforms the AUC value 0.848 of the winning team at Web Spam Challenge [19] and 0.854 obtained by [20]. Therefore, our EOM cascade detection mechanism can improve the performance of Web spam detection.

Table 6. Introduction of 3 Experiments

	Experiment objectives	Data set
Experiment 1	(1) Comparison between the EOM cascade detection with the methods of the winning team at [19] and that of [20] (2) Parameter selection	Samples from WEBSPAM-UK2007 are shown in Table 5
Experiment 2	The detection performance evaluation based on the new feature set integrating with Web quality features and semantic features	
Experiment 3	The performance comparison between EOM algorithm and three classic classifiers	

Table 7. Results of EOM cascade detection

	Precision	Recall	F1_measure	AUC	Accuracy
Spam is positive	0.304	0.752	0.433	0.867	0.886
Nonspam is positive	0.983	0.894	0.937	0.867	0.886

In order to eliminate the randomness, five-fold cross validation (three training sets, one validation set, and one test set) are conducted. Our EOM cascade detection is proved to be well-performed and the results are shown in Table 8.

Table 8. Five-fold cross validation results

	Precision	Recall	F1_measure	AUC	Accuracy
Spam is positive	0.27	0.595	0.37	0.81	0.886
Nonspam is positive	0.974	0.903	0.937	0.81	0.886

5.2.2 Experiment 2

The effectiveness of quality features and semantic features is proved in this experiment. In addition, the contrast experiment between one stage detection and the cascade detection is carried out.

Integrating either quality or semantic features can improve the performance of spam detection as shown in Table 9. Cascade detection outperforms one-stage detection even if the same features are used. This is because cascade detection can better play to the role of all kinds of features. 3-fold filtering can clean out much more spam and becomes

Table 9. Results of experiment 2(spam is positive)

Method	Feature	AUC
EOM one-stage detection (K = 120)	Content + link	0.762
	Content + link + quality	0.776
	Content + link + quality + semantics	0.78
EOM cascade detection (K1 + K2 + K3 = 280)	Content + link + quality + semantics	0.867

more accurate by narrowing the detection range step by step. Integrating all features together at one stage detection may result in that each feature cannot realize its full effectiveness.

5.2.3   Experiment 3

The Experiment 3 is designed to compare the EOM algorithm with SVM, C4.5 and Random Forest for dealing with the unbalance issue of a few spammed pages and huge normal pages. We just use the content-based features and the link-based features provided by WEBSPAM-UK2007 to perform one-stage detection, in order to conduct the reasonable comparison on a common platform.

Due to the performance of SVM are greatly influenced by the tradeoff factor C and kernel function, we utilize validation set for parameter selection. The tradeoff factor C is chosen from {0.001, 0.01, 0.1, 1, 5, 10, 15, 20, 25, 30} so as to achieve the highest AUC against the validation set. Similarly, the kernel function is chosen from four types, that is linear, polynomial, radial basis function and sigmoid. With parameter optimization, the performance of SVM is better than the original one.

As shown in Table 10, the AUC of the EOM algorithm proposed in this paper is the highest when the data set is unbalanced. EOM algorithm is a kind of outlier mining methods which are appropriate for unbalanced data distribution naturally, while the classic classifiers (e.g., SVM and C4.5 in this experiment) are easily affected when data set is extremely unbalanced. The AUC of Random Forest is the second highest one, because Random Forest is an ensemble classification method by integrating the advantages of multiple sub-classifiers and can correct some errors caused by unbalanced data set.

Table 10. Comparison of EOM algorithm and classifiers (Spam is positive)

Algorithm	Balance method	AUC
SVM (C = 1, linear function)	No balancing	0.534
C4.5		0.588
Random forest		0.742
EOM one-stage detection (Our method K = 120)		**0.762**
SVM (C = 25, linear function)	K-Means under-sampling	0.665
C4.5		0.634
Random forest		0.577
SVM (C = 5, linear function)	SMOTE oversampling	0.722
C4.5		0.635
Random forest		0.779

Some methods adopt under-sampling and oversampling before classification for performance improvement. Our EOM approach outperforms SVM and C4.5 no matter which balance method they have used, and can reach the similar effectiveness of Random Forest with SMOTE balancing method. The performance of classifiers is easily affected by the data balancing method as in Table 10, while the EOM algorithm doesn't need to balance two types of data and still performs well. The EOM algorithm is more stable in the case of unbalance data distribution.

6 Conclusions

A Web spam cascade detection mechanism is introduced in the paper. The main novelties include (1) integrating quality features from the dimensions of Web source, content and information usage-specific; (2) extracting semantic features to quantify Web page semantics; (3) using EOM (a kind of outlier mining) algorithm instead of classification algorithm to deal with the performance degradation caused by data imbalance; (4) designing a cascade detection mechanism based on the EOM algorithm to discover Web spam more correctly.

Most of spam pages are of poor quality and have short life cycle, because the spammers want to avoid various spam filtering techniques and are not able to organize and maintain their pages carefully. Besides, spam pages have harmful content more or less and unordinary topical characteristics. Therefore, it is proven in this paper that integrating discriminative features, such as quality and semantic metrics, can significantly improve the performance of spam detection.

The EOM approach can resist data imbalance and save the cost for balancing. The cascade detection mechanism based on EOM algorithm outperforms one-stage detection, but it needs more time and space in the exchange procedure for the higher performance.

In the future, our work can be further improved in the following ways.

1. The semantic features only refer to the content harmfulness. Integration of link harmfulness characteristics (e.g. the target of an outgoing link, anchor text of links) could be helpful for discovering more spam.
2. More effective syntax and semantics analysis of Web content should be done to improve feature selection. For example, to detect the spelling and grammatical errors in those automatically generated spam pages by using linguistic models.
3. The performance of EOM algorithm is easily affected by the parameter K. Even though we have introduced a method for parameter selection, more feasible methods may be needed.
4. Discretization plays an important role in EOM algorithm and it directly affects the performance of minimum entropy exchanging. More effectively discretized methods should be studied in the future.
5. The performance of the proposed approach in a practical environment will be studied.

Acknowledgements. This work was supported by the Academic and Technological Leadership Training Foundation of Sichuan Province, China [WZ0100112371601/004, WZ0100112371408, YH1500411031402].

References

1. Spirin, N., Han, J.: Survey on web spam detection: principles and algorithms. ACM **13**(2), 50–64 (2012)

2. Cheng, Z., Gao, B., Sun, C., Jiang, Y., Liu, T.: Let web spammers expose themselves. In: Proceedings of the 4th ACM International Conference on Web Search and Data Mining, New York, pp. 525–534 (2011)
3. Wei, X., Li, C., Chen, H.: Content and link based web spam detection with co-training. J. Frontiers Comput. Sci. Technol. **4**, 899–908 (2010)
4. Wang, W., Zeng, G., Tang, D.: Using evidence based content trust model for spam detection. Expert Syst. Appl. **37**(8), 5599–5606 (2010)
5. Lee, P.Y., Hui, S.C., Fong, A.C.M.: Neural Networks for Web Content Filter. IEEE Intell. Syst. **17**, 48–57 (2002)
6. Dong, C., Zhou, B.: Effectively detecting content spam on the web using topical diversity measures. In: IEEE/WIC/ACM International Conference on Web Intelligence and Intelligent Agent Technology, pp. 1115–1123 (2012)
7. Suhara, Y., Toda, H., Nishioka, S., Susaki, S.: Automatically generated spam detection based on sentence-level topic information. In: Proceedings of the 22nd International Conference on World Wide Web Companion, pp. 1157–1160 (2013)
8. Fang, X., Tan, Y., Zheng, X., Zhuang, H., Zhou, S.: Imbalanced web spam classification using self-labeled techniques and multi-classifier models. In: Proceedings of International Conference on Knowledge Science, Engineering and Management, pp. 663–668 (2015)
9. Bhowan, U., Johnston, M., Zhang, M.: Developing new fitness functions in genetic programming for classification with unbalanced data. IEEE Trans. Syst. Man Cybern. **42**, 406–421 (2012)
10. Bhattacharya, G., Ghosh, K., Chowdhury, A.S.: Outlier detection using neighborhood rank difference. Pattern Recogn. Lett. **60**(C), 24–31 (2015)
11. Daneshpazhouh, A., Sami, A.: Entropy-based outlier detection using semi-supervised approach with few positive examples. Pattern Recogn. Lett. **49**, 77–84 (2014)
12. Fayyad, U.M., Irani, K.B.: Multi-interval discretization of continuous-valued attributes for classification learning. In: Proceedings of the 13th International Joint Conference on Artificial Intelligence, pp. 1022–1027 (2010)
13. Zhao, B., Zhu, Y.: Formalizing and validating the Web quality model for Web source quality evaluation. Expert Syst. Appl. **41**, 3306–3312 (2014)
14. Wei, S., Zhu, Y.: Combining topic similarity with link weight for Web spam ranking detection. J. Comput. Appl. **36**(3), 735–739 (2016). (in Chinese)
15. Goh, K.L., Patchmuthu, R.K., Singh, A.K.: Link-based web spam detection using weight properties. J. Intell. Inf. Syst. **43**(1), 129–145 (2014)
16. Krishnan, V., Raj, R.: Web spam detection with anti-trust rank. In: Proceedings of the Second International Workshop on Adversarial Information Retrieval on the Web, Seattle, Washington, USA, pp. 37–40 (2006)
17. McAfee Inc. TrustSource Web Database Reference Guide (category set 4). https://support.mcafee.com/ServicePortal/faces/knowledgecenter. Accessed 29 Nov 2016
18. Standardization Administration of the People's Republic of China (SAC). Information security technology—Guidelines for the category and classification of information security incidents. GB/Z 20986-2007 (2013)
19. Web Spam Challenge: Results. http://webspam.lip6.fr/wiki/pmwiki.php?n=Main.PhaseIII. Accessed 29 Nov 2016 (2008)
20. Bíró, I., Siklósi, D., Szabó, J, Benczúr, A.: Linked latent Dirichlet allocation in web spam filtering. In: Proceedings of the 5th International Workshop on Adversarial Information Retrieval on the Web, Madrid, Spain, pp. 37–40 (2009)

Logical Schema for Data Warehouse on Column-Oriented NoSQL Databases

Mohamed Boussahoua[✉], Omar Boussaid, and Fadila Bentayeb

ERIC EA 3083, Universite Lumiere Lyon 2,
5 avenue Pierre Mendes-France, 69676 Bron Cedex, France
{mohamed.boussahoua,omar.boussaid,fadila.bentayeb}@univ-lyon2.fr

Abstract. The column-oriented NoSQL systems propose a flexible and highly denormalized data schema that facilitates data warehouse scalability. However, the implementation process of data warehouses with NoSQL databases is a challenging task as it involves a distributed data management policy on multi-nodes clusters. Indeed, in column-oriented NoSQL systems, the query performances can be improved by a careful data grouping. In this paper, we present a method that uses clustering techniques, in particular *k-means*, to model the better form of column families, from existing fact and dimensional tables. To validate our method, we adopt TPC-DS data benchmark. We have conducted several experiments to examine the benefits of clustering techniques for the creation of column families in a column-oriented NoSQL HBase database on Hadoop platform. Our experiments suggest that defining a good data grouping on HBase database during the implementation of a data warehouse increases significantly the performance of the decisional queries.

Keywords: Data warehouses · NoSQL databases · Columns family

1 Introduction

Data warehouses play an important role in collecting and archiving large amounts of data for decision support. Usually, they are implemented as traditional relational databases management systems (RDBMS). These systems have their weaknesses, they are not suitable for distributed databases *(scalability problems, join operation problems, mass data storage and access problems)* as argued in [1, 2]. Then, it is necessary to use new reliable storage solutions with lower cost. Among these solutions, there is the Hadoop platform[1], which comprises different modules such as Apache Hive[2], Apache Pig ...etc, and new data models named NoSQL *(Not Only SQL)*[3] are used, supported by the major platforms Web such as *Google, Yahoo, Facebook, Twitter* and *Amazon*. The NoSQL databases, rely on the CAP theorem (*Consistency, Availability* and *Partition Tolerence*) by Brewer [3].

---

[1] http://hadoop.apache.org/.
[2] https://hive.apache.org/.
[3] http://nosql-database.org/.

© Springer International Publishing AG 2017
D. Benslimane et al. (Eds.): DEXA 2017, Part II, LNCS 10439, pp. 247–256, 2017.
DOI: 10.1007/978-3-319-64471-4_20

They offer great flexibility of data representation and allows management of large amounts of data storage in distributed servers. NoSQL databases can be classified into at lest 4 categories that correspond to different modeling types: *Key/Value Store, Column Store, Document Store* and *Graph Store*. Their degree of performance strongly depends on the suitability to a use case. Among the major types of NoSQL systems, we chose to implement the data on a column oriented databases. These provide variable-width tables that can be partitioned vertically and horizontally across multiple nodes [4]. Other advantage of this databases their ability to store any data structures, high performance on aggregation queries and highly efficient data compression, which offer a more appropriate model for data warehouse storage.

In this paper, we address the storage and implementation process of data warehouses with column-oriented NoSQL database. In fact, a good way to design the column families in the most popular NoSQL databases (HBase, Cassandra, Hypertable...) is more challenging compared to that in design principle relational database. Effectively, there are no rules of normalization for column-oriented NoSQL databases, these databases break the rules of normalization by denormalizing. So, to take advantage of this denormalized databases, the question is how to organize data in column families to serve effectively specific queries, in our particular case the OLAP queries, where the read-load is more than the write-load. In this case, we propose the application of clustering techniques *k-means* to determine which attributes (frequently used by on set of queries) should be grouped together. In order, to obtain a better design of column families, these column families form a data model for the data warehouse in column-oriented NoSQL Databases. Several tests are made to evaluate the effectiveness of the proposed method. We adopt TPC-DS data benchmark[4]. For designing the columnar NoSQL data warehouse (*CN-DW*) for TPC-DS benchmarking database, we used 3 different methods, first by our method, and then by 2 other methods that have been already tested and implemented successfully in [5,6]. We proceed to perform a query workload on different schemas upon *CN-DW* built over TPC-DS. It has been found that, the application of clustering techniques for designing a data model of *CN-DW*, effectively improve the queries execution time, compared to the two other methods. The remainder of the paper is organized as follows. Section 2 provides a state of the art on columnar NoSQL data warehouses. Section 3 describes the problem we address in this paper. Section 4, presents the proposed approach. Section 5 evaluates our approach. Conclusion and future works are given in Sect. 6.

2   Related Work

Using column oriented NoSQL databases for data warehouse solutions has been debated within the scientific community. Several works have treat the problem of modeling and implementing the data warehouse according to these models. These works can be classified into two main categories:

---

[4] Benchmark (TPC-DS) v2.0.0, http://www.tpc.org/tpcds/.

1. Denormalized approaches: the aim of these works is to propose a storage schema that combines the fact and dimension tables into the same big table, using different ways to create the column families. In this context, in [7], the authors convert a relational database schema to column oriented data-schema based on HBase DBMS. The relational database schema is converted into a large table with several column families, one for each relational table. In [8] the authors propose a method to construct an OLAP cube from a data warehouse implemented in big table on HBase and one family of columns has been defined for each column. To make direct access to the data blocks, the authors required to add the index upon HBase table. In [5], the authors propose two methods. The first one, consist of storing the fact and dimension tables into one table with a single column family for all attributes. The second method stores the fact and dimension tables into one table, the fact table is mapped into one column family, and the attributes belonging to the same dimension table are gathered in one column family. In [6], the authors propose a set of rules to convert a multidimensional conceptual model into two different NoSQL models (column-oriented or document-oriented). Then, in the context of the column oriented, they propose two column oriented models to implement a data warehouse in HBase. The first model, for each fact all related dimension attributes and all measures are combined in one table with one column family. In the second model, the fact and dimension tables are combined in one table, but one column family for each dimension table and one column family dedicated for the fact table. In [9], the authors tackled the problem of distributing attributes between column families. They implemented the data warehouse in HBase table with multiple column families (CFs), one (CF) groups some of the more frequently used dimensions to the fact table. The others (CFs), one (CF) for the fact's attributes and one (CF) for each less frequently used dimension table.

2. Normalized approaches: In [5,6], the authors propose to split data into multiple tables, where the fact table is stored into one table with one column family, each dimension table is stored into one table with one column family. This implementation reduces the redundancy of the dimensions data, but requires the using of a special join between the fact table and dimension tables.

3   Problem Statement

The implementation of a data warehouse with column-oriented NoSQL databases, which takes into account the specific characteristics of the column-oriented storage environment. Thus, all data are stored in the same table consisting of one or more column families. Where each column family is composed of a set of attributes, that implies a principle of vertical partitioning data. The Fig. 1 shows an example of data storage and data distribution of a table T, with 2 column families. Each column family CF_i consists of a set of attributes, each attribute having a value, each row of data is referenced by a *Row-key* (R_i).

In reality, all data referring to the same *Row-key* (R_i) are stored together. The column family name is acting as a key to each of its columns, the *Row-key* as key of all attributes. It may be noted in this storage technique, that

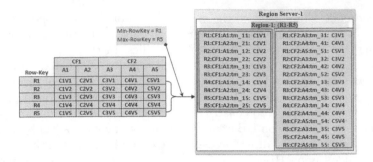

Fig. 1. Data storage using a column oriented model - HBase storage

two elements influence on the execution speed of queries: the choice of columns and the choice of rows. To make it simple, let consider a query that involves the attributes values *(A3, A4, A5)*. To respond to this query, the search will be done on data partitions $CF2$ only. Comparing this query with another query, which access the attributes values *(A1, A4)*, the column families table is scanned two times (the search will be done on two different data partitions) $CF1$ and $CF2$ with a complete scan of the values of the concerned attribute, which penalizes the execution time of this query.

Overall, we found that, until now, the implementation of data warehouses based on the columnar NoSQL systems usually utilizes the denormalization approaches. On the other hand, the first works don't take in consideration the clustering techniques to create the column families that contain the required data for processing a query or multiple queries. Also, these methods could not control the number of column families targeted by the complex queries. This number could be very large and would make the data warehouse management very complex, and therefore, limits the advantage of clustering of attributes into column families. For these reasons, to enhance the load balancing between column families, the implementation of the data warehouses on column-oriented NoSQL databases requires an appropriate design technique. This is what we propose in our approach that we detail in the following section.

4 The Proposed Approach

Our strategy, to modeling the form of Columnar NoSQL Data Warehouse (CN-DW), is subdivided into four steps that are detailed in the following sub-sections. Its general principle is summarized in (Fig. 2).

1. extracting the set of attributes of the fact table and dimension tables
2. grouping of attributes and constructing column families by *k-means*
3. generating a logical schema of the *CN-DW*.
4. preparing and loading the data into *CN-DW*.

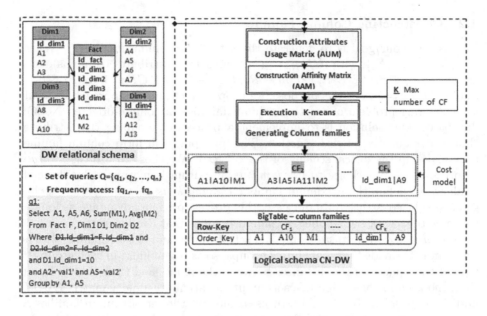

Fig. 2. *k-means* method to design columnar NoSQL data warehouse

4.1   Extracting the Set of Attributes of the Fact and Dimension Tables

Let $T = \{t_1, t_2, ..., t_m\}$ be the set of the warehouse tables (fact and dimension tables), where each table $(t_j)_{j=1,...,m}$ is composed of set of attributes. The workload which consists of a set of most frequent queries $Q = \{q_1, q_2, .., q_n\}$ that access the set of attributes R, where $R = \{\forall a \in R, \exists t_j \in T : a \in t_j\}$.

In this first step, we treat all attributes found in workload's queries, we taken into account all the attributes present in each query, in particular those that appear in the *Select* and *Where* clauses, except for the attributes of the join predicates. Our objective is to construct the Attribute Usage Matrix (AUM) and the Attribute Affinity Matrix (AAM). To do this, we were inspired by the works of Navathe [10]. They use the principle of affinity between attributes to design the vertical fragments. The matrix (AAM) denotes which query uses which attribute $use(q_i, a_j) = 1$ if q_i uses a_j, and to 0 otherwise. The Attribute Affinity Matrix (AAM) tells how closely related attributes are, this matrix which is a n x n symmetric matrix, where n is the number of different attributes in R. Each element (aa_{ij}) in (AAM) equals the sum of the access frequency of the queries simultaneously involves two attributes a_i and a_j, this measure is called affinity, is defined in [11] as follows: $Aff(a_i, a_j) = \sum_{h|use(q_h,a_i)=1 \wedge use(q_h,a_j)=1} \sum_{\forall S_l} ref_l(q_h)Acc_l(q_h)$,

where $ref_l(q_h)$ is the number of accesses to attributes (a_i, a_j) for each execution of query q_h at site s_l, and $acc_l(q_h)$ is the access frequency. In our case, for reasons of simplification and experimentation, we consider $(ref_l(q_h) = 1$ for all q_h and the number of sites $(l = 1)$).

4.2   Constructing Column Families

In this step, our goal is to generate the set of the column families S. Where $S = \{CF_1, ..., CF_k\}$, with $(CF_i)_{i=1,..,k}$ are subsets of attributes (columns), such as: (1) $\forall CF_i \in S, CF_i \subset R$; (2) $\forall a \in R, \exists CF_i \in S : a \in CF_i$; (3) $\forall CF_i, CF_h \in S, CF_i \cap CF_h = \varnothing$. For this purpose, we propose a clustering approach to tackle a challenging problem of creating column families in column-oriented NoSQL databases. Our solution is to implement a process of grouping the attributes that are frequently queried together, This grouping will form column families that make up the logical schema of the $CN\text{-}DW$. We chose to use the $k\text{-}means$ algorithm [12], our choice to use $k\text{-}means$ algorithm is motivated by the fact, it allows to define the maximum number *(having k as an input parameter)* of groups to be constructed. In our case, this proves to be an advantage as long as we want to control the number of column families that can be created.

$k\text{-}means$ algorithm takes as input a set of points and cluster number k. The problem is to divide the points into k groups, so as to minimize in each group the sum of the squares of the distances between the points and their center. In our case, the $k\text{-}means$ algorithm takes as input the attribute affinity matrix (AAM) and the cluster number k, and returns as output, the set of column families S. To measure the quality of the obtained schema S, we adapt a cost model based on the works of Derrar et al. [13]. Initially, this model is computed using the *Square Error (E^2)*, taking account of the access frequency (f_q) of queries. The (E^2) of the grouping attributes schema is calculated as follow:

$$E_S^2 = \sum_{i=1}^{k} \sum_{l=1}^{n} [(f_{q_l})^2 \times \alpha_i^{q_l} (1 - \frac{\alpha_i^{q_l}}{\beta_i})] \tag{1}$$

$(\alpha_i^{q_l})$ is the number of attributes in $(CF_i)_{i=1,..,k} \in S$ accessed by the query q_l, (β_i) is the number of attributes in column family CF_i. Also, more the E_S^2 value approaches zero (0), more optimum is this grouping schema.

5   Implementation, Experiments and Results

To validate our $k\text{-}means$ method for designing the column families, we developed a software tool named $(RDW2CNoSQL: Relational Data warehouse to Columnar NoSQL)$ with Java programming language.

1. Dataset: To evaluate our approach, we used the TPC-DS benchmark[5]. The TPC-DS uses a constellation schema which consists of 17 dimension tables and 7 fact tables. In our case, we used the STORE_SALES fact table and its 9 dimension tables (CUSTOMER, CUSTOMER_DEMOGRAPHICS, CUSTOMER_ADDRESS, ITEM, TIME, DATE, HOUSEHOLD_DEMOGRAPHICS, PROMOTION, STORE). The DSD-GEN data generator of TPC-DS allows to generate data files in a *(file.data)* format with different sizes according to a *Scale Factor (SF)*. We set SF to 100 which produces in STORE_SALES fact table (28,799,7024 tuples).

---

[5] Benchmark (TPC-DS) v2.0.0, http://www.tpc.org/tpcds/.

2. Query workload: The TPC-DS benchmark offers 99 queries. We selected 19 separate queries that access 67 attributes, which exploit the entire schema of the STORE_SALES fact table and its dimension tables, using the operations *(selection, join, aggregate, projection)*. These queries compute the OLAP cubes with a gradually increasing number of dimensions. The degree of this dimensionality is divided into 3 levels: *(small: SD)*, *(medium: MD)* and *(large: LD)*, according to: (1) The number of tables used by a query; (2) The number of attributes and predicates for each query. It should be noted that our objective is to use TPC-DS benchmark to evaluate the performance of our technique when forming column families. Due to, some requirements are not feasible with Apache Phoenix[6] (on query read capabilities) and HBase databases, these queries would require some modifications (syntax changes). Table 1 describe the used queries.

Table 1. Queries characteristics

	(SD)			(MD)									(LD)						
	q1	q2	q3	q4	q5	q6	q7	q8	q9	q10	q11	q12	q13	q14	q15	q16	q17	q18	q19
Tables	1	2		3							4		5						6
Attributes	4	9	9	4	4	6	6	6	6	9	6	7	8	10	10	11	11	12	14
Predicates	4	3	10	4	5	12	7	7	6	6	5	7	21	13	10	8	23	15	15

3. Experimental configuration: To achieve our evaluation goals, we setup two storage environments. The first one is relational non-distributed with intel-core machine TMi7-4790S CPU@3.20 GHZ with 8 GB of RAM, and a 500 GB disk. It runs under the 64-bit Ubuntu-14.04 LTS operating system, which is used as a PostgreSQL server dedicated to the storage of the relational data warehouse. The second is a distributed NoSQL storage environment. It is a cluster of computers consisting of 1 master server *(NameNode)* and 3 slave machines *(Data Nodes)*. The *(NameNode)* has an Intel-Core TMi5-3550 processor CPU@3.30 GHZx4 with 16 GB RAM, and a 1TB SATA drive. Each of the *(DataNodes)* has an Intel-Core TMi5-3550 processor CPU@3.30 GHZx4 with 16 GB RAM and 500 GB of disk space. These machines run on 64bit Ubuntu-14.04 LTS and Java JDK 8. We used Hadoop (v2.6.0), MapReduce for processing, HBase (v0.98.8), ZooKeeper for track the status of distributed data in the *Region-Servers (DataNodes)*, Phoenix (v4.6.0) and SQuirreL SQL Client to simplify data manipulation and increase the performance of the HBase. In order to efficiently transfer PostgreSQL data to HBase, we integrated all the features of Sqoop[7] into our *(RDW2CNoSQL)* tool.

4. Tests and results: We choose three different methods, 2 already existing approaches in addition to our method, for implementing the data warehouses *CN-DW* in HBase DBMS: (1) in the first one, the STORE_SALES fact table and its 9 dimension tables are stored into one HBase table with only one column family

---

[6] https://phoenix.apache.org/.
[7] https://sqoop.apache.org.

for all attributes (called *Flat schema*); (2) in the second, the STORE_SALES fact table and its 9 dimension tables are stored into one HBase table with 10 column families, one for each table (called *Naïve schema*); (3) the last one consists of *CN-DW*, built according to our method with in addition three different schemas, i.e. all data are stored in one HBase table, we varied the number of the column families ($k = 4, k = 11, k = 13$) corresponding to (*schema $k = 4$*, with a square error value $E^2_{k=4} = 0.437$), (*schema $k = 11$*, with $E^2_{k=11} = 0.197$) and (*schema $k = 13$*, with $E^2_{k=13} = 0.213$). We executed all queries presented in Table 1, on the five configurations described above. Note that, in this experiment, we do not want to make a performance comparison between the relational DBMS and NoSQL databases. Our primary focus is to seek the main elements that have an impact on query execution time in a Columnar NoSQL Data Warehouse (CN-DW). The obtained results are presented in the Figures from 3, 4, 5 and 6.

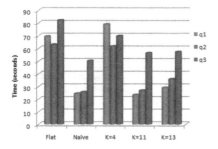

Fig. 3. SD Queries execution time

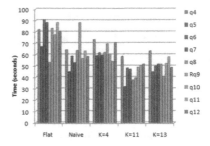

Fig. 4. MD Queries execution time

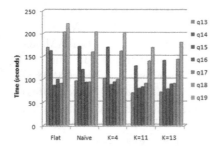

Fig. 5. LD Queries execution time

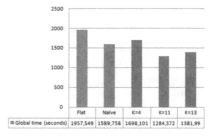

Fig. 6. Global queries response time

5. Discussion: We discus our results in this sub-section.

(a) Impact of the data size on query execution time: As shown in Fig. 3, the queries execution time increases significantly, when using the (*Flat schema*) or the (*schema $k = 4$*). These schemas records poor results compared with other schemas *(Naïve, schema k = 11, schema k = 13)*. In the *(schema k = 4)*, these results are due to the fact that poor quality of this schema (the attributes in column families aren't well grouped, having a greater value of the $E^2_{k=4} = 0.437$), the performance can degrade in situations where evil choice of the number of column families k. But, in the *(Flat schema)*, these results are due to the pressure on the memory caused by the large amounts of data coming from the same column family *(Flat* method: all attributes of the fact and dimension tables are combined in one column family). Indeed, In HBase, the data from a single column family are stored in set of *HFiles*. It is important to note that the number of *HFiles* depends to the size of the data of a column family. For read data in *(Flat schema)*, HBase will automatically solicit and loaded into memory a large number of *HFiles*, this offer the possibility of performing multiple processing at the memory. As such, it is clear that this process results in a significant increase in the cost of input/output, which in turn results in increased execution time and decreased system performance. On the other hand, we observe a slight variation between the queries execution times, run on the schemas *(Naïve, schema k = 11, schema k = 13)*. In the *(Naïve schema)* the queries frequent 1 to 2 column families, in the *(schema k = 11)* and *(schema k = 13)* the queries use 2 to 3 column families. To respond to $q1$, $q2$ and $q3$ in these 3 schemas, HBase exploits column families having small data sizes. This, allows it to considerably reduces the number of *HFiles* in the memory (fewer *HFiles* of data are scanned during the query execution), which enables efficient query execution.

(b) Impact of the number of column families on query execution time: The objective of this experiment is to examine the scalability of the *k-means* method, when faced with variations in the number of dimensions. To do this, we executed 16 queries $q4$ to $q19$ presented in Table 1. These queries compute the OLAP cubes with a gradually increasing number of dimensions. In Figs. 4 and 5, we observed that the proposed method gives better performance for all queries, whatever the number of dimensions involved, when using the *(schema k = 11, schema k = 13)*, in these schemas the queries can be recall (3 to 4 column families). On the other hand, from Fig. 5 it can be seen that *(Naïve schema)* and the *(Flat schema)* are equivalent, in terms of responses times for some complex queries, this was foreseeable. Indeed, To respond of these queries in the *(Naïve schema)*. HBase system solicits a very large number of column families (5 to 6 column families), which generates a high cost of combinations and reconstruction of the intermediate results. Recall that the *Naïve* approach constructs the column families according to the principle where each dimension of the relational model must be transformed into a column family.

Finally, by analyzing these preliminary results, we observe the query runtime is dependent on the way to model the form of column families. Figure 6 shows the *k-means* method in the *(schema k = 11)* has lowered global queries execution time, up to 19.20 % and 34.38 % compared to *Naïve* approach and *Flat* approach, respectively. In general, to improve query run time on the data warehouse implemented on HBase database, It's readily apparent that: (1) limit the number of columns in families; (2) define the good number of column families, it is not advisable to create too many column families.

6   Conclusion

The major contribution of this work is the transformation of relational data warehouse into a columnar NoSQL data warehouse (*CN-DW*). We have first proposed a method that uses clustering techniques in particular, *k-means*, to create column families, according to our specific necessities, from existing fact and dimension tables. These column families form an effective logical model. This latter is easily converted into an appropriate *CN-DW*. Experiments are carried out using the TPC-DS benchmark, several tests are made to evaluate the effectiveness of the our approach. The results we obtain confirm the benefits of clustering techniques for the creation of column families to increase the performance of the decisional queries. In future work, it we plan to consider all changes and configuration parameters related to the data warehouse environment.

References

1. Padhy, R.P., Patra, M.R., Satapathy, S.C.: RDBMS to NoSQL reviewing some next-generation non-relational database's. IJSEAT **11**(1), 15–30 (2011)
2. Song, J., Chaopeng, G., Wang, Z., et al.: HaoLap: a Hadoop based OLAP system for big data. J. Syst. Softw. **102**, 167–181 (2015)
3. Brewer, E.A.: Towards robust distributed systems. In: Proceedings of the Nineteenth Annual ACM Symposium on Principles of Distributed Computing, PODC 2000, p. 7. ACM, New York (2000)
4. Stonebraker, M., Cattell, R.: 10 rules for scalable performance in simple operation' datastores. Commun. ACM **54**(6), 72–80 (2011)
5. Dehdouh, K., Boussaid, O., et al.: Using the column oriented NoSQL model for implementing big data warehouses. In: International Conference on PDPTA, pp. 469–475 (2015)
6. Chevalier, M., Malki, M.E., Kopliku, A., Teste, O., Tournier, R.: Implementation of multidimensional databases in column-oriented NoSQL systems. In: Morzy, T., Valduriez, P., Bellatreche, L. (eds.) ADBIS 2015. LNCS, vol. 9282, pp. 79–91. Springer, Cham (2015). doi:10.1007/978-3-319-23135-8_6
7. Li, C.: Transforming relational database into HBase: a case study. In: IEEE International Conference on Software Engineering and Service Sciences, pp. 683–687 (2010)
8. Abell, A., Ferrarons, J., Romero, O.: Building cubes with MapReduce. In: Proceedings of the ACM 14th International Workshop on Data Warehousing and OLAP, pp. 17–24 (2011)
9. Scabora, L.C., Brito, J.J., et al.: Physical data warehouse design on NoSQL databases OLAP query processing over HBase. In: ICEIS, pp. 111–118 (2016)
10. Navathe, S., Ceri, S., et al.: Vertical partitioning algorithms for database design. ACM Trans. Database Syst. **9**, 680–710 (1984)
11. Tamerözsu, M., Valduriez, P.: Principles of Distributed Database Systems. Springer Science & Business Media, New York (2011)
12. MacQueen, J., et al.: Some methods for classification and analysis of multivariate observations. In: Proceedings of the Fifth Berkeley Symposium on Mathematical Statistics and Probability, Oakland, CA, USA, vol. 1, pp. 281–297 (1967)
13. Derrar, H., Boussaid, O., Ahmed-Nacer, M.: An objective function for evaluation of fragmentation schema in data warehouse. In: Encyclopedia of Information Science and Technology, 3rd edn. pp. 1949–1957. IGI Global (2015)

Data Mining and Machine Learning

PALM: A Parallel Mining Algorithm for Extracting Maximal Frequent Conceptual Links from Social Networks

Erick Stattner$^{(\boxtimes)}$, Reynald Eugenie, and Martine Collard

LAMIA Laboratory, University of the French West Indies, Pointe-á-Pitre, France
{estattne,reugenie,mcollard}@univ-ag.fr

Abstract. Numerous methods have been proposed in order to perform clustering from social networks. While significant works have been carried out on the design of new approaches, able to search for various kinds of clusters, a major challenge concerns the scalability of these approaches. Indeed, given the mass of data that can now be collected from online social networks, particularly from social platforms, it is important to have efficient methods for exploring and analyzing these very large amount of data. One of the recent social network clustering approaches is the extraction of conceptual links, a new approach that performs link clustering by exploiting both the structure of the network and attributes of nodes to identify strong links between groups of nodes in which nodes share common attributes. In this paper, we focus on the optimization of the search for conceptual links. In particular, we propose PALM, a parallel algorithm that aims to improve the efficiency of the extraction by simultaneously exploring several areas of the search space. For this purpose, we begin by demonstrating that the solution space forms a concept lattice. Then, we propose an approach that explores in parallel the branches of the lattice while reducing the search space based on various properties of conceptual links. We demonstrate the efficiency of the algorithm by comparing the performances with the original extraction approach. The results obtained show a significant gain on the computation time.

1 Introduction

The study of social networks has become one of the most active research areas of the 21st century, which is found in the literature as the "*new science of networks*" [1]. We can explain this growing interest in network analysis by two factors. Firstly more and more data are now available, especially through online social platforms which allows to collect data on users as well as the content they exchange and requires efficient methods for exploring and analyzing these very large amount of data. Secondly the network formalism allows to address various kinds of real world phenomena such as diffusion problems [2], questions of influence [3], purchasing behaviours [4] or the prediction of various events [5].

One of the very active lines of research of network science concerns the knowledge extraction from networks, known as "*social network mining*" or more simply

© Springer International Publishing AG 2017
D. Benslimane et al. (Eds.): DEXA 2017, Part II, LNCS 10439, pp. 259–274, 2017.
DOI: 10.1007/978-3-319-64471-4_21

"*link mining*" [6,7]. In particular, the identification of clusters is the most common task of social network mining [8]. Indeed in many systems, whether natural or social, the entities involved often tend to organize into groups. Identifying these groups proves to be an important challenge to understand the structures emerging from the interactions between entities, to study the mechanisms that take place and to determine the role of agents within these kinds of systems. Thus many clustering methods can be found in the literature. If traditional clustering methods dedicated to networks only focused on the network structure for extracting clusters called *communities* [9,10] (groups of nodes densely connected), more recent approaches have focused on the extraction of new kinds of clusters, more complex, defined by the network structure and the attributes of nodes [11]. Thus a major challenge concerns the scalability of these new approaches.

In this paper, we focus on the optimization of a recent network clustering approach called *conceptual links*, that performs clusters of links by exploiting both the network structure and node attributes to identify strong links between groups of nodes in which nodes share common attributes. This work is motivated by the observation that the original extraction algorithm [12] performs a sequential search of these clusters, without taking into account possible parallelisms. Thus in this paper we present PALM, a parallel algorithm that aims to improve the efficiency of the extraction process of conceptual links by simultaneously exploring several areas of the search space. For this purpose, we begin by demonstrating that the solution space forms a concept lattice. Then, we propose an approach that explores in parallel the branches of the lattice while reducing the search space based on various properties of conceptual links. The efficiency of the algorithm is demonstrated by comparing the performances with the original extraction algorithm on a telecommunication network. The results obtained show a significant gain on the runtime.

The paper is organized as follows. Section 2 reviews the main descriptive approaches proposed to extract clusters from social networks. Section 3 formally describes conceptual links and shows how the solution space forms a concept lattice. Section 4 details the parallel algorithm we propose. Section 5 is devoted to experiments conducted on a telecommunication network. Finally, Sect. 6 concludes and presents our future directions.

2   Related Works

In recent years, numerous methods have been proposed to extract knowledge from social networks. As for the classical data mining area, these methods can be classified according to two main families. (1) Approaches that rely on *predictive modelling*, that cover all methods focusing on historical and current data in order to formulate hypotheses about future or unknown events [13,14]. (2) Approaches based on *descriptive modeling*, that cover the set of methods that aim to summarize data by identifying some relevant hidden patterns to describe how the components of a system are organized, work and interact [15,16]. A recent survey on descriptive modeling appraoches on social networks can be found in [8].

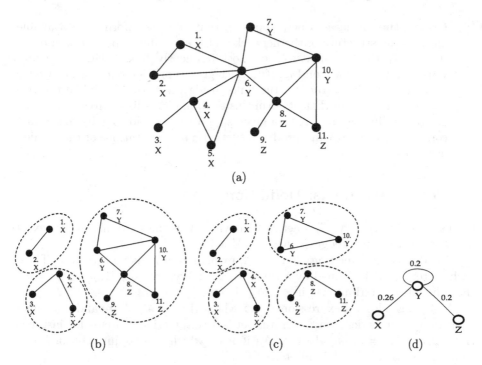

Fig. 1. Examples of (b) communities, (c) hybrid clusters and (d) conceptual links extracted from the reference network (a)

The search for clusters in a social network, also called *network clustering*, is one of the best known descriptive modeling approaches. The objective is to analyze the network in order to identify groups that satisfy some specific criteria. Main network clustering methods can be classified as follows.

(i) **Link-based clustering** refers to a family of methods that search for node partitions (also called "*communities*") by only taking into account the structure of the network [9,10] as shown on Fig. 1(b). The objective is to decompose the network into several *communities*, namely groups of nodes with a high density of connections. The algorithms attempt to identify node groups that maximize the intra-community links while minimizing inter-community ones [17]. Two kinds of methods are commonly distinguished, aggregative methods that merge nodes iteratively, and separative methods which start with a single group containing all nodes and divide it iteratively.

(ii) **Hybrid clustering** is a network clustering approach that aims to take into account the attributes of the nodes during the extraction phase of communities [11,18,19] as shown on Fig. 1(c). Indeed, in numerous applications the classical definition of *a community* does not allow to fully understand the studied structures. Thus, hybrid clustering techniques attempt to identify densely connected groups of nodes, which have in addition a high homogeneity in their attributes.

(iii) **Conceptual links** is a new approach that exploits information available on both the structure of the network and the attributes of nodes, in order to identify the set of attributes most frequently linked within the network [12, 20] as shown on Fig. 1(d). Such groups provide relevant knowledge on attributes that structure the links inside the network. Unlike the two previous approaches that perform clusters of nodes, this approach identifies clusters of links according to a given support threshold. It relies on formal concept analysis [15] and produces synthetic representations of the studied networks.

3   Conceptual Links: Definition

Let us introduce $G = (V, E)$ a network, in which V is the set of nodes (vertexes) and E the set of links (edges) with $E \subseteq V \times V$.

V is defined as a relation $R(A_1, ..., A_p)$ where each A_i is an attribute. Thus, each vertex $v \in V$ is defined by a tuple $(a_1, ..., a_p)$ where $\forall q \in [1..p], v[A_q] = a_q$, the value of the attribute A_q in v and $|R| = p$.

An item is a logical expression $A = x$ where A is an attribute and x a value. The empty item is denoted $\emptyset$. An itemset is a conjunction of items for instance $A_1 = x$ and $A_2 = y$ and $A_3 = z$. An itemset which is a conjunction of k non empty items is called a k-itemsets.

Let m and sm be two itemsets. If $sm \subseteq m$, we say that sm is a sub-itemset of m and m is a super-itemset of sm. For instance $sm = xy$ is a sub-itemset of $m = xyz$.

Any itemset is a sub-itemset of itself.

We denote I_V the set of all itemsets built from V.

Let us consider G as a *unipartite directed graph*. Thus, for any itemset m in I_V, we denote V_m the set of nodes in V that satisfy m and we define:

- the *m-left-hand linkset* LE_m as the set of links in E that start from nodes satisfying m i.e
  $LE_m = \{e \in E \; ; \; e = (a, b) \quad a \in V_m\}$
- the *m-right-hand linkset* RE_m as the set of links in E that arrive to nodes in V_m i.e
  $RE_m = \{e \in E \; ; \; e = (a, b) \quad b \in V_m\}$

Definition 1 (Conceptual link). For any two elements m_1 and m_2 in I_V, the *conceptual link* (m_1, m_2) of G is the set of links connecting nodes in V_{m_1} to nodes in V_{m_2} (as shown on Fig. 2).

Obviously, conceptual links are defined on various kinds of networks: oriented, non-oriented, unipartite or bipartite networks.

For instance, if m_1 is the itemset cd and m_2 is the itemset efj, the *conceptual link* $(m_1, m_2) = (cd, efj)$ includes all links in E between nodes in V that satisfy the property cd with nodes in V that satisfy the property efj.

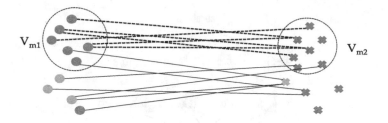

Fig. 2. Example of conceptual link extracted between m_1 and m_2

Let L_V be the set of conceptual links of $G = (V, E)$ and (m_1, m_2) be any element in L_V.

$$(m_1, m_2) = LE_{m_1} \cap RE_{m_2}$$
$$= \{e \in E \ ; \ e = (a, b) \quad a \in V_{m_1} \ and \ b \in V_{m_2}\}$$

Definition 2 (Support of conceptual link). We call *support* of any element $l = (m_1, m_2)$ in L_V, the proportion of links in E that belong to l.

$$supp(l) = \frac{|(m_1, m_2)|}{|E|}$$

For an itemset m and a conceptual link l, if $l = (\emptyset, m)$ or $l = (m, \emptyset)$ then $supp(l) = 0$.

Definition 3 (Frequent Conceptual Link). Given a real number $\beta \in [0..1]$, a conceptual link l in L_V is *frequent* if its support is greater than a minimum link support threshold β,

$$supp(l) > \beta$$

Let FL_V be the set of frequent conceptual links (FCL) in $G = (V, E)$ according to a given link support threshold β.

$$FL_V = \bigcup_{m_1 \in I_V, m_2 \in I_V} \{ (m_1, m_2) \in L_V \ ; \ \frac{|(m_1, m_2)|}{|E|} > \beta \}$$

Definition 4 (Conceptual sub-link). Let two any itemsets sm_1 and sm_2 be respectively sub-itemsets of m_1 and m_2 in I_V. The conceptual link (sm_1, sm_2) is called conceptual *sub-link* of (m_1, m_2).

Similarly, (m_1, m_2) is called conceptual *super-link* of (sm_1, sm_2).

We write $(sm_1, sm_2) \subseteq (m_1, m_2)$

Property 1 (Downward-closure property). If a conceptual link l is frequent then all its sub-links are frequent. Thus if a link is infrequent then all its super-links are infrequent.

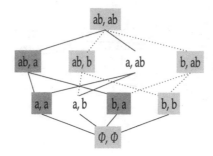

Fig. 3. Example of concept lattice formed by conceptual links with conceptual link (ab;b) frequent and conceptual link (b;b) infrequent (irrelevant links are pruned)

Proof. Let sm_1 and sm_2 be respectively sub-itemsets of m_1 and m_2. The properties $V_{m_1} \subseteq V_{sm_1}$ and $V_{m_2} \subseteq V_{sm_2}$ hold.

Therefore $(m_1, m_2) \subseteq (sm_1, sm_2)$.

Thus $|(m_1, m_2)| \leq |(sm_1, sm_2)|$.

Definition 5 (Maximal frequent conceptual link). Let β be a given link support threshold, we call *maximal frequent conceptual link* (MFCL), any frequent conceptual link l such as, there exists no super-link l' of l that is also frequent.

More formally, $\nexists l' \in FL_V$ such as $l \subset l'$.

Thus, the relation $\subseteq$ that defines a partial order on I_V can be extended to L_V. Thus as shown on Fig. 3 $(L_V, \subseteq)$ induces a concept lattice that can be used to extract *maximal frequent conceptual links*. On example of Fig. 3 when a MFCL like (ab, a) is found, all its sub-links may be pruned since they are not maximal. Similarly, when an infrequent conceptual link like (b, b) is found, all its super-links may be pruned since they cannot be frequent.

Maximal frequent conceptual links provide a conceptual view of the social network about groups of nodes that share common *internal* properties (or concepts) and are the most connected into the network [20]. This knowledge can be summarized into a graph structure in which each node is associated to an itemset, and each link corresponds to a MFCL.

4   PALM Algorithm

In this section, we describe the parallelized algorithm PALM that we propose. In Sect. 4.1 we begin by detailing MFCLMin, the original algorithm proposed to extract maximal frequent conceptual links. Then, Sect. 4.2 introduces the improvements introduced in PALM to reduce the search space of solutions and to parallelize its exploration.

4.1 Original Algorithm MFCLMin

The original algorithm can be divided in 4 parts, with the last two parts that are iteratives:

(i) **Generation of all 1-itemsets candidate**: Browse all nodes of the graph and then build the list of the left hand 1-itemsets LI_1 and the right hand 1-itemsets RI_1 by extracting all attributes from the network nodes.

(ii) **Generation of the 1-frequent conceptual links**: Evaluate all conceptual links (m_1, m_2) whose left-hand m_1 is in LI_1 and right-hand m_2 is in RI_1. When the support value is greater than the support threshold β, the conceptual link (m_1, m_2) is stored in $FL_{V_{max}}$.

(iii) **Generation of the t-itemsets candidate at iteration $t > 1$**: Compare all elements of LI_{t-1} and generate t-itemsets candidate by joining those that share $(t - 2)$ attributes. New itemsets are added in a list LI_{cand}. In the same way, this task is done for RI_{t-1} and a list RI_{cand} is generated.

(iv) **Generation of the t-frequent conceptual links at iteration $t > 1$**: Check all conceptual links l composed of a t-itemset and a k-itemset, with $k \leq t$. If the conceptual link l has a support value higher than the support threshold β and has non super itemsets in FLV_{max}, all sub-links of l are removed from FLV_{max} and l is added to FLV_{max}.

Tasks (iii) and (iv) are repeated until no new candidate is generated. While this algorithm allows to extract maximal frequent conceptual links from social networks, the search process is conducted in a sequential way. Moreover it does not optimize the exploration of the search space, which makes it difficult to extract conceptual links on large datasets.

Thus, in order to optimize the search for these clusters of links, we propose the algorithm PALM that introduces two kinds of improvements. First, it reduces the number of computation made by limiting the exploration of the search space, and secondly it parallelizes some parts of the extraction process to explore simultaneously several areas of the search space.

4.2 Improvements Provided in PALM

The first improvements that have been introduced in PALM concerns the candidate generation process. In part (iii) of MFCLMin, in order to generate the t-itemsets at iteration t, it was trying to join all m_1, m_2 as $(m_1, m_2) \in LI_{t-1}^2$ or $(m_1, m_2) \in RI_{t-1}^2$. Only those that share $(t - 2)$ items could be join. However, 2 itemsets share k items if and only if they have at least one common subitemset. So in PALM, we save the structure of the concept lattice and reduce the time spent for the creation of new t-itemsets at each iteration (see lines 22 and 23 of Algorithm 1) by exploring the $(t - 2)$ itemsets and trying to join their own superitemsets that are respectively in LI_{t-1} and RI_{t-1}.

In part (iv) of MFCLMin, the conceptual links that are tested are made from all possible pairs with an itemset from LI_{cand} and an itemset of RI_{cand}. Then each conceptual link will be compared with all edges of the network to evaluate its frequency. This part is the most time consuming part. Thus in order to improve this treatment two enhancements have been applied. Indeed, using the property 1 and the structure of the concept lattice saved, we reduce the number of conceptual links that have to be tested at each iteration. As the superlinks of frequent conceptual links from the previous step could be frequent as well, we reduce the links that have to be checked at each step in the Algorithm 4 by selecting only super conceptual links (see line 2). Furthermore, by keeping edges that match each conceptual links, we can reduce the number of edges that have to be compared as well for each conceptual link to the number of edges that match its superconceptual link.

Many parts of the process are composed by subtasks that are entirely independent. Then a second way to reduce the computation time is to perform these tasks simultaneously by using multithreading.

In part (i), all nodes of the network are explored to extract 1-itemsets. However this task may be performed simultaneously by dividing the network nodes

---

Algorithm 1. PALM

Require: Network : $G = (V, E)$; Threshold : β; Integer : nbCore
1: $FL_{V_{max}}$: **empty list of conceptual links**
2: LI_{cand} : **empty list of itemsets**
3: RI_{cand} : **empty list of itemsets**
4: **for** $threadIndice \leftarrow 1$ **to** $nbCore$ **do**
5:    **run** $threadGeneration1ItemSet(G, threadIndice, LI_{cand})$
6:    **run** $threadGeneration1ItemSet(G, threadIndice, RI_{cand})$
7: **end for**
8: $iteration \leftarrow 1$
9: LI_n with $n \geq 0$: **Empty lists of Itemsets**
10: RI_n with $n \geq 0$: **Empty lists of Itemsets**
11: $LI_0.addEmptyItemset()$
12: $RI_0.addEmptyItemset()$
13: $LI_0.firstItemset().addSuperItemsets(LI_{cand})$
14: $RI_0.firstItemset().addSuperItemsets(RI_{cand})$
15: **for** $threadIndice \leftarrow 1$ **to** $nbCore$ **do**
16:    **run** $threadGen1CL(G, \beta, threadIndice, FL_{V_{max}}, LI_1, RI_1, LI_{cand}, RI_{cand})$
17: **end for**
18: $iteration \leftarrow iteration + 1$
19: **while** $(LI_{t-1} \neq \emptyset)$ **and** $(RI_{t-1} \neq \emptyset)$ **do**
20:    **Generate** LI_{cand} from LI_{t-2} and LI_{t-1} **and** update lattice structure
21:    **Generate** RI_{cand} from RI_{t-2} and RI_{t-1} **and** update lattice structure
22:    **for** $threadIndice \leftarrow 1$ **to** $nbCore$ **do**
23:       **run** $threadGenCL(G, \beta, threadIndice, FLV_{max}, LI_t, RI_t, LI_{cand}, RI_{cand})$
24:    **end for**
25:    $iteration \leftarrow iteration + 1$
26: **end while**

Algorithm 2. threadGeneration1ItemSet

Require: Network : $G = (V, E)$; Integer : tIndex; $List \langle ItemSet \rangle$: Cand
1: $tCand \leftarrow \emptyset$: **Empty list of ItemSet**
2: **for all node** v_i **with** $i \in \left] \frac{(tIndex-1) \times |V|}{nbCore}, \frac{tIndex \times |V|}{nbCore} \right]$ **do**
3:     $tCand \leftarrow tCand \cup v_i.extractAll1ItemSets()$
4: **end for**
5: $Cand \leftarrow Cand \cup tCand$

Algorithm 3. threadGen1CL

Require: Network : $G = (V, E)$; Threshold β; Integer tIndex; $List \langle ConceptualLink \rangle$
     : $FL_{V_{max}}$; $List \langle ItemSet \rangle$: $LI_1, RI_1, LI_{cand}, RI_{cand}$
1: **for all Itemset** $m_i \in LI_{cand}$ **with** $i \in \left] \frac{(tIndex-1) \times |LI_{cand}|}{nbCore}, \frac{tIndex \times |LI_{cand}|}{nbCore} \right]$ **do**
2:     **for all Itemset** $n_j \in RI_{cand}$ **do**
3:         **if** $|(m_i, n_j)| > \beta \times |E|$ **then**
4:             $FL_{V_{max}}.addConceptualLink(m_i, n_j)$
5:             $LI_1.add(m_i)$
6:             $RI_1.add(n_j)$
7:         **end if**
8:     **end for**
9: **end for**

---

on **T** threads that will extract a part of the 1-itemsets before merge them in a global list, as shown in Algorithm 2.

In part (ii), all elements of LI_{cand} are combined with those of RI_{cand} in order to create new conceptual links candidate that will be checked and stored if their support is greater than the support threshold β. In the same way, the frequency of each conceptual link could be calculated in different threads as well. Therefore, in the Algorithm 3, we divided the LI_{cand} on **T** threads that aim to create the candidate conceptual links using their fraction of LI_{cand} and all itemsets of RI_{cand}.

Finally, in part (iv), each element of FVL_{max} is used in order to find all conceptual links at current iteration that could be frequent. Algorithm 4, a distributed version of this task, splits the current FVL_{max} so we can attribute each part to **T** threads that will generate the superconceptual links and test them.

5   Experimental Results

This section is devoted to the performances of the PALM algorithm. Section 5.1 presents the test environment used for experiments. Section 5.2 describes the behavior of the algorithm when the size of the network (expressed in number of links) and the number of attributes evolve. Finally, Sect. 5.3 details the gain on the computation time provided by PALM with respect to the original algorithm MFCLMin.

Algorithm 4. threadGenCL

Require: Network : $G = (V, E)$; Threshold β; Integer i; $List \langle ConceptualLink \rangle$:
$FL_{V_{max}}$; $List \langle ItemSet \rangle$: $LI_t, RI_t, LI_{cand}, LI_{cand}$

1: **for all** C.Link $L_i \in FL_{V_{max}}$ **with** $i \in \left] \frac{(tIndex-1) \times |FL_{V_{max}}|}{nbCore}, \frac{tIndex \times |FL_{V_{max}}|}{nbCore} \right]$ **do**
2:     **for all** C.Link l in $L_i.listSuperConceptualLink()$ **do**
3:         $m_l \leftarrow l.leftItemSet()$
4:         $m_r \leftarrow l.rightItemSet()$
5:         **if** $(m_l \in LI_{cand}$ **or** $m_r \in RI_{cand})$ **and** $|(m_l, m_r)| > \beta \times |E|$ **then**
6:             $FL_{V_{max}}.remove(L_i)$
7:             $FL_{V_{max}}.add(l)$
8:             **if** $|m_l| = t$ **then** $LI_t.add(m_l)$
9:             **if** $|m_r| = t$ **then** $RI_t.add(m_r)$
10:        **end if**
11:    **end for**
12: **end for**

5.1   Test Environnement

The dataset used is a telecommunication network provided by a local mobile phone operator. This network represents the calls made by subscribers on June 1st 2009 from 5 am to 3 pm. In a such a network, nodes are subscribers and links are calls made on the study period. The network has a scale-free structure and is composed of about 246 000 nodes and 510 000 links collected over the 10 h of study. Figure 4 describes the evolution of the main properties of the network on the study period.

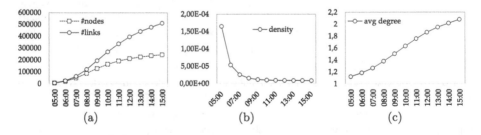

Fig. 4. Evolution of the main properties of the network on the study period: (a) nodes and links, (b) density and (c) average degree

Each node of the network is characterized by an ID and 9 attributes: (1) localization (Martinique, Guadeloupe or Guyana), (2) time slot on which it is most active, (3) type of package (Limited, Young or Professional), (4) average number of calls made, (5) average duration of calls made, (6) average number of received calls, (7) average duration of received calls, (8) number of sms sent, (9) number of sms received. Except for attributes (1) and (3), all attributes have been pre-processed in order to discretize them on five classes having equivalent sizes.

Our objective was to evaluate the performances of PALM algorithm. For this purpose, we studied for several support thresholds, the evolution of the runtime according to (i) the size of the network and (ii) the number of attributes. We vary the size of the network by capturing its state on each hour of the study. In same way, we vary the number of attributes by reducing the attributes on nodes. Results have been compared to those obtained by the original algorithm MFCLMin, to understand how PALM behaves compared to MFCLMin and what is the gain on runtime provided by our approach.

Experiments have been conducted with the following simulation environment: Intel Core i7-4770 3.40 GHz, 8 cores, 32Go Ram, Linux Ubuntu 14.04 64 Bits, Java JDK 1.8. The algorithm PALM have been implemented in Java and all tests have been conducted by parallelizing the program on 8 cores.

5.2 Algorithm Behavior

As a first approach, we have studied the evolution of the runtime (in seconds) with some support thresholds ($\beta = 0.3$ and $\beta = 0.4$) for both algorithms. Figure 5 shows the results when (a) the number of links evolves and (b) the number of attributes varies.

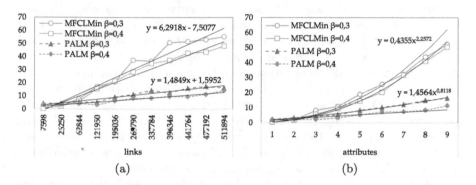

Fig. 5. Evolution of runtime (sec.) of MFCLMin and PALM according to (a) the number of links and (b) the number of attributes

First of all, we observe that for the two thresholds used, the runtime is always lower for the PALM algorithm we propose. For instance with 9 attributes, the runtime is about 55 s with MFCLMin for the two thresholds used, while it is about 15 s for $\beta = 0.3$, and 10 s for $\beta = 0.4$ with PALM. Nevertheless, in a more general way, we can observe common tendencies. Indeed, the runtime increases linearly with the size of the network, while it can be approximated by a power function when the number of attributes is increased, as shown on some equations approximated by regression displayed on Fig. 5.

These trends have been observed for several support thresholds used. Thus, to better understand the evolution of the runtime, we focused on the evolution of (a) the slope of the runtime curve when the number of links varies and (b) the exponent of the power function according to the number of attributes. Figures 6 presents these results for various support threshold $\beta \in [0.15..0.5]$.

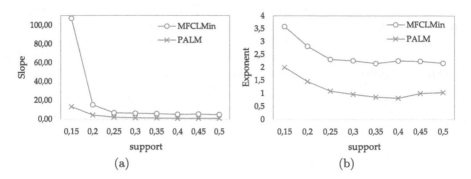

Fig. 6. Evolution of (a) slope and (b) exponent of the runtime curve according to the support threshold

If we focus on the evolution of the slope (see Fig. 6(a)), we can observe that for both algorithms, the slope of the runtime curve increases when the support threshold decreases. Thus, although the runtime increases linearly with the number of links (observed on Fig. 5(a)), we note that the slope of the curve is stronger for small support thresholds. This can be explained by the fact that for small support thresholds the solution space become larger, since many conceptual links are identified and have to be evaluated at the next iteration. However, the algorithm PALM explores more quickly this solution space, especially for small thresholds. Indeed, the slope is always lower PALM algorithm whatever is the support threshold. In particular, when the search space becomes larger, namely with small support thresholds, the difference is significant, which demonstrate that PALM explores the search space more efficiently.

A similar behavior is observed if we focus on the exponent of the curve (see Fig. 6(b)). Indeed although the runtime increases with the number of attributes according to a power function (observed on Fig. 5(b)) we note that the exponent of the curve is always lower for the PALM algorithm whatever is the support thresholds used. This also demonstrates an improvement in computing time.

Thus, if the behaviors are similar when the size of the network evolves or when the number of attributes varies, we have observed that the parallelization performed by the algorithm PALM allows to explore more efficiently the solution space, which significantly reduces the runtime.

5.3   Gain Provided by PALM

To go further, we sought to understand what was the gain on the runtime provided by PALM. In our context, the gain is defined as the fraction of time saved

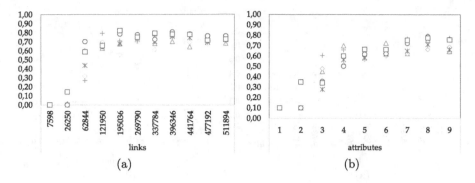

Fig. 7. Gain on the runtime (sec.) compared to original algorithm for various support thresholds according to (a) the number of links and (b) the number of attributes

from the original algorithm, namely $\frac{runtime(MFCLMin)-runtime(PALM)}{runtime(MFCLMin)}$. Figure 7 shows for various support thresholds how evolves the gain when we vary (a) the number of links and (b) the number of attributes.

First of all, we note that the gain is negligible when the number of links is very low (see Fig. 7(a)). This can be easily explained by the fact that the original algorithm also extract the conceptual links relatively quickly on small networks. However, the gain provided by the PALM algorithm grows quickly with the size of the network and seems to stabilize, even with high number of links. Indeed, it is interesting to observe that for all support thresholds used the gain is about 75% and remains stable when the number of links is high.

Regarding the results obtained according to the number of attributes (see Fig. 7(b)), similar trends can be observed. Indeed, the gain is low when nodes have few attributes and increases with attributes. For instance, the gain on the runtime is about 70% when nodes have more than 5 attributes. Moreover, as previously observed it seems to remain relatively stable for high numbers of attributes.

Thus, the results obtained confirm the efficiency of the PALM algorithm, since the gain on the runtime compared to the original algorithm is significant and remains stable when the number of links evolves or when the number of attributes is varying.

Finally, to complete these results, we sought to understand the impact of the number of cores on the gain. For this purpose, as we observed that the gain remains rather stable regardless the β support thresholds used (see Fig. 7), we studied the impact of the number of cores on the average gain on the runtime when (a) the number of links evolves and (b) the number of attributes varies. Figure 8 presents these results with 8 cores, 12 cores and 16 cores.

As observed previously, we observed that the gain is not important when the number of links is low or when the number of attributes is reduced. This is due to the fact that on small networks, or on networks with very few attributes, the original algorithm identifies the clusters relatively quickly. However we note that

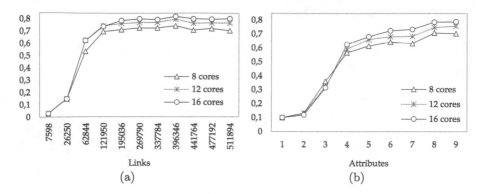

Fig. 8. Impact of the number of cores on the average gain according to (a) the number of links and (b) the number of attributes

the gain on the runtime is significant when the size of the network increases or when the number of attributes evolves. For instance with 9 attributes, the gain is about 80% when the parallelization is made on 16 cores. Finally, we can observe that the parallelization on more cores actually accelerates the computation time, since the average gain increases with the number of cores used. For instance, when the number of links is high the gain is about 70% with 8 cores while it is about 80% with 16 cores.

Thus, the results obtained have shown the efficiency of the PALM algorithm in the search for maximal frequent conceptual links. In particular, the improvements provided by the parallelization of the treatments and by the reduction of the research space have allowed to obtain gains on the computation time of up to 80%.

6   Conclusion and Future Works

In this paper, we have addressed the problem of the extraction of clusters from social networks. Unlike many other studies that focus on the search for different types of clusters, we have addressed here the major problem of scaling up existing methods. In particular, we are interested in the optimization of the search for maximal conceptual links, which is a new approach that performs link clustering by exploiting both the structure of the network and the attributes of nodes to identify strong links between groups of nodes in which nodes share common attributes. Our contributions can be summarized as follows.

- We have formally described the notion of conceptual links and shown that the solution space forms a lattice concept. This formalization have allowed us to extract some properties on conceptual links and these properties have been used to reduce the search space.
- We have proposed the algorithm PALM that aims to improve the efficiency of the extraction by simultaneously exploring several areas of the search space.

Moreover, PALM reduces the search space based on some properties highlighted on the solution space.

- Finally, we have implemented the algorithm in Java and we have evaluated its performances on a telecommunication datasets. The efficiency of the algorithm has been demonstrated by comparing the performances with the original extraction algorithm. The results obtained have shown a significant gain on the computation time.

As short term perspectives, the work conducted in this paper will allow to extract conceptual links on larger social networks and in less time. To go further, we plan to extend this work of parallelization by distributing computations on machine clusters and to use the concepts of big data in order to adapt the proposed algorithm to big data framework such as Hadoop or Spark.

References

1. Barabasi, A., Crandall, R.: Linked: the new science of networks. Am. J. Phys. **71**, 409 (2003)
2. Fowler, J., Christakis, N.: The dynamic spread of happiness in a large social network. BMJ Br. Med. J. **337**, a2338 (2008)
3. Tang, J., Sun, J., Wang, C., Yang, Z.: Social influence analysis in large-scale networks. In: Proceedings of the 15th ACM SIGKDD International Conference on Knowledge Discovery and Data Mining, KDD 2009, pp. 807–816. ACM, New York (2009)
4. Leskovec, J., Adamic, L.A., Huberman, B.A.: The dynamics of viral marketing. ACM Trans. Web **1**, 228–237 (2007)
5. Tumasjan, A., Sprenger, T.O., Sandner, P.G., Welpe, I.M.: Predicting elections with twitter: what 140 characters reveal about political sentiment. ICWSM **10**(1), 178–185 (2010)
6. Getoor, L., Diehl, C.P.: SIGKDD Explor. **7**, 3–12 (2005)
7. Kleinberg, J.M.: Challenges in mining social network data: processes, privacy, and paradoxes. In: Proceedings of the 13th ACM SIGKDD International Conference on Knowledge Discovery and Data Mining, pp. 4–5. ACM (2007)
8. Stattner, E., Collard, M.: Descriptive modeling of social networks. Procedia Comput. Sci. **52**, 226–233 (2015)
9. Blondel, V., Guillaume, J.L., Lambiotte, R., Lefebvre, E.: Fast unfolding of communities in large networks. J. Stat. Mech: Theory Exp. **2008**, P10008 (2008)
10. Fortunato, S.: Community detection in graphs. Phys. Rep. **486**, 75–174 (2010)
11. Zhou, Y., Cheng, H., Yu, J.: Graph clustering based on structural/attribute similarities. VLDB Endowment **2**(1), 718–729 (2009)
12. Stattner, E., Collard, M.: *MAX-FLMin*: an approach for mining maximal frequent links and generating semantical structures from social networks. In: Liddle, S.W., Schewe, K.-D., Tjoa, A.M., Zhou, X. (eds.) DEXA 2012. LNCS, vol. 7446, pp. 468–483. Springer, Heidelberg (2012). doi:10.1007/978-3-642-32600-4_35
13. Liben-Nowell, D., Kleinberg, J.: The link-prediction problem for social networks. J. Am. Soc. Inf. Sci. Technol. **58**, 1019–1031 (2007)
14. Heatherly, R., Kantarcioglu, M., Li, X.: Social network classification incorporating link type. In: Proceedings of IEEE Intelligence and Security Informatics, Dallas, Texas, June 2009

15. Gaume, B., Navarro, E., Prade, H.: A parallel between extended formal concept analysis and bipartite graphs analysis. In: Hüllermeier, E., Kruse, R., Hoffmann, F. (eds.) IPMU 2010. LNCS, vol. 6178, pp. 270–280. Springer, Heidelberg (2010). doi:10.1007/978-3-642-14049-5_28

16. Kuramochi, M., Karypis, G.: Discovering frequent geometric subgraphs. Inf. Syst. **32**(8), 1101–1120 (2007)

17. Newman, M.E.: Modularity and community structure in networks. Proc. Natl. Acad. Sci. **103**(23), 8577–8582 (2006)

18. Yoon, S.H., Song, S.S., Kim, S.W.: Efficient link-based clustering in a large scaled blog network. In: Proceedings of the 5th International Conference on Ubiquitous Information Management and Communication, ICUIMC 2011, pp. 71:1–71:5. ACM, New York (2011)

19. Bothorel, C., Cruz, J.D., Magnani, M., Micenkova, B.: Clustering attributed graphs: models, measures and methods. Netw. Sci. **3**(3), 408–444 (2015)

20. Stattner, E., Collard, M.: Social-based conceptual links: conceptual analysis applied to social networks. In: International Conference on Advances in Social Networks Analysis and Mining (2012)

Learning Interactions from Web Service Logs

Hamza Labbaci[1,2(✉)], Brahim Medjahed[2], and Youcef Aklouf[1]

[1] USTHB University, Algiers, Algeria
{hlabbaci,yaklouf}@usthb.dz
[2] University of Michigan - Dearborn, Dearborn, USA
brahim@umich.edu

Abstract. Web services are typically involved in various types of interaction during their lifespan. They may participate as components in more
complex services (composition) or replace unavailable services (substitution). Identifying the invocations that are part of the same interaction
relationship and the nature of these relationships provides support for
mashup developers. In this paper, we propose a novel approach for discovering composition and substitution relationships from service logs.
We introduce a technique to correlate events that are part of the same
relationship. We use association rule algorithms to determine the most
frequent item-sets of correlated events. We infer composition and substitution relationships from these item-sets and derive a multi-relation
network of Web services. Experiments show that 80% of the interaction
relationships can be learned with 70% precision.

Keywords: Web service interactions · Service logs analysis · FP-
Growth algorithm

1 Introduction

The tremendous recent technological advances such as cloud computing, Internet of Things (IoT), mobile computing, and social media have motivated organizations to export their applications as Web services. These services do not
operate in silos; they usually participate in several kinds of interaction relationships such as *composition* and *substitution*. *Composition* denotes collaboration
between services with complementary functionalities to provide an added-value
services known as *composite service* [16]. We consider two composition patterns [19]: *orchestration* and *composability*. *Orchestration* refers to an executable
Web service process and is controlled by a single party. The mashups listed in
`programmableweb.com` exemplify the orchestration pattern where independent
APIs are combined to provide new services. BPEL-like composition is another
example of orchestration [19]. *Composability* ascertains that Web services can
safely be combined in a composite service, hence avoiding unexpected failures
at runtime. It refers to the ability of services to be combined together (both
syntactically and semantically), hence being part of the same composite service
[11]. *Substitution* [9] is defined as the possibility of replacing the invocation of

© Springer International Publishing AG 2017
D. Benslimane et al. (Eds.): DEXA 2017, Part II, LNCS 10439, pp. 275–289, 2017.
DOI: 10.1007/978-3-319-64471-4_22

a service (at compile or run-time) by another service. It may reflect competition between services offering comparable functionalities (e.g., Bing Maps and Google Maps) or potential collaboration. For instance, a tax filing service may be unable to cope with the increasing number of requests before the tax filing deadline. Instead of delaying these requests and penalizing customers, it may delegate some requests to partner services.

As services interact with each other, the hosting servers record the history of such interactions as events in service logs. Service logs contain raw data about past service invocations. Some of these invocations may be part of the same interaction relationship (e.g., composition, substitution). Service logs may be leveraged to learn about past interaction relationships. This provides useful information to mashup developers such as (i) which services can be combined together to build new mashups? (ii) which services are likely to be composable? (iii) which services can be used to replace failing ones?

The learned relationship offers three major to mashup developers. First, service composition involves multiple time-consuming phases such as identifying the business tasks that make-up the composite service, discovering partner services, comparing competitors and selecting the "best ones", checking the composability of the selected services, and orchestrating those services. This process gets even more tedious as developers combine a larger number of services. For instance, the **Better Home**[1] home finder mashup integrates seven services: **Zillow**, **Walk Score**, **Google Maps**, **geocoder**, **Trulia**, **Factual**, **Yelp**, **Socrata Open Data**. Learning from mashups that have been successfully used in the past may assist developers in building new ones. Unfortunately, very few programmers devote time to share information about their mashups as they are generally busy with their programming and maintenance duties.

Second, Web services are subject to unexpected events during their lifetime such as changes (e.g., new parameters added to messages) and failures (e.g., server shutdown). Such events may impact the composite services that use those services. For example, **MapsKrieg**[2] mashes-up **Craigslist**'s apartment and housing listings with **Google Maps** to provide a visual way to find places to rent. We noticed that **MapsKrieg** was not working on 1/29/2017 because of **Craigslist**'s unavailability. Substitution relationships provide support for mashup developers in handling exceptions (e.g., failed invocation); developers can replace unavailable partners by others using the substitution relationships discovered from service logs.

Third, mashup developers have to sift through a large service space to discover the services that meet their needs and requirements, hence exacerbating attempts to compose services. One solution to alleviate this problem is to organize services into smaller cluster also referred to as *communities*. The different types of interactions relationships learned from service logs can be modeled as a *multi-relation service network* [7] where nodes represent services and edges denote relationships among those services. In our previous work [7],

---

[1] https://boiling-eyrie-10872.herokuapp.com.

[2] http://www.mapskrieg.com/.

we introduced an approach for clumping services into communities of services that are more likely to interoperate. The relationships learned through our approach can be used as input for building communities of interoperable Web services.

Mining Web service logs has been subject of past research in the literature. [4] proposes a statistical iterative approach for discovering compositions from service logs. [12,14] study correlation of events across different logs to find the ones that belong to the same business process. However, the previous techniques do not consider learning the different types of interaction relationships which may help developers identify potential collaborators and competitors. Furthermore, we provide an end-to-end approach for learning interactions starting from cleaning service logs to generating multi-relation networks of services. [13,17] mine logs to build networks that depict the causal relationships between service execution events. They do not identify which causal relationships are part of the same composition, composability, or substitution relationship. [10,18] analyze Web service interactions for trust assessment. However, they do not consider the nature of the relationships services are engaged in and which services are part of the same relationship.

In this paper, we propose a new approach for learning interaction relationships from service log files. We derive a multi-relation network of services using the inferred relationships. The main contributions of this paper are:

- We define a three-phase approach for learning interaction relationships from service logs. The *pre-processing* phase cleans log files from irrelevant data and prepares data structures for the next phases. The *interaction mining* is the core of our learning process. It generates itemsets of services that are likely to participate in composition and substitution relationships. The *relationship generation* phase returns a multi-relation network of services that depicts services and their relationships.
- We introduce an event-correlation technique and algorithms for mining service logs to identify orchestration, composability, and substitution relationships. An event in the log denotes a client's invocation to another service. The proposed technique groups correlated events and their services into *interaction windows*. We use the parallel FP-Growth [8], an efficient association rule algorithm, to determine the most frequent interaction windows. We refer to these interaction windows as *frequent* or *learned*.
- We conduct extensive experiments to assess the efficiency of the proposed approach. The experiments show that 80% of the interaction relationships can be learned with 70% precision.

The rest of this paper is organized as follows. Section 2 reviews related work. Section 3 describes the proposed approach for learning relationships and building the multi-relation services network. Section 4 is devoted to experiments and performance analysis. We provide concluding remarks in Sect. 5.

2   Related Work

Machine learning has been applied to extract insights from log files for various purposes. [3,13] mine software and service logs for failure detection. [3] aims at predicting failures in software via logs. The different changes of state that software experience during their lifetime are recorded in logs as messages (operations). Random Indexing is used to represent sequences of those messages; it is combined with Support Vector Machines to classify a sequence of messages as a failure or non-failure. [13] proposes a diagnosis-based approach for identifying service execution events suspected to cause failures. It uses FP-Growth to build a causality graph of events detailing service execution. The events make up the graph nodes while the edges depict the rules that present the causality between the events. Based on such a graph, a diagnosis is performed to identify the top-k events suspected to raise a failure. [2] mines the traces of software components interaction to apply dynamic changes to a running software system (such as replacing a running component) without creating inconsistencies. The mining allows also the identification of potentially malicious (abnormal) behavior. [17] models dependencies among the activities relating to software execution by using a Bayesian network. The Bayesian network depicts the causal relationships between the recorded log events. Such a network is used to predict the lateness probability of the process managed by the software. The aforementioned approaches mine logs for different purposes such as failure prediction or process lateness probability computation. Our approach aims at learning the different possible interactions among services (mainly composition and substitution) and the services involved in those interactions. We also model the learned interactions using a multi-relation network. [4] proposes a statistical incremental approach for discovering composition patterns from service logs. [12,14] study events correlation across a set of service logs. [12] proposes heuristics and algorithms for identifying the events that are part of the same process execution. However, it does not consider the type of the relationships these events are part of. [14] proposes parallel algorithms that use map-reduce for discovering event correlations over big event datasets. In our work we use the parallel FP-Growth and Apache Spark for performing parallel computations during the learning process. [15] proposes an FPGrowth-based approach for ranking Web services after semantically modeling their execution logs. The focus of our work is on correlating log events that are part of the same composition or substitution whereas this approach considers ranking services. In another hand, [6] uses collaborative filtering to recommend APIs (items) for mashups by comparing with similar mashups. Once similar mashups are identified, their preferences in term of APIs are added to the desired mashup. Our approach identifies more opportunities for compositions and substitutions, and hence we allow improving mashups make-up and offer alternatives when a particular API is no longer available. By doing so, our network allows mashups self-healing, self-adapting and self-configuring. [1] proposes an unsupervised semantic method to learn profiles of Web services from semantically rich documents shared by IoT devices. Although existing work tried to leverage data mining to address various challenges related to Web services, the originality of our contribution is in mining service logs to derive a network of Web services along with their interaction relationships. The generated network provides support for building flexible, reliable, dynamic service compositions.

3   The Proposed Approach for Learning Interactions

Each service S_i is identified by an IP address IP_i and has a service log L_i that records all invocations to S_i. S_i's invocation of S_j is recorded in S_j's log file (i.e., L_j). An interaction window $IW(S_u, \delta)$ refers to the set of services that have been invoked by a certain service S_u during a given timeframe δt. For example, if $IW(S_u, \delta t = 5s) = \{L_2, L_4, L_5\}$ then services S_u invoked S_2, S_4, and S_5 in a 5-second timeframe.

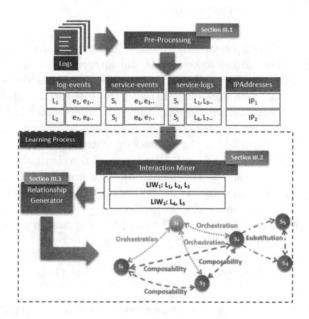

Fig. 1. Approach overview

Figure 1 depicts the main steps of the proposed approach for learning interaction relationships. The *pre-processing* phase (Sect. 3.1) cleans the service log files from irrelevant data. It returns the list of events recorded in each log file, the list of events raised by each Web service, the list of log files where such events have been recorded, and the list of service IP addresses. An event in the log denotes a client's invocation to another service. The following two phases constitute the learning process of our approach. The *interaction miner* (Sect. 3.2) uses the aforementioned lists to identify the interaction windows IWs for each service. If the same IP address is observed in the *clientIP* entry of many service logs during a given timeframe δt, then those services are candidate to correlate.

An Interaction window IW that includes those services is generated. The generated IWs are grouped into itemsets. We use FP-Growth [8], an association rule algorithm, to determine the most frequent interaction windows. We refer to these interaction windows as *frequent* or *learned*. The *relationship generator* (Sect. 3.3) uses the frequent interaction windows to infer orchestration, composability, and substitution relationships. The outcome is a multi-relation network of services that depicts services and their relationships. In the rest of this section, we give details about the three phases.

3.1 Pre-processing Log Files

The aim of this initial phase is twofold. First, we clean log files from irrelevant data (data cleaning). Second, we prepare the cleaned logs to be used by the subsequent phases (data preparation).

Data Cleaning. This step cleans log files from irrelevant data (e.g., data needed by robots) and derives the logs to be used by our approach. We need the following three pieces of information for each service invocation: (i) IP address of the client service (ii) invocation date and time (iii) and the invocation outcome (success or failure). Such information is usually scattered across two files: `Server.log` and `Access.log`. `Server.log` contains the invocation date and time, details of the invoked server, duration of the related operation, and level. `Access.log` includes the invocation date and time, IP address of the client service, and other details about the client such as the client's operating system. To clean the log files, we first perform a join operation between `Server.log` and `Access.log` on the attribute `Date-Time`. Then, we project the results over the attributes (`ClientIP`, `Date-Time`, `Outcome`). The algebraic expression given below summarizes the previous operations:

$$\Pi_{(ClientIP,Date-Time,Outcome)}(Server.log \bowtie_{(Date-Time)} Access.log)$$

Data Preparation. The goal of this second step it to prepare the data-structures to be used in the learning process, namely $IPAddresses$, $log-events$, $service-events$, and $service-logs$ (Fig. 1). $IPAddresses$ is a list containing each service IP address. $log-events$ is a dictionary where each key is a log identifier (such as L_i). Each value is a list of events recorded in L_i (i.e., the invocations targeting S_i). $service-events$ is a dictionary where each key is a service identifier (such as IP_i). Each value is a list of events service S_i has raised (i.e., the invocations made by S_i). $service-logs$ is a dictionary where each key is a service identifier (such as IP_i). Each value is a list of logs where the events raised by S_i have been observed.

An event denotes the invocation of a client service to another service and is defined by the quadruplet $< ClientIP, DateTime, Status, CurrentIP >$. $ClientIP$ and $CurrentIP$ refer to the IP addresses of the client service and service being invoked, respectively. $DateTime$ represents the event's date and time.

Status refers to the event outcome. *Status* may take multiple possible values such as warning, success, and failure. For simplicity, we only consider *Success* and *Failure* as values. Creating the data structures listed above is crucial for learning the service relationships. For the composition relationship, we look at the successful events that have been recorded almost at the same date and time and issued by the same client service. For the substitution relationship, we focus on the events that raised failures and have been followed by successful invocation in a short time range and from the same client service.

3.2   Mining Service Interactions

In this section, we illustrate our techniques for discovering composition and substitution relationships from service logs. For that purpose, we define the following two heuristics for inferring orchestration, composability, and substitution relationships:

- **Heuristic 1** - If a service S_i successfully invokes S_m, S_n, S_p within a timeframe δt and with a high frequency, then we consider an orchestration relationship between S_i and S_m, S_i and S_n, and S_i and S_p. Moreover, we infer that S_m is composable with S_n, S_n is composable with S_p, and S_m is composable with S_p.
- **Heuristic 2** - If the unsuccessful invocation of S_i to S_m is followed by the successful invocation of S_i to S_n with a high frequency, then we infer a substitution relationship between S_m and S_n.

The intuition behind the first heuristic is that the different executions of an orchestration process generally involve the invocation of a given set of services during a similar period of time. If we notice that a given service S_i invokes S_m, S_n, S_p during a timeframe δt and if this same invocation pattern is observed frequently enough (above a support threshold defined by the user) in the service logs, then we assume that S_i is a process that orchestrates S_m, S_n, S_p. The rationale behind the second heuristic is that the substitution of S_m by S_n generally occurs as a result of the failure of S_m invocation by S_i. The following sequence of events are then observed (i) S_i invokes S_m, (ii) a failure of S_m invocation is recorded in L_m, (iii) S_i invokes S_n. If this sequence of events is recorded frequently enough (above a support threshold defined by the user) in the service logs, then we assume a substitution relationship between S_m and S_n.

Algorithm 1 returns (i) *serviceIWs*, a dictionary whose keys are IP addresses and values are the related interaction windows; and (ii) *frequentIWs*, a set of frequent interaction windows. The latter are also called *learned* interaction windows ($LIWs$). The algorithm first identifies the interaction windows IWs for each service. An interaction window IW can be either a composition or a substitution. It is a set of services that have been invoked by a certain service S_u during a given timeframe δt. For each event e raised by S_i, a new interaction window IW is initialized with the log containing e. We look for the logs that contain events that *correlate* with e. For each log file L_j that contains an event f that correlates with e, we append L_j to IW.

Algorithm 1. Learning Algorithm

Input: log-events, service-events, service-logs, IPAddresses, min-support, δt.
Output: serviceIWs, frequentIWs
1: $itemsetAllIWs \leftarrow serviceIWs \leftarrow frequentIWs \leftarrow null$
2: **for** each $IP_i \in IPAddresses$ **do**
3:     $events \leftarrow service - events[IP_i]$
4:     $logs \leftarrow service - logs[IP_i]$
5:     $IW \leftarrow null$
6:     $used \leftarrow null$
7:     **for** each $e_k \in events$ & $e_k \notin used$ **do**
8:         $IW.append(L_k)$
9:         $used.append(e_k)$
10:         **for** each $L_j \in logs$ **do**
11:             **if** $(f \in log - events[L_j]$ & $f \notin used$ & $CorrelationCheck(e_k, f, \delta t))$ **then**
12:                 $IW.append(L_j)$
13:                 $used.append(f)$
14:             **end if**
15:         **end for**
16:         $serviceIW[IP_i].append(IW)$
17:         $itemsetAllIWs.append(IW)$
18:     **end for**
19: **end for**
20: $frequentIWs \leftarrow FP\text{-}Growth(itemsetAllIWs, min - support)$

Two events e and f recorded in L_k and L_j, respectively, *correlate for composition* iff (i) they have been recorded during a given time frame δt (i.e., $e.dateTime - f.dateTime <= \delta t$); (ii) they have been raised by the same client service S_i (i.e., $e.clientIP = f.clientIP = IP_i$); (iii) and they have been achieved successfully (i.e., $e.status = f.status = success$). Algorithm 2 describes the correlation condition for composition relationships.

Similarly, two events e and f recorded in L_k and L_j, respectively, *correlate for substitution* iff (i) they have been recorded during a given time frame δt; (ii) they have been raised by the same client service S_i; (iii) the invocation of S_i to S_k failed and has been followed by a successful invocation of S_i to S_j. Checking if two services correlate in the same substitution is given by Algorithm 3.

Algorithm 2. *CorrelationCheck* Algorithm for Composition Relationships

---

Input: e, f, δt.
Output: $result$.
1: $result \leftarrow false$
2: **if** $(e.clientIP == f.clientIP$ & $((e.DateTime - f.DateTime) \leq \delta t)$ & $(e.status == \text{"Success"}$ & $f.status == \text{"Success"}))$ **then**
3:     $result \leftarrow true$
4: **end if**

Each log that contains an event that correlates with e is added to IW. Once all logs are checked, we save IW in the list of interaction windows IWs for S_i (i.e., $serviceIWs$). We also save it in the list of all interaction windows (i.e., $itemsetAllIWs$). We iterate the previous process for the remaining events of S_i to generate new interaction windows IWs. To avoid generating redundant IWs, we do not consider the events processed in previous iterations. The whole process is then repeated for each Web service. Finally, we run the FP-Growth association rule algorithm [8], to determine the learned interaction windows $LIWs$. FP-Growth uses as inputs $itemsetAllIWs$ and the support; it returns the set $frequentIWs$ of learned interaction windows.

Algorithm 3. $CorrelationCheck$ Algorithm for Substitution Relationships

---

Input: e, f, δt.
Output: $result$.
1: $result \leftarrow false$
2: **if** $(e.clientIP == f.clientIP$ & $((e.DateTime - f.DateTime) \leq \delta t)$ & $(e.status == \text{``Success''}$ & $f.status == \text{``Failure''}))$ **then**
3:    $result \leftarrow true$
4: **end if**

Example 1: The example in Fig. 2a illustrates our technique. We want to learn the frequent composition and substitution interaction windows for a set of services S_i, S_1, S_2, S_3. The events raised by S_i are e_1, e_2, e_3, e_4, e_5, e_8, e_9, e_{10}, and e_{11}. We assume $\delta t = 2s$ and $8s$ for the composition and substitution correlation checking, respectively. First, we start by identifying the composition interaction windows for S_i. Algorithm 1 appends L_1 (i.e., the log where e_1 has been recorded) to a new interaction window IW_1^C. Then, it reviews the events recorded in L_2 and L_3 looking for the ones that correlate with e_1. Since e_5 in L_2 and e_9 in L_3 satisfy the composition correlation condition, L_2 and L_3 are added to IW_1^C. IW_1^C is saved in the list of the composition interaction windows for S_i and the list of all composition interaction windows. Similarly, the composition interaction window IW_2^C is identified. This process is iterated for S_1, S_2, and S_3. For simplicity, we do not show information about invocations made by these three services in Fig. 2.a. As a result, we generate $IW_1^C = \{L_1, L_2, L_3\}$, $IW_2^C = \{L_1, L_2\}$, $IW_3^C = \{L_1, L_2\}$, and $IW_4^C = \{L_3\}$ (IW_3^C and IW_4^C not shown in Fig. 2a). The generated IWs are used as input by the FP-Growth algorithm with $support = 0.75\%$. FP-Growth returns the learned interaction windows (i.e., frequent ones) namely $LIW_1^C = \{L_1\}$, $LIW_2^C = \{L_2\}$, and $LIW_3^C = \{L_1, L_2\}$. Similarly to the composition learning process, we learn the most frequent substitution interaction windows. We use the substitution correlation condition to identify substitution interaction windows. After iterating the previous process for S_1, S_2, and S_3, the generated substitution interaction windows are $IW_1^S = \{L_1, L_3\}, IW_2^S = \{L_1, L_3\}$, and $IW_3^S = \{L_1, L_3\}$. FP-Growth returns $LIW_1^S = \{L_1\}$, $LIW_2^S = \{L_3\}$, and $LIW_3^S = \{L_1, L_3\}$.

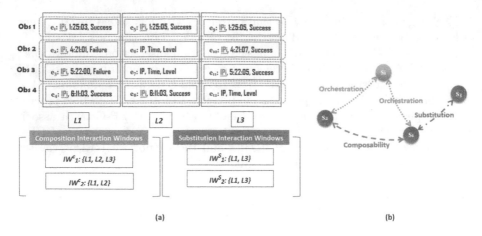

(a)   (b)

Fig. 2. Interaction windows and Multi-relation graph generation

3.3 Generating Relationships

The third phase uses the most frequent interaction windows $LIWs$ returned by Algorithm 1, namely frequentIWs, to infer the orchestration, composability, and substitution relationships among services. We use these relationships to build a graph or network of services, with services as nodes and relationships as edges. As services have multiple types of relationships, we adopt *multiplex graphs* (also known as *multi-relation graphs*) as a model [7]. This is a particular kind of complex networks defined over a set of nodes from the same type linked by different types of relations. A multi-relation network of Web services is a directed graph $G = (V, R_1, R_2, R_3)$ where V is a set of services, and $R_k = (S_u, S_v) \in V * V$ is the set of edges of the k-th relationship. Precisely, R_1, R_2, and R_3 are the set of edges of the orchestration, composability, and substitution relationships, respectively. For example, an edge $(S_i, S_j) \in R_3$ indicates that there is a relationship of type substitution between Web services S_i and S_j,. It states that S_j is a substitute of S_i and clients can invoke S_j instead of S_i if S_i fails. Algorithm 4 describes the process of building the graph.

Algorithm 4 checks whether the interaction windows $LIWs$ learned in Algorithm 1 (i.e., itemsets form frequentCompositionIWs or frequentSubstitutionIWs) match the generated ones (i.e., serviceCompositionIWs and serviceSubstitutionIWs). If a learned composition interaction window LIW^C is part of the initial composition interaction windows IWs generated for S_i. then we add the orchestration relationships among S_i and all services in LIW_C to the network. In addition, we add the composability relationships among services in LIW^C to the network. This process is iterated for all the learned composition interaction windows. Similarly, If a learned substitution interaction window LIW^S is part of the initial substitution IWs generated for S_i, we add the substitution relationship among the services in LIW^S to the network. This process is iterated for all the learned substitution interaction windows.

Algorithm 4. Network Building Algorithm

Input: frequentCompositionIWs, frequentSubstitutionIWs, serviceCompositionIWs, serviceSubstitutionIWs.

Output: The resulting network

1: **for** each $IW \in frequentCompositionIWs$ **do**
2:      **for** each $L_j \in serviceCompositionIWs.keys()$ **do**
3:          **if** $(serviceCompositionIWs[L_j]).contains(IW))$ **then**
4:              **for** each $L_k \in IW$ **do**
5:                  **if** $(Orchestration(L_j, L_k) \notin network))$ **then**
6:                      $network.addOrchestration(L_j, L_k)$
7:                  **end if**
8:              **end for**
9:              **for** each $(L_{k1}, L_{k2}) \in IW$ **do**
10:                  **if** $(Composability(L_{k1}, L_{k2}) \notin network))$ **then**
11:                      $network.addComposability(L_{k1}, L_{k2})$
12:                  **end if**
13:              **end for**
14:          **end if**
15:      **end for**
16: **end for**
17: **for** each $IW \in frequentSubstitutionIWs$ **do**
18:      **for** each $(L_{k1}, L_{k2}) \in IW$ **do**
19:          **if** $(Substitution(L_{k1}, L_{k2}) \notin network)$ **then**
20:              $network.addSubstitution(L_{k1}, L_{k2})$
21:          **end if**
22:      **end for**
23: **end for**

Example 2:   Example 2: Let us consider the following composition IWs learned in Example 1: $LIW_1^C = \{L_1\}$, $LIW_2^C = \{L_2\}$, and $LIW_3^C = \{L_1, L_2\}$; the most frequent substitution interaction windows were $LIW_1^S = \{L_1\}$, $LIW_2^S = \{L_3\}$, and $LIW_3^S = \{L_1, L_3\}$. Algorithm 5 checks that $LIW_3^C = \{L_1, L_2\}$ is part of the initial composition interaction windows generated for S_i. Therefore, it establishes orchestration relationships between S_i and S_1, and between S_i and S_2. Moreover, the composability relationship is established among S_1 and S_2. Additionally, Algorithm 4 checks that $LIW_3^C = \{L_1, L_3\}$ is part of the initial substitution interaction windows generated for S_i. Thus, it establishes the substitution relationship among S_1 and S_3. The generated multiplex graph is showed in Fig. 2b.

4 Experimental Study

We conducted experiments to evaluate the performance of our approach. We first overview our experimental set-up. Then, we discuss our experiment results.

4.1   Experimental Set-Up

We ran our experiments on a 64-bit Windows 10 environment, in a machine equipped with an intel i7 and 12 GO RAM. We used Java 8 and leveraged the powerful Java streams while processing log files. We used the parallel FP-Growth algorithm [8] from the machine learning library MLIB of Apache Spark. Parallel FP-Growth is a faster version of the initially proposed FP-Growth [5]. It is more convenient for processing big amounts of data. It leverages parallel computation algorithms and splits the computation task among several processors. We simulated between 425 and 18757 compositions and substitutions for a number of services varying between 100 to 2000 services. A composition is simulated by the successful invocation of the same service S_i to some other services S_j, S_k,.. in a given time frame. A substitution is simulated by an unsuccessful invocation of S_i to S_j which is followed by a successful invocation of S_i to S_k in a given time frame. Each service has been assigned a log file and an IP address. We recorded the details of the simulated interactions as events in logs in a such way that each service S_i's invocation of service S_j is recorded in S_j's log file. The goal of the experiments is to assess the ability of the approach to learn the initial compositions and substitutions from a set of logs.

4.2   Results and Discussion

In this section, we first assess the ability of the proposed approach at learning the initially generated compositions and substitutions. In this regard, we compare the learned interaction windows with the initial ones.

A composition denoted by the interaction window (IW_i) is *successfully learned* by our technique if and only if there is a learned interaction window (LIW_j) that contains l services from (IW_i), where l is the learning threshold. For instance, let us consider $l = 0.5$. Assume that $LIW_1^C = \{L_1, L_9, L_{12}\}$ and the intially executed composition involve the following sets of services: $S_1 = \{L_1, L_2, L_9, L_{12}\}$, $S_2 = \{L_2, L_3, L_{10}, L_{11}\}$, and $S_3 = \{L_3, L_4 L_{15}\}$. Since LIW_1 contains 75% of S_1 services, we conclude that IW_1 is learned and the learning is valid at 75%. We call this a *learning ratio*. Similarly, S_2 is learned and the learning is valid at 75%, while S_3 has not been learned. Additionally, we compute the *average learning validation* (i.e., how valid are the learned compositions). We calculate the average of the learning ratios of the successfully learned interaction windows. In this example, the average validation of the learning equals $\frac{0.75+0.75}{2} = 0.75$. We proceed similarly for learning substitutions.

We ran our algorithms using the service logs generated as explained in Sect. 4.1 (simulated compositions and substitutions). The average learning validation approximates 98%. It means that initial compositions and substitutions are almost entirely retrieved among the learned interaction windows.

We also computed the *recall* and *precision* of the learning process. *Recall* is the number of interaction windows that are succcessfully learned divided by the number of all correct interaction windows (i.e., initial ones). It is the fraction of

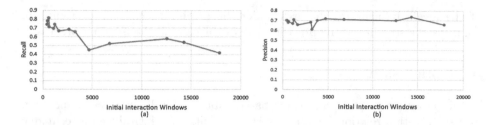

Fig. 3. Precision and recall with respect to the number of initial interaction windows

relevant interaction windows that are retrieved. It can be also seen as the percentage of initial interaction windows that have been learned. Figure 3a depicts the recall with respect to the number of initial interaction windows. It shows that up to 80% of interaction windows are recalled (i.e., learned). However, this number decreases for very high numbers of initial interaction windows. The justification is that as the number of interaction windows increases, it becomes harder to identify frequent interaction windows that satisfy the minimum support. *Precision* is the number of successfully learned interaction windows divided by the number of all learned interaction windows. It is the fraction of the retrieved results that are relevant. It can also be seen as the percentage of the learned interaction windows that are correct among all learned interaction windows. Figure 3b compares the variation of the learning precision with respect to the number of initial interaction windows. Contrarily to the recall, the precision is stable even when the number of initial interaction windows increases. The justification is that the number of the correctly learned interaction windows is proportional to the number of the initial interaction windows.

We finally calculated the *F-measure* and *average-precision* of our approach (Fig. 4). The *F-measure* is a trade-off of precision and recall. It is the harmonic mean of precision and recall. Figure 4a. shows that the F-measure is impacted by the increase in the number of initial interaction windows which is predictable since the recall is also impacted by the variation of the number of initial interaction windows. The obtained results are promising. However, we believe that they can be improved by leveraging in the learning process expert opinions on the identified relationships. The *average-precision* shows the precision as a function

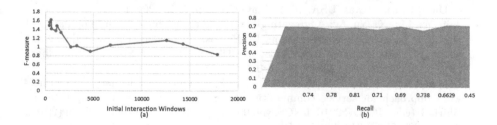

Fig. 4. Average precision and F-measure

of the recall. It measures the area under the precision-recall curve. Figure 4b. confirms the stability of the precision values through all the learning process.

5   Conclusion

In this paper, we proposed a learning-based approach for discovering composition and substitution relationships from service log files. We introduced a technique to correlate invocations that are part of the same composition or substitution. Identifying the invocations that are part of the same interaction relationship and the nature of these relationships can be helpful for developers in composing and substituting APIs. We used the association rule algorithm FP-Growth [8] to determine the most frequent correlated invocations and identify orchestration, composability, and substitution relationships. We derived a multidimensional network of services using the inferred relationships. We developed a tool that implements the proposed approach and allows generating a multidimensional services network from log files. We conducted extensive experiments. The obtained results are promising since 80% of the relationships can be learned with 70% precision. We believe that such results could even be improved further by including feedback from experts such as mashup and service developers (a.k.a, crowd-sourcing) in the learning process.

References

1. Antunes, M., Gomes, D., Aguiar, R.: Semantic features for context organization. In: 2015 3rd International Conference on Future Internet of Things and Cloud (FiCloud), pp. 87–92 (2015)
2. Esfahani, N., Yuan, E., Canavera, K.R., Malek, S.: Inferring software component interaction dependencies for adaptation support. TAAS 10(4), 26:1–26:32 (2016)
3. Fronza, I., Sillitti, A., Succi, G., Terho, M., Vlasenko, J.: Failure prediction based on log files using random indexing and support vector machines. J. Syst. Soft. 86(1), 2–11 (2013)
4. Gaaloul, W., Baïna, K., Godart, C.: Log-based mining techniques applied to web service composition reengineering. Serv. Oriented Comput. Appl. 2(2–3), 93–110 (2008)
5. Han, J., Pei, J., Yin, Y.: Mining frequent patterns without candidate generation. In: Proceedings of the 2000 ACM SIGMOD International Conference on Management of Data, Dallas, Texas, USA, 16–18 May 2000, pp. 1–12 (2000)
6. Jain, A., Liu, X., Yu, Q.: Aggregating functionality, use history, and popularity of apis to recommend mashup creation. In: Barros, A., Grigori, D., Narendra, N.C., Dam, H.K. (eds.) ICSOC 2015. LNCS, vol. 9435, pp. 188–202. Springer, Heidelberg (2015). doi:10.1007/978-3-662-48616-0_12
7. Labbaci, H., Medjahed, B., Aklouf, Y., Malik, Z.: Follow the leader: A social network approach for service communities. In: Sheng, Q.Z., Stroulia, E., Tata, S., Bhiri, S. (eds.) ICSOC 2016. LNCS, vol. 9936, pp. 705–712. Springer, Cham (2016). doi:10.1007/978-3-319-46295-0_50

8. Li, H., Wang, Y., Zhang, D., Zhang, M., Chang, E.Y.: Pfp: parallel fp-growth for query recommendation. In: Proceedings of the 2008 ACM Conference on Recommender Systems (RecSys 2008), Lausanne, Switzerland, 23–25 October 2008, pp. 107–114 (2008)

9. Maamar, Z., Santos, P., Wives, L., Badr, Y., Faci, N., de Oliveira, J.P.M.: Using social networks for web services discovery. IEEE Internet Comput. **15**(4), 48–54 (2011)

10. Malik, Z., Medjahed, B.: Trust assessment for Web services under uncertainty. In: Maglio, P.P., Weske, M., Yang, J., Fantinato, M. (eds.) ICSOC 2010. LNCS, vol. 6470, pp. 471–485. Springer, Heidelberg (2010). doi:10.1007/978-3-642-17358-5_32

11. Medjahed, B., Malik, Z., Benbernou, S.: On the composability of semantic web services. In: Web Services Foundations, pp. 137–160 (2014)

12. Nezhad, H.R.M., Saint-Paul, R., Casati, F., Benatallah, B.: Event correlation for process discovery from web service interaction logs. VLDB J. **20**(3), 417–444 (2011)

13. Nie, X., Zhao, Y., Sui, K., Pei, D., Chen, Y., Qu, X.: Mining causality graph for automatic web-based service diagnosis. In: 2016 IEEE 35th International Performance Computing and Communications Conference (IPCCC), pp. 1–8. IEEE (2016)

14. Reguieg, H., Benatallah, B., Nezhad, H.R.M., Toumani, F.: Event correlation analytics: Scaling process mining using mapreduce-aware event correlation discovery techniques. IEEE Trans. Serv. Comput. **8**(6), 847–860 (2015)

15. Shafiq, M.O., Alhajj, R., Rokne, J.G.: Reducing search space for web service ranking using semantic logs and semantic FP-tree based association rule mining. In: Proceedings of the 9th IEEE International Conference on Semantic Computing (ICSC 2015), Anaheim, CA, USA, 7–9 February 2015, pp. 1–8 (2015)

16. Singh, M.P.: Physics of service composition. IEEE Internet Comput. **5**(3), 6 (2001)

17. Sutrisnowati, R.A., Bae, H., Song, M.: Bayesian network construction from event log for lateness analysis in port logistics. Comput. Industr. Eng. **89**, 53–66 (2015)

18. Wahab, O.A., Bentahar, J., Otrok, H., Mourad, A.: Towards trustworthy multicloud services communities: A trust-based hedonic coalitional game. IEEE Trans. Serv. Comput. (2017). http://ieeexplore.ieee.org/document/7445255/

19. Yu, Q., Liu, X., Bouguettaya, A., Medjahed, B.: Deploying and managing web services: issues, solutions, and directions. VLDB J. **17**(3), 537–572 (2008)

Data Driven Generation of Synthetic Data with Support Vector Data Description

Fajrian Yunus$^{(\boxtimes)}$, Ashish Dandekar, and Stéphane Bressan

Department of Computer Science, School of Computing,
National University of Singapore, Singapore, Singapore
{fajrian.yunus,steph}@nus.edu.sg, ashishdandekar@u.nus.edu

Abstract. We propose a method to generate synthetic data by Support Vector Data Description. Support Vector Data Description is a variant of Support Vector Machine for one-class classification problem. Our method assumes that an observed data is a sample of a random variable which satisfies an unknown membership decision function. The unknown membership decision function is to be learned by Support Vector Data Description based on the training data.

By using the learned membership decision function, we perform rejection sampling. Firstly, we generate a random data point. Secondly, we test the data point against the membership decision function. Lastly, if the data point fails the test, we repeat from the first step.

However, in some cases, the rejection sampling approach runs slowly. Therefore, we also propose another approach. The approach works by using a heuristic to find a good starting point and then performs gradient descent to gradually move the data point into inside the positive region boundary while maintaining randomness of the generated data. This approach runs noticeably faster than rejection sampling when rejection sampling runs slowly.

Keywords: Machine learning · Novelty detection · One-class classification · Support Vector Data Description · Generative model · Data generation · Synthetic data

1 Introduction

We propose a method based on Support Vector Data Description [16] to generate synthetic data. Support Vector Data Generation is a one-class classifier. Our method assumes that an observed data is a sample of a random variable which satisfies an unknown membership decision function. The unknown membership decision function is to be learned by Support Vector Data Description.

Support Vector Data Description is a variant of Support Vector Machine for one-class classification. Unlike the standard Support Vector Machine which uses hyperplane, Support Vector Data Description uses hypersphere. The hypersphere envelopes the positive region. Similar to the case of standard Support

© Springer International Publishing AG 2017
D. Benslimane et al. (Eds.): DEXA 2017, Part II, LNCS 10439, pp. 290–298, 2017.
DOI: 10.1007/978-3-319-64471-4_23

Vector Machine, we can employ kernel trick in Support Vector Data Description. Kernel trick works by mapping the data into a higher dimensional space, in which the data obeys separability criterion. The effect is the decision boundary in the normal space is no longer regularly shaped. In the case of Support Vector Data Description with a kernel, the decision boundary is no longer a hypersphere. This effect makes Support Vector Data Description more flexible [16]. It should be noted that polynomial kernel cannot be used in Support Vector Data Description [16]. In this work, we focus on Support Vector Data Description with Gaussian kernel.

In this work, we use the term "inside the hypersphere" and "positive data" interchangeably. We will also use the term "hypersphere", "membership decision function", and "positive region" interchangeably. We explain the related work in Sect. 2 and the Support Vector Data Description in Sect. 3. We describe our proposed method in Sect. 4. We conclude our work in Sect. 6. Interested readers can refer to [17] for our experiment result and more detailed discussion.

2   Related Work

Support Vector Data Description [16] is a one-class classifier. One-class classification is also called novelty detection, anomaly detection, or outlier detection. Support Vector Data Description is classified as a domain-based one-class classifier [12]. It means that the classifier only cares about the boundary of the classes and ignores the probability density [12].

Support Vector Data Description has been applied to, among others, the problems of land-cover classification [14], defect identification of TFT-LCD manufacturing [8], chillers fault detection [18], diagnosing circuit faults [9], and target detection in hyperspectral imagery [13].

There are some existing popular generative models which are used to generate data. Among them are Latent Dirichlet Allocation [1], Restricted Boltzmann Machine [15], and Generative Adversarial Network [4]. Latent Dirichlet Allocation and Restricted Boltzmann Machine assume the observed variables to be the manifestations of the latent factors. These algorithms learn the latent factors and then generate the data based on the learned latent factors. Generative Adversarial Network, on the other hand, works by having one neural network generates synthetic data, one other neural network distinguishes whether its input is a genuine or synthetic data, and both of them compete until the generated synthetic data is no longer distinguishable from the genuine data.

Among the uses of data generation are for semi-supervised learning [2,7], generation of synthetic handwritings [5], and image generation [6]. Data generation has also been used to simulate some specific problems [3,10,11,19].

3   Support Vector Data Description

Support Vector Data Description [16] works by finding the smallest hypersphere which envelopes the positive examples while keeping the negative examples outside the hypersphere. Similar to many other machine learning algorithms, Support Vector Data Description also tolerates misclassification to some extent.

Formally, Support Vector Data Description minimizes $F(R, a)$, which is the volume of the hypersphere as given in function 1, subject to inequality 2.

$$F(R, a) = R^2 + C \sum_i \xi_i \tag{1}$$

$$\forall i : \|x_i - a\| \leq R^2 + \xi_i \tag{2}$$

In function 1 and inequality 2, R is the radius of the hypersphere, a is the centroid of the hypersphere, C is the slack coefficient parameter, and ξ are slack variables, x are the positive training data. The minimization problem is about finding the values of R and a which minimize function 1.

We can set the misclassification tolerance by setting the value of the slack coefficient C. Large C value means there is a large penalty for every misclassification. This minimization problem is solved by expressing the problem in Lagrangian form and setting the derivative to 0. We then apply kernel trick to the resulting equation. In this case, we apply the Gaussian kernel, which is stated in Eq. 3.

$$K(x_1, x_2) = e^{-\gamma \|x_1 - x_2\|^2} \tag{3}$$

After applying the Gaussian kernel, we obtain the membership decision function which is expressed in formula 4 and 5.

$$k = 1 + \sum_{i,j} \alpha_i \alpha_j e^{-\gamma \|x_i - x_j\|^2}$$
$$f(z) = k - 2 \sum_i \alpha_i e^{-\gamma \|z - x_i\|^2} \tag{4}$$

$$f(z) \leq R^2 \iff z \mapsto + \tag{5}$$

The x_is in formula 4 are the support vectors. Support vectors are the subset of the training data which affect the boundary. In other words, if we keep the support vectors and eliminate all other training data, the boundary will still be the same. The support vectors are located on the boundary or near the boundary of the hypersphere. The α_is in formula 4 are Lagrange multipliers. There is one Lagrange multiplier for every support vector. The Lagrange multipliers satisfy the conditions stated in formula 6.

$$\sum_i \alpha_i = 1 \qquad \forall i : \alpha_i > 0 \tag{6}$$

The intuitive explanation of Eq. 4, formula 5, and 6 is that Support Vector Data Description with Gaussian kernel models the membership decision function as a high-dimensional hypersphere. An arbitrary data point z is considered as a positive data point if and only if the data maps to a high-dimensional point located inside the (high-dimensional) hypersphere. The hypersphere itself is defined by a subset of the training data which are called "support vectors". Each of these support vectors has a coefficient, and these coefficients are the Lagrange multipliers.

Interested readers can refer to [16] for further details.

4   Proposed Method

We train the Support Vector Data Description model by feeding the training data to the Support Vector Data Description algorithm along with the γ and C parameters. Parameter C affects the size of the hypersphere. Larger C creates a model with larger hypersphere. On the other hand, parameter γ affects the complexity of the function shape. As can be seen in Fig. 1, $f(z)$ with small γ has the shape of a simple convex function while $f(z)$ with large γ has many convex curves.

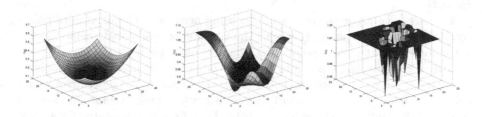

Fig. 1. Examples of $f(z)$ plots where z is a 2D vector. $f(z)$ is the vertical axis and the remaining axes are z. The plots have different γ values. From left to right, the corresponding γ is small, medium, and large.

We choose the parameters based on our knowledge about the data. If the data is simple, we should use small γ. On the other hand, if the data is "complex", we should use large γ. Parameter C depends on how "average" we want our generated data to be. Small C corresponds to small hypersphere. Therefore, the data we generate is similar to the typical training data points. On the other hand, large C corresponds to large hypersphere. Therefore, some of the data points we generate may be similar to the outliers in the training data.

After the training, we obtain the support vectors, the Lagrange multipliers, and the radius of the hypersphere. These, along with the γ parameter defines our membership decision function (Eq. 4 and formula 5). After that, we sample this model to generate synthetic data. We will denote the generated random data point as z.

4.1 Sampling Approaches

Rejection Sampling. An obvious way to generate a random synthetic point which is inside the hypersphere is by using rejection sampling. Firstly, we generate a random data point. Secondly, we test the data point against the membership decision function (formula 4 and 5). Lastly, if the data point fails the test, we repeat from the first step. It should be noted that the runtime of the membership decision function is constant with respect to the location of data point we test.

If the hypersphere is large, then the rejection sampling approach will run fast. This is because of many of the randomly generated data points fall inside the hypersphere.

Heuristic and Gradient Descent (Save Our Samples!). Unfortunately, if the hypersphere is small, many randomly generated data points will fall outside the hypersphere, and thus the rejection sampling approach fails to work efficiently.

Therefore, we use gradient descent to bring negatively classified data points into inside the hypersphere. However, the runtime of gradient descent depends on the location of the data point. If the data point is far from the hypersphere, we will need many steps to move the data point into inside the hypersphere. Therefore, we use a heuristic to choose our starting point. We generate our data point to be near one of the support vectors because support vectors are located on or near the boundary of the hypersphere. Therefore, the generated data point is located near the boundary of the hypersphere.

An important intuition from Eq. 4 is that $f(z)$ is an inverted Gaussian mixture model where each of the support vectors are the centers of each Gaussian curve. This can be seen in Fig. 1. Therefore, we can always slide from any of the support vectors to a local minimum. Moreover, if the γ is big, then the local minimum will be near the support vectors because the contour will be full of deep "holes" (see Fig. 1).

Formally, we initialize our random data point z as one of the support vectors (here, denoted as x_{sv}), plus noise ϵ to add some randomness.

$$z_0 = x_{sv} + \epsilon \tag{7}$$

After initializing z, we perform gradient descent to move z to the desirable range.

$$z_{t+1} = z_t - \eta \frac{df'(z)}{dz} \tag{8}$$

where $\eta > 0$. η is the learning rate.

From formula 4, we calculate the gradient of $f(z)$, and then we plug it to the gradient descent function 8, and we then obtain the update function 9.

$$z_{t+1} = z_t - 4\eta\gamma \sum_i \alpha_i e^{-\gamma\|z - x_i\|^2}(z - x_i) \tag{9}$$

Our aim is to update z such that eventually z falls within a desired range from the centroid.

$$f(z) \leq rand(0,1) \times R^2 \tag{10}$$

The purpose of $rand(0,1)$ in inequality 10 is to spread out the generated data points because having the data points concentrated on the boundary of the hypersphere is not an ideal outcome.

Note that we do not guarantee that condition in inequality 10 can certainly be achieved. One of the problems is that the gradient descent might get trapped in a saddle point or a bad local minimum. There is also an issue of numerical precision. The gradient might be almost zero and is thus represented as zero, before achieving inequality 10. In any of these situations, we simply reset z according to formula 7.

The pseudocode of the method is given in Algorithm 1.

Algorithm 1. Data generation algorithm by Support Vector Data Description with Gaussian kernel, initialization near a support vector, and gradient descent

---

Input: R, γ, α (Lagrange multipliers), and x (support vectors)
Output: z

1: **while** true **do**
2:     $dist\_target \leftarrow R^2 * rand(0,1)$
3:     initialize data point z according to Eq. 7 (requires x)
4:     $flag \leftarrow false$
5:     calculate $dist$ by Eq. 4 (requires γ, α, and x)
6:     **while** $dist > dist\_target$ **do**
7:         $pre\_update\_z \leftarrow z$
8:         update z by Eq. 9
9:         **if** $z = pre\_update\_z$ **then**
10:           $flag \leftarrow true$
11:           $break$
12:         **else**
13:           calculate $dist$ by Eq. 4 (requires γ, α, and x)
14:         **end if**
15:     **end while**
16:     **if** $\neg flag$ **then**
17:         $return\ z$
18:     **end if**
19: **end while**

Constrained Data Generation. We extend the idea of gradient descent into the case where the random data point we want to generate is required to have one feature to be within a certain interval.

We use the same initialization formula 7, except for the feature we want to constrain. For the feature, we initialize it to a random value within the permitted range. We also use the same update function 9 for the gradient descent, except that we only allow the update of the restricted feature if and only if the new value of the feature is within our constraint. Otherwise, we simply do not update that particular feature.

However, there is no guarantee that a data point which both satisfies the prescribed constraint and is inside the hypersphere exists at all, and we do not know whether such data point exists. In that situation, the program will run continuously without getting any result.

5   Experiments

We perform several experiments to compare the runtime and agreement of different γ and C combinations. Interested readers may refer to [17] for the details.

6   Conclusion

Our proposed method has two approaches, namely rejection sampling approach and heuristic and gradient descent approach.

The rejection sampling approach works in three steps. Firstly, we generate a random data point. Secondly, we test the data point against the membership decision function. Lastly, if the data point fails the test, we repeat from the first step. The runtime of the membership decision function itself is constant with respect to the location of the data point. If the hypersphere is large, the rejection sampling approach works efficiently. However, if the hypersphere is small, this approach does not work efficiently because many data points fall outside the hypersphere. To ameliorate this weakness, we propose the heuristic and gradient descent approach.

Gradient descent allows a randomly generated data point which falls outside the hypersphere to be brought into inside the hypersphere. However, if the data point is far from the hypersphere, it will take many steps to bring the data point into inside the hypersphere. Therefore, we employ a heuristic to find a good starting point which is near the hypersphere boundary so that we do not need many steps to bring the data point into inside the hypersphere. However, we do not want the gradient descent to always stop immediately after reaching the hypersphere lest the data points will be concentrated on the boundary. Therefore, we vary the objective distance between the generated data point and the hypersphere's centroid. Each data point generation attempt has a different objective distance. Therefore, the generated data points have a variety of distances to the centroid, which effectively randomizes the generated data set.

Acknowledgement. This work is supported by the National Research Foundation, Prime Minister's Office, Singapore under its Campus for Research Excellence and Technological Enterprise (CREATE) programme.

References

1. Blei, D.M., Ng, A.Y., Jordan, M.I.: Latent Dirichlet Allocation. J. Mach. Learn. Res. **3**, 993–1022 (2003)
2. Cozman, F.G., Cohen, I., Cirelo, M.C.: Semi-supervised learning of mixture models. In: Fawcett, T., Mishra, N. (eds.) Proceedings of the 20th International Conference on Machine Learning (ICML-03), pp. 99–106 (2003). http://www.aaai.org/Papers/ICML/2003/ICML03-016.pdf
3. Eshky, A., Allison, B., Ramamoorthy, S., Steedman, M.: A generative model for user simulation in a spatial navigation domain. In: EACL, pp. 626–635 (2014)
4. Goodfellow, I., Pouget-Abadie, J., Mirza, M., Xu, B., Warde-Farley, D., Ozair, S., Courville, A., Bengio, Y.: Generative adversarial nets. In: Ghahramani, Z., Welling, M., Cortes, C., Lawrence, N.D., Weinberger, K.Q. (eds.) Advances in Neural Information Processing Systems, vol. 27, pp. 2672–2680. Curran Associates Inc. (2014). http://papers.nips.cc/paper/5423-generative-adversarial-nets.pdf
5. Graves, A.: Generating sequences with recurrent neural networks. CoRR abs/1308.0850 (2013). http://arxiv.org/abs/1308.0850
6. Gregor, K., Danihelka, I., Graves, A., Rezende, D., Wierstra, D.: Draw: A recurrent neural network for image generation. In: Blei, D., Bach, F. (eds.) Proceedings of the 32nd International Conference on Machine Learning (ICML-15), JMLR Workshop and Conference Proceedings, pp. 1462–1471 (2015). http://jmlr.org/proceedings/papers/v37/gregor15.pdf
7. Kingma, D.P., Mohamed, S., Jimenez Rezende, D., Welling, M.: Semi-supervised learning with deep generative models. In: Ghahramani, Z., Welling, M., Cortes, C., Lawrence, N.D., Weinberger, K.Q. (eds.) Advances in Neural Information Processing Systems, vol. 27, pp. 3581–3589. Curran Associates Inc. (2014). http://papers.nips.cc/paper/5352-semi-supervised-learning-with-deep-generative-models.pdf
8. Liu, Y.H., Lin, S.H., Hsueh, Y.L., Lee, M.J.: Automatic target defect identification for tft-lcd array process inspection using kernel fcm-based fuzzy SVDD ensemble. Exper. Syst. with Appl. **36**(2, Part I), 1978–1998 (2009). http://www.sciencedirect.com/science/article/pii/S0957417407006240
9. Luo, H., Wang, Y., Cui, J.: A SVDD approach of fuzzy classification for analog circuit fault diagnosis with FWT as preprocessor. Expert Syst. with Appl. **38**(8), 10554–10561 (2011). http://www.sciencedirect.com/science/article/pii/S0957417411002934
10. McGrath, R., Pozdnukhov, A.: A generative model of urban activities: simulating a population. In: Proceedings of the 3rd ACM SIGKDD International Workshop on Urban Computing (2014)
11. Perry, G.L.W., Wainwright, J., Etherington, T.R., Wilmshurst, J.M.: Experimental simulation: Using generative modeling and palaeoecological data to understand human-environment interactions. Front. in Ecol. Evol. **4**, 109 (2016). http://journal.frontiersin.org/article/10.3389/fevo.2016.00109
12. Pimentel, M.A., Clifton, D.A., Clifton, L., Tarassenko, L.: A review of novelty detection. Signal Process. **99**, 215–249 (2014). http://www.sciencedirect.com/science/article/pii/S016516841300515X
13. Sakla, W., Chan, A., Ji, J., Sakla, A.: An svdd-based algorithm for target detection in hyperspectral imagery. IEEE Geosci. Remote Sens. Lett. **8**(2), 384–388 (2011)
14. Sanchez-Hernandez, C., Boyd, D.S., Foody, G.M.: One-class classification for mapping a specific land-cover class: Svdd classification of fenland. IEEE Trans. Geosci. Remote Sens. **45**(4), 1061–1073 (2007)

15. Smolensky, P.: Information Processing in Dynamical Systems: Foundations of Harmony Theory. Technical report, DTIC Document (1986)
16. Tax, D.M., Duin, R.P.: Support vector data description. Mach. Learn. **54**(1), 45–66 (2004). http://dx.doi.org/10.1023/B:MACH.0000008084.60811.49
17. Yunus, F., Dandekar, A., Bressan, S.: Data driven generation of synthetic data with support vector data description. Technical Report TRA6/17, National University of Singapore. https://dl.comp.nus.edu.sg/jspui/handle/1900.100/6428
18. Zhao, Y., Wang, S., Xiao, F.: Pattern recognition-based chillers fault detection method using support vector data description (svdd). Appl.D Energ. **112**, 1041–1048 (2013). http://www.sciencedirect.com/science/article/pii/S0306261912009348
19. Zheng, A., Goldenberg, A.: A generative model for dynamic contextual friendship networks. Technical Report, Machine Learning Department. Carnegie Mellon University (2006). http://ra.adm.cs.cmu.edu/anon/usr/ftp/anon/ml/CMU-ML-06-107.pdf

Conversion Rate Estimation in Online Advertising via Exploring Potential Impact of Creative

Junxiang Jiang[✉] and Huiqiang Jiang

Software School, South China University of Technology, Guangzhou, China
junxiang_jiang@163.com, seblock@mail.scut.edu.cn

Abstract. As a key criterion to measure ads performance, CVR quantitatively describes the proportion of users who take a desirable action (such as purchasing an item, adding to a cart, adding favorite items, etc.) on given ads in the ads ecosystem. Therefore, it is a critical issue to allocate ads-budget and increase advertisers profits. Focusing on improving the accuracy of CVR prediction in online advertising, this paper firstly analyzes and reveals the correlation underlying creatives associated with ads and CVR, which is excluded by most state-of-the-arts in this literature. Furthermore, we propose a novel LR+ model to utilize the potential impacts of creatives on predicting CVR. Experimental results and analysis on two public real-world datasets (REC-TMALL dataset and Taobao Clothes Matching dataset) validate the effectiveness of the proposed LR+ and demonstrate that the proposed LR+ outperforms typical models (e.g., LR, GBDT and linear SVR) in term of root mean square of error (RMSE).

Keywords: Online advertising · Machine learning · Conversion rate prediction

1 Introduction

With the proliferation of the Internet and e-commerce, digital marketing has become an effective method to promote brand and sell items. Therefore, the performance of online advertising is a particularly concerned issue for advertisers. In general, the criteria of online advertising performance include click through rate (CTR) [11], CVR [4], dwell time [3], bounce rate [13] and the user post-clicked behaviors [8]. Especially, CVR refers to the proportion of audiences who take a predefined, desirable action (such as purchasing an item, adding to a cart, adding favorite items, etc.) on given ads in the ads ecosystem. Estimating CVR accurately not only helps advertisers allocate budget and increase profits, but also improves the experience of audiences on the website so that it attracts more users to visit the website. Therefore, CVR prediction turns into an essential issue in online advertising.

© Springer International Publishing AG 2017
D. Benslimane et al. (Eds.): DEXA 2017, Part II, LNCS 10439, pp. 299–309, 2017.
DOI: 10.1007/978-3-319-64471-4_24

(a) creative 1   (b) creative 2

Fig. 1. Two different creatives.

People generally believe that the CVR of the ad is affected by several main factors which include: (1) the correlation between the audience and the ad; (2) the position of the ad on the web page; (3) the attribution of the item; (4) the quality of the ad. Researchers have studied CVR prediction and the correlation between CTR and creative [2], but there is no researchers who pay attention to the potential impact of creatives on predicting CVR. Figure 1 illustrates two creatives of one item. These two creatives have different layout, background and tones. In the same situation (such as the context of the ad), we present these two creatives to the same audiences, the CVRs are 0% and 2.27%. This brief comparison demonstrates that creative 2 makes more audiences to buy this item. From this perspective, the design of creatives has a potential impact on audiences' purchase behaviors. Therefore, in such situation, how to quantitatively analyze the potential impact of creatives on predicting CVR and improve the prediction accuracy of CVR by utilizing this impact, is the main point of this paper.

Against this background, we perform data analysis on Taobao[1] which is the famous e-commerce website in China. The analysis results demonstrate that the creative affects audiences' purchase behavior given the same audience and the same item. Therefore, we justify the creative has a potential impact on predicting CVR. However, there is no work to analyze and incorporate this factor into CVR prediction. This paper proposes a novel model called LR+ (logistic regression +) to incorporate the potential impact of creatives. In summary, the main contributions of this paper are two-fold:

- We quantitatively analyze the correlation between creatives and CVR, and then justify that the potential impact of the creative provides more cues to improve the prediction accuracy of CVR.
- The proposed CVR estimation algorithm can easily incorporate the potential impact of creative into CVR prediction so that it not only outperforms many state-of-the-art models, but also can be used to find out which creative attracts audiences to purchase.

The rest of this paper is organized as follows. In Sect. 2, we give a brief overview of the related work and limitations. Section 3 analyzes the correlation

---

[1] https://www.taobao.com.

between creatives and CVR so that the impact of creative on predicting CVR can be justified. By doing so, we describe the proposed model in detail. Section 4 performs a set of experiments and comparisons. Finally, Sect. 5 concludes this paper with future work.

2   Related Works

In most cases, the techniques used for CTR and CVR prediction share some commonalities. Therefore, we introduce the works of CTR/CVR prediction at the same time. Azimi et al. [2] firstly extracted some visual features from creatives and then incorporated these features into some state-of-the-art (such as LR and SVR), which is the most relevant work to ours. However, the conversion behavior is more complex than click behavior so that their work is not suitable for CVR prediction. Agarwal et al. [1] estimated CVR by utilizing the dependency between audience information and ad-publisher pair. Menon et al. [10] used data hierarchical constraints to estimate CVR/CTR via using collaborative filtering. In contrast with them, Lee et al. [9] proposed hierarchies contain user information and the structure becomes more complicated, and the ordering of hierarchies is not straightforward in some cases. Rosales et al. provided a detailed analysis of conversion rates in the context of non-guaranteed delivery targeted advertising [12]. Chapelle et al. [4] observed that the time delay between impression and click is so short, while the time delay between impression and conversion is so much longer, maybe days or weeks. Therefore, they proposed a CVR prediction model by utilizing the time delay. Richardson et al. [11] utilized LR to estimate CTR via using relative term CTR, ad quality, and other features for search engine advertising. Furthermore, Dave et al. [7] learned the CTR for rare/new ads from similar ads, Chapelle et al. [5] proposed a Dynamic Bayesian Network to estimate the CTR.

In summary, CVR prediction has been widely studied, but few researchers pay attention to the potential impact of creative on predicting CVR. Therefore, this paper analyzes and justifies that the creative affects CVR, and we propose a novel model to incorporate with the potential impact of creative.

3   Conversions

In this section, we present the potential impact of creatives on predicting CVR via conducting data analysis on Taobao. By doing so, we propose a novel model called LR+ to incorporate into the potential impact of creative.

3.1   The Impact of Creative on Predicting CVR

To analyze the potential impact of creatives on predicting CVR, we perform experiments on Taobao. Due to the page limitation, we select two types of items as an example. Figures 2 and 3 illustrate two groups of creatives of items. Keeping

(a) creative 1   (b) creative 2   (c) creative 3   (d) creative 4   (e) creative 5

Fig. 2. The first group of selected creatives. The price of this item is 239 yuan.

(a) creative 1   (b) creative 2   (c) creative 3   (d) creative 4   (e) creative 5

Fig. 3. The second group of selected creatives. The price of this item is 999 yuan.

Table 1. The data analysis on Taobao.

Item	Creative	CVR(%)
The price of this item is 239 yuan.	1	1.27
	2	1.34
	3	1.16
	4	1.70
	5	1.05

Item	Creative	CVR(%)
The price of this item is 999 yuan.	1	0
	2	0.12
	3	0.17
	4	0.11
	5	0.15

the condition (e.g., price, audience and item) unchanged, we adopt 5 creatives per item to experiment. By analyzing Table 1, we find that some creatives make ads gain high CVR but some creatives are the opposite. Obviously, the CVR of the ad is affected by creatives. As discussed above, we justify the potential impact of creatives on predicting CVR. From this perspective, incorporating such impact into CVR prediction provides more cues to improve prediction accuracy.

3.2   The Proposed Method

Motivated by the finding in Sect. 3.1, in this section, we propose a novel model by utilizing the potential impact of creative on predicting CVR.

Probabilistic Inference. This paper aims to estimate the CVR of the ad $Pr(conversion|click)$ by utilizing the potential impact of creatives on predicting CVR. Taking the purchase behavior as an example, audiences always choose one item to purchase from several similar items, and the creative of the ad is one of the most important criteria to measure whether the item satisfies audiences. Therefore, by analyzing the different response of audiences to creative, the conversion of the ad can be summarized into two parts: when the creative of the ad satisfies audiences, CVR is equal to $Pr(conversion|click, creative)$ which means the proportion of audiences who take a conversion action after they click the ad and the creative of the ad satisfies them; (2) otherwise, CVR is equal to Pr($conversion|click, \neg creative$) which stands for the proportion of audiences who take a conversion action after they click the ad but doesn't satisfy with the creative. In summary, the formulation of CVR is transformed into the following equation.

$$\Pr(conversion|click) = \Pr(conversion|click, creative)$$
$$\times \Pr(creative|click) + \Pr(conversion|click, \neg creative)$$
$$\times \Pr(\neg creative|click). \tag{1}$$

where Pr($creative|click$) refers to the proportion of audiences who satisfy with the creative after they clicked the ad, and Pr($\neg creative|click$) refers to proportion of audiences who dissatisfy with creatives after they clicked the ad.

Considering that Pr($conversion|click, \neg creative$) $\times$ Pr($\neg creative|click$) is always equal to zero in practice, therefore, to simplify the formulation of CVR, we change the formulation as follows

$$\Pr(conversion|click) = \Pr(conversion|click, creative)$$
$$\times \Pr(creative|click). \tag{2}$$

Notations. Suppose we have extremely sparse instances $\mathcal{X} = \{x_i\}_{i=1:N}$, creatives $\mathcal{S} = \{s_i\}_{i=1:N}$ and labels $\mathcal{Y} = \{y_i\}_{i=1:N}$. Besides, N is the number of instances, x_i represents the features characterizes the i-th instance, s_i represents the features characterizes the i-th creative, and y_i denotes the CVR associated with x_i. Given instances $\mathcal{X}$ and creatives $\mathcal{S}$, the target of the proposed model is to find out the most suitable labels $\mathcal{Y}$ which satisfies the following condition.

$$\arg\min \ \mathcal{L}(\mathcal{X}, \mathcal{S}, \mathcal{Y}). \tag{3}$$

where $\mathcal{L}(\mathcal{X}, \mathcal{S}, \mathcal{Y})$ is the loss function of the proposed model.

Model Formulation. By modeling the Eq. (2), it transforms into the following formulation, which is the formulation of the proposed model.

$$\mathcal{H}(x, s) = \mathcal{G}(x)\mathcal{T}(s), \tag{4}$$

where $H(x, s)$, $G(x)$ and $T(s)$ refer to the probabilities Pr($conversion|click$), Pr($conversion|click, creative$) and Pr($creative|click$).

Both models are generalized standard LR since LR is more suitable to solve the regression problem and always achieve good performance. These two models are shown in following equations.

$$\mathcal{G}(x) = \frac{1}{1 + exp(-\theta^T x)}, \tag{5}$$

$$\mathcal{T}(s) = \frac{1}{1 + exp(-\omega^T s)}, \tag{6}$$

where θ and ω are the parameter need to learn.

We consider that the audiences' satisfaction with creatives is closely related to the aesthetics of the creative. Therefore, firstly, we clawed some rated images from flickr[2]; secondly, we extract nine visual features from these rated images and creatives based on artistic intuition, such as average pixel density, tones, saturation hue and the rule of thirds [6]; thirdly, using these features of the rated images and corresponding labels to train a SVM-based assessment model [6][3]; finally, we use the trained assessment model to rate the creative as the input of $T(s)$.

Optimization. The parameters associated with the proposed model are learned by solving the regularized least squares problem

$$\mathcal{L}(\mathcal{X}, \mathcal{Y}, \mathcal{S}) = \frac{1}{2N} \sum_{i=1}^{N} (\mathcal{H}(x, s) - y_i)^2 + \frac{\lambda}{2N} \theta^T \theta + \frac{\lambda}{2N} \omega^T \omega \qquad (7)$$

Here, we put L2-norm for regularization to avoid over fitting on the loss function, λ is the regularization constant of parameters θ and ω.

The optimization method that we implement the experiments in this paper is a gradient descent algorithm. We estimate the model parameters by minimizing the regularized squared error function.

$$\frac{\partial \mathcal{L}(\mathcal{X}, \mathcal{Y}, \mathcal{S})}{\partial \theta} = \frac{1}{N} \sum_{i=1}^{N} (\mathcal{H}(x, s) - y_i) \mathcal{H}(x, s)(1 - \mathcal{G}(x)) + \frac{\lambda}{N} \theta \qquad (8)$$

$$\frac{\partial \mathcal{L}(\mathcal{X}, \mathcal{Y}, \mathcal{S})}{\partial \omega} = \frac{1}{N} \sum_{i=1}^{N} (\mathcal{H}(x, s) - y_i) \mathcal{H}(x, s)(1 - \mathcal{T}(s)) + \frac{\lambda}{N} \omega \qquad (9)$$

Since the optimization problem is unconstrained and twice differentiable, it can be solved with any gradient based optimization technique. We use limited-memory BFGS (LBFGS) algorithm to optimize the parameters, a state-of-the-art optimizer.

Complexity Analysis. Let q denote the total number of parameters in θ and ω. In the training procedure, we need $O(qN)$ time to compute the gradient, $O(q^2)$ time to approximately compute the Hessian matrix. Therefore, the total time complexity is $O(qN + q^2)\mu$ for μ iterations.

---

[2] www.flickr.com.
[3] Source Code: http://www.cs.unc.edu/~vicente/code.html.

4   Experiments

In this section, we introduce the experiments on two public real-world datasets: REC-TMALL dataset[4] and Taobao Clothes Matching dataset[5]. By comparing with three state-of-the-art (LR, GBDT and linear SVR), we verify the effectiveness of the proposed model.

4.1   Datasets and Features

REC-TMALL dataset contains several types of behavior logs (click, adding to favorite, adding to cart and the purchase), but it only has 549 ads in total and the number of purchase logs is small. Therefore, we treat the behaviors except the click as the conversion behavior.

Taobao Clothes Matching dataset only contains purchase behavior logs and 462008 ads in total, and the majority of items are purchased less than 100. Therefore, we assume a hyper parameter (which is equal to 100) instead of the clicks. Furthermore, we randomly sample 10000 ads from the dataset to experiment.

Besides, we split each dataset into the training set and the test set in the proportion of 9:1. Moreover, to analyze the robustness of the proposed model and make a comparison between LR+ and LR, we also perform experiment in the proportion 8:2 of each dataset.

Because datasets only have a few features to use, we adopt the title of the ad as features to experiment. The title of the ad is informative, including time, brand, gender and etc. But they contain plenty of useless features, such as punctuation, conjunction and some rarely used words. Thus, we select the words whose frequency is more than 1% and less than 50% size of the dataset.

4.2   Baseline Model

This paper adopts three state-of-the-art methods as baselines, i.e., LR, GBDT (Gradient Boosting Decision Tree), linear SVR (Support Vector Regression). Due to the page limitation, we do not introduce baselines in detail. Specifically, the target of ours is a regression problem; therefore, we optimize the model parameters of LR by minimizing the regularized squared error function through LBFGS algorithm.

4.3   Metrics

There are two metrics to measure the performance of the proposed model and baselines, i.e., Root Mean Squared Error (RMSE) and relative improvement (RelaImpr).

---

[4] https://tianchi.shuju.aliyun.com/datalab/dataSet.htm?spm=5176.100073.888.15.VlfDKh&id=2.

[5] https://tianchi.shuju.aliyun.com/datalab/dataSet.htm?spm=5176.100073.888.29.U0CN9W&id=13.

RMSE is often used to evaluate the performance of regression models. The formulation of RMSE is formed: $RMSE = \sqrt{\frac{1}{N}\sum_{i=1}^{N}(\widehat{y_i} - y_i)^2}$ where N is the size of instances, $\widehat{y_i}$ is produced by the prediction model and y_i refers to ground truth. The smaller the value, the better the performance is.

To compare the proposed model with baselines more intuitively, we introduce the metric RelaImpr. The formulation of RelaImpr as follow: $RelaImpr = \frac{x_b - x}{x_b}$ where x_b means RMSE of the baseline, and x means RMSE of the proposed model.

4.4   Performance Comparisons

We make a comparison between LR+ and baselines on REC-TMALL dataset and Taobao Clothes Matching dataset. By analyzing the experimental results, we demonstrate the advantages of our model in improving the accuracy of CVR prediction. Due to the proposed model LR+ is based on LR, thus the comparison between LR+ and LR can verify the significance of the potential impact of creative.

Comparison on REC-TMALL Dataset. The experimental results on REC-TMALL dataset are provided in Table 2. All results reported in the baselines and the proposed model use the same visual features and scores as side information whenever needed. The proposed model LR+ outperforms other models and the relative improvement reaches 7% at least, which verifies the effectiveness of the proposed model. Furthermore, LR+ outperforms LR, and this comparison indicates that involving the impact of creative in such non-linear manner is a significant advantage. By comparing with directly using visual features, we find that the score of creative leads to better performance of the proposed model.

Comparison on Taobao Clothes Matching Dataset. To verify the effectiveness of the proposed model sufficiently, we also perform experiments on Taobao Clothes Matching dataset. Table 2 details the results. As we can see, the proposed model LR+ is able to produce better predictions compared with all previously reported results. And the proposed model achieves 3% relative improvement at least except for GBDT which performs excellently in this dataset. We qualitatively observe that the prediction performance clearly benefit from the potential impact of creative. By comparing the performance of LR and LR+ after using visual features, these experiments produce a same conclusion: LR+ capture the potential impact of creative on predicting CVR better than LR, and the score of creative is more suitable than the visual features.

The Robustness Analysis of the Proposed Model. To analyze the robustness of LR+, we perform experiments on REC-TMALL dataset and Taobao Clothes Matching dataset. We vary the values of the hyper parameter λ, and draw the influence on the performance in Fig. 4. We can find that the best performance can be achieved when λ is equal to 0.5 both on REC-TMALL dataset and Taobao Clothes Matching dataset. Since the proposed method LR+ is based on LR, we make a comparison with a different ratio between the training set and the

Table 2. Comparison of different models on REC-TMALL dataset and Taobao Clothes Matching dataset. The letters "V" and "S" refer to 9 visual features which extracted from creative and the score which rated by image aesthetics assessment model. "+" means concatenate operator. For example, +V means that directly adding visual features into models. The RelaImpr reflects the difference between (LR+) + S and other method.

<table>
<tr><td colspan="3" align="center">REC-TMALL Dataset</td><td colspan="3" align="center">Taobao Clothes Matching Dataset</td></tr>
<tr><td>Model</td><td>RMSE</td><td>RelaImpr(%)</td><td>Model</td><td>RMSE</td><td>RelaImpr(%)</td></tr>
<tr><td>Linear SVR</td><td>0.1440</td><td>33.82</td><td>Linear SVR</td><td>0.1758</td><td>7.28</td></tr>
<tr><td>Linear SVR + V</td><td>0.1482</td><td>35.70</td><td>Linear SVR + V</td><td>0.1746</td><td>6.64</td></tr>
<tr><td>Linear SVR + S</td><td>0.1410</td><td>32.41</td><td>Linear SVR + S</td><td>0.1758</td><td>7.28</td></tr>
<tr><td>GBDT</td><td>0.1029</td><td>7.39</td><td>GBDT</td><td>0.1639</td><td>0.55</td></tr>
<tr><td>GBDT + V</td><td>0.1042</td><td>8.54</td><td>GBDT + V</td><td>0.1639</td><td>0.55</td></tr>
<tr><td>GBDT + S</td><td>0.1029</td><td>7.39</td><td>GBDT + S</td><td>0.1639</td><td>0.55</td></tr>
<tr><td>LR</td><td>0.1190</td><td>19.92</td><td>LR</td><td>0.1756</td><td>7.18</td></tr>
<tr><td>LR + V</td><td>0.1168</td><td>18.41</td><td>LR + V</td><td>0.1682</td><td>3.09</td></tr>
<tr><td>LR + S</td><td>0.1189</td><td>19.85</td><td>LR + S</td><td>0.1752</td><td>6.96</td></tr>
<tr><td>(LR+) + V</td><td>**0.0976**</td><td>2.36</td><td>(LR+) + V</td><td>**0.1634**</td><td>0.24</td></tr>
<tr><td>(LR+) + S</td><td>**0.0953**</td><td>-</td><td>(LR+) + S</td><td>**0.1630**</td><td>-</td></tr>
</table>

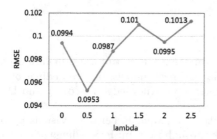

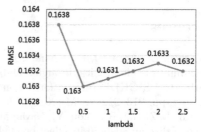

(a) REC-TMALL Dataset.   (b) Taobao Clothes Matching Dataset.

Fig. 4. The influence of hyper parameter λ.

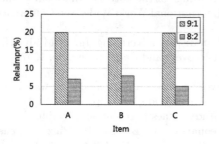

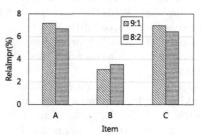

(a) REC-TMALL Dataset.   (b) Taobao Clothes Matching Dataset.

Fig. 5. Comparison between LR+ and LR on REC-TMALL dataset and Taobao Clothes Matching dataset. A, B and C stand for (LR+)+S vs. LR, (LR+)+S vs. LR+V and (LR+)+S vs. LR+S.

test set. The results are shown in Fig. 5. As we can see, there is much difference on REC-TMALL dataset while it is the converse on Taobao Clothes Matching dataset. The reason is that the scale of REC-TMALL is small, thus 80% of the dataset cannot train models sufficiently.

5 Conclusions

In this work, we propose a novel method called LR+, which allows to include the potential impact of creative in helping the CVR prediction task. Compared to the traditional algorithm which explores the utilization of different features, data hierarchical information and etc., the major novelty of our work is that we focus on the unexplored area: the potential impact of creative on predicting CVR, and we demonstrate the effectiveness of the proposed model via experimenting on two public real-world datasets. For the future work, we hope to explore more suitable methods which quantify the impact of creative on predicting CVR. Besides, we also hope to explore broader CVR prediction methods.

Acknowledgements. This research was supported by the National Natural Science Foundation of China (7167010139).

References

1. Agarwal, D., Agrawal, R., Khanna, R., Kota, N.: Estimating rates of rare events with multiple hierarchies through scalable log-linear models. In: Proceedings of the 16th ACM SIGKDD International Conference on Knowledge Discovery and Data Mining, pp. 213–222. ACM (2010)
2. Azimi, J., Zhang, R., Zhou, Y., Navalpakkam, V., Mao, J., Fern, X.: Visual appearance of display ads and its effect on click through rate. In: Proceedings of the 21st ACM International Conference on Information and Knowledge Management, pp. 495–504. ACM (2012)
3. Barbieri, N., Silvestri, F., Lalmas, M.: Improving post-click user engagement on native ads via survival analysis. In: Proceedings of the 25th International Conference on World Wide Web, pp. 761–770. International World Wide Web Conferences Steering Committee (2016)
4. Chapelle, O.: Modeling delayed feedback in display advertising. In: ACM SIGKDD International Conference on Knowledge Discovery and Data Mining, pp. 1097–1105 (2014)
5. Chapelle, O., Zhang, Y.: A dynamic Bayesian network click model for web search ranking. In: International Conference on World Wide Web, pp. 1–10 (2009)
6. Datta, R., Joshi, D., Li, J., Wang, J.Z.: Studying aesthetics in photographic images using a computational approach. In: Leonardis, A., Bischof, H., Pinz, A. (eds.) ECCV 2006. LNCS, vol. 3953, pp. 288–301. Springer, Heidelberg (2006). doi:10.1007/11744078_23
7. Dave, K.S., Varma, V.: Learning the click-through rate for rare/new ads from similar ads. In: Proceeding of the International ACM SIGIR Conference on Research and Development in Information Retrieval, SIGIR 2010, Geneva, Switzerland, pp. 897–898, July 2010

8. Lalmas, M., Lehmann, J., Shaked, G., Silvestri, F., Tolomei, G.: Promoting positive post-click experience for in-stream Yahoo Gemini users. In: Proceedings of the 21th ACM SIGKDD International Conference on Knowledge Discovery and Data Mining, pp. 1929–1938. ACM (2015)

9. Lee, K., Orten, B., Dasdan, A., Li, W.: Estimating conversion rate in display advertising from past performance data. In: Proceedings of the 18th ACM SIGKDD International Conference on Knowledge Discovery and Data Mining, pp. 768–776. ACM (2012)

10. Menon, A.K., Chitrapura, K.P., Garg, S., Agarwal, D., Kota, N.: Response prediction using collaborative filtering with hierarchies and side-information. In: Proceedings of the 17th ACM SIGKDD International Conference on Knowledge Discovery and Data Mining, pp. 141–149. ACM (2011)

11. Richardson, M., Dominowska, E., Ragno, R.: Predicting clicks: estimating the click-through rate for new ads. In: Proceedings of the 16th International Conference on World Wide Web, pp. 521–530. ACM (2007)

12. Rosales, R., Cheng, H., Manavoglu, E.: Post-click conversion modeling and analysis for non-guaranteed delivery display advertising. In: Proceedings of the Fifth ACM International Conference on Web Search and Data Mining, pp. 293–302. ACM (2012)

13. Sculley, D., Malkin, R.G., Basu, S., Bayardo, R.J.: Predicting bounce rates in sponsored search advertisements. In: Proceedings of the 15th ACM SIGKDD International Conference on Knowledge Discovery and Data Mining, pp. 1325–1334. ACM (2009)

Recommender Systems and Query Recommendation

A NoSQL Data-Based Personalized Recommendation System for C2C e-Commerce

Tran Khanh Dang[1(✉)], An Khuong Vo[1(✉)], and Josef Küng[2]

[1] HCMC University of Technology, Vietnam National University, Ho Chi Minh City, Vietnam
khanh@hcmut.edu.vn, voankhuong14@gmail.com
[2] FAW Institute, Johannes Kepler University Linz, Linz, Austria
jkueng@faw.jku.at

Abstract. With the considerable development of customer-to-customer (C2C) e-commerce in the recent years, there is a big demand for an effective recommendation system that suggests suitable websites for users to sell their items with some specified needs. Nonetheless, e-commerce recommendation systems are mostly designed for business-to-customer (B2C) websites, where the systems offer the consumers the products that they might like to buy. Almost none of the related research works focus on choosing selling sites for target items. In this paper, we introduce an approach that recommends the selling websites based upon the item's description, category, and desired selling price. This approach employs NoSQL data-based machine learning techniques for building and training topic models and classification models. The trained models can then be used to rank the websites dynamically with respect to the user needs. The experimental results with real-world datasets from Vietnam C2C websites will demonstrate the effectiveness of our proposed method.

Keywords: C2C e-commerce · Recommendation system · Ensemble learning · Topic modeling

1 Introduction

With the increasing popularity of the Internet, e-commerce markets are developing rapidly these days. Among types of e-commerce, the B2C (Business-to-Customer) websites in which manufactures or retailers sell their products to consumers are occupying the largest market share. However, with the trend towards more economical and convenient transaction between consumers, many people are now shifting to using the C2C (Customer-to-Customer) websites. The C2C model brings lower costs and higher profits for buyers and sellers. Therefore, C2C websites are currently growing strongly in quantity as well as in quality. In Vietnam, as a proof for this trend, there are now hundreds of active C2C marketplaces with some big names such as *Cho Tot* [20], *Nhat Tao* [21], *Vat Gia* [22], etc. This poses problems known as information overload that it is difficult for users to choose the suitable websites to sell their items. However, most of the current decision support systems, specifically e-commerce recommendation systems, only directly benefit the buyers [1, 2].

© Springer International Publishing AG 2017
D. Benslimane et al. (Eds.): DEXA 2017, Part II, LNCS 10439, pp. 313–324, 2017.
DOI: 10.1007/978-3-319-64471-4_25

In this paper, we are going to design a recommendation system which suggests suitable C2C websites to sell an item by applying machine learning techniques. The system is built in the context that the statistics, facts, and sale data is not provided. In this regard, the simplest way is selecting websites based on the category of the item. However, such category-based recommendation is not enough to model a specific item. After deeply studying the common features on some popular C2C websites, we focus on three features for processing: the description, the category, and the price of the item. Specifically, when a user provides the system those item's information, the system will recommend a ranked list of appropriate websites. In the construction of the C2C recommendation engine, one of the challenges we need to deal with is the unstructured text data on C2C websites, since the descriptions are usually written freely in user styles. To overcome this, we apply topic modeling [3–5], a form of text mining for identifying patterns in a corpus. We compute with topic modeling techniques to learn the latent subtopics in the descriptions to get the distribution of topics in each description and use the results as semantic features for further processing.

To personalize the recommendation system that can adapt to seller needs. We give two options for ranking websites. First, the websites are ranked in order of a specific criterion. For example, the websites are either ranked by the quantity or the average price of the similar items that are posted on each site. Second, the websites are ranked based on the whole input: the description, the category and desired selling price. For the first option, we measure the description similarity on the pre-processed semantic features and accumulate the results. For the second option, we build a classification model to predict a set of target websites for an item. The model must learn on single-label dataset but produce multiple labels. Moreover, the model must handle the data in which the same instance could belong to different labels since the same stuff could be posted to many similar websites. To solve this, we utilize the ensemble learning methods [6, 7], which have lots of learners that each learns a different aspect of the dataset and put them together to aggregate the results. The trained classifier, based on the voting scheme [8], ranks the websites in order of most voted.

The rest of the paper is organized as follows. We first introduce the related works in the Sect. 2. Section 3 goes into details of the proposed method. In Sect. 4, we describe our experimental setup and present the results. Finally, Sect. 5 concludes the paper and provides some potential ideas for future work.

2   Related Works

There are a few research works about C2C e-commerce recommendation systems have been conducted. In [9], Guangyao proposed a recommendation model for C2C online trading. The method calculates the similarity between current active users and other users based on a preference matrix. The matrix is constructed by user's behaviors on C2C context in four criteria: browse, attention, bidding, and purchase. Then, a recommendation list is built for each user based on most important items for his neighbors. Also, three-dimensional collaborative filtering technique which is extended from traditional two-dimensional collaborative filtering is an approach for C2C recommendation system [10]. The method firstly calculates seller similarities using seller features, and fills the rating matrix based on sales relations and seller similarities. Next, it calculates the buyer similarities using historical ratings, and defines neighbors of each user to predict unknown ratings. Lastly, the system recommends the seller and product combinations with the highest prediction ratings to the target user. In [11], Bahabadi et al. proposed a solution based on social network analysis. The research focused on users and transactions, which formed the map and the network to incorporate link prediction techniques for building recommendation system using prior transactions of users, categories of items, rating of users and reputation of sellers.

To the best of our knowledge, this is the first attempt to construct C2C e-commerce recommendation system aiming at choosing appropriate websites for target items. Besides, this paper uses machine learning techniques such as topic modeling and ensemble learning, an approach that has not been systematically studied in previous research works about C2C e-commerce recommendation system.

3   Proposed Approach

In this section, we will describe in detail: (i) how the C2C e-commerce database is constructed for further data processing, (ii) how the unstructured text data, the item's description, is represented, (iii) how the websites are ranked based on specific criteria such as quantity or price using document similarity, (iv) how the websites are ranked based on the voting principle of the classifier.

3.1   The C2C e-Commerce Database

The local database of C2C e-commerce is constructed with the aim to provide an extensive and reliable representation of items that can be queried in a reasonable time. Our database for recommendation engine is illustrated in Fig. 1. First, the data is collected from many C2C e-commerce websites. Then, it is extracted and stored in a local database. Since there is a vast volume of data, local repository needs to perform queries in a short time. Besides, the database must allow flexibility in various document structures in many websites, as well as allow an easy adaption to integrate new dataset into the system, this database is in form of NoSQL database. We choose Elasticsearch [12], a distributed, open-source full-text search and analytical engine which can serve as a schema-less database. It provides horizontal scalability and automatically maintains fault tolerance and load distribution. The efficiency of Elasticsearch is due to its quick searching on inverted index [13] instead of searching the text directly. As Elasticsearch does not enforce document structure, single index can store documents with different fields. Based on this, we create each index for each website.

Once the data is in indexed, we run searches and aggregations to mine the data needed for learning algorithms in the recommendation engine. After the models are trained, they are used to compute the relevance of documents as well as to rank the websites for user's queries. Additionally, the trained models can help to organize, search, and classify web documents in the database more effectively.

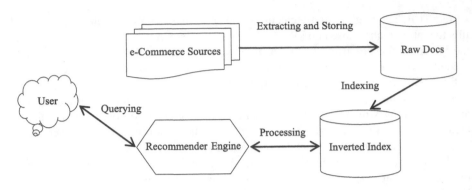

Fig. 1. The local C2C e-commerce database

3.2 Item's Description Representation

We first created a corpus using the descriptions of items from C2C e-commerce websites. First of all, each description is lowercased and tokenized, all the common words, as well as words that only appear once in all the descriptions are removed. Then descriptions are represented in bag-of-words (BOW) model [13]. Each document is represented by one vector where each vector element is represented by term frequency–inverse document frequency (tf-idf) weighting [13], denoted as:

$$tf * idf = f_{t,d} \log \frac{N}{n_t} \tag{1}$$

where $f_{t,d}$ is the number of occurrences of term t in document d, N is the total number of documents in a corpus and n_t is the number of documents where term t appears. The tf-idf value increases proportionally to the number of times a word appears in the document, but is compensated by the frequency of the word in the corpus, which reduces the importance of terms that occur very frequently. However, this BOW model alone is not enough to depict the underlying semantic meanings or concepts of documents. For example, the phrases "Samsung S6 32gb like new" and "Galaxy S6 32gb 99%" are the same in meaning but barely identical vectors in BOW model. Also, the model generates very high dimensional features, over 10000 dimensions.

To improve efficiency, topic modeling [3–5] technique is proposed to fit the constructed tf-idf corpus. Topic modeling is an unsupervised approach, aiming at discovering the hidden thematic structure within text body. Using themes to explore the content of the description, we can reveal various aspects of it such as brand, status, hardware specification, etc. A topic is considered as a cluster of words that tend to contextually co-occur together. It connects words with similar meanings and distinguish the uses of words with multiple meanings. Besides, text representation with topic semantic space can greatly reduce feature dimension comparing to BOW feature. This paper compares two most common techniques, the probabilistic model, Latent Dirichlet Allocation (LDA) [14], and more recently, the non-probabilistic model, Non-Negative Matrix Factorization (NMF) [15]. Both algorithms take as input the bag-of-words matrix

$A \in R^{n \times m}$. Each algorithm attempts to produce two smaller matrices: a word to topic matrix $W \in R^{n \times k}$ and a document to topic matrix $H \in R^{k \times m}$, that when multiplied together reproduce approximately matrix A, as illustrated in Fig. 2.

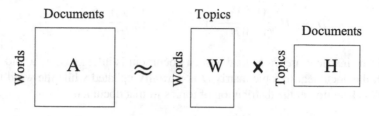

Fig. 2. Matrix decomposition for LDA and NMF

Latent Dirichlet Allocation (LDA) learns the relationships between words, topics, and documents by the assumption that documents are generated by a probabilistic model. Let K be the specified number of topics, N is the number of words in the corpus and two hyper parameters α and η that guide the distributions. Each topic z is associated with a multinomial distribution over the vocabulary β_k, drawn from $Dir(\eta)$. A given document d_i in total D documents, is then generated by the following process:

- Choose $\Theta_d \sim Dir(\alpha)$, a topic distribution for d_i
- For each word
  - Select a topic $z_{d,n} \sim \Theta_d$
  - Select the word $w_{d,n} \sim \beta_{z_{d,n}}$

In LDA model, the β and Θ distributions represent the probability of words used in each topic and the probability of each topic appearing in each document, which correspond to word to topic matrix W and document to topic matrix H, respectively. The process above defines a generative model corresponds to the following joint distribution, specifies a number of dependencies of the hidden and observed variables:

$$\prod_{i=1}^{K} p(\beta_i|\eta) \prod_{d=1}^{D} p(\Theta_d|\alpha) \left(\prod_{n=1}^{N} p(z_{d,n}|\Theta_d)p(w_{d,n}|\beta_{1:K}, z_{d,n}) \right) \tag{2}$$

Non-Negative Matrix Factorization (NMF) also factorizes A into two lower rank matrices W and H with the goal to reproduce the original matrix with the lowest error, with only one constraint: the decomposed matrices consist of only non-negative values. Using the Frobenius norm for matrices (a distance measure between two given matrices), the approximation of A is achieved by minimizing the error function:

$$\min_{W,H} \|A - WH\|_F, \quad \text{s.t } W \geq 0, H \geq 0 \tag{3}$$

NMF algorithm has fewer parameter choices involved in the modeling process than those of LDA. NMF uses an iterative process to update the initial values of W and H. The process stops when the approximation error converges or it reaches the specified number of iterations. The algorithm randomly initializes each matrix with non-negative values, iterate for each element index c, j, and i, and modify them as the following multiplicative update rule:

$$H_{cj} \leftarrow H_{cj}\frac{(W^TA)_{cj}}{(W^TWH)_{cj}};\ W_{ic} \leftarrow W_{ic}\frac{(AH^T)_{ic}}{(WHH^T)_{ic}} \tag{4}$$

After the matrices are trained, an unseen document is inferenced by a fold-in step, in which the document to topic matrix H is partially updated while the word to topic matrix W is kept, to get the distribution of topics in that document.

3.3   Description Similarity Based Ranking

As previously described by using LDA or NMF, the bag-of-words matrix is decomposed into two lower rank matrices. The document to topic matrix H is used to represent the descriptions and to compute the similarity score between two descriptions. Based on this and the category of the item, the websites are either ranked by the quantity or average price of similar items that are posted on each site.

The cosine similarity [13] is used as a distance metric to calculate the similarity score between two documents. Given two k dimension vectors d_i and d_j, which represent the distribution of k topics in two different descriptions, the value of the cosine angle between them is a real number in the range [0, 1]. Two vectors with the same orientation have a cosine similarity of 1 and two vectors which are perpendicular have a similarity of 0. The cosine angle between them can be calculated as follows:

$$cos(d_i, d_j) = \frac{\sum_k (d_i[k] \cdot d_j[k])}{\sqrt{\sum_k d_i[k]^2} \cdot \sqrt{\sum_k d_j[k]^2}} \tag{5}$$

For calculating the average price, after having a list of the items which have similar descriptions, we use the measures of data dispersion to get the range of price data. Let $x_1, x_2, \ldots, x_n$ be a set of ordered observations for the price attribute. The lower quartile Q_1, which cuts off the lowest 25% of the data and the upper quartile Q_3, which cuts off the lowest 75% of the data. We use the distance between the first and third quartiles which cover the middle half of the data to calculate the average price of the items whose prices fall into this range.

3.4   Voting Scheme Based Ranking

Since users may not only want to know the selling websites by the statistics of similar descriptions but they also want other aspects such as desired selling price to be taken into account, the recommendation system also has to deal with this scenario. Based on the pre-processed features: computed semantic features, along with category feature and price feature of the training dataset, the system learns to recommend a ranked list of suitable websites for users to sell their items. This problem can be considered as classification process. However, this is a non-standard classification task. The classifier must learn how to rank a list of labels for each of the input based on a single-label training dataset. Besides, the training dataset is informal in the way that the same instance could belong to different corresponding labels, due to the fact that a user may sell his item on more than one website. We propose the ensemble learning methods [6–8], specifically Random Forests method [16], where we utilize the voting principle of the algorithm to handle the recommendation.

The basic idea of ensemble learning methods is that by combining lots of learners, some learns certain things well and some learns others so that they perform in different ways, the generated results will be significantly better than any single learner. One of the approach to combine learners is bagging [16], which stands for bootstrap aggregating. Bagging is sampling the original dataset with replacement, so that some data points in the original dataset may be replicated in the sample, whereas other points may be absent from the sample. The bootstrap sample is the same size as the original, and many of the samples are taken. This method get each learner performs slightly differently, which achieves the requirement of ensemble methods. This is also what we desire, each learner learns a different aspect and to avoid overfitting on such informal dataset.

Random Forests is a meta-learner, belongs to the ensemble learning methods, which consists of many individual decision trees [16], that are used for classification using the information entropy concept. This tree based learner requires no input preparation which is suitable to handle our mixed-type data: categorical features and numerical features in different scales. In Random Forests, there are two different sources of randomness involved in the process of building trees. First, it uses bagging, each tree is trained on different random subset of data. Second, the random selection of the attributes to split at each node, which is only the subset of the whole set of features. Commonly, the subset size is equal to the square root of the number of all features. The randomness reduces the variance without effecting the bias. Also, it speeds up the training when at each node, there is fewer features to compute. The training algorithm is described as follows:

- For each of the N trees
  - Draw a bootstrap sample from the training data to train a decision tree
  - At each node, select a subset of m features
  - Pick the best split point among m features and split the node into child nodes
  - Repeat until the tree in complete

The algorithm works well on large dataset since each tree is independent of each other, trees can be trained in parallel. After the trees are trained, the output of the forest is the majority vote, which resemble the governments election system, a candidate must receive a majority of electoral votes to win the election. On this basis, the websites are ranked in order of decreasing number of votes.

4   Experiments and Results

4.1   The Dataset and Evaluation Method

We used the data from Vietnam C2C e-commerce, including 7 websites published by these companies: *Cho Tot* [20], *Nhat Tao* [21], *Vat Gia* [22], *Kypernet Viet Nam* [23], *Truyen Thong So* [24], *Viet Giang* [25] and *Mua Ban* [26]. The data collected in the first three websites have items of various categories, whereas the last four websites specialize in cars, motorbikes, cameras and luxury phones, respectively. In this experiment, we get from database about 55,000 pre-processed posts in which the poor quality data [17] are removed. Each post is a record with attributes containing its product's information (title description, category and price). We also get other 1,000 posts to evaluate the proposed approaches. We asked real users to judge the ranking of the websites for each of the item. The first set is ranked based on the specific criteria: the quantity or the average price of similar items, as the test set for description similarity based ranking. The second set is ranked based on the whole input information: description, category and desired selling price, to use as the test set for voting scheme ranking.

In this paper, we used evaluation metric which is specifically designed to measure the performance of a ranking system, the Normalized Distance-based Performance Measure (NDPM) [18]. Let C^- be the number of contradictory preference relations which happen when the system's ranking is opposite to the reference user's ranking for a pair, C^u be number of compatible preference relations which happen when the system preferred an item to another item but user sees them equal, and C^i be the total number of preferred relations, in same order, of the pairs between the system's and user's ranking. The NDPM is then given by:

$$NDPM = \frac{C^- + 0.5C^u}{C^i} \tag{6}$$

The NDPM measure gives a perfect score of 0 to systems where there is total agreement between system's ranking and user's ranking. Contradicting every reference preference relation gives the worst score of 1. Contradicting a reference preference is penalized twice as much as not predicting it.

4.2   Results and Interpretations

We used the NDPM metric to evaluate the performance of both the description similarity based ranking and voting scheme based ranking. The goal here is to minimize the average NDPM score on the test set.

In Fig. 3, we measured the performance of two topic modeling techniques, NMF and LDA, and simple BOW model on the recommendation task in which users want to sort the websites in order of a specific criterion, which is the quantity or the average price of similar items in a category. The experiment is conducted with different number of topics $n_{topic} \in \{50, 100, 150, 200, 250, 300, 350\}$ and we found that when the number of topics is increased to 250, the NDPM score for NMF method obtains the lowest score compared to both LDA and BOW in both criteria. It reaches approximately 0.16 in the quantity ranking and 0.18 in the average price ranking. On the other hand, LDA method performed worse on tests than both BOW and NMF model.

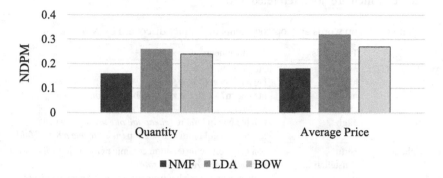

Fig. 3. NDPM score for three text presentation models, NMF, LDA and BOW for C2C website recommendation based on the quantity or the average price of similar items

Figure 4 shows the performance of the C2C website recommendation system based on voting principle of Random Forest. The experiment is conducted on BOW method with feature selection [19] for dimensionality reduction, NMF method with 250 topics and LDA method with 100 topics, where larger numbers of topics do not significantly improve the performance of each topic model. As the graph shows, the performance of NMF method far exceeded that of LDA and BOW method. As the number of trees is increased to about 100 trees, NDPM score reaches roughly 0.13 for NMF method. Again, LDA method gave the worst performance.

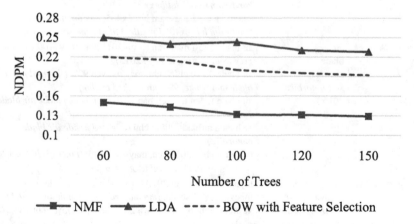

Fig. 4. NDPM score for C2C website recommendation based on voting rule of Random Forests

As shown in Table 1, we select 5 interpretable topics in 250 topics generated by NMF method as examples. We manually name each topic based on highest weighted words and highest weighted documents, which representing categories including *HTC One M smartphone, luxury phone, Sony Handycam camera, Air Blade motorbike* and

Toyota Innova car. These topics are about categorical groups of the items of several brands without going to details about specifications of the items. Besides, there are also others topics which are not interpreted well.

Table 1. Top words and top documents in 5 topics discoverd by NMF method

Topic	Top words	Top documents
Smartphone, HTC One M	one htc m7 e8 32gb	htc m8 gold 32gb htc one m8 gold new 99% htc one m7 - nguyên zin – fullbox *(htc one m7 - original – fullbox)* bán htc one m7 giá rẻ *(cheap htc one m7 for sale)* htc one m8 gold mới nguyên hộp *(new htc one m8 gold fullbox)*
Luxury phone	vertu constellation ti ascent goldvish	bán máy vertu constellation ceramic nguyên zin *(original vertu constellation ceramic for sale)* vertu ascent ti chính hãng giá rẻ *(cheap and genuine vertu ascent ti)* vertu constellation ayxta black alligator đẹp quý phái *(luxuriously beautiful vertu constellation ayxta black alligator)* vertu constellation pink nguyên zin cần bán *(original vertu constellation pink for sale)* goldvish-vertu các loại *(goldvish-vertu series)*
Camera, Sony Handycam	sony hadycam hd hdr-gw77 cx405	bán máy quay handycam hdr-gw77 99% *(99% handycam hdr-gw77 camera for sale)* sony handycam hdr-gw77 còn mới *(like new sony handycam hdr-gw77)* bán nhanh sony handycam hdr-gw77 *(sony handycam hdr-gw77 for quick sale)* sony handycam cx405 xách tay fullbox *(hand-carried sony handycam cx405 fullbox)* cần bán sony handycam hdr cx405, đầy đủ phụ kiện *(sony handycam hdr cx405 for sale, full of accessories)*
Motorbike, Air Blade	air blade zin xe *(motorbike)* 2015	air blade 125fi màu trắng xám 2015 zin 100% *(100% original grey white air blade 125fi 2015)* air blade đỏ xe chính chủ máy nguyên 99% *(licensed red air blade motorbike, 99% untouched engine)* air blade fi đen nguyên bản 2010 *(untouched black air blade fi 2010)* xe honda airblade chính chủ *(licensed honda air blade motorbike)* bán airblade fi 2015 đen, máy zin giá tốt *(black air blade fi 2015 for sale, untouched engine, good price)*
Car, Toyota Innova	toyota innova số *(transmission)* sàn *(manual)* chủ *(licensed)*	toyota innova g đời 2010 *(toyota innova g 2010 series)* innova g 2011 số sàn *(manual transmission innova g 2011)* xe chính chủ toyota innova 2011 *(licensed toyota innova 2011 car)* toyota innova 2016 xe gia đình còn zin 100% *(100% original toyota innova 2016 family car)* xe innova 2k15 2.0G màu bạc *(silver innova 2k15 2.0G car)*

5   Conclusion and Future Work

In this work, we explored the NoSQL data-based machine learning techniques in C2C e-commerce recommendation system for suggesting appropriate websites for target items, to solve the lack of similar systems. The unstructured data is first stored and indexed, then applied machine-learned models for feature representation and label ranking. We used topic modeling methods for text presentation, and utilized the voting scheme of ensemble learning methods to rank the websites. The experimental results showed that our proposed approach is effective. Using Non-Negative Matrix Factorization method for representing the distribution of topics in documents gained the best performance in both ranking based on similar descriptions and ranking based on votes of Random Forests. Specifically, the NDPM score reaches approximately 0.16, 0.18 and 0.13 in ranking according to quantity, to average price, and to whole item's information, respectively. This also showed that Random Forests, widely known for very good performance on classification task, can be used to rank labels efficiently. Further studies on applying these techniques to larger scale dataset and improving the performance by more sophisticated topic models and ranking models should be necessary.

References

1. Konstan, J.A., Riedl, J., Schafer, J.B.: E-commerce recommendation applications. Data Min. Knowl. Discov. **5**(1–2), 115–153 (2001)
2. Choi, I.Y., Kim, H.K., Kim, J.K., Park, D.H.: A literature review and classification of recommender systems research. Expert Syst. Appl. **39**(11), 10059–10072 (2012)
3. Blei, D.M., Carin, L., Dunson, D.B.: Probabilistic topic models. IEEE Sig. Process. Mag. **27**(6), 55–65 (2010)
4. Arora, S., Ge, R., Moitra, A.: Learning topic models–going beyond SVD. In: 2012 IEEE 53rd Annual Symposium on Foundations of Computer Science, pp. 1–10 (2012)
5. Andrzejewski, D., Buttler, D., Kegelmeyer, W.P., Stevens, K.: Exploring topic coherence over many models and many topics. In: 2012 Joint Conference on Empirical Methods in Natural Language Processing and Computational Natural Language Learning, pp. 952–961 (2012)
6. Dietterich, T.G.: Ensemble methods in machine learning. In: Kittler, J., Roli, F. (eds.) MCS 2000. LNCS, vol. 1857, pp. 1–15. Springer, Heidelberg (2000). doi:10.1007/3-540-45014-9_1
7. Shi, C., Kong, X., Yu, P.S., Wang, B.: Multi-label ensemble learning. In: Gunopulos, D., Hofmann, T., Malerba, D., Vazirgiannis, M. (eds.) ECML PKDD 2011. LNCS, vol. 6913, pp. 223–239. Springer, Heidelberg (2011). doi:10.1007/978-3-642-23808-6_15
8. Brown, G., Kuncheva, L.I.: "Good" and "Bad" diversity in majority vote ensembles. In: Gayar, N., Kittler, J., Roli, F. (eds.) MCS 2010. LNCS, vol. 5997, pp. 124–133. Springer, Heidelberg (2010). doi:10.1007/978-3-642-12127-2_13
9. Guangyao, C.: Research on the recommending method used in C2C online trading. In: 2007 IEEE/WIC/ACM International Conferences on Web Intelligence and Intelligent Agent Technology-Workshops, pp. 103–106 (2007)
10. Ai, D.X., Zuo, H., Yang, J.: C2C e-commerce recommender system based on three-dimensional collaborative filtering. Appl. Mech. Mater. **336**, 2563–2566 (2013)
11. Bahabadi, M.D., Golpayegani, A.H., Esmaeili, L.: A novel C2C e-commerce recommender system based on link prediction: applying social network analysis. CoRR, abs/1407.8365 (2014)

12. Kononenko, O., Baysal, O., Holmes, R., Godfrey, M.W.: Mining modern repositories with elasticsearch. In: 11th Working Conference on Mining Software Repositories, pp. 328–331 (2014)
13. Manning, C.D., Raghavan, P., Schütze, H.: Introduction to Information Retrieval. Cambridge University Press, Cambridge (2008)
14. Blei, D.M., Ng, A.Y., Jordan, M.I.: Latent Dirichlet allocation. J. Mach. Learn. Res. **3**, 993–1022 (2003)
15. Lee, D.D., Seung, H.S.: Learning the parts of objects by non-negative matrix factorization. Nature **401**(6755), 788–791 (1999)
16. Breiman, L.: Random forests. Mach. Learn. **45**(1), 5–32 (2001)
17. Dang, T.K., Ho, D.D., Pham, D.M.C., Vo, A.K., Nguyen, H.H.: A cross-checking based method for fraudulent detection on e-commercial crawling data. In: 2016 International Conference on Advanced Computing and Applications, pp. 32–39 (2016)
18. Gunawardana, A., Shani, G.: Evaluating recommender systems. In: Ricci, F., Rokach, L., Shapira, B. (eds.) Recommender Systems Handbook, pp. 265–308. Springer, Boston (2015). doi:10.1007/978-1-4899-7637-6_8
19. Yang, Y., Pedersen, J.O.: A comparative study on feature selection in text categorization. In: 14th International Conference on Machine Learning, pp. 412–420 (1997)
20. Cho Tot Co., Ltd. https://www.chotot.com
21. Nhat Tao E-Commerce JSC. https://nhattao.com
22. Viet Nam Price JSC. http://www.vatgia.com
23. Kypernet Viet Nam JSC. https://bonbanh.com
24. Truyen Thong So Co., Ltd. http://www.2banh.vn
25. Viet Giang Co., Ltd. http://mayanhcu.vn
26. Mua Ban JSC. https://muaban.net

Recommending Diverse and Personalized Travel Packages

Idir Benouaret[(✉)] and Dominique Lenne[(✉)]

Sorbonne Universités, Université de Technologie de Compiègne,
CNRS, Heudiasyc UMR 7253, Compiègne, France
{idir.benouaret,dominique.lenne}@hds.utc.fr

Abstract. The success of recommender systems has made them the focus of a massive research effort in both industry and academia. The aim of most recommender systems is to identify a ranked list of items that are likely to be of interest to users. However, there are several applications such as trip planning, where users are interested in package recommendations as collections of items. In this paper, we consider the problem of recommending the top-k packages to the active user, where each package is constituted with a set of points of interest (POIs) and is under user-specified constraints (time, price, etc.). We formally define the problem of top-k composite recommendations and present our approach which is inspired from composite retrieval. We introduce a scoring function and propose a ranking algorithm for solving the top-k packages problem, taking into account the preferences of the user, the diversity of recommended packages as well as the popularity of POIs. An experimental evaluation of our proposal, using data crawled from Tripadvisor demonstrates its quality and its ability to provide relevant and diverse package recommendations.

Keywords: Recommender systems · top-k packages · Diversity · Tourism

1 Introduction

With the beginning of the Web 2.0 era, the Internet began growing up and developing with massive speed. The amount of information in the tourism field available in the World Wide Web and its number of users have noticed an enormous increase in the last decade. All this information may be useful for those users who plan to visit a new city. Information about travel destinations and their associated resources is commonly searched for tourists in order to plan their trip. However, the amount of information and items get extremely huge, leading to an information overload. It becomes a big problem to find what the user is actually looking for. Search engines partially solves that problem, however personalization of information was not given. Recommender systems (RS) have emerged as a popular paradigm for enabling users to find what they might be interested in, complementing search engines. Recommender systems are tools for filtering

© Springer International Publishing AG 2017
D. Benslimane et al. (Eds.): DEXA 2017, Part II, LNCS 10439, pp. 325–339, 2017.
DOI: 10.1007/978-3-319-64471-4_26

and sorting items and information. They generally use opinions of a community of users to help individuals in that community to more effectively identify content of interest from a potentially overwhelming set of choices [11]. There is a huge diversity of algorithms and approaches that help creating personalized recommendations. Two of them are very popular: collaborative filtering and content-based filtering. They are used as a base of most modern recommender systems. However, classical recommender systems are confined to recommending *individual* items, e.g., books or DVDs. But, there are several applications where users are interested in packages, i.e., *sets* of items. Examples of such applications may include music [8] (e.g., play lists of songs on Last.fm), education [9] (e.g., course combination) and trip planning [14] (e.g., sets of Points of Interest).

During the last decades, recommender systems have found their way in the context of travel planning to help tourists finding relevant Points of Interest (POIs) that might be interesting for them. In trip planning, a user is interested in suggestions for POIs that could be very heterogeneous, e.g., museums, parks, restaurants, etc. Tourism recommender systems can benefit from a system capable of recommending POIs organized in packages rather than ranked lists, which will constitute an improved exploratory experience for the visitor. Therefore, there is a need to recommend for the user the top-k packages that match his preferences. Furthermore, there may be a cost and a time needed for visiting each POI, which the user may want to constraint with a budget. The budget may also simply be the number of POIs per package. Independently of the budget, there may be a notion of compatibility among items in a packages, that can be modeled as constraints that the user may specify: e.g., "no more than three museums in a package", "no more than two restaurants", "the total distance covered for visiting all POIs in a package should be less than 20 km", etc. Some so-called "Third generation" travel planning web sites, such as *YourTour*[1] and *nileGUIDE*[2], aim at assisting the user with suggestions of POIs integrating these kinds of features, but the suggestions are based only on the well-known POIs and neglect the personalization aspect. Thus, the use of these web sites is very limited.

The majority of the existing work in recommender systems for tourism focus on modeling users' profiles and the representation of Points of Interest, in order to get a ranking of the most relevant POIs according to the user preferences. However, the diversity of recommendations has never been the main focus of the proposed approaches. Nevertheless, it has been suggested that diversification has a positive effect on the satisfaction of users of recommender systems [5].

Our contribution in this paper is the design and implementation of a suggestion model promoting diversity and inspired from composite retrieval [2]. The approach we propose is to group recommendations in different packages, where each package is constituted with a set of POIs. The set of all package recommendations covers a wide diversity of themes. Each POI has a time and a price associated with it, and the user specifies a maximum total value for price and

---

[1] http://www.yourtour.com/.
[2] https://www.nileguide.com/.

time (budgets) for any recommended package. POIs in each package are chosen using a scoring function that takes into account the preferences of the user, the diversity of items including the package, and the popularity of the items in the package. The evaluation of our proposed system using a real dataset with data crawled from the website Tripadvisor, shows that our system is competitive, it can improve the diversity of recommendations without deteriorating their relevance.

The road map of the paper is as follows. In Sect. 2, we present the architecture of our system and give a formalization of the problem. We then describe our model and define the quality criteria of packages (Sect. 3). In Sect. 4, we describe our algorithm for computing the top-k packages. In Sect. 5, we subject our system to experimental analysis using a real dataset and investigate the quality of the recommended packages. In Sect. 6, we present related works. Finally, we conclude the paper in Sect. 7.

2   System Architecture and Problem Statement

2.1   System Architecture

In a traditional recommender system, users give ratings to items based on their personal interest. These ratings are then used by the system to predict ratings for items not rated by the active user. The predicted ratings can be used to give the user a ranked recommendation list. However, as we discussed it above, these ranked lists are not suitable for applications such as trip planning which deal with heterogeneous items.

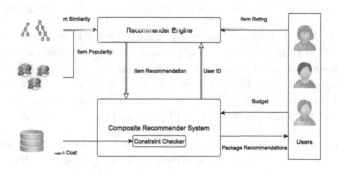

Fig. 1. Architecture of our system

As shown in Fig. 1, our system is composed of two main components: the recommender engine and the composite recommender module. The recommender engine captures users item ratings based on their preferences, and these ratings are used to predict an appreciation score for POIs not rated yet by the active user. The estimated appreciation score of a POI is calculated using items similarities, which are computed using the taxonomy presented in Fig. 2. In our system, each candidate POI has also a popularity assigned to it, which can be calculated using the number of users positive reviews, the number of visits per period or the number of "likes" via social networks, etc. Then, the composite recommender component receives information about the predicted ratings and the popularity of the candidate POIs. From this information, the role of the composite recommender system is then to provide the active user with the top-k packages. This component also includes a constraint checker module which checks whether a package satisfies the constraints specified by the user. The user interacts with the system by specifying his budgets in term of the total price and the total visiting time for each package, an integer k which is the number of packages and optionally compatibly constraints such as "no more than three museums in a package".

2.2   Problem Statement

Given a set I of POIs, a set U of users, an active user $u \in U$ and a POI $i \in I$. We denote by $c(i)$ the cost of POI i and by $t(i)$ the average time needed for visiting POI i. Given a set of POIs $P \subset I$, we define $Score(P)$ the score of a package P, which estimates the quality of a package (see more details in Sect. 3), $c(P) = \sum_{i \in P} c(i)$ the total cost of a package P and $t(P) = \sum_{i \in P} t(i)$ the time to visit POIs in package P. Given a cost budget B_c and a time budget B_t, a package P is said *Valid* iff $c(P) \leq B_c$ and $t(P) \leq B_t$.

We note that in this paper, we restrict attention to the problem of recommending packages where no compatibly constraint is imposed. We discuss in Sect. 4.2 how to extend our algorithms to take into account compatibility constraints. We also assume that the cost budget B_c and the time budget B_t do not include the cost and the time associate with traveling between POIs.

Problem 1. *Top-k Composite recommendations Given a set I of POIs, an active user u with his preferences background, a cost budget B_c, a time budget B_t and an integer k, a top-k composite recommender system has to determine the top-k packages $P_1, P_2, ..., P_k$ such that each P_i has $c(P_i) \leq B_c$, $t(P_i) \leq B_t$, and among all Valid packages, $P_1, P_2, ..., P_k$ have the k highest scores, i.e. $Score(P) \leq Score(P_i)$ for all Valid packages $P \notin \{P_1, P_2, ..., P_k\}$*

The complexity of the top-k composite recommendations problem has been studied in [1] and was proved to be NP-hard. More precisely, it has been shown that the problem can be reduced to the well known *Maximum Edge Subgraph* problem, which is NP-complete. It is therefore necessary to design efficient algorithms for the recommendation of top-k packages.

3   Model

3.1   Topical Distance and Similarity Between Points of Interest

Our distance between POIs is based on a taxonomy of hierarchical topic categories organized in a tree structure. Formally, we used a domain ontology developed by [6] to represent these categories. Figure 2 shows a part of this taxonomy.

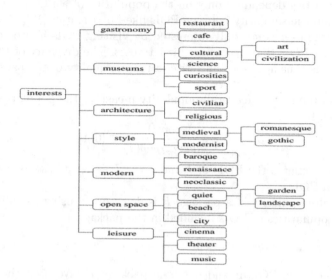

Fig. 2. Part of the taxonomy [6] used to represent the categories of POIs

Let be I the set of all possible POIs for potential suggestions. Each POI in
I is associated to one or more categories in the taxonomy, e.g. museum, park,
building, etc. We define the topical distance $dist_t : I \times I \rightarrow \mathbb{N}$ between two
POIs i and j as the length of the shortest path in the taxonomy between the
two closest categories of i and j in the taxonomy. Formally, we denote by C_i
and C_j the set of topical categories of POI i, j respectively. The topical distance
between the POIs i and j is computed using the following formula:

$$dist_t(i,j) = min_{c_i \in C_i, c_j \in C_j} dist_{edge}(c_i, c_j) \qquad (1)$$

where $dist_{edge}$ is the function computing the shortest path between two nodes
in the taxonomy.

The topical similarity between two POIs i and j can then be calculated as
a function of their topical distance. The similarity evaluates to what extent two
POIs i and j deal with the same thematic. The greater the topical distance
between two POIs, the smaller the similarity between these two POIs and vice
versa. Formally, the similarity between two POIs i and j is calculated as follows:

$$sim(i,j) = \frac{1}{1 + dist_t(i,j)} \qquad (2)$$

3.2 Packages Quality Criteria

In order to create the top-k packages of POIs for a specific user, it is necessary to have a criteria estimating how "good" a package P is, according to a user u. We denote it by the score of a package. To this end, we need to measure how a new POI not visited yet by the user would be interesting for him. The popularity is also an important factor in the general appreciation of a POI. In fact, the properties depending only on the popularity of a POI are often more important than the similarity between POIs liked by the user [12]. Moreover, we assume that the user is not only interested by visiting POIs he would like, but instead visiting POIs that best cover his interests. The diversity of POIs in the same package is thus an important criteria for the quality of the package.

Overall Popularity: The overall popularity measures the popularity of a POI i. It is defined by:

$$opop(i) = \frac{pop(i)}{max_{j \in I} pop(j)} \in [0, 1] \tag{3}$$

where j designates the POIs of I and $pop : I \to \mathbb{N}$ represents a popularity indicator of a POI.

The overall popularity of a package P is determined using the min aggregation of the popularities of POIs included in the package:

$$opop(P) = Min_{i \in P} opop(i) \in [0, 1] \tag{4}$$

Where i are the POIs included in the package P. We used the minimum operator as an aggregation function rather than the mean operator to ensure a high overall popularity of the packages.

Intra Package Diversity: Most of travel recommender systems focus on the modeling of user preferences and representation of POIs in order to get a ranking of the most pertinent POIs for a user. However, the diversity of suggestions in travel planning applications has never been the focus point. Nevertheless, it has been suggested that the diversity of the recommendations has a large positive effect on the satisfaction of the user [15]. In order to take into account the diversity of POIs in a package, we adapt the intra list diversity introduced by [15]. For a package of POIs P, we define the intra package diversity:

$$ipd(P) = \frac{\sum_{i,j \in P} 1 - sim(i, j)}{|P|^2} \tag{5}$$

Estimated Appreciation: The estimated appreciation evaluates to what extent a POI i that a user has not yet rated or visited would be interesting for a user u. This estimation is based on the preferences of the user and is calculated using ratings that he gave to a sample of similar POIs S_i pondered by the similarity between the POI to estimate and POIs of the sample. The estimated appreciation of a user $u \in U$ for a POI $i \in I$ is defined by:

$$eapp_u(i) = \frac{\sum_{j \in S_i} rating_u(j) \times sim(i,j)}{\sum_{j \in S_i} sim(i,j)} \tag{6}$$

where j designates POIs of the sample S_i, the set of similar POIs rated by the user u, and $rating_u \rightarrow [0,1]$ associates for a POI the rating given by the user u, divided by the maximum rate.

Note that our algorithm, that we describe in the Sect. 4 do not depend on a specific recommendation algorithm, we used a simple memory-based item-item collaborative filtering approach to generate predicted ratings for each user. This method is chosen for its simplicity and low computational cost.

The estimated appreciation for a user u for a package P is computed using the min aggregation of the estimated appreciations of POIs included in the package:

$$eapp_u(P) = Min_{i \in P} eapp(i) \in [0,1] \tag{7}$$

where i are the POIs included in the package P.

Score of a package: The score of a package evaluates the quality of POIs that form a package for a user u according to the overall popularity, the diversity and the estimated appreciation. The score of a package P for a user u is calculated by:

$$Score_u(P) = C_{eapp} \times eapp_u(P) + C_{opop} \times opop(P) + C_{div} \times ipd(P) \tag{8}$$

where $C_{eapp}, C_{opop}, C_{div}$ are positive Coefficients that modulate respectively, the importance of the estimated appreciation, the overall popularity and the diversity in the score function.

4   Construction of the top-k Packages

Composite recommendations can be solved by generating a set of candidate packages and then selecting the best possible subset. This approach is called Produce-and-choose. We choose it as a fundamental paradigm for solving our problem. The construction of top-k packages is done in two steps: first, a set of valid packages are produced in large quantities with a cardinality $c >> k$, packages are formed by aggregation around a pivot POI and taking into account the quality criteria. After that, the packages are ranked according to their respective score to recommend the top-k packages.

4.1   Creating Good Packages

Our approach for forming a set of good valid packages is inspired from the algorithm "BOBO" (*Bundles One-By-One*) introduced by Amer-Yahia *et al.* [2]. We adopted this algorithm to take into account the quality criteria of the packages (estimated appreciation, popularity and diversity) defined in Sect. 3.2. The goal of this algorithm is to create c valid packages that respect the budget constraints. It is inspired from $k - nn$ clustering. At each step a POI is chosen

---

Algorithm 1. BOBO

Input: I, B_c, B_t, number of packages c
Output: a set c of packages
1   $Packages \leftarrow \emptyset$
2   $Pivots \leftarrow Descending\_sort(I, opop)$
3   **while** $(Pivots \neq \emptyset)$ **and** $|Packages| < c$ **do**
4     $w \leftarrow Pivots[0]$
5     $Pivots \leftarrow Pivots - \{w\}$
6     $P \leftarrow$ Pick_bundle(w, I, B_c, B_t)
7     $Pivots \leftarrow Pivots - P$
8     $Packages \leftarrow Packages \cup P$
9   **end**
10 **return** $Packages$

Algorithm 2. PICK_BUNDLE

Input: pivot w, I, B_c, B_t
Output: a package P
1   $S \leftarrow w$
2   $active \leftarrow I - \{w\}$
3   $cost \leftarrow c(w)$
4   $time \leftarrow t(w)$
5   **while** $(not\ finish)$ **do**
6     $i \leftarrow argmax_{i \in active} Score_u(S \cup \{i\})$
7   **if** $(cost + c(i) \leq B_c)$ **and** $(time + t(i) \leq B_t)$ **then**
8       $S \leftarrow S \cup \{i\}$
9       $cost \leftarrow cost + c(i)$
10       $time \leftarrow time + t(i)$
11   **end**
12   **else**
13       $finish \leftarrow true$
14   **end**
15     $active \leftarrow active - i$
16   **end**
17 **return** S

---

as pivot, and a valid package with maximum score is built around that pivot. The pseudo code is described in Algorithm 1.

BOBO starts with an empty set of packages (line 1). Then, a list of candidates pivots POIs is constituted (line 2), this list ranks the list of potential suggestions (I) in a decreasing order of their overall popularity (*opop*).

As long as the number of formed packages is less than the number of required packages c, at each iteration the first POI is taken from the set of *Pivots* (line 4), and a package is built around it (line 6). This is done by the routine *Pick_bundle* described in Algorithm 2. This routine greedily keeps picking the next POI that maximizes the score of the package formed around the pivot (line 6), as far as the budget constraints are satisfied (line 7). If the selected POI respects the budget constraints then it is added to the package (line 8), its cost is added to the cost of the package (line 9) and its time is added to the whole time of the package (line 10). It is then discarded from the *active* POIs (line 15) so that it will not appear in any other package. Note that, without loss of generality, we assume that all POIs have a smaller cost than the cost budget B_c and have a smaller visiting time than the time budget B_t. Let us go back to BOBO's main loop: once a candidate package is created, it is added to *"Packages"* (line 8) and its elements are removed from *"Pivots"* (line 7) so that they are not longer used.

Once the required number of packages has been created, they are ranked following their respective scores (Sect. 3.2). Afterwards, the k packages having the best scores are selected as the top-k packages.

4.2 Discussion

As mentioned in Sect. 2.1, our system optionally includes the notion of a package satisfying compatibility constraints. E.g., the user may require the recommended package to contain no more than 3 museums. To handle these kind of constraints in our algorithms, we can add a definition of a Boolean compatibility function over the packages. Given a package P, $C(P) = true$ iff all constraints on POIs included in P are satisfied. Then, we can easily add a call to C in BOBO algorithm after each candidate package has been generated. If the package fails the compatibility check, we just discard it and search for the next candidate package.

5 Evaluation

5.1 Data Set

The goal of our experiments were: (1) evaluate the relevance of the packages recommended by our algorithm, and (2) evaluate their diversity as well. In order to have a set of POIs constituting potential recommendations, we have taken advantage of the well known travel website Tripadvisor, where users can share and explore travel information. Tripadvisor provides a list of POIs, each POI is associated with a thematic category. These categories are organized in a tree structure, which allows us to build the semantic similarity measure introduced in Sect. 3.1. In addition, Tripadvisor provides for a POI a set of social indicators, such as the average rating for all users, or the number of users that have rated it, etc. These information constitute an important popularity index. In particular, we used the number of users positive ratings for a POI as an indicator of its popularity, for estimating the function $pop : I \rightarrow \mathbb{N}$ defined in Sect. 3.2.

For our experiments, we crawled user rating information from POIs in the five most popular cities in France. We exclude POIs that have very few or no reviews. The dataset contains 40635 ratings for 1183 POIs by 18227 users, so as we see the data is very sparse. We associate with each POI its cost and its average time of visit crawled also from Tripadvisor. The average price of POIs was close to €7 and the average time for visiting a POI ranges from 30 min to 3 h. Because of the large sparsity of the underlying user rating matrix, we exclude users that have rated less than 30 POIs.

5.2 Evaluation Metrics

The goal of our experiments was to test the relevance and the diversity of the recommended packages of POIs. For this purpose we used two metrics. The first one is *Precision*, widely used in the evaluation of the relevance of recommended items. Precision is calculated as the ratio of recommended POIs that are relevant to the total number of recommended POIs. In our settings, we consider that a POI is relevant for a user if it was rated with 4 or 5 stars.

Precision: is calculated as the ratio of recommended POIs that are relevant to the total number of recommended POIs.

$$Precision = \frac{|relevant\ recommended\ POIs|}{|recommended\ POIs|} \tag{9}$$

The second used metric is for the purpose of evaluating the diversity of the recommended packages. We extend the intralist similarity introduced by *Ziegler et al.* [15] to a set of k packages $\{P_1, ..., P_k\}$. The Mean Intralist Diversity (MILD) is defined as bellow:

Diversity: we extend the intralist diversity introduced by *Ziegler et al.* [15] to a set of k packages $\{P_1, ..., P_k\}$. The Mean Intralist Diversity (MILD) is defined.

$$MILD(\{P_1, ..., P_k\}) = \frac{\sum_{i=1}^{k} ILD(P_i)}{k} \tag{10}$$

We used a third metric in order to compare approaches within a better compromise between precision and diversity. F_{PD} is the harmonic mean of precision and diversity:

$$F_{PD} = \frac{2 \times precision \times diversity}{precision + diversity} \tag{11}$$

5.3 Experimental Protocol

Our goal was to test the impact of personalization (per), popularity (pop) and diversity (div) on the quality of recommendations. To this end, we compared several versions of our system, corresponding to different possible combinations of the factors of C_{eapp}, C_{opop} C_{div}. Each version corresponds to a different combination of the parameters. Versions we tested are summarized in the Table 1. The name of each version indicates the use or not of the different aspects when constituting the set of packages.

Table 1. Different versions of our system

Versions of our system	C_{eapp}	C_{opop}	C_{div}
per (personalization)	1	0	0
pop (popularity)	0	1	0
div (diversity)	0	0	1
per + pop	1/2	1/2	0
per + div	1/2	0	1/2
pop + div	0	1/2	1/2
per + pop + div	1/3	1/3	1/3

Note that in our experiments, we do not seek to achieve the optimal combination of the weights C_{eapp}, C_{opop} C_{div}. Our goal is not to find the best combination of weights for each criteria, but rather to test the impact of the different criteria on the quality of recommendations. This justifies the simple values we have assigned to the different weights C_{eapp}, C_{opop} C_{div}. For each version, we wanted to test the impact of the presence or absence of each criteria.

To evaluate also the effectiveness of the proposed system, we compare our results with the package recommendation method proposed by *Xie et al.* [13], which is the closest work to ours. The authors explored approximate solutions to composite recommendations using a Fagin-style algorithm. The authors also computed the estimated appreciation of POIs using an item-item collaborative filtering approach, without taking in consideration the popularity of POIs and the diversity aspect of recommendations.

We tested our system varying the number of returned packages k, we vary it between four values: $\{5, 10, 15, 20\}$. The cost budget is fixed to €60 and the time budget fixed to 300 min. We tested our algorithms under various cost and time budgets with very similar results, so other budgets are not presented for lack of space. The variation of cost and time budget does not really affects the precision and diversity of recommendations, but rather affects the POIs that will be selected into each package, which means, if the cost budget is very small, this will basically limits the budget to only free attractions (POIs). As well, if the time budget is very small, the algorithm will tend to create packages which may contain very few POIs in a package or even empty packages. We fixed the cost budget to €60 and the time budget 300 min, which are medium budgets given the distribution of costs and times of all POIs.

We used the standard approach which consists of dividing the data into two categories randomly. We performed 5-fold cross validation: approximatively 80% of ratings were assigned to the training set and the remaining 20% to the test set. Ratings are translated into binary data to be able to use the precision metric. POIs ratings in the set $\{1, 2, 3\}$ are converted to "0" (irrelevant), and POIs ratings in the set $\{4, 5\}$ are converted to "1" (relevant). For each combination of the weights C_{eapp}, C_{opop} and C_{div} our algorithm uses the training set to produce a list of recommended k packages for each user. Our goal is to calculate the relevance of the POIs that are built in the k packages using the Formula 9. For this we must compare the recommendations with the truth values of the test set. Therefore, a POI recommendation is relevant if and only if the value of this POI in the test set is relevant. Moreover, for the recommended packages we compute the mean intralist diversity according to the Formula 10, as well as the harmonic mean between precision and diversity according to the Formula 11. The obtained results are presented and discussed in the following section.

5.4 Results

Results of our versions compared to the competitive approach according to precision, diversity and F_{PD} are reported in Table 2. In all our versions, we can notice a high influence of the popularity of POIs with respect to the precision. It is important to underline that the popularity is a significant factor as well as the personalization. In fact, in most cases, the "pop" version leads to a better precision than the "per" version and the "pop + div" version better than the "per + div" version. These results are in accordance with [12], which highlights the importance of the popularity and its effect on the relevance of recommendations. The div version performs the worst precision, since it ignores the personalization aspect and the popularity of POIs.

Table 2. Comparison of our different versions with the competitive approach

k=5

Version	P	D	F_PD
per	0.5073	0.3902	0.4411
pop	0.5875	0.4889	0.5336
div	0.4101	**0.6196**	0.4935
pop+div	0.5411	0.5533	0.5471
per+div	0.4836	0.5403	0.5103
per+pop	**0.5978**	0.4225	0.4950
per+pop+div	0.5589	0.5825	**0.5704**
Xie et al.	0.5724	0.4358	0.4948

k=10

Version	P	D	F_PD
per	0.5106	0.4112	0.4555
pop	0.5692	0.4133	0.5398
div	0.4301	**0.6039**	0.5023
pop+div	0.5381	0.5804	0.5584
per+div	0.4938	0.5784	0.5327
per+pop	**0.5403**	0.5057	0.5224
per+pop+div	0.5394	0.5965	**0.5665**
Xie et al.	0.5332	0.4811	0.5058

k=15

Version	P	D	F_PD
per	0.5263	0.4274	0.4771
pop	0.5236	0.4876	0.5049
div	0.4268	**0.5921**	0.4960
pop+div	0.5186	0.5606	0.5387
per+div	0.4839	0.5698	0.5233
per+pop	**0.5368**	0.4943	0.5146
per+pop+div	0.5191	0.5719	**0.5442**
Xie et al.	0.5165	0.5081	0.5122

k=20

Version	P	D	F_PD
per	0.4935	0.4285	0.4587
pop	0.5126	0.4792	0.4953
div	0.3808	**0.5781**	0.4591
pop+div	0.4967	0.5542	0.5238
per+div	0.4803	0.5316	0.5046
per+pop	**0.5121**	0.4825	0.4968
per+pop+div	0.5041	0.5602	**0.5306**
Xie et al.	0.5045	0.5285	0.5162

Varying the number of packages, the "per + pop" version always performs the best precision and outperform the algorithm of *Xie et al.*, due to combining the personalization and the popularity.

Concerning the diversity of recommendations, without surprise, the "div" version is the one that achieves the best diversity compared to all others. However, it has also the worst values for precision. In all cases, the "pop + div" performs a better diversity than "pop" and "per + div" better than "per". But we highlight that the "pop + div" and "per + div" versions performs a precision which is not close to the best precision score performed by the version "per + pop".

Let us now study the case of F_{PD} to analyze the compromise between precision and diversity. we notice that the "per + pop + div" realizes the highest values of F_{PD}, and out performs the competitive algorithm. This version tends to promote a large diversity, performs better than *Xie et al.*, and is not significantly different in precision comparing to the optimal "per + pop" version. Thus, the "per + pop + div" is the best approach when considering both precision and diversity.

Through this analysis, we argued on the quality of our recommended packages for the different settings. We especially confirmed our hypothesis on the importance of taking into account the popularity and the diversity in addition to the personalization aspect when constructing the top-k packages. We proved that combining these three aspects for the task of composite recommendations leads to a better results on the F-measure compared to other approaches.

6   Related Work

In [3], authors are interested in finding the top-k tuples of entities. Examples of entities could be cities, hotels and airlines, while packages are tuples of entities, they query documents using keywords in order to determine entity scores. A package in their framework has a fixed size, e.g., one city, one hotel and one airline. Instead, in our work, we allow packages (composite recommendations) of variable size, subject to a budget constraint specified by the user.

CARD [4] is a framework for finding top-K recommendations of composite products or services. A language similar to SQL is proposed to specify user requirements as well as how atomic costs are combined. However as in [3], recommended packages have a fixed size, making the problem simpler.

CourseRank [9] is a project motivated by a course recommendation for helping students planning their academic program at Stanford University. The recommended set of courses must also satisfy constraints (e.g., take two out of a set of five math courses). Similar to our work, each course is associated with a value that is calculated using an underlying recommender engine. Formally, they use the popularity of courses and courses taken by similar students. Given a number of constraints, the system finds a minimal set of courses that satisfy the requirements of the user and has the highest score.

The same authors in [10] extend Courserank with prerequisite constraints, and propose several approximation algorithms that return high-quality course recommendations satisfying all prerequisites. Like in our work, such suggestions of packages are not in a fixed size. However, [9, 10] do not consider the cost of items (i.e. courses), while we capture POI costs and user budget, which are essential features in the application of trip planning that we consider.

Other closely related work is [7] where a framework is proposed to automatically recommend travel itineraries from online user-generated data, like picture uploads using social websites such as Flickr. They formulate the problem of recommending travel itineraries that might be interesting for users while the travel time is under a given time budget. However, in this work, the value (score) of each POI is only determined by the number of times it was mentioned by other users in the social network, whereas in our work, the importance of a POI is determined not only by the popularity of the POI but also with a personalized score depending on user's preferences and his ratings for other POIs.

Finally, the closest work to our's is [13], where the authors explore returning approximate solutions to composite recommendations. The focus of the work is on using a Fagin-style algorithm for variable size packages and proving its optimality. The same authors further develop the idea into a prototype of recommender system for travel planning (CompRec) [14]. However, the score of an item is just the predicted rating of a user, while we believe that using also the popularity of items improves the relevance of the recommended packages. Furthermore, none of these works accounts for the diversity in packages which leads to a better satisfaction of the user.

7   Conclusion

Motivated by applications of trip planning, we studied the problem of recommending packages consisting of sets of POIs. Each POI is associated to a category, where categories are organized in a hierarchical tree structure, which allowed us to define a semantic similarity measure between POIs. Our composite recommendation system consists of ranking packages according to a score function, where the score of a package depends on the estimated appreciation, the overall popularity and the diversity of POIs constituting the package. We formalized the problem of generating top-k packages recommendations that are under cost and time budgets, where a cost and a time of visit are incurred by visiting each recommended POI and the budgets are user specified. We developed an algorithm for retrieving the top-k packages with best scores. The evaluation of our system using a real world dataset crawled from the website Tripadvisor demonstrates its quality and its ability to improve both the relevance and the diversity of recommendations. We plan now to realize a study of the proposed system with real users on a situation of mobility, where the localization context will take an important role on the recommendation process, the task will be to recommend the best packages for a given user provided that the recommended POIs are close to the position of the user. Furthermore, it will be interesting to compare between a recommender system providing classical ranked lists and our composite recommender system.

References

1. Amer-Yahia, S., Bonchi, F., Castillo, C., Feuerstein, E., Méndez-Díaz, I., Zabala, P.: Complexity and algorithms for composite retrieval. In: Proceedings of the 22nd International Conference on World Wide Web, pp. 79–80. ACM (2013)
2. Amer-Yahia, S., Bonchi, F., Castillo, C., Feuerstein, E., Mendez-Diaz, I., Zabala, P.: Composite retrieval of diverse and complementary bundles. IEEE Trans. Knowl. Data Eng. **26**(11), 2662–2675 (2014)
3. Angel, A., Chaudhuri, S., Das, G., Koudas, N.: Ranking objects based on relationships and fixed associations. In: Proceedings of the 12th International Conference on Extending Database Technology: Advances in Database Technology, EDBT 2009, pp. 910–921. ACM, New York (2009)
4. Brodsky, A., Henshaw, S.M., Whittle, J.: Card: a decision-guidance framework and application for recommending composite alternatives. In: Proceedings of the 2008 ACM Conference on Recommender Systems, RecSys 2008, pp. 171–178. ACM, New York (2008)
5. Candillier, L., Chevalier, M., Dudognon, D., Mothe, J.: Diversity in recommender systems: bridging the gap between users and systems. Centrics (2011). To appear
6. Castillo, L., Armengol, E., Onaindía, E., Sebastiá, L., González-Boticario, J., Rodríguez, A., Fernández, S., Arias, J.D., Borrajo, D.: Samap: an user-oriented adaptive system for planning tourist visits. Expert Syst. Appl. **34**(2), 1318–1332 (2008)
7. De Choudhury, M., Feldman, M., Amer-Yahia, S., Golbandi, N., Lempel, R., Yu, C.: Automatic construction of travel itineraries using social breadcrumbs. In: Proceedings of the 21st ACM Conference on Hypertext and Hypermedia, pp. 35–44. ACM (2010)
8. Hansen, D.L., Golbeck, J.: Mixing it up: recommending collections of items. In: Proceedings of the SIGCHI Conference on Human Factors in Computing Systems, pp. 1217–1226. ACM (2009)
9. Parameswaran, A., Venetis, P., Garcia-Molina, H.: A course recommendation perspective. Technical report, Recommendation systems with complex constraints (2009)
10. Parameswaran, A.G., Garcia-Molina, H.: Recommendations with prerequisites. In: Proceedings of the Third ACM Conference on Recommender Systems, pp. 353–356. ACM (2009)
11. Resnick, P., Varian, H.R.: Recommender systems. Commun. ACM **40**(3), 56–58 (1997)
12. Steck, H.: Item popularity and recommendation accuracy. In: Proceedings of the Fifth ACM Conference on Recommender Systems, RecSys 2011, pp. 125–132. ACM, New York (2011)
13. Xie, M., Lakshmanan, L.V.S., Wood, P.T.: Breaking out of the box of recommendations: from items to packages. In: Proceedings of the Fourth ACM Conference on Recommender Systems, RecSys 2010, pp. 151–158. ACM, New York (2010)
14. Xie, M., Lakshmanan, L.V.S., Wood, P.T.: Comprec-trip: a composite recommendation system for travel planning. In: 2011 IEEE 27th International Conference on Data Engineering (ICDE), pp. 1352–1355. IEEE (2011)
15. Ziegler, C.-N., McNee, S.M., Konstan, J.A., Lausen, G.: Improving recommendation lists through topic diversification. In: Proceedings of the 14th International Conference on World Wide Web, pp. 22–32. ACM (2005)

Association Rule Based Approach to Improve Diversity of Query Recommendations

M. Kumara Swamy[1]($\boxtimes$), P. Krishna Reddy[1], and Subhash Bhalla[2]

[1] Kohli Center on Intelligent Systems (KCIS), International Institute of Information Technology-Hyderabad (IIIT-H), Gachibowli, Hyderabad 500032, India
kumaraswamy@research.iiit.ac.in, pkreddy@iiit.ac.in
[2] Department of Computer Software, University of Aizu, Aizuwakamatsu, Japan
bhalla@u-aizu.ac.jp

Abstract. Query recommendation (QR) support search engine to provide alternative queries as a recommendation using similarity-based approaches. In the literature, orthogonal query recommendation (OQR) has been proposed to compute the diversity of QR when the user does not formulate proper queries. The OQR uses dissimilarity measure in QR to recommend completely different queries. In this paper, we propose an approach in QR by extending association rules, diverse patterns, and unbalanced concept hierarchy of search terms. We conceptualize association rules based QR, and order the rules based on *confidence* and *diversity*. Subsequently, the high ranked rules based on *confidence* and *diversity* are provided in QRs. The experimental results on real world AOL click-through dataset show that the diverse QRs improve the performance significantly.

Keywords: Query recommendation · Diverse query recommendations · Diversity · Diverse rank

1 Introduction

The search engines (google.com, yahoo.com, bing.com, etc.) employ the query recommendations (QRs) to improve the user satisfaction. The QR provides alternate set of queries incase of unsatisfactory query results. The search engine expects the users to click on the queries in recommendation to find the required information. Efforts have been made in the literature [2,10] to develop QR approaches in search engines using similarity measures. These approaches may fail, when the original query has not formulate using proper keywords. Orthogonal query recommendation (OQR) approach [11] has been proposed to recommend terms that are syntactically different from the original query and semantically similar. We observe that OQR generates different queries that may not be relevant to the user intent and also degrades the accuracy performance. The research issue is to generate the different queries as recommendation that are relevant to the user intent without degrading the accuracy performance.

© Springer International Publishing AG 2017
D. Benslimane et al. (Eds.): DEXA 2017, Part II, LNCS 10439, pp. 340–350, 2017.
DOI: 10.1007/978-3-319-64471-4_27

In this paper, we propose an improved approach using diverse QRs to generate potential queries that are relevant to the user intent without degrading accuracy performance. We extend *association rules, diverse patterns*, and *unbalanced concept hierarchy* of search terms for QRs. In this approach, we refine ranking mechanism over the approach proposed in [8] to rank rules based on the diversity of the queries using the unbalanced concept hierarchy. We divide the rules considering confidence, diversity, and combining both confidence and diversity measures. The experimental results on real world AOL click-through dataset show that the proposed approach with high confidence and high diversity is able to provide diverse QRs relevant to the user intent and improve the performance significantly over OQR approach.

The paper is organized as follows. Section 2 discusses the background. The proposed approach to compute the diversity is explained in Sect. 3. Proposed association rule based diverse query recommendation is discussed in Sect. 4. Section 5 presents experimental results. Last section contains summary and conclusions.

2   Background

In this section, we explain association rule based recommender system, and overview of diversity.

2.1   Association Rule Based Recommender System

Association Rules (ARs): The association among the items in a transactional data is identified by AR mining algorithms [1]. A rule is in the form $X \rightarrow Y$, where X, Y are itemsets and $X \cap Y = \emptyset$, states that X and Y present together in a transactional data. The rule is measured by a fraction of transactions that contains both X and Y at minimum support (*minsup*), indicates how frequently the itemset appears, and minimum confidence (*minconf*), indicates how often the rule is found, thresholds. Diversity is a new interestingness measure which is different from Lift, Conviction, AllConfidence, etc.

AR based Recommendations: In AR based recommender system (RS) [9], the top-N recommendations are generated. The items purchased by n users form transactional data. The AR mining algorithm generate rules satisfying *minsup* and *minconf* thresholds. For all the rules, the items purchased by a user are compared with the LHS of the rule. If they match, the item(s) on the RHS are recommended.

2.2   Overview of Diversity

We present an overview of concept hierarchy, diversity of a pattern and approach to compute the diversity of patterns using balanced concept hierarchy.

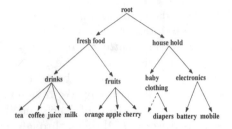

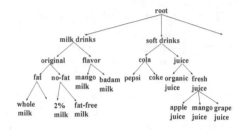

Fig. 1. An example of a balanced concept hierarchy (BCH)

Fig. 2. An example of an unbalanced concept hierarchy (UCH)

Concept Hierarchy (CH): A CH is a tree in which the data items are organized in an hierarchical manner. In this tree, all the leaf nodes represent the *items*, the internal nodes represent the *categories* and the top node represents the *root* (a virtual node). We consider that a lower-level node is mapped to only one higher-level node. The CHs can be balanced or unbalanced. In a balanced concept hierarchy (BCH) (refer Fig. 1), the height of all leaf level nodes is the same. In an unbalanced concept hierarchy (UCH) (refer Fig. 2), the height of at least one of the leaf level node is different from the height of other leaf level nodes.

Diversity of Pattern: The diversity of a pattern is based on the category of the items within it. If the items of a pattern are mapped to the same/few categories in a CH, we consider that the pattern has low diversity. Relatively, if the items are mapped to multiple categories, we consider that the pattern has more diversity.

Approach to Compute the Diversity of Patterns Using BCH: An approach was proposed in [7] to compute the diversity of a pattern. Given a BCH and a pattern, we assign diverse rank (*drank*) by analyzing the portion of BCH formed by the items in a pattern. The number of nodes vary based on the patterns, if the items in the pattern merge quickly, the corresponding projection has less number of nodes as compared to the projection of a pattern in which the items merge slowly.

We explain the process of calculating *drank* (0 to 1) of balanced pattern (BP). A pattern is called BP, if the height of all the items in a pattern is equal, i.e., at the same level of BCH. Let Y be a pattern with n items, C be a concept hierarchy of height h and $Y(\Pi(Y/C))$ be a projection of Y. The projection, $Y(\Pi(Y/C))$, is a sub-tree which represents a CH concerning to the items in Y. To compute the *drank*, we need two terms called *maximal* ($max\Pi(Y/C)$) and *minimal* ($min\Pi(Y/C)$). The $max\Pi(Y/C)$ is a projection of C for Y by considering all the leaf-level items merge at the root through distinct intermediate nodes and it is equal to ($|Y| \times h$). The $min\Pi(Y/C)$ is a projection of C for Y

by considering all the items merge at the immediate parent node and it is equal to $(|Y| + h - 1)$, where h is the height of C and $|Y|$ is the number of items Y.

Computing the drank of BP: The ratio of $|\Pi(Y/C)|$ to $|max\Pi(Y/C)|$ is called *drank* which is equal to $\frac{|\Pi(Y/C)|}{|max\Pi(Y/C)|}$. The minimum value of this ratio is equal to $\frac{|min\Pi(Y/C)|}{|max\Pi(Y/C)|}$ and the maximum value of this ratio is equal to $\frac{|max\Pi(Y/C)|}{|max\Pi(Y/C)|}$ (which is equal to 1). We replace the $|min\Pi(Y/C)|$ with $(|Y| \times h)$ and $|max\Pi(Y/C)|$ with $(|Y| + h - 1)$. The final formula after min-max normalization is shown in 1. The more explanation can be found in [7].

$$drank(Y) = \frac{(|\Pi(Y/C)|) - (|Y| + h - 1)}{(h - 1)(|Y| - 1)} \tag{1}$$

where, $|\Pi(Y/C)|$ is the number of nodes in $\Pi(Y/C)$, $|Y|$ is the number of items in a pattern, and h is the height of $\Pi(Y/C)$.

Fig. 3. The query recommendations in search engine, (a) query, (b) search results and (c) query recommendations.

3 Proposed Approach to Compute Diversity Using UCH

In this section, we explain problem description and basic idea, and approach to compute the diversity using unbalanced concept hierarchy (UCH).

3.1 Problem Description and Basic Idea

We illustrate the QR in Fig. 3, contains three parts: 'part a' is a search query, 'part b' is the search results, and 'part c' is the QRs. The QR may be inappropriate, if the user uses wrong key words. In similarity-based approaches, the variety/diversity is not considered, and in orthogonal query recommendation (OQR), completely different queries are recommended that may not be relevant to the user. The issue is to recommend queries that are relevant and variety/diversity without loosing the accuracy.

The *basic idea* is as follows. By processing click-through data, we identify click-through transactions of queries. There is an opportunity to improve diversity as well as accuracy, if high confidence and high diversity rules are generated. Such an approach will be able to recommend queries with high accuracy, reasonable diversity and nearer to user intent even though the user may input inappropriate query.

3.2 Approach to Compute the Diversity Using UCH

We propose the refinements to the approach proposed in [8] by capturing the diversity based on the ratio of number of edges of projection and maximal projection of a pattern. We define unbalanced pattern and extended UCH as follows.

Definition 1. *Unbalanced Pattern (UP)*: *Let Y be a pattern Y and U be a UCH of height 'h'. A pattern is called an UP, if the height of at least one of the item in Y is less than 'h'.*

Definition 2. *Extended Unbalanced Concept Hierarchy (E)*: *For a given UCH (U) with height 'h', we convert U into extended U, say E, by adding dummy nodes and edges such that the height of each leaf level item is equal to 'h'.*

The notion of unbalanced-ness depends on how the heights of the nodes in CH are distributed. Suppose, all the items of a pattern are at the height h. The Y is unbalanced, if h is less than the height of BCH.

The basic idea to compute *drank* of UP is as follows. We first convert the UCH to BCH called, "extended UCH" by adding dummy nodes and edges. We calculate the *drank* of UP with Eq. 1 by considering the "extended UCH." Next, we reduce the effect of dummy nodes and edges from *drank*. So, the *drank* of UP is relative to the *drank* of the same pattern computed by considering all of its items that are at the leaf level of the extended UCH. Given UP and the corresponding UCH (U), we compute *drank* of UP using the following steps: (i) Convert the U to the corresponding extended U, (ii) Compute the effect of the dummy nodes and edges, and (iii) Compute the *drank*.

(i) Convert the U to the corresponding extended U: We explain the extended UCH through the example as follows.

Example 1. Consider UCH in Fig. 4(i). Figure 4(ii) is the extended UCH of Fig. 4(i). In Fig. 4(ii), '∗' indicates dummy node and dotted line indicates dummy edge. In Fig. 4(i), the items *a, b, c,* and *d* are located at different levels. Considering the height *h* of $\langle root, l, k, a \rangle$, the extended UCH is generated adding dummy nodes and edges as in Fig. 4(ii). The projections of patterns {a,b}, {b,c}, {b,d}, and {c,d} are shown from Fig. 4(iii) to (vi) respectively.

(ii) Compute the Effect of the Dummy Nodes and Edges: We define the notion of adjustment factor to compute the effect of dummy nodes and edges.

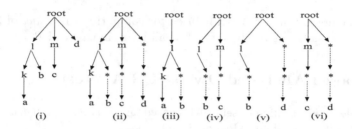

Fig. 4. (i) UCH, (ii) Extended UCH of (i). The projections of patterns {a, b}, {b, c}, {b, d}, and {c, d} are shown in (iii), (iv), (v), and (vi) respectively in Extended UCHs.

Adjustment Factor (AF): The *AF* reduce the *drank* by measuring the contribution of dummy edges/nodes relative to the original edges/nodes. The *AF* for *Y* should depend on the ratio of number of real edges formed with the children of the real nodes in $\Pi(Y/E)$ versus total number of edges formed with the children of dummy nodes in $\Pi(Y/E)$. The value of *AF* at a given height should lie between 0 and 1. If the number of real edges is equals to zero, *AF* is zero. If the pattern does not contain dummy nodes/edges, the value of *AF* is 1. The *AF* for *Y* is denoted as $AF(Y, \Pi(Y/E))$ and is calculated as follows.

$$AF(Y, \Pi(Y/E)) = \left[\frac{(|\Pi(Y/C)|) - (|Y| + h - 1)}{(h-1)(|Y| - 1)} \right] \left[\frac{\# \ of \ Dummy \ Edges \ of \ UP}{\# \ of \ Total \ Edges \ of \ UP} \right] \quad (2)$$

We compute the total number of edges in UP including actual and dummy using $|\Pi(Y/E)|$ and we replace '*# of Total Edges of UP*' with '$|\Pi(Y/E)|$'. On simplification, we get the following equation.

$$AF(Y, \Pi(Y/E)) = \frac{[(|\Pi(Y/E)|) - (|Y| + h - 1)] \ [\# \ of \ Dummy \ Edges \ of \ UP]}{(h-1)(|Y| - 1) \ (|\Pi(Y/E)|)} \quad (3)$$

where *numerator* is the number of edges formed with the children of the real nodes and *denominator* is the number of edges formed with the children of both real and dummy nodes in $\Pi(Y/E)$.

Example 2. Consider a frequent pattern $Y = \{whole\ milk, pepsi, coke, shampoo\}$ in Fig. 2. The $|\Pi(Y/E)| = 14$, $|Y| = 4$, $h = 4$ and '*# of Dummy Edges of UP*' = 4. The $AF(Y, \Pi(Y/E)) = \frac{[14-(4+4-1)]\times[4]}{(4-1)(4-1)(14)} = \frac{28}{126} = 0.2222$.

(iii) Computing the *drank*: The *drank* of UP is defined as follows.

Definition 3. *Diverse rank of a pattern* Y *(drank(Y)): Let* Y *be a pattern and* U *be a UCH of height 'h'. The drank of* Y *(drank(Y)) is given Eq. 4.*

$$drank(Y) = drank(Y) - AF(Y, \Pi(Y/E)) \tag{4}$$

where, $drank(Y)$ is the *drank* of a BP and $AF(Y, \Pi(Y/E))$ is the AF of Y.

Example 3. In continuation of the Example 2, the $drank(Y) = 0.7778$ and the $AF(Y, \Pi(Y/E)) = 0.2222$. The final $drank(Y) = 0.5556 (= 0.7778 - 0.2222)$.

It can be noted that Eq. 4 can be used for computing the *drank* of both BP and UP as the values of AF becomes 1 at all levels in case of BP.

4   Proposed AR Based Diverse QR Approach

It is possible to select different set of ARs based on $confidence$ and $drank$ values. As a result, different sets of ARs could be formed.

- HCLD: Top-N rules with high $confidence$ and low $drank$.
- HCHD: Top-N rules with high $confidence$ and high $drank$.
- LCLD: Top-N rules with low $confidence$ and low $drank$.
- LCHD: Top-N rules with low $confidence$ and high $drank$.

The rules with LCLD are not interesting due to low confidence and diversity. The rules with HCLD, HCHD and LCHD represent different types of interestingness with variations in accuracy and diversity. The framework is as follows.

i. **Data set preparation**
   (a) **Preparation of web click-through data:** Convert click-through data to transactional data. For each session, the queries executed by a user forms a transaction. Set of such transactions form transactional data.
   (b) **Formation of CH:** The queries in the transactional data are organized in hierarchical manner.
ii. (a) **Generation of ARs:** The ARs [1] are generated at $minsup$ and $minconf$ from transactional data. Top-N high confidence rules are selected for QR.
   (b) **Computing the $drank$:** For each set of items in the rules, we compute $drank$ using the Eq. 4.
iii. **Recommendations:** Top-N rules are used for QR. The rules are selected using following approaches.
   (a) HCLD: Improving the accuracy for QR.
   (b) HCHD: Improving both accuracy and diversity for QR.
   (c) LCHD: Improving diversity for QR.

5   Experimental Results

In this section, we explain preparation of dataset and methodology, and results.

5.1   Preparation of Dataset and Methodology

We carry the experiments on AOL click-through dataset [4]. The dataset consists of ≈20 million web queries from ≈650000 users over a period of three months.

Table 1. DMOZ hierarchy levels and their topic count.

Hierarchy level	Topic count	Hierarchy level	Topic count	Hierarchy level	Topic count	Hierarchy level	Topic count	Hierarchy Level	Topic count
0	1	3	7764	6	109847	9	107407	12	3906
1	17	4	39946	7	167528	10	56265	13	648
2	656	5	89934	8	165460	11	15903	Total	765282

Concept Hierarchy: We generate CH using the hierarchical structure of the open web directory - DMOZ [5], a comprehensive human-edited directory for all the queries from leaf level to 'top'. ≈765282 topics are made into 17 categories [6]. The categories at each level are shown in Table 1. The CH is a UCH, we compute the diversity using UP. Table 2 shows the paths of sample queries. For each query in the pattern, the paths are extracted. We observe that DMOZ gives multiple paths for a query. Finding a suitable path is one of the issues. We select the suitable path considering "the high support patterns have low diversity and low support patterns have high diversity [8]".

Table 2. Sample queries and their paths.

SNo	Query	Path
1	Animation World Network	Top/Arts/Animation/
2	Hitoshi Doi	Top/Arts/Animation/Anime/Collectibles/

Methodology: Using Apriori algorithm [1], we generate ARs at $minsup = 0.01$ and $minconf = 0.10$. The transactional data is considering session = 30 min. The dataset is divided into five distinct splits of training and test data. The training set is used to generate the ARs, and the test set is used for QRs. The ARs extracted from training set, called top-N rules. We then look into the test set and match items with top-N rules. The items that appear in the LHS of the rule is matched to the test set, if they match, the RHS item is included in *Hit* set considering match as a *Hit*. The *Hit* set is used for testing accuracy. The experiments are performed on five splits, and average is reported. We employ Precision, Recall, F1-metric [3], and Diversity (refer Eq. 4) as performance metrics. We also compute performance of OQR using Jaccard Coefficient similarity measure. In all the experiments, we evaluate the performance for 1 to 10 QRs.

5.2 Results

Figure 5 shows the precision of HCHD, HCLD, LCHD, and OQR. The HCHD out performs the other approaches. As the ARs in HCHD contains both high confidence and high diverse queries which are used by large number of users. It can be observed that the performance of LCHD have significantly low as compared to HCHD and HCLD. This is due to the fact that LCHD does not consider support and confidence values. Still, the performance of LCHD does not reach 0. This is due to the appearance of few high confidence rules. The performance OQR is lower than the other three approaches and observe that it is also not reached to 0, as some users have used these QRs. Similar trends can be observed in recall and F1-metric as shown in Figs. 7 and 8 respectively.

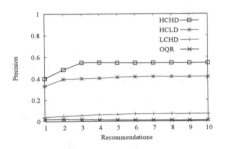

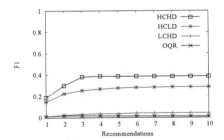

Fig. 5. Precision   Fig. 6. Diversity

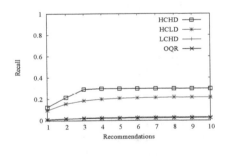

Fig. 7. Recall   Fig. 8. F1-Score

In Fig. 6, we show the diversity performance. The results show that HCHD is returning the results with high drank as compared to HCLD. The performance of LCHD is high as compared to HCHD, HCLD, and OQR. Ideally, the OQR is expected to perform better than other approaches, it happened at one QR only and in other QRs the performance is lower than the LCHD. The reason could be that the completely different recommendations that are not relevant to user intent. The performance of HCHD is higher than HCLD and is nearer to LCHD. From 4 QRs onwards, the HCHD approach got high performance than OQR approach. The reason is HCHD approach has both high confidence and high diversity which are liked by the users.

From the results, it can be concluded that the performance of HCHD approach gives recommendations with high diversity nearer to the user intent and significantly improves the precision performance.

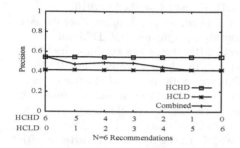
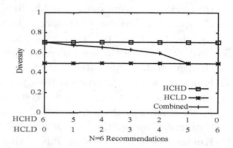

Fig. 9. Precision of HCLD and HCHD

Fig. 10. Diversity of HCLD and HCHD

We have also conducted experiment by combining the rules of HCLD and HCHD. Selecting x rules from HCHD and $(6 - x)$ rules from HCLD, where x is varied from 1 to 6. From both Figs. 9 and 10, the following are the observations.

– By selecting 2 rules from HCLD and 4 rules from HCHD high precision which is equal to HCHD could be achieved by improving diversity considerably.
– By selecting 4 rules from HCHD and 2 rules from HCLD high diversity which is equal to HCHD could be achieved by improving the precision considerably.

From the results, it can be consider that the performance of the accuracy and diversity can be adjusted based on the user requirements.

6   Summary and Conclusion

In search engine, QR guides the users when search results do not satisfy the user's requirement. It is important to build better QR. Normally, similarity-based QRs fail, if the user input an inappropriate query. In this paper, we propose an approach called diverse QR by exploiting ARs to generate potential QRs that are relevant to the user intent without compromising accuracy. We conduct the experiments on real world dataset and show that the proposed approach is able to provide relevant diverse QRs with high accuracy. As a part of future work, we are planning to propose an approach in network data and social networks.

References

1. Agrawal, R., Imieliński, T., Swami, A.: Mining association rules between sets of items in large databases. In: The SIGMOD, pp. 207–216. ACM (1993)
2. Baeza-Yates, R., Hurtado, C., Mendoza, M.: Query recommendation using query logs in search engines. In: Lindner, W., Mesiti, M., Türker, C., Tzitzikas, Y., Vakali, A.I. (eds.) EDBT 2004. LNCS, vol. 3268, pp. 588–596. Springer, Heidelberg (2004). doi:10.1007/978-3-540-30192-9_58
3. Basu Roy, S., Amer-Yahia, S., Chawla, A., Das, G., Yu, C.: Constructing and exploring composite items. In: The SIGMOD, pp. 843–854. ACM (2010)
4. Campos, R.: Analysis of temporal data in explicit temporal queries. aol dataset (aol_ds) (2011). http://www.ccc.ipt.pt/ricardo/experiments/AOL_DS.html
5. DMOZ: DMOZ - the open directory project (2016). https://www.dmoz.org/
6. Henderson, L.: Automated text classification in the dmoz hierarchy. TR (2009)
7. Kumara Swamy, M., Krishna Reddy, P.: Improving diversity performance of association rule based recommender systems. In: Chen, Q., Hameurlain, A., Toumani, F., Wagner, R., Decker, H. (eds.) DEXA 2015. LNCS, vol. 9261, pp. 499–508. Springer, Cham (2015). doi:10.1007/978-3-319-22849-5_34
8. Kumara Swamy, M., Reddy, P.K., Srivastava, S.: Extracting diverse patterns with unbalanced concept hierarchy. In: Tseng, V.S., Ho, T.B., Zhou, Z.-H., Chen, A.L.P., Kao, H.-Y. (eds.) PAKDD 2014. LNCS (LNAI), vol. 8443, pp. 15–27. Springer, Cham (2014). doi:10.1007/978-3-319-06608-0_2
9. Sarwar, B., Karypis, G., Konstan, J., Riedl, J.: Analysis of recommendation algorithms for e-commerce. In: Conference on e-Commerce, pp. 158–167. ACM (2000)
10. Silverstein, C., Marais, H., Henzinger, M., Moricz, M.: Analysis of a very large web search engine query log. SIGIR Forum 1, 6–12 (1999)
11. Vahabi, H., Ackerman, M., Loker, D., Baeza-Yates, R., Lopez-Ortiz, A.: Orthogonal query recommendation. In: 7th RecSys, pp. 33–40. ACM (2013)

How to Find the Best Rated Items on a Likert Scale and How Many Ratings Are Enough

Qing Liu[1]([✉]), Debabrota Basu[1], Shruti Goel[2], Talel Abdessalem[3],
and Stéphane Bressan[1]

[1] School of Computing, National University of Singapore, Singapore, Singapore
{liuqing,debabrota.basu}@u.nus.edu, steph@nus.edu.sg
[2] Jacobs School of Engineering, University of California San Diego,
La Jolla, CA, USA
s2goel@eng.ucsd.edu
[3] LTCI/IPAL CNRS, Télécom ParisTech, Université Paris-Saclay, Paris, France
talel.abdessalem@telecom-paristech.fr

Abstract. The collection and exploitation of ratings from users are
modern pillars of collaborative filtering. Likert scale is a psychomet-
ric quantifier of ratings popular among the electronic commerce sites.
In this paper, we consider the tasks of collecting Likert scale ratings
of items and of finding the n-k best-rated items, i.e., the n items that
are most likely to be the top-k in a ranking constructed from these ratings. We devise an algorithm, Pundit, that computes the n-k best-rated
items. Pundit uses the probability-generating function constructed from
the Likert scale responses to avoid the combinatorial exploration of the
possible outcomes and to compute the result efficiently. Selection of the
best-rated items meets, in practice, the major obstacle of the scarcity
of ratings. We propose an approach that learns from the available data
how many ratings are enough to meet a prescribed error. We empirically
validate with real datasets the effectiveness of our method to recommend
the collection of additional ratings.

1 Introduction

The collection and exploitation of ratings from users are modern pillars of collaborative filtering [6,8]. Likert scale is an ordinal rating scale popular among
the electronic commerce sites and crowdsourced information systems such as
TripAdvisor. Each value in the scale gauges the degree of satisfaction of the user
towards a particular item, e.g., product or service.

Ranking the items based on Likert scale ratings is not always as obvious as it
seems to be. For instance, ranking items using the expectations of the Likert scale
responses can yield incorrect results [5]. Thus, we consider the task of finding
the list of n items that are most likely to be the top-k in a ranking constructed
from the ratings as our target recommendation task. For the sake of simplicity,
we refer to this list as the n-k *best-rated items*. We define the problem of finding
the n-k best-rated items as a probabilistic one. The uncertainty arises not from

© Springer International Publishing AG 2017
D. Benslimane et al. (Eds.): DEXA 2017, Part II, LNCS 10439, pp. 351–359, 2017.
DOI: 10.1007/978-3-319-64471-4_28

the unreliability of users but from the unavailability of ratings by all the users. We assume that the ratings from all users are correct and exact.

In this paper, we represent the scores of the items as discrete distributions on an L-valued Likert scale. We develop a polynomial-time algorithm, Pundit, that computes the n-k best-rated items. Pundit exploits the probability-generating functions of the discrete distributions of the items to avoid the combinatorial exploration of all possible outcomes and to compute the result efficiently. Our method is exact whereas the other methods like ranking by mean ratings or Monte Carlo sampling [15] are approximate algorithms.

In practice, the problem is not solved yet. Since selection of the n-k best-rated items is constrained by insufficient ratings. Corresponding discrete distributions are not 'true' representations of the score of the items. We devise a score distribution error model based on KL-divergence to answer the question "How many ratings are enough". This error model estimate the deviation of the discrete distribution formed with the available data from the 'true' universal distribution. As we empirically evaluate this model on the Amazon review dataset, we observe the KL-divergence based model follows inverse law. Following this we use the inverse law for KL-divergence based error to recommend how many additional ratings should be proactively sought to reach a certain error threshold.

2   Related Work

The problem of finding the n-k best-rated items is related to the probabilistic threshold top-k query [4] that returns the items having a probability of being in the top-k over a user specified threshold. [10] proposes a unified way to summarize a category of probabilistic top-k queries. Though the problem definitions seem similar, the category of probabilistic top-k queries is applicable to the scenario where the existence of an item is uncertain. Each item has a fixed known score representing its quality. However, the uncertainty modeled in our problem emerges from the unavailability of the ratings by the universal user-pool and is expressed as an evolving distribution over a Likert scale. [9] studies the problem of ranking continuous probabilistic data, where the score of each item is modelled as a continuous probability distribution. The authors focus on the probability of an item being ranked at a certain position. This result cannot be applied directly to our problem since the probability of an item being in top-k is not simply the sum over the probabilities of it being at different positions in the ranking. Another variant of these queries is UTop-Rank query [16]. This query searches for the item that has the highest probability of being ranked within a certain range of positions. They solve UTop-Rank query based on Monte Carlo sampling techniques which produce an approximate result. We construct a polynomial time exact algorithm for our problem which is more effective and efficient than the Monte Carlo method.

Crowdsourcing-based approaches have been proposed for the ranking and top-k problems in recent years. For example, [2,3] study the problem of finding the 'max' item or ranking the items by asking the crowd to compare pairs of items. Then, heuristic algorithms or learning approaches are proposed to aggregate the opinions collected from the crowd and to find the item with the maximal score. Beside these, [18] provides a thorough experimental study of the crowdsourced top-k queries. Most of the works in crowdsourcing use the *preference judgement* scheme which is based on the pairwise comparisons results from the crowd for inferring the global ranking. Hybrid approaches, such as [14,17], combine preference judgement and *absolute judgement*, like ratings, to infer the ranking. These approaches either transform the absolute judgement into the preference judgement [14] or use the parametric analysis [17] which may not be suitable for the ordinal data [5]. In this paper, we adopt the absolute judgement in form of the correct and exact ratings to infer the ranking of the items. The score of the item is modeled as a discrete distribution over a L-valued Likert scale.

3 How to Find the n-k Best-Rated Items?

3.1 Problem Definition

We use similar notations as in [11]. Consider a set of N items, $\mathcal{O} = \{o_1, \cdots, o_N\}$. A *scoring function* s maps the set of items $\mathcal{O}$ to a totally ordered domain $\mathcal{D}$, i.e. $s : \mathcal{O} \to \mathcal{D}$. $(\mathcal{D}, \geq)$ denotes a total order and $(\mathcal{D}, >)$ is the corresponding strict total order of $\mathcal{O}$ induced by s. We call the image $s(o)$ of an item $o \in \mathcal{O}$ by the function s the score of the item. A *ranking* $r : \mathcal{O} \to S_N$ is an indexing function induced on $\mathcal{O}$ by the total order $(\mathcal{D}, \geq)$. Here, S_N denotes the permutation group on $\{1, \ldots, N\}$. It is the set of all possible rankings of N items. For any two items o_i and $o_j \in \mathcal{O}$, if score of o_i is greater than or equal to that of o_j, i.e., $s(o_i) \geq s(o_j)$, we say that o_i is ranked equally with or above o_j, i.e., $r(o_i) \leq r(o_j)$.

In our problem, the score of each item is constructed from a collection of ratings. This epistemic uncertainty introduced by insufficiency of ratings prohibits existence of a deterministic score. Thus, we model the score $s(o_i)$ of an item $o_i \in \mathcal{O}$ as a random variable X_i with a probability mass function f_i. We define the score function as $s : \mathcal{O} \to \{f : \mathcal{L} \to [0,1]\}$. Here, $\mathcal{L} \triangleq \{1, \ldots, L\}$ is the L-valued Likert scale and f is a probability mass function defined over the support $\mathcal{L}$. For example, $\mathcal{L}$ is $\{1, \ldots, 5\}$ for a 5-valued Likert scale. We call f a *score distribution*.

If $x_1, \cdots, x_N \in \mathcal{L}$ are the observed ratings for the N items correspondingly, the probability of an item o_i to be ranked in top-k is expressed in Eq. 1.

$$\mathbb{P}(r(o_i) \leq k) = \sum_{\{x_1, \cdots, x_N\} \in S_i^k} f_1(x_1) \cdots f_N(x_N). \tag{1}$$

Here, S_i^k is the set of all N-tuples $\{x_1, \cdots, x_N\}$, such that for each $\{x_1, \cdots, x_N\}$ there exist at least $(N - k + 1)$ number of x's which are less than or equal to x_i. We call $\mathbb{P}(r(o_i) \leq k)$ the *positional probability* of o_i.

Example 1. Suppose there are three items, o_1, o_2 and o_3. If $k = 1, i = 2$, $\{x_1 = 5, x_2 = 1, x_3 = 5\}$ is not in S_2^1. Because it consists no rating lower than or equal to x_2. But $\{x_1 = 1, x_2 = 1, x_3 = 1\}$ is in S_2^1. Because ratings of o_1 and o_3 are equal to the rating of o_2.

We are looking for the list of n items $\Omega = [o_1, \ldots, o_n]$ that are most likely to be the top-k. That is, $\mathbb{P}(r(o_1) \le k) \ge \cdots \ge \mathbb{P}(r(o_n) \le k)$, and $\mathbb{P}(r(o_n) \le k) \ge \mathbb{P}(r(o_{i'}) \le k)$ for all $o_{i'} \notin \Omega$. This means that the items in Ω are ranked according to their positional probability and probability of other $N - n$ items to be in top-k is less than that of any item in Ω.

3.2 An Exact Algorithm for Finding the n-k Best-Rated Items

Approaches like ranking by the mean scores and Monte Carlo approaches give approximate results. Here, we develop an exact algorithm, Pundit, that finds the n-k best-rated items in polynomial time. The idea is to construct a degree N polynomial such that its coefficients are dependent on the positional probability. In the following, we will explain how to construct such a polynomial and then how to compute the coefficients. Once we can compute the positional probabilities efficiently, the n items with the highest positional probabilities are the result.

Construction of the Polynomial. We observe that by construction "rank of o_i is higher than or equal to k, i.e., $r(o_i) \le k$" is equivalent to the fact that "at least $N - k$ items other than o_i have scores lower than or equal to score of item o_i, i.e., $s(o_i)$". This fact includes k mutually exclusive cases. Case $j \in \{1, \cdots, k\}$ occurs if exactly $N - k + j - 1$ items other than o_i have scores lower than or equal to $s(o_i)$ and other $k - j$ scores are higher than $s(o_i)$. Thus, if we can calculate the probability for each of the k cases, the positional probability is the sum of the probabilities of these k cases.

In order to calculate the probability of each of the k cases, we construct a probability-generating function as shown in Eq. 2. This construction connects the probability of each of the k cases to the coefficients of the polynomial.

$$\mathcal{F}_i(x, l) \triangleq \prod_{j \ne i} (\mathbb{P}(s(o_j) \le l) + \mathbb{P}(s(o_j) > l)x) \tag{2}$$

In Eq. 2, $\mathbb{P}(s(o_j) \le l)$ denotes the probability that score of o_j is lower than or equal to l. For a given l, $\mathcal{F}_i(x, l)$ is a polynomial of x. In particular, the coefficient of the term x^k equals to $\mathbb{P}(\sum_{j \ne i} \mathbb{I}(s(o_j) > l) = k)$ [10]. Here, $\mathbb{I}(s(o_j) > l)$ is the indicator function that returns 1 or 0 depending on whether $s(o_j) > l$ is true or not. This implies that the coefficient of the term x^k is the probability that there are exactly k items having scores higher than l. If $s(o_i) = l$, the coefficient of the term x^{k-j} is the probability that there are exactly $k - j$ items having scores higher than $s(o_i)$. Thus, the coefficient of x^{k-j} exactly quantifies the jth ($j \in \{1, \cdots, k\}$) case.

Now, we just need to think about how to compute the coefficients in Eq. 2 efficiently.

Coefficients Calculation. We reconstruct the generating function of Eq. 2 into the polynomial expression of x.

$$\mathcal{F}_i(x, l) = c_0(l)x^0 + \cdots + c_{N-1}(l)x^{N-1} \tag{3}$$

where $c_q(l)$ represents the qth coefficient. The coefficients $c_0(l), \cdots, c_{N-1}(l)$ can be computed in $O(N^2)$ time by expanding Eq. 2 into Eq. 3.

We propose an efficient divide-and-conquer algorithm which applies Fast Fourier transform (FFT) to compute the coefficients $c_0(l), \cdots, c_{N-1}(l)$ more efficiently. Time complexity of this divide-and-conquer algorithm is $O(Nlog^2N)$. Due to the limitation of space, we refer the readers to our technical report [1] for more details of the efficient computation of the coefficients.

Pundit: The Algorithm. Once $c_0(l), \cdots, c_{N-1}(l)$ are computed, the positional probability for a L-valued Likert scale is calculated using $\mathbb{P}(r(o_i) \leq k) = \sum_{l=1}^{L} (c_0(l) + \cdots + c_{k-1}(l)) \mathbb{P}(s(o_i) = l)$. Once the positional probabilities for all the N items are computed, the n items that have the highest positional probabilities are the n-k best-rated items. Calculating the positional probability $\mathbb{P}(r(o_i) \leq k)$ for each item takes $O(Nlog^2N)$ time, it would take $O(N^2log^2N)$ time for all the items. Here, we propose two techniques to accelerate the computation. The first technique is to pre-compute the coefficient C^l of $\mathcal{F}' = \prod_{o_j \in \mathcal{O}} (\mathbb{P}(s(o_j) \leq l) + \mathbb{P}(s(o_j) > l)x)$ for all $1 \leq l \leq L$. Using the shorthand notation, we get $\mathcal{F}_i(x, l) = \mathcal{F}' \left[p_i^l + (1 - p_i^l)x \right]^{-1}$, where $p_i^l = \mathbb{P}(s(o_i) \leq l)$. Thus, we need to compute the set of coefficients once for each l and all the coefficients can be deduced correspondingly. Secondly, we observe that explicit calculation of all the coefficients is not needed, we calculate only the first k coefficients $c_0(l), \cdots, c_{k-1}(l)$. Time complexity of Pundit reduces to $O(Nlog^2N + Nk)$.

4   How Many Ratings Are Enough?

Though we have formulated an exact algorithm, Pundit, for finding n-k best-rated items, the problem is not solved yet. For real applications, ratings of some items are either missing or insufficient. For example, more than 30000 books in our Amazon book dataset have only one rating while the entire population of our datasets is 8726569. If we try to find the 10 best-rated books from this dataset, we would get 10 books which are rated as 5-star by only one user. This result is statistically insignificant and probably biased. Thus, the question that naturally appears is – "how many ratings are enough to construct the score of an item?" We investigate error of the score distribution of an item if we have a finite number of ratings. This model allows us to set a threshold in the required number of ratings for ranking the items without introducing remarkable error.

4.1   Score Distribution Error Model

We represent the 'true' score of an item by the *oracle score distribution* f^* constructed with all the ratings from the universal user pool while the *observed* *score distribution* f is constructed with a limited number of ratings. For brevity, we call f^* and f the *oracle distribution* and the *observed distribution* respectively.

The Optimization Problem. Consider the scenario when m ratings of an item are collected in the form of L-valued Likert scale. Suppose $z_1, \cdots, z_L$ are the number of ratings for each of the L values, such that $\sum_{i=1}^{L} z_i = m$. Such a rating pool can be represented by a multinomial distribution, $\mathbb{P}(z_1, \cdots, z_L) = \frac{m!}{z_1! \cdots z_L!} p_1^{*z_1} \cdots p_L^{*z_L}$. Here, $\{p_1^*, \cdots, p_L^*\}$ is the oracle distribution f^* of this item. The observed distribution f^m based on m ratings is $\{\frac{z_1}{m}, \cdots, \frac{z_L}{m}\}$.

In order to model the information gap between the observed distribution f^m and the oracle distribution f^*, we define the *expected score distribution error* as $E_m^{\mathrm{KL}} \triangleq \sum \mathbb{P}(z_1, \cdots, z_L) \, Dist(f^m, f^*)$ with a distance function $Dist(f^m, f^*)$. The sum is calculated over all $\{z_1, \cdots, z_L\}$ in the set of all possible L-partitions of m, $P(m, L)$. Thus, the expected error depends on three factors – the number of ratings m, the oracle distribution f^* and the distance function $Dist$. As m is given at an instance and the oracle distribution f^* is constructed with the universal review pool, modeling the expected error reduces to choice of the distance function. Since KL-divergence [7] quantifies the expected information per sample to discriminate between the uncertainty encompassed by one distribution against the other, we choose KL divergence as the eligible choice of distance function between the oracle and the observed score distributions. Thus, the expected error can be written as in Eq. 4.

$$E_m^{\mathrm{KL}} = \sum_{\{z_1, \cdots, z_L\} \in P(m.L)} \left(\frac{m! \, p_1^{*z_1} \cdots p_L^{*z_L}}{z_1! \cdots z_L!} \sum_{i=1}^{L} \left(\frac{z_i}{m} \log \frac{z_i}{m p_i^*} \right) \right) \tag{4}$$

We want to find the minimal number of ratings m^* such that the expected error between the oracle and the observed distribution is less than a predefined threshold ϵ. Our objective is mathematically expressed in Eq. 5.

$$m^* = \arg\min_{m} m \quad \text{such that, } E_m^{\mathrm{KL}} \le \epsilon. \tag{5}$$

Efficient Solution. In order to compute m^* in Eq. 5, we need to compute E_m^{KL}. E_m^{KL} depends on the oracle distribution, which is a choice, and the observed distribution f^m, which is observable. As we focus on the method for efficient calculation of the error, let us assume f^* is either given as a model parameter or constructed from the user-pool of a given dataset. But even when the oracle distribution is given, the expected error is not easy to compute. Because we sum over the set $P(m, L)$ that contains $\binom{m+L-1}{L-1}$ elements. It makes exact calculation of E_m^{KL} combinatorially expensive.

Thus, we propose a sampling approach to estimate the expected error based on the Ergodic Theorem [15]. Due to the limitation of space, we refer readers to our technical report [1] for more details of the computation of the expected error. Once we are able to calculate a sufficient approximation of the expected error E_m^{KL} efficiently, we can formulate the relation between E_m^{KL} and m. This allows us to find the minimal number of ratings required (m^*) to reach a prescribed error.

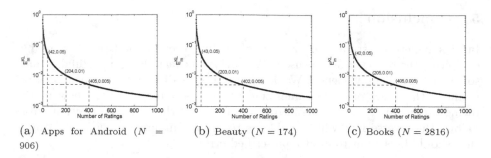

Fig. 1. Expected Error E_m^{KL} with Different Number of Ratings

4.2   Experimental Investigation of Error Models

Dataset and Set-up. We use Amazon review dataset[1] [12,13] with six categories of products – 'Apps for Android', 'Beauty', 'Books', 'Cell phones and Accessories', 'Electronics' and 'Movies and TVs'. Each review contains a rating for an item collected using a 5-valued Likert scale. This dataset is collected from May 1996 to July 2014. We consider only the items with more than 500 reviews. The remaining number of items is summarized in Fig. 1. We only show the results on three datasets in Fig. 1 due to the limitation of space, the results on other datasets are similar to those in Fig. 1 [1]. We aggregate the ratings for each item to construct the oracle distributions of the items. Once we obtain the oracle distributions, we focus on uncovering the relation between the score distribution error and the number of ratings m. In the experiments, we increase the number of ratings accumulated for the items and then observe evolution of the error.

Score Distribution Error and Number of Ratings. In Fig. 1, we present a smooth curve that fits the evolution of the error. We observe that the score distribution error decreases with increase in the number of ratings. This observation proves that this error model is consistent. Because the observed score distribution would converge to the oracle distribution with accumulation of more ratings, i.e., information about the item.

We also observe that decay of the expected score distribution error follows the inverse law, it fits with the hyperbolic equation $E_m^{KL} = \frac{c}{m}$. For the six categories, c is a constant between 2.01 and 2.024. Also, evolution of the error is almost category independent as it quantifies the evolution of observed distribution with accumulation of ratings. Thus, the score distribution error depends on the accumulation of ratings but not on the exact object names or categories. Now, we are able to answer the question "How many ratings are enough". For example, in order to restrict the score distribution error to a prescribed value 0.005, we need around 405 ratings for each item.

---

[1] http://jmcauley.ucsd.edu/data/amazon/.

5 Conclusion

In this paper, we study the problem of finding the n-k best-rated items by exploiting the ratings from the users. We devise an exact algorithm, Pundit, that solves this problem efficiently. We develop the score distribution error model to quantify the effect of the accumulation of ratings and to answer "how many ratings are enough". Then, we uncover the fact that the score distribution error follows the inverse law, which enable us to predict minimal number of ratings that should be sought to meet a prescribed error.

Acknowledgement. This work is supported by the National University of Singapore under a grant from Singapore Ministry of Education for research project number T1 251RES1607 and is partially funded by the Big Data and Market Insights Chair of Télécom ParisTech.

References

1. http://www.comp.nus.edu.sg/~liuqing/tech-reports/TRB6-17-likert.pdf
2. Chen, X., Bennett, P.N., Collins-Thompson, K., Horvitz, E.: Pairwise ranking aggregation in a crowdsourced setting. In: WSDM, pp. 193–202 (2013)
3. Guo, S., Parameswaran, A., Garcia-Molina, H.: So who won?: dynamic max discovery with the crowd. In: SIGMOD, pp. 385–396 (2012)
4. Hua, M., Pei, J., Zhang, W., Lin, X.: Ranking queries on uncertain data: a probabilistic threshold approach. In: SIGMOD, pp. 673–686 (2008)
5. Jamieson, S., et al.: Likert scales: how to (ab)use them. Med. Educ. **38**(12), 1217–1218 (2004)
6. Jin, R., Si, L., Zhai, C., Callan, J.: Collaborative filtering with decoupled models for preferences and ratings. In: CIKM, pp. 309–316. ACM (2003)
7. Kullback, S., Leibler, R.A.: On information and sufficiency. Ann. Math. Stat. **22**(1), 79–86 (1951)
8. Lee, J., Lee, D., Lee, Y.C., Hwang, W.S., Kim, S.W.: Improving the accuracy of top-n recommendation using a preference model. Inf. Sci. **348**, 290–304 (2016)
9. Li, J., Deshpande, A.: Ranking continuous probabilistic datasets. VLDB **3**(1–2), 638–649 (2010)
10. Li, J., Saha, B., Deshpande, A.: A unified approach to ranking in probabilistic databases. VLDB **2**(1), 502–513 (2009)
11. Liu, Q., Basu, D., Abdessalem, T., Bressan, S.: Top-k queries over uncertain scores. In: Debruyne, C., et al. (eds.) OTM 2016. LNCS, vol. 10033, pp. 245–262. Springer, Cham (2016). doi:10.1007/978-3-319-48472-3_14
12. McAuley, J., Pandey, R., Leskovec, J.: Inferring networks of substitutable and complementary products. In: SIGKDD, pp. 785–794 (2015)
13. McAuley, J., Targett, C., Shi, Q., van den Hengel, A.: Image-based recommendations on styles and substitutes. In: SIGIR, pp. 43–52 (2015)
14. Niu, S., Lan, Y., Guo, J., Cheng, X., Yu, L., Long, G.: Listwise approach for rank aggregation in crowdsourcing. In: WSDM, pp. 253–262 (2015)
15. Robert, C.P., Casella, G.: Monte Carlo Statistical Methods. Springer, New York (2004). doi:10.1007/978-1-4757-4145-2
16. Soliman, M.A., Ilyas, I.F., Ben-David, S.: Supporting ranking queries on uncertain and incomplete data. VLDB **19**(4), 477–501 (2010)

17. Ye, P., Doermann, D.: Combining preference and absolute judgements in a crowd-sourced setting. In: ICML, pp. 1–7 (2013)
18. Zhang, X., Li, G., Feng, J.: Crowdsourced top-k algorithms: an experimental evaluation. VLDB **9**(8), 612–623 (2016)

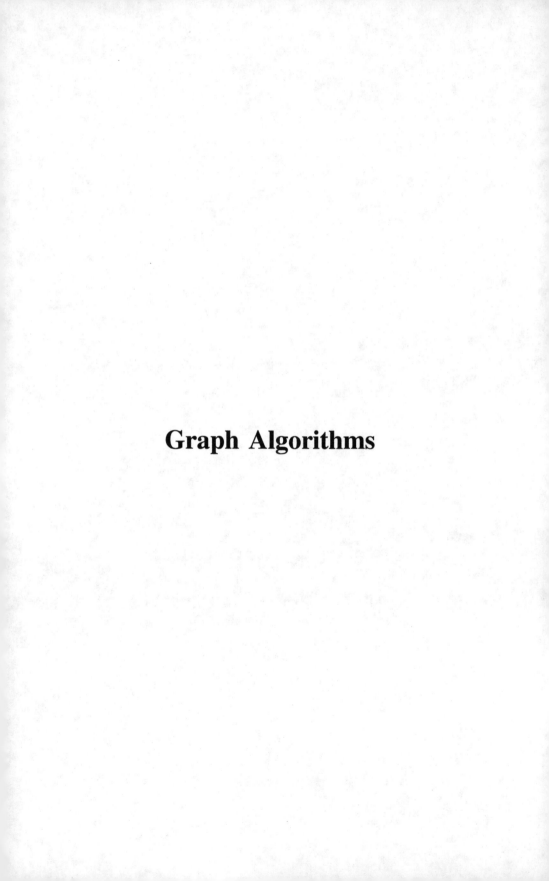

Graph Algorithms

Itinerary Planning with Category Constraints Using a Probabilistic Approach

Paolo Bolzoni, Fabio Persia, and Sven Helmer[✉]

Free University of Bozen-Bolzano, Bozen-Bolzano, Italy
paolo.bolzoni@stud-inf.unibz.it, {fabio.persia,sven.helmer}@unibz.it

Abstract. We propose a probabilistic approach for finding approximate solutions to rooted orienteering problems with category constraints. The basic idea is to select nodes from the input graph according to a probability distribution considering properties such as the reward of a node, the attractiveness of its neighborhood, its visiting time, and its proximity to the direct route from source to destination. In this way, we reduce the size of the input considerably, resulting in a much faster execution time. Surprisingly, the quality of the generated solutions does not suffer significantly compared to the optimal ones. We illustrate the effectiveness of our approach with an experimental evaluation also including real-world data sets.

1 Introduction

Finding itineraries for domains such as tourist trip planning and logistics often involves solving an orienteering problem, as those tasks are not about determining shortest paths but routes covering the most attractive points of interests or the most needy customers. In principle, the orienteering problem (OP) is about determining a path from a starting node to an ending node in an edge-weighted graph with a score for each node, maximizing the total score while staying within a certain time budget. We focus on a variant that assumes that every point of interest and customer has a category. This categorization helps a user in expressing preferences, e.g. a tourist may only be interested in certain types of venues, such as museums, galleries, and cafes, while certain vehicles can only supply particular customers.

In general, orienteering is an NP-hard problem, adding categories does not change this fact [2]. While an algorithm computing the optimal solution is able to utilize pruning to filter out partial solutions that cannot possibly result in an optimal route, the run time is still exponential in the worst case. If we want to have a chance to find an answer within an acceptable time frame, we need to decrease the size of the problem instance in some way. In our previous approach [2] we accomplished this by clustering the nodes in a graph and then generating paths containing clusters rather than individual nodes. In a second phase, nodes were selected from the clusters to determine the actual path. Here, we propose a different approach: we decrease the size of a problem instance by

© Springer International Publishing AG 2017
D. Benslimane et al. (Eds.): DEXA 2017, Part II, LNCS 10439, pp. 363–377, 2017.
DOI: 10.1007/978-3-319-64471-4_29

randomly selecting nodes from a graph according to probabilities assigned to these nodes. Basically, the chance of a node to be chosen depends on its score, the attractiveness of its surroundings, its visiting time, and its proximity to the direct route between the source and destination node. After removing nodes from a graph, we run an optimal algorithm on the reduced graph obtaining the best possible solution for this particular subgraph.

The randomized approach also gives us a mechanism with which to trade accuracy for execution speed. The more nodes we remove from the graph, the faster the algorithm will run, but the worse the quality of the found solution will be. At the extreme ends we have the optimal algorithm (removing no nodes), which is very slow, and the worst possible one (removing all the nodes), very quickly computing an empty path between the source and destination node.

In summary, we make the following contributions:

- We present variants of a probabilistic algorithm based on reducing the size of the graph for approximately solving orienteering problems with category constraints.
- In an experimental evaluation we demonstrate that our approach generates itineraries close to optimal ones, but does so much faster than the state-of-the-art algorithm.
- While the main bulk of our evaluation relies on experiments, we also provide a theoretical analysis for the score and run time of our algorithm.

2   Related Work

Introduced by Tsiligrides in [18], orienteering is an NP-hard problem and algorithms computing exact solutions using branch and bound [6,11] as well as dynamic programming techniques [10,13] are of limited use, as they can only solve small problem instances. Consequently, there is a body of work on approximation algorithms and heuristics, most of them employing a two-step approach of partial path construction [8,18] and (partial) path improvement [1,3,14]. Metaheuristics, such as genetic algorithms [17], neural networks [19], and ant colony optimization [9] have also been tested. For a recent overview on orienteering algorithms, see [5]. However, none of the approaches investigate OP generalized with categories.

There is also work on planning and optimizing errands, e.g., someone wants to drop by an ATM, a gas station, and a pharmacy on the way home. The generalized traveling salesman version minimizes the time spent on this trip [12], while the generalized orienteering version maximizes the number of visited points of interest (POIs) given a fixed time budget. However, as there are no scores, no trade-offs between scores and distances are considered.

Adapting an existing algorithm for OP would be a natural starting point for developing an approximation algorithm considering categories. However, many of the existing algorithms have a high-order polynomial complexity or no implementation exists, due to their very complicated structure. Two of the most promising approaches we found were the segment-partition-based technique by Blum et al. [1] and the method by Chekuri and Pál, exploiting properties of submodular functions [4]. The latter approach, a quasi-polynomial algorithm, is still too slow for practical purposes. Nevertheless, Singh et al. modified the algorithm by introducing spatial decomposition for Euclidean spaces in the form of a grid, making it more efficient [16]. It has been adapted by us for OPs on road networks with category constraints [2].

3   Problem Formalization

We assume a set $\mathbf{P}$ of points of interest (POIs) $p_i, 1 \leq i \leq n$. The POIs, together with a starting and a destination node, denoted by s and d, respectively, are connected by a complete, metric, weighted, undirected graph[1] $G = (\mathbf{P} \cup \{s, d\}, \mathbf{E})$, whose edges, $e_l \in \mathbf{E} = \{(x, y) \mid x, y \in \mathbf{P} \cup \{s, d\}\}$ connect them. Each edge e_l has a cost $c(p_i, p_j)$ that signifies the duration of the trip from p_i to p_j, while every node $p_i \in \mathbf{P}$ has a cost $c(p_i)$ that denotes its visiting time. Each POI belongs to a certain category, such as *museums*, *restaurants*, or *galleries*. The set of m categories is denoted by $\mathbf{K}$ and each POI p_i belongs to exactly one category $k_j, 1 \leq j \leq m$. Given a POI p_i, $\texttt{cat}(p_i)$ denotes the category p_i belongs to and $\texttt{score}(p_i)$ denotes its score or reward, with higher values indicating higher interest to the user. Finally, users have a certain maximum time in their budget to complete the itinerary, denoted by $t_{\max}$.

Definition 1 (Itinerary). *An itinerary $\mathcal{I}$ starts from a starting point s and finishes at a destination point d (s and d can be identical). It includes an ordered sequence of connected nodes $\mathcal{I} = \langle s, p_{i_1}, p_{i_2}, \ldots, p_{i_q}, d \rangle$, each of which is visited once. We define the* cost *of itinerary $\mathcal{I}$ to be the total duration of the path from s to d passing through and visiting the POIs in $\mathcal{I}$, $\texttt{cost}(\mathcal{I}) = c(s, p_{i_1}) + c(p_{i_1}) + \sum_{j=2}^{q}(c(p_{i_{j-1}}, p_{i_j}) + c(p_{i_j})) + c(p_{i_q}, d)$, and its* score *to be the sum of the scores of the individual POIs visited, $\texttt{score}(\mathcal{I}) = \sum_{j=1}^{q} \texttt{score}(p_{i_j})$.*

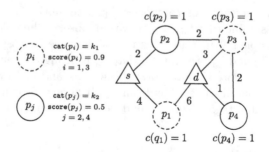

Fig. 1. Itinerary including $n = 4$ POIs

---

[1] The algorithm can be adapted to directed graphs.

Example 1. Figure 1 shows an example with four POIs, p_1, p_2, p_3, and p_4, along with their distances, visiting times, scores, and categories. We simplify the graph slightly to keep it readable: all POIs of the same category have the same score and we also omit some edges. Three example itineraries with one, two, and three POIs, respectively, are: $\mathcal{I}_1 = \langle s, p_1, d \rangle$, $\mathcal{I}_2 = \langle s, p_2, p_3, d \rangle$, and $\mathcal{I}_3 = \langle s, p_2, p_3, p_4, d \rangle$. Their costs and scores are as follows:

- $\mathcal{I}_1 = \langle s, p_1, d \rangle$: $\mathtt{cost}(\mathcal{I}_1) = 4 + 1 + 6 = 11$, $\mathtt{score}(\mathcal{I}_1) = 0.9$;
- $\mathcal{I}_2 = \langle s, p_2, p_3, d \rangle$: $\mathtt{cost}(\mathcal{I}_2) = 2 + 1 + 2 + 1 + 3 = 9$, $\mathtt{score}(\mathcal{I}_2) = 0.5 + 0.9 = 1.4$;
- $\mathcal{I}_3 = \langle s, p_2, p_3, p_4, d \rangle$: $\mathtt{cost}(\mathcal{I}_3) = 2 + 1 + 2 + 1 + 2 + 1 + 1 = 10$, $\mathtt{score}(\mathcal{I}_3) = 0.5 + 0.9 + 0.5 = 1.9$

Given traveling and visiting times as well as scores, we need to build an itinerary starting at s and ending at d from a subset P of $\mathbf{P}$ with duration smaller than $t_{\max}$ and maximum cumulative score. We introduce an additional constraint specifying the number of POIs per category that can be included in the final itinerary. More precisely, we introduce a parameter $\max_{k_j}$ for each category k_j that is set by the user to the maximum number of POIs in a category that he or she prefers to visit during the trip. We are now ready to define the *Orienteering Problem with Maximum Point Categories (OPMPC)*.

Definition 2 (OPMPC). *Given a starting point s, a destination point d, n points of interest $p_i \in \mathbf{P}$, each with $score(p_i)$, visiting times $c(p_i), 1 \leq i \leq n$, traveling times $c(x, y)$ for $x, y \in \mathbf{P} \cup \{s, d\}$, categories $k_j \in \mathbf{K}, 1 \leq j \leq m$, and the following two parameters: (a) the maximum total time t_{max} a user can spend on the itinerary and, (b) the maximum number of POIs $\max_{k_j}$ that can be used for the category k_j ($1 \leq j \leq m$), a solution to the OPMPC is an itinerary $\mathcal{I} = \langle s, p_{i_1}, p_{i_2}, \ldots, p_{i_q}, d \rangle, 1 \leq q \leq n$, such that*

- *the total score of the points, $score(\mathcal{I})$, is maximized;*
- *no more than $\max_{k_j}$ POIs are used for category k_j;*
- *the time constraint is met, i.e., $cost(\mathcal{I}) \leq t_{max}$.*

Example 2. In the presence of categories k_1 with $\max_{k_1} = 1$ and k_2 with $\max_{k_2} = 1$, and assuming that $t_{\max} = 10$, we can observe the following about the itineraries in Example 1: Itinerary $\mathcal{I}_1$ is infeasible since its cost is greater than $t_{\max}$. Comparing $\mathcal{I}_2$ and $\mathcal{I}_3$, we can see that $\mathcal{I}_3$ is of higher benefit to the user, even though it takes more time to travel between s and d. However, it cannot be chosen since it contains two POIs from k_2. Itinerary $\mathcal{I}_2$ contains two POIs, each from a different category and it could be one recommended to the user.

4   Our Approach

Our approach is based on the following observations. Generally, POIs with a higher reward or score are better. However, it is not as simple as that: if we have a POI with a long visiting time, the returns are significantly diminished.

We need to look at the overall utility of a node, considering its score in relation to its visiting time, rather than its raw score. This is not a particular new insight, the utility of a POI was already used in some of the earliest heuristics for orienteering [18]. While the distance to the source s and destination d has previously been used for pruning nodes that are too far away to be reached, its use in heuristics is much more recent and not as widespread [7]. The idea here is that a POI at a long distance from s and d tends to incur a high travel cost, which has a negative impact on the quality of an itinerary. Nevertheless, this is not always the case. Clearly, an isolated node far away from s and d is only worth traveling to if it offers a very high reward. Again, we should be using a utility-based function rather than raw values. However, while the visiting time of a POI can be determined fairly accurately, traveling times may fluctuate widely, depending on our point of origin. As we are not visiting nodes individually, going back to s in between, but on an itinerary, it may well be worth going there if we can visit a lot of other high-scoring nodes in the vicinity as well. One novelty of our approach is to implicitly consider the neighborhood of a POI when determining its utility.

4.1   Reducing the Input Graph

In a first step, we remove from $\mathbf{P}$ all POIs p that belong to categories excluded by the user and in a second step all those that cannot be reached, meaning the length of the path $\langle s, p, d \rangle$ is longer than t_{max}. These two steps remove nodes that are not part of any feasible solution. From now on, when referring to $\mathbf{P}$, we mean all the nodes that can potentially be part of a solution and disregard those removed in the first two steps described above.

In the following, we visualize the main idea of our probabilistic algorithm. We snap every POI in the graph onto a grid, more specifically we overlay the road network with a grid of points and place a POI at the position of the closest grid point (see Fig. 2). Additionally, we surround every POI with an area, covering neighboring grid points (the areas covered by POIs can overlap). The extent of the region around a POI is determined by its quality in terms of the score, visiting time, centrality, and category: the better a POI, the larger its surrounding area. For selecting POIs, we randomly determine a point in the grid and all POIs that cover this point are added to the subgraph used for computing the solution. Choosing multiple POIs connected to a single grid point is done deliberately. In this way we consider the neighborhood of a POI: if there is a lot of overlap, i.e., a particular neighborhood is rich in valuable POIs, chances are high that we select multiple POIs in a single step.

We now have a closer look at the criteria determining the quality of a POI. First we have the score: the higher the score, the better. We also normalize the score: we subtract the overall minimal score and divide by the difference of the maximal and minimal score obtaining

$$\text{nscore}(p) = \frac{\text{score}(p) - \min_{q \in \mathbf{P}} \text{score}(q)}{\max_{q \in \mathbf{P}} \text{score}(q) - \min_{q \in \mathbf{P}} \text{score}(q)}$$

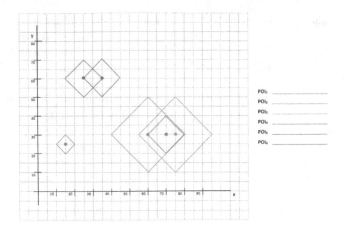

Fig. 2. The POIs and their areas

Second, we look at the visiting time. Here, the opposite is true: the longer the visiting time, the worse, since it takes away time we could spend on other POIs or for traveling. Consequently, we subtract the normalized visiting time from 1:

$$\text{nvisit}(p) = 1 - \frac{c(p) - \min_{q \in \mathbf{P}} c(q)}{\max_{q \in \mathbf{P}} c(q) - \min_{q \in \mathbf{P}} c(q)}$$

Third, we look at the centrality of a POI. The larger the distance of a POI to s and d, the harder it is to reach it when traveling from s to d. Again, we subtract the normalized value from 1:

$$\text{centrality}(p) = 1 - \frac{c(s,p) + c(p,d) - \min_{q \in \mathbf{P}} \big(c(s,q) + c(q,d)\big)}{\max_{q \in \mathbf{P}} \big(c(s,q) + c(q,d)\big) - \min_{q \in \mathbf{P}} \big(c(s,q) + c(q,d)\big)}$$

Finally, we look at the category of a POI. The higher the maximum constraint of a category, the more POIs of this category we can fit into an itinerary.

$$\text{constraint}(p) = \frac{\max_{\text{cat}(q)} - \min_{k \in K} \max_k}{\max_{k \in K} \max_k - \min_{k \in K} \max_k}$$

If the denominator of any of the previous fractions is equal to zero, it means that the corresponding criterion does not provide any useful information and we set its value to one. The overall quality γ_p of a POI p is computed by summing up the weighted values for the different criteria:

$$\gamma_p = w_1 \cdot \text{nscore}(p) + w_2 \cdot \text{nvisit}(p) + w_3 \cdot \text{centrality}(p) + w_4 \cdot \text{constraint}(p)$$

In order to determine the area around a POI, we use its overall quality γ_p and multiply it with a constant r, which is the largest area a POI can potentially cover.

Algorithm 1. ConstructSubgraph $(\mathbf{P}, o)$
Input: set of POIs $\mathbf{P}$, size L of output graph
Output: subgraph Q

```
1  generate grid g
2  for all p ∈ P do
3  │   find point gp in grid closest to p
4  │   assign p to gp
5  │   compute γp · r
6  │   determine ⌈γp · r⌉ points in grid covered by p
7  │   assign p to these points
8  while L > 0 do
9  │   select random point rp in grid g
10 │   for all p associated with rp do
11 │   │   if  p not in Q yet  and L > 0 then
12 │   │   │   L ← L − 1
13 │   │   └   Q.add(p)
14 │   remove rp from g
15 return Q
```

Algorithm 1 illustrates the construction of the subgraph used for generating solutions. In the first part, we assign the POIs to points in the grid. Each POI is assigned to the closest grid point and then, depending on its quality, it is also assigned to neighboring grid points. We use a forest-fire flood fill algorithm to determine the neighboring grid points [15]. In the second part we randomly select grid points and add the associated POIs to the subgraph Q until the size of Q reaches L.

4.2   Computing the Optimal Solution

Algorithm 2 builds itineraries that are feasible solutions in a systematic way by extending already computed itineraries with POIs not yet visited. An itinerary $\mathcal{I}$ is a *feasible solution* if $\mathsf{cost}(\mathcal{I}) \leq t_{max}$ and no more than max_{k_j} POIs are used for category k_j. When generating new solutions we discard any itineraries that violate the time or any category constraints. We also prune itineraries $\mathcal{I}_j$ that are dominated by another itinerary $\mathcal{I}_i$, which means that the extensions of $\mathcal{I}_i$ will always lead to a better solution than the extensions of $\mathcal{I}_j$.

Generating Feasible Solutions. When starting, we compute the all-pairs shortest paths between POIs (e.g., with the Floyd-Warshall algorithm). While not strictly necessary, this will improve the performance. At the core of our algorithm is a loop that generates feasible solutions in a systematic way. For each feasible solution computed so far, we store a tuple containing the itinerary $\mathcal{I}$, its score, its cost, and an array keeping track of the number of categories of POIs

---

Algorithm 2. DynamicProg

Input: n POIs with categories, constraints, and scores, time budget t_{max}
Output: $\mathcal{I}_{opt}$, the optimal itinerary under the given constraints

1   $\mathcal{I}_{opt} \leftarrow \langle \rangle$ // empty path
2   $s_{opt} \leftarrow 0$
3   compute all-pairs shortest paths on $\mathbf{P}$ // optional
4   **if** $cost(\langle s, d \rangle) \leq t_{max}$ **then**
5   enqueue($queue$, $[\langle s, d \rangle, 0, cost(\langle s, d \rangle), [0, .., 0]]$)
6   **while** $queue$ *not empty* **do**
7          $t_{cur} \leftarrow$ dequeue($queue$)
8   **if** $t_{cur}.score > s_{opt}$ **then**
9             $\mathcal{I}_{opt} \leftarrow t_{cur}.\mathcal{I}$
10             $s_{opt} \leftarrow t_{cur}.score$

11   **for** *(every $p_i \in \mathbf{P}$)* **do**
12   **if** $p_i \notin t_{cur}.\mathcal{I} \wedge t_{cur}.count[cat(p_i)] < max_k[cat(p_i)] \wedge$
                $cost(t_{cur}.\mathcal{I} \oplus p_i) \leq t_{max}$ **then**
13                 $t_{new} \leftarrow [t_{cur}.\mathcal{I} \oplus p_i, t_{cur}.score + score(p_i),$
                  $cost(t_{cur}.\mathcal{I} \oplus p_i), count[cat(p_i)] + 1]$
14   **if** *there is no t_{sim} in queue with* $\{p_q | p_q \in t_{sim}.\mathcal{I}\} =$
                  $\{p_q | p_q \in t_{new}.\mathcal{I}\} \wedge t_{sim}.\mathcal{I}[-2] = t_{new}.\mathcal{I}[-2]$ **then**
15   enqueue($queue$, t_{new})
16   **else**
17   **if** $t_{sim}.cost > t_{new}.cost$ **then**
18   enqueue($queue$, t_{new})
19   remove($queue$, t_{sim})

20 **return** $\mathcal{I}_{opt}$

---

visited so far: $[\mathcal{I}, score(\mathcal{I}), cost(\mathcal{I}), count[1 \ldots m]]$. For example, we represent $\mathcal{I}_2$ in Example 1 by $[\langle s, p_2, p_3, d \rangle, 1.4, 9, [1, 1]]$.

Given a tuple $t_j = [\mathcal{I}_j = \langle s, p_{j_1}, p_{j_2}, \ldots, p_{j_r}, d \rangle, score(\mathcal{I}_j), cost(\mathcal{I}_j), count[1 \ldots m]]$ and the set $\mathbf{P}_{\overline{\mathcal{I}_j}} = \{p_i | p_i \in \mathbf{P}, p_i \notin \mathcal{I}_j\}$ of all POIs not in the itinerary $\mathcal{I}_j$, we extend t_j by adding each POI p_q in $\mathbf{P}_{\overline{\mathcal{I}_j}}$ to $\mathcal{I}_j$, creating new tuples of the form $t_{jq} = [\mathcal{I}_{jq} = \langle s, p_{j_1}, p_{j_2}, \ldots, p_{j_r}, p_q, d \rangle, score(\mathcal{I}_j) + score(p_q), cost(\mathcal{I}_j) - cost(\langle p_{j_r}, d \rangle) + cost(\langle p_{j_r}, p_q, d \rangle), count[cat(p_q)] + 1]$, where $cat(p_q)$ is the index of the category of p_q.

We start the search for the optimal itinerary by inserting $t_{init} = [\mathcal{I}_{init} = \langle s, d \rangle, 0, cost(\mathcal{I}_{init}), [0, \ldots, 0]]$ into an empty first-in first-out (FIFO) queue. The algorithm removes the first item from the queue and extends its itinerary as described above. Before re-inserting a newly created tuple t_{jq} we check whether it is a feasible solution or not, i.e., it is discarded if $cost(\mathcal{I}_{jq}) > t_{max}$ or one of the category counts exceeds its maximum value. In each step we compare the item removed from the queue to the best solution found so far, keeping track of the highest scoring itinerary. The process continues until the queue runs empty; we then return the optimal solution to the user.

Pruning Dominated Itineraries. When comparing two tuples $t_i = [\mathcal{I}_i = \langle s, p_{i_1}, p_{i_2}, \ldots, p_{i_r}, d\rangle, \texttt{score}(\mathcal{I}_i), \texttt{cost}(\mathcal{I}_i), count[1 \ldots m]]$ and $t_j = [\mathcal{I}_j = \langle s, p_{j_1}, p_{j_2}, \ldots, p_{j_r}, d\rangle, \texttt{score}(\mathcal{I}_j), \texttt{cost}(\mathcal{I}_j), count[1 \ldots m]]$, both of length $r + 2$, with $p_{i_r} = p_{j_r}$, and containing the same set of POIs ($\mathbf{P}_{\mathcal{I}_i} = \{p_q | p_q \in \mathcal{I}_i\} = \mathbf{P}_{\mathcal{I}_j} = \{p_q | p_q \in \mathcal{I}_j\}$), we say that t_i dominates t_j, iff $\texttt{cost}(\mathcal{I}_i) < \texttt{cost}(\mathcal{I}_j)$ (we break ties by lexicographical order of the POIs in the itineraries). In that case we can drop t_j, since any extended path starting with the itinerary $\mathcal{I}_j$ can always be improved by replacing $\mathcal{I}_j$ with $\mathcal{I}_i$. Thus, when generating a new tuple, in addition to verifying that it is a feasible solution we also search for tuples in the queue that either dominate or are dominated by the newly generated tuple.[2]

5   Theoretical Analysis

Dynamic Programming. Algorithm 2 avoids having to generate all possible paths by employing pruning and exploiting the category constraints. We have m categories k_j, $1 \le j \le m$, each with the constraint max_{k_j} and the set $K_{k_j} = \{p_i | p_i \in P, cat(p_i) = k_j\}$ containing the POIs belonging to this category. The set of relevant POIs R is equal to $\bigcup_{j=1, max_{k_j}>0}^{m} K_{k_j}$, we denote its cardinality by $|R|$. We now derive an upper bound for the number of paths we need to generate.

For the first POI we have $|R|$ choices, for the next one $|R| - 1$, and so on until we have created paths of length $\lambda = \sum_{j=1, max_{k_j}>0}^{m} max_{k_j}$. We cannot possibly have itineraries containing more than λ POIs, as this would mean violating at least one of the category constraints. So, in a first step this gives us $\prod_{i=0}^{\lambda-1}(|R|-i)$ different itineraries. This is still an overestimation, though, as it assumes that we can extend every path to a length of λ, since we neither consider the time constraint t_{max} nor the individual category constraints.

For itineraries including at least three POIs we can start pruning dominated paths. For example, for the itineraries containing p_1, p_2, and p_3 (visiting p_3 last) we have $\{\langle s, p_1, p_2, p_3, d\rangle, \langle s, p_2, p_1, p_3, d\rangle\}$, for POIs p_1, p_2, p_3, and p_4 (visiting p_4 last) we get $\{\langle s, p_1, p_2, p_3, p_4, d\rangle, \langle p_1, p_3, p_2, p_4, d\rangle, \langle s, p_2, p_1, p_3, p_4, d\rangle, \langle s, p_2, p_3, p_1, p_4, d\rangle, \langle s, p_3, p_1, p_2, p_4, d\rangle, \langle s, p_3, p_2, p_1, p_4, d\rangle\}$. From each of these sets we only have to keep a single itinerary, the one with the lowest cost. In general, for all the paths containing l POIs, $l \ge 3$, we only need to keep $\frac{1}{(l-1)!}$ paths. Combining this with the earlier result, we obtain $|R| \cdot (|R| - 1) \cdot \frac{(|R|-2)}{2} \cdot \frac{(|R|-3)}{3} \cdot \ldots \cdot \frac{(|R|-\lambda+1)}{\lambda-1}$.

For step l we have an increase by a factor of $\frac{|R|-(l-1)}{l-1}$ as the $(l-2)!$ other plans are not included anymore, having already been considered in the previous steps. In summary, we have to generate at most $\lambda \binom{|R|}{\lambda}$ itineraries. This illustrates

---

[2] In the code we use the Ruby notation [-2] for accessing the last but one element of an array.

why reducing the input size will have a significant impact on the run time, as $\lambda\binom{|R|}{\lambda}$ grows exponentially in $|R|$.

Graph Reduction. We introduce a few simplifications, as it is very hard to model path and category constraints accurately (thus, we fix $w_2 = w_3 = w_4 = 0$). First of all, we assume that our graph consists of n nodes and that the highest-scoring node has a score of s_{max} and the lowest-scoring one has a score of s_{min}. Furthermore, we suppose that the scores of the nodes are uniformly distributed among the n nodes, meaning, for example, that the $n/2$-th node has a score of $s_{min} + 1/2(s_{max} - s_{min})$ and the $3n/4$-th node has one of $s_{min} + 1/4(s_{max} - s_{min})$. Assuming that the optimal itinerary consists of m POIs, we can give an upper bound for the score of the optimal path (by disregarding path and category constraints):

$$\sum_{i=0}^{m-1} \left(\frac{(n-i)-1}{n-1} \cdot (s_{max} - s_{min}) + s_{min} \right) \tag{1}$$

In our reduced graph we keep a proportion of k $(0 < k \le 1)$ of the nodes in **P**. Selecting the nodes for the reduced graph completely randomly and choosing among them the m highest-scoring ones, we arrive at an upper bound for the total score of the best itinerary in the reduced subgraph:

$$\sum_{i=0}^{m-1} \left(\frac{(n-\frac{i}{k})-1}{n-1} \cdot (s_{max} - s_{min}) + s_{min} \right) \tag{2}$$

Figure 3(a) shows an example of applying Eqs. (1) and (2) to a graph consisting of 1400 nodes with $s_{min} = 50$, $s_{max} = 99$, and a path length $m = 6$. The constant black line represents the optimal solution for the full graph and the blue line shows the score for different sizes of the reduced graph (varying the parameter k). As we can see, the score of the best itinerary in the reduced graph increases quickly at the beginning and then slowly approaches the score for the full graph. Figure 3(b) gives a worst-case approximation for the number

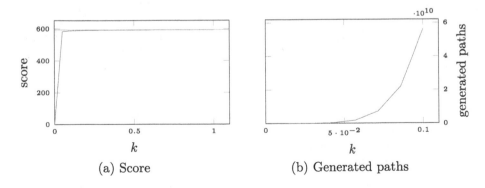

(a) Score   (b) Generated paths

Fig. 3. Effect of removing nodes (Color figure online)

of paths generated during the search, which grows exponentially. Actually, the theoretical results in Fig. 3 resemble those we get in our experimental evaluation.

6   Experimental Evaluation

We executed our experiments on a Linux machine (Archlinux kernel version 4.9.6) with a i7-4800MQ CPU with a clock speed of 2.7 Ghz and 24 Gb of RAM (of which up to 15 were used). The code is written in C++ and compiled with g++ version 6.3.1. For the artificial networks we used a grid and a spider network with 10,000 nodes each. The grid is formed by a regular 100×100 node square and each edge has a weight of 60 s. The spider network is composed of 100 rings with 100 nodes each. The length of an edge depends on its position. Edges connecting rings are 300 s long. Edges located on a ring gradually increase as we move outwards: edges on the innermost ring have a weight of 19 s, those on the outermost ring have one of 1885 s. The scores of the POIs are uniformly distributed between 49 and 100, their visiting times are uniformly distributed between three minutes and three hours. Due to space constraints, we only show the results for the grid network here. The results for the spider network are similar, if not slightly better for our algorithm. Additionally, we use two real-world data sets: one for the city of Bolzano, Italy, with a total of 1830 POIs and one for San Francisco with a total of 1983 POIs.

6.1   Tuning the Parameters

Area of POIs. The first parameter we need to tune is r, the potential area surrounding a POI. Figure 4 shows the results for different values of r on the grid data set (the different criteria all had the same weight, we show several curves for different values of k: 5%, 10%, 20%, 30%, and 50%). If the value we choose for r is too small, there is hardly any overlap, which means we will not find promising neighborhoods. If it is too large, a considerable number of POIs have a high likelihood of getting picked, which leads to a more randomized selection. In the following experiments we set r to 240, as there is a sharp rise in score (from 20 to 240), but no significant increase in run time. The value for r is calibrated for a grid granularity of $\frac{1}{2000}$ of a degree using the SRID 4326 coordinate system.

Weighting of Different Criteria. Next, we investigated the impact of choosing different values for the weights w_i, identifying their importance when choosing POIs. The weight r was set to 240 in all cases, we average the values for subgraph sizes of 5%, 10%, 20%, 30%, and 50% of the total graph size. For the weights, we distinguish nine different configurations: full: setting $w_i = 1$ ($1 \le i \le 4$) and $w_j = 0$ ($1 \le j \le 4, i \ne j$); favored: $w_i = 0.4$ ($1 \le i \le 4$) and $w_j = 0.2$ ($1 \le j \le 4, i \ne j$); balanced: $w_i = 0.25$ for all i ($1 \le i \le 4$). For every variant we measured its utility compared to the greedy algorithm: (score of our algorithm − score of greedy)/run time. Then we determined how far away every

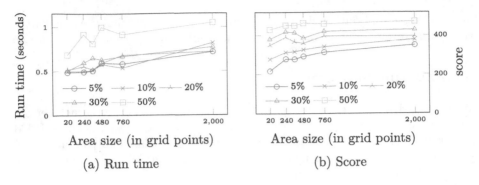

Fig. 4. Determining the area r

configuration is from the best performer on average (the smaller the value, the better), Table 1 summarizes the results for different values of t_{max}.

The greedy algorithm works as follows. Given a partial itinerary $\mathcal{I} = \langle s, p_1, p_2, \ldots, p_i, d \rangle$ (starting with the empty itinerary), the greedy strategy adds to this path a POI $p \in \mathbf{P}_{rem(\mathcal{I})}$ such that its $\mathtt{utility}(p) = \frac{\mathtt{score}(p)}{c(p_i,p)+c(p)+c(p,d)}$ is maximal and no constraints are violated. We repeat this until no further POIs can be added to the itinerary.

Table 1. Performance versus greedy

t_{max}	$w_1 = 1$ $w_j = 0$	$w_2 = 1$ $w_j = 0$	$w_3 = 1$ $w_j = 0$	$w_4 = 1$ $w_j = 0$	$w_1 = .4$ $w_j = .2$	$w_2 = .4$ $w_j = .2$	$w_3 = .4$ $w_j = .2$	$w_4 = .4$ $w_j = .2$	$w_i = .25$
3600	0.108	0.095	0.037	0.174	0.095	0.075	0.079	0.134	0.101
5400	0.029	0.036	0.029	0.067	0.032	0.029	0.043	0.035	0.030
7200	0.004	0.012	0.014	0.023	0.012	0.015	0.017	0.013	0.011
9000	0.004	0.002	0.004	0.007	0.003	0.003	0.004	0.006	0.002

There is a general tendency that the difference to the best performer gets smaller as t_{max} increases.[3] Emphasizing the weight on the category constraint ($w_4 = 1$ or $w_4 = .4$) produces the worst results overall, so we focus on the other weights. Concentrating on centrality ($w_3 = 1$ or $w_3 = .4$) is a good strategy for small values of t_{max}, where it becomes important to not waste a lot of time traveling between POIs. We would have expected a similar effect for visiting times ($w_2 = 1$ or $w_2 = .4$), as shorter visiting times are more crucial when time is scarce. However, the impact is not as distinct as for centrality, it seems that the actual location is more important. Emphasizing score ($w_1 = 1$ or $w_1 = .4$) has an opposite effect on the quality: the larger t_{max}, the better it becomes. When

---

[3] Run time increases considerably for large t_{max} values, decreasing the utility for all variants, making the differences smaller.

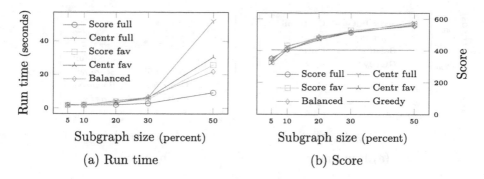

(a) Run time

(b) Score

Fig. 5. Subgraph size

a wider area can be reached, the actual score has a greater impact. Finally, the balanced approach ($w_i = .25$ for all i), while only being a top contender for t_{max} = 9000 s, performs well overall.

Size of Subgraph. Figure 5 shows (a) the run time and (b) the score of executing our algorithm, selecting 5%, 10%, 20%, 30%, and 50% of the whole graph as a subgraph, using the following parameters: grid network, $t_{max} = 7200\,s$, $r = 240$, and different weights w_i (due to space constraints, we only show the better performing configurations). The different curves represent different ways to balance the weights related to the four criteria described in Sect. 4. Clearly, the larger the chosen subgraph, the better the score. However, we have to pay a price: the run time also increases. More importantly, the run time grows super-linearly, while the score goes up sublinearly. Consequently, we have to strike a balance between run time and score. Still we get a better score than greedy with only 10% of the graph.

6.2 Comparison with State-of-the-Art

In this section we compare our probabilistic approach with the state-of-the-art algorithm for itinerary planning with category constraints: CLuster Itinerary Planning (CLIP) [2]. For these experiments we use the real-world data sets of Bolzano and San Francisco, selecting the following values for the parameters: $r = 240$ and a constant size of 200 POIs (Bolzano) and 150 POIs (San Francisco) for the randomly chosen subgraph. Figure 6 shows the results for the Bolzano network, Fig. 7 those for the San Francisco network, both of them varying the overall trip time t_{max}. Again, due to space constraints, we only show the configurations performing best.

As can be clearly seen in Fig. 6 for the Bolzano network, in terms of score the probabilistic approach lies right in the middle between the greedy strategy and CLIP. However, when we look at the run time, we notice that the probabilistic approach is only slightly slower than the greedy algorithm, whereas it takes

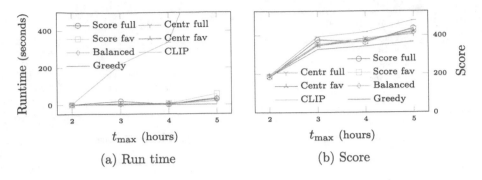

(a) Run time   (b) Score

Fig. 6. Bolzano network

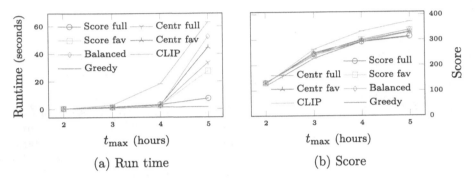

(a) Run time   (b) Score

Fig. 7. San Francisco network

CLIP significantly more time to come up with its solution. For the San Francisco network (Fig. 7), our algorithm has a similar performance, while CLIP exhibits a better run time compared to the Bolzano network.

7   Conclusion and Future Work

We developed a probabilistic algorithm for solving the orienteering problem with category constraints (OPMPC), selecting only a small part of the original graph probabilistically based on the quality of the points of interest. In an experimental evaluation we have shown that our approach can stand its ground. It produces better results than a greedy strategy, incurring only a slightly longer run time. While it does not quite reach the level of the state-of-the-art algorithm CLIP in terms of the achieved score, it generates solutions faster.

The selection of a subgraph is basically a filtering step before running the actual algorithm (in our case a dynamic programming algorithm), which means that it can be applied to any other algorithm used for solving OPMPC to reduce its run time. As demonstrated, this can be done without a significant loss in quality and in the future we want to investigate combining the probabilistic filtering step with other approaches. Additionally, it would be very interesting to tune the input parameters w_i using a machine learning approach.

References

1. Blum, A., Chawla, S., Karger, D.R., Lane, T., Meyerson, A., Minkoff, M.: Approximation algorithms for orienteering and discounted-reward TSP. SIAM J. Comput. **37**(2), 653–670 (2007)
2. Bolzoni, P., Helmer, S., Wellenzohn, K., Gamper, J., Andritsos, P.: Efficient itinerary planning with category constraints. In: Proceedings of the 22nd International Conference on Advances in Geographic Information Systems (SIGSPATIAL 2014), Dallas, Texas, pp. 203–212 (2014)
3. Chekuri, C., Korula, N., Pál, M.: Improved algorithms for orienteering and related problems. In: SODA 2008, pp. 661–670 (2008)
4. Chekuri, C., Pál, M.: A recursive greedy algorithm for walks in directed graphs. In: FOCS 2005, pp. 245–253 (2005)
5. Gavalas, D., Konstantopoulos, C., Mastakas, K., Pantziou, G.: A survey on algorithmic approaches for solving tourist trip design problems. J. Heuristics **20**(3), 291–328 (2014)
6. Gendreau, M., Laporte, G., Semet, F.: A branch-and-cut algorithm for the undirected selective traveling salesman problem. Networks **32**(4), 263–273 (1998)
7. Kanza, Y., Safra, E., Sagiv, Y., Doytsher, Y.: Heuristic algorithms for route-search queries over geographical data. In: ACM GIS 2008, pp. 11:1–11:10 (2008)
8. Keller, C.: Algorithms to solve the orienteering problem: a comparison. Eur. J. OR **41**, 224–231 (1989)
9. Liang, Y.-C., Kulturel-Konak, S., Smith, A.: Meta heuristics for the orienteering problem. In: CEC 2002, pp. 384–389 (2002)
10. Lu, E.H.-C., Lin, C.-Y., Tseng, V.S.: Trip-mine: an efficient trip planning approach with travel time constraints. In: MDM 2011, pp. 152–161 (2011)
11. Ramesh, R., Yoon, Y.-S., Karwan, M.H.: An optimal algorithm for the orienteering tour problem. INFORMS J. Comput. **4**(2), 155–165 (1992)
12. Rice, M.N., Tsotras, V.J.: Parameterized algorithms for generalized traveling salesman problems in road networks. In: ACM GIS 2013, Orlando, Florida, pp. 114–123 (2013)
13. Righini, G., Salani, M.: Decremental state space relaxation strategies and initialization heuristics for solving the orienteering problem with time windows with dynamic programming. Comput. OR **36**(4), 1191–1203 (2009)
14. Sevkli, Z., Sevilgen, F.E.: Variable neighborhood search for the orienteering problem. In: Levi, A., Savaş, E., Yenigün, H., Balcısoy, S., Saygın, Y. (eds.) ISCIS 2006. LNCS, vol. 4263, pp. 134–143. Springer, Heidelberg (2006). doi:10.1007/11902140_16
15. Shane, T.: Applied Computer Science. Springer, Cham (2016). doi:10.1007/978-3-319-30866-1
16. Singh, A., Krause, A., Guestrin, C., Kaiser, W.J., Batalin, M.A.: Efficient planning of informative paths for multiple robots. In: IJCAI 2007, pp. 2204–2211 (2007)
17. Tasgetiren, F., Smith, A.: A genetic algorithm for the orienteering problem. In: IEEE Congress on Evolutionary Computation (2000)
18. Tsiligrides, T.A.: Heuristic methods applied to orienteering. J. Oper. Res. Soc. **35**(9), 797–809 (1984)
19. Wang, Q., Sun, X., Golden, B.L., Jia, J.: Using artificial neural networks to solve the orienteering problem. Ann. OR **61**, 111–120 (1995)

Minimizing Negative Influence in Social Networks: A Graph OLAP Based Approach

Zakia Challal[1(✉)], Omar Boussaid[2], and Kamel Boukhalfa[1]

[1] University of Science and Technology Houari Boumediene,
BP 32 El Alia, 16111 Bab Ezzouar, Algiers, Algeria
{zchallal,kboukhalfa}@usthb.dz
[2] University of Lyon 2, 5, avenue Pierre Mondès France, 69676 Bron Cedex, France
omar.boussaid@univ-lyon2.fr

Abstract. Users of social networks (SN) interact continuously and are easily influenced. Research on influence analysis in SN gives much attention to positive influence. Few works on negative influence minimization make a restricted analysis limited to the influence degree between users ignoring the context and more importantly the users' opinion which may reflect the positive or negative aspect. None of the works do an OLAP style analysis of the influence to enable a multidimensional and a multi-level view of the network. In this paper, we propose an influence analysis graph OLAPing framework which offers a more complete, flexible and dynamic analysis. We then apply it to minimize negative influence.

Keywords: Social networks · Influence analysis · Graph OLAP

1 Introduction

Social influence analysis receives a particular attention due to its applications in important area such as marketing and politics. It studies how people interact and influence each other, and how this influence spreads in SN [12]. A large number of researchers works on identifying influential users [13] and influence maximization [6] problems. Few studies [2] have been interested in influence minimization considering the negative influence. The problem of minimizing the negative influence in a SN consists in selecting a minimum number of nodes or links to be blocked or protected in order to minimize the propagation of the influence. Most of the existing works are based on greedy algorithms which work well in terms of influence minimization but are time consuming [11,16]. Furthermore, we have identified some limitations in the description of the problem. The first concerns the influence probabilities: the arcs of the social graph are labeled by probabilities indicating the influence strength of one user on another. However, the question of how these probabilities can be computed from real SN data has been largely ignored [5]. The second concerns the negative influence: in all works, it is assumed that a negative influence is launched in the network without explaining what this means concretely. Finally, most of the works do not

© Springer International Publishing AG 2017
D. Benslimane et al. (Eds.): DEXA 2017, Part II, LNCS 10439, pp. 378–386, 2017.
DOI: 10.1007/978-3-319-64471-4_30

consider the subject and time of the influence, whereas the influence changes with change of context and time [12]. In this work, we propose a more complete analysis covering different aspects reflecting the context and the nature of the influence and allowing to have different views of the network according to different dimensions and different granularities. We aim to answer questions like "Which users have a high influence degree and a strong negative opinion on a given topic at a given time?". With such queries, we can extract the sub-graphs that participate massively in the diffusion of the negative influence for different subjects and different times according to different influence degrees. To this end, we propose a multidimensional analysis of the network based on graph OLAPing [3]. The rest of the paper is organized as follows: Sect. 2 gives a background on influence minimization problem in SN and graph OLAPing. Section 3 presents related works. Sections 4 and 5 introduce our approach and its application to the influence minimization. The evaluation and the results are presented in Sect. 6. We conclude in Sect. 7.

2   Definitions

2.1   Minimizing Negative Influence in Social Network

Let $G(V, E)$ be a graph representing a SN, M a propagation model [6], I a set of initially influenced nodes, and $\sigma_M(I)$ the function which estimates the propagation of the influence. The question is how to minimize $\sigma_M(I)$? [14–16].

2.2   Graph OLAP

A graph OLAP is a collection of network snapshots $S = \{S_1, \cdots, S_N\}$. Each snapshot is defined as $S_i = \{I_{1,i}, \cdots, I_{k,i}; G_i\}$ where $I_{k,i}$ are informational attributes describing the snapshot. $G_i(V_i, E_i)$ is a graph. V_i represent nodes and E_i represent edges. Attributes may be associated to V_i and E_i [3].

3   Related Works

We presented in [2] a survey on the minimization of negative influence in SN. We have classified the works according to their approach in three classes: (1) Blocking nodes: A minimum number of nodes are blocked to minimize the propagation of the negative influence in the network [14,15]. (2) Blocking links: A minimum number of links are blocked to minimize the propagation of the negative influence in the network [7–9,16]. Our work adopt this approach. (3)Competitive influence: A minimum number of nodes are selected to adopt a counter-campaign in order to limit the diffusion of negative influence [1,4,10,11]. Table 1 summarizes the related works. Analyzing these works allowed us to draw the following limitations: (1) Most of the works propose greedy algorithms and compare it to centrality measures. The greedy algorithms always give better results in terms of influence minimization but are clearly worse in term of execution time.

Table 1. Related works on negative influence minimization in social networks

Approach	Reference	Algorithms					Results	
		Proposed		Tested with			Influence minimization	Time
		Geedy	Others	Degree	Betweenness	Others		
Blocking nodes	[14]	x		x	x	x	+	
	[15]		x	x	x		+	
Blocking links	[9]	x		x	x	x	+	
	[8]	x		x	x		+	
	[7]	x		x	x	x	+	
	[16]	x		x	x		+	-
Competitive influence	[1]	x	x	x		x	+	
	[11]	x		x		x	+	-
	[10]	x		x		x	+	
	[4]	x		x		x	+	

(2) Most of the works analyze the influence with no regard to subject and time while the influence depends strongly on these two elements. (3) All the works do not define clearly two key elements of the problem: influence probabilities and the negative influence. In this paper, we answers these limits through a multidimensional model. To the best of our knowledge, no work does an OLAP analysis of the influence in SN. [3] propose a graph OLAP general framework to analyze information networks.

4   Graph OLAP for Influence Analysis in Social Networks

Our work aim to examine the influence in a SN through four dimensions: subject, time, influence degree and opinion value in a multidimensional model.

4.1   Model Definition

Given a SN represented by a graph $G(V, E)$. V is the set of nodes representing users. E is the set of links representing relations between users. s is a subject of analysis and t is the time of analysis. The influence probability between two users, referred as the influence degree di, represents the strength of interaction between two users. $DI_{s,t} : V * V \rightarrow [0,1]$ is the function which calculates the influence degree of a node u on a node v according to formula (1). The nature of the influence, positive or negative, depends on the user's opinion. We represent it by vo. $VO_{s,t} : V \rightarrow [-1,1]$ is the function that computes the opinion value of a node v according to the formula (2).

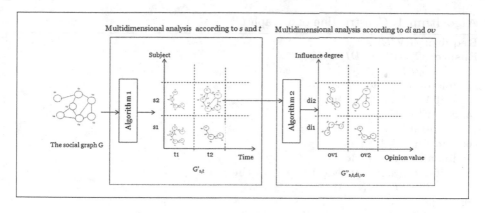

Fig. 1. Graph data warehouse construction for influence analysis

To build our data warehouse, we proceed in two steps (see Fig. 1): (1) On a the social graph G, we make an analysis according to s and t. We calculate di and vo for each subject and instant. We obtain the influence graphs $G'_{s,t}$. (2) On an influence graph $G'_{s,t}$, we make an analysis according to di and vo. We obtain the graph $G''_{s,t,di,vo}$ which constitute the final cube of our data warehouse. Next, we detail the two steps.

4.2 Multidimensional Analysis According to Subject and Time

Influence degrees and opinion values have different values for different subject and instant. Thus, we initially, analyze the social graph according to these two dimensions. The result is a set of oriented graphs $G'_{s,t}(V', E', DI_{s,t}, VO_{s,t})$ extracted from the global graph $G(V, E)$. V' is the set of nodes that participate in the subject s at time t. E' is the set of arcs (u, v) such that u influences v. $DI_{s,t}$ associates for each arc (u, v) the influence degree of u on v. $VO_{s,t}$ associates for each node v of V' its opinion value. We also calculate the global influence in each graph $G'_{s,t}$ according to formula (3). The Algorithm 1 describes the construction of a graph $G'_{s,t}$.

Learning Influence Degrees. Goyal and al. [5] propose models based on the user's action log. We refer to their discrete model by enriching the definition of an action with the subject.

$$DI_{s,t}(u, v) = A^{s,t}_{u2v} / A^{s,t}_u \qquad (1)$$

$A^{s,t}_{u2v}$ is the number of actions propagated from u to v.
$A^{s,t}_u$ is the number of actions of u.

Algorithm 1. Construction of a graph $G'_{s,t}$

Require: $G(V, E), s, t$
Ensure: $G'_{s,t}(V', E', DI_{s,t}, VO_{s,t})$
1: $V' \leftarrow \oslash;\ E' \leftarrow \oslash$
2: **for all** $v \in V$ **do**
3:     **if** $VO_{s,t}(v) \neq 0$ **then**
4:         $V' \leftarrow V' \cup \{v\}$
5:     **end if**
6: **end for**
7: **for all** $(u, v) \in E$ **do**
8:     **if** $DI_{s,t}(u, v) \neq 0 \wedge u \in V' \wedge v \in V'$ **then**
9:         $E' \leftarrow E' \cup \{(u, v)\}$
10:     **end if**
11:     **if** $DI_{s,t}(v, u) \neq 0 \wedge u \in V' \wedge v \in V'$ **then**
12:         $E' \leftarrow E' \cup \{(v, u)\}$
13:     **end if**
14: **end for**

Learning Opinion Values. We calculate the opinion value of a user on the basis of his text messages published in the network on the subject s until time t. We refer to the work of [17]. The messages are decomposed into phrases. The opinion value is first calculated in each phrase p.

$$VO_{s,t}(u) = (\sum_p (\sum_{w_j : w_j \in p \wedge w_j \in WL} \frac{OriSem(w_j)}{dist(w_j, s)})/n)/m \tag{2}$$

w_j is a word designating an opinion. WL is the Wilson Lexicon dictionary which contains all English words designating an opinion with their semantic orientation. $OriSem(w_j)$ is the function which returns the semantic orientation of w_j $(-1, -0.5, 0.0.5, 1)$. $dist(w_j, s)$ is the distance between the word w_j and the subject s. n is the number of words w_j in the phrase p. m is the number of phrases.

Learning the Global Influence in the Network. The global influence in each graph $G'_{s,t}$ depends on the influence degrees and opinion values of $G'_{s,t}$.

$$inf_{G'_{s,t}} = (\sum_{(u,v) \in E'} DI_{s,t}(u, v) * VO_{s,t}(v))/|E'| \tag{3}$$

4.3 Multidimensional Analysis According to Influence Degree and Opinion Value

Each influence graph $G'_{s,t}$ constitutes a new graph to analyze, this time, according to the two dimensions: influence degree and opinion value. For a given influence degree di and a given opinion value vo, we obtain a sub-graph $G''_{s,t,di,vo}$ extracted from $G'_{s,t}$. E'' constitutes the set of arcs (u,v) such that the degree of influence of the arc multiplied by the opinion value of its source node belongs to an interval $[\theta 1, \theta 2]$ defined in terms of $di * vo$. $[\theta 1, \theta 2]$ is equal to $[-1, di * vo]$ if we want to analyze the values below $di * vo$, it is equal to $[di * vo, 1]$ if we want to analyze the values above $di * vo$ and it is equal to $[di * vo - \alpha, di * vo + \alpha]$ if we want to analyze the values around $di * vo$ ($\alpha \in [0,1]$). The nodes u and v of the arcs (u,v) of E'' constitute the set V''. We have analyzed the graph according to $di * vo$ instead of the two dimensions di and vo separately because we consider that a user with a high influence degree and a low opinion value has the same effect as a user having a low influence degree and a strong opinion. The Algorithm 2 describes the construction of a graph $G''_{s,t,di,vo}$.

Algorithm 2. Construction of the graph $G''_{s,t,di,vo}$

Require: $G'_{s,t}(V', E', DI_{s,t}, VO_{s,t}), di, vo$
Ensure: $G''_{s,t,di,vo}(V'', E'', DI_{s,t}, VO_{s,t})$
 1: Define $[\theta 1, \theta 2]$ in terms of $di * vo$
 2: $V'' \leftarrow \oslash$; $E'' \leftarrow \oslash$
 3: **for all** $(u,v) \in E'$ **do**
 4:    **if** $DI_{s,t}(u,v) * VO_{s,t}(u) \in [\theta 1, \theta 2]$ **then**
 5:        $V'' \leftarrow V'' \cup \{u,v\}$; $E'' \leftarrow E'' \cup \{(u,v)\}$
 6:    **end if**
 7: **end for**

---

5 Application to Negative Influence Minimization

The graph $G''_{s,t,di,vo}$ can informs us about the arcs and nodes which strongly participate in the propagation of the negative influence if we choose the appropriate values of the dimensions di and vo. For example, a high value of di and a strong negative value of vo would respond to this request. After selecting the suspicious arcs or nodes, we can apply the minimization influence techniques like blocking nodes or links. In our case, we decide to block arcs instead of blocking nodes. Indeed, when we block a node, all the arcs attached to this node will be blocked, including arcs that do not rise the negative influence. Once the selected arcs are blocked, the global influence in the graph $G'_{s,t}$ is again measured. If it still negative, we change the values of di and vo and launch the request again. The process is repeated until the global influence becomes positive or reaches a defined threshold θ. Algorithm 3 summarizes the process.

Algorithm 3. Negative influence minimization

Require: $G'_{s,t}, \theta$
Ensure: $G'_{s,t}$ Where some arcs may be blocked
1: Calculate $inf_{G'_{s,t}}$
2: **while** $inf_{G'_{s,t}} < \theta$ **do**
3:   Read (di, vo)
4:   Extract $G''_{s,t,di,vo}$
5:   Block arcs returned by $G''_{s,t,di,vo}$ in $G'_{s,t}$
6:   Calculate $inf_{G'_{s,t}}$
7: **end while**

---

6   Experiments

For this first evaluation, we mainly aim at testing the negative influence minimization algorithm. We randomly generated a graph of 32767 nodes and 1.62
million arcs representing the graph $G'_{s,t}$ with random influence degree and opinion value. The global influence in the network is -0.256. We apply our proposed
Algorithm 3 by varying the values of di and vo. We block arcs whose influence
degree multiplied by the opinion value of the source node is less than $di * vo$
(i.e. $[\theta_1, \theta_2] = [-1, di * vo]$) and then calculate the global influence. Figure 2 illustrate the variation of the global influence in terms of the number of blocked arcs
for our approach, the greedy algorithm and Weights centrality measure [7]. Our
approach achieve the same influence minimization as greedy algorithm in less
iterations. In the greedy algorithm, at each iteration, we block the arc which
best improves the global influence by checking all the arcs of the graph. We stop
when k arcs are blocked. This require to browse all the graph k times which is
very time consuming. In our approach, we also block arcs that best improves the
global influence but in just one iteration. This is achieved by guiding the selection
of the arcs with the values of the dimensions vo and di. In the Weights centrality
measure, the top-k arcs according to their influence degree are blocked without
considering the opinion values. Our approach gives better results in terms of
influence minimization which shows the importance of taking into account the
user's opinion value.

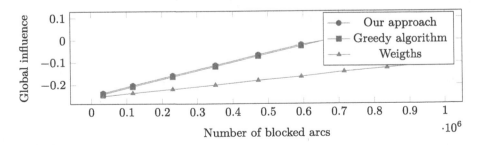

Fig. 2. Minimizing negative influence by blocking arcs: approaches comparison

7   Conclusion

Influence analysis in SN is so important when social media become the main way for information diffusion. In this paper, we proposed a multidimensional analysis of the influence in a SN which we apply to resolve the minimization of negative influence problem. In the future, we intend to work on three main contributions to offer a complete OLAP framework for influence analysis in SN. First, the ETL phase including influence degrees and opinion values computation, should be implemented with regard of scalability. Second, we need to guess for an efficient storage model for the graph data warehouse. Finally, we will develop an analytic tool with OLAP operations like roll up and drill down.

References

1. Budak, C., Agrawal, D., El Abbadi, A.: Limiting the spread of misinformation in social networks. In: Proceedings of the 20th International Conference on World Wide Web, WWW 2011, pp. 665–674. ACM, New York (2011)
2. Challal, Z., Boukhalfa, K.: Minimisation de l'influence négative dans les réseaux sociaux. In: Actes du XXXIVème Congrès INFORSID, Grenoble, France, pp. 117–130, 31 May–3 June 2016
3. Chen, C., Yan, X., Zhu, F., Han, J., Yu, P.S.: Graph OLAP: towards online analytical processing on graphs. In: 2008 Eighth IEEE International Conference on Data Mining, pp. 103–112, December 2008
4. Fan, L., Lu, Z., Wu, W., Thuraisingham, B., Ma, H., Bi, Y.: Least cost rumor blocking in social networks. In: 2013 IEEE 33rd International Conference on Distributed Computing Systems, pp. 540–549, July 2013
5. Goyal, A., Bonchi, F., Lakshmanan, L.V.: Learning influence probabilities in social networks. In: Proceedings of the International Conference on Web Search and Data Mining, WSDM 2010, pp. 241 250. ACM, New York (2010)
6. Kempe, D., Kleinberg, J., Tardos, E.: Maximizing the spread of influence through a social network. Theor. Comput. **11**(4), 105–147 (2015)
7. Khalil, E., Dilkina, B., Song, L.: Cuttingedge: influence minimization in networks. In: Workshop on Frontiers of Network Analysis: Methods, Models, and Applications at NIPS (2013)
8. Kimura, M., Saito, K., Motoda, H.: Solving the contamination minimization problem on networks for the linear threshold model. In: Ho, T.-B., Zhou, Z.-H. (eds.) PRICAI 2008. LNCS, vol. 5351, pp. 977–984. Springer, Heidelberg (2008). doi:10.1007/978-3-540-89197-0_94
9. Kimura, M., Saito, K., Motoda, H.: Blocking links to minimize contamination spread in a social network. ACM Trans. Knowl. Discov. Data **3**(2), 9:1–9:23 (2009)
10. Luo, C., Cui, K., Zheng, X., Zeng, D.: Time critical disinformation influence minimization in online social networks. In: 2014 IEEE Joint Intelligence and Security Informatics Conference, pp. 68–74, September 2014
11. Nguyen, N.P., Yan, G., Thai, M.T., Eidenbenz, S.: Containment of misinformation spread in online social networks. In: Proceedings of the 4th Annual ACM Web Science Conference, WebSci 2012, pp. 213–222. ACM, New York (2012)
12. Peng, S., Wang, G., Xie, D.: Social influence analysis in social networking big data: opportunities and challenges. IEEE Netw. **31**(1), 11–17 (2017)

13. Tan, C.W., Yu, P.D., Lai, C.K., Zhang, W., Fu, H.L.: Optimal detection of influential spreaders in online social networks. In: Proceedings of Annual Conference on Information Science and Systems (CISS), pp. 145–150, March 2016
14. Wang, S., Zhao, X., Chen, Y., Li, Z., Zhang, K., Xia, J.: Negative influence minimizing by blocking nodes in social networks. In: Proceedings of the 17th AAAI Conference on Late-Breaking Developments in the Field of Artificial Intelligence, AAAIWS 2013, pp. 134–136. AAAI Press (2013)
15. Yao, Q., Shi, R., Zhou, C., Wang, P., Guo, L.: Topic-aware social influence minimization. In: Proceedings of the 24th International Conference on World Wide Web, WWW 2015 Companion, pp. 139–140. ACM, New York (2015)
16. Yao, Q., Zhou, C., Xiang, L., Cao, Y., Guo, L.: Minimizing the negative influence by blocking links in social networks. In: Lu, Y., Wu, X., Zhang, X. (eds.) ISCTCS 2014. CCIS, vol. 520, pp. 65–73. Springer, Heidelberg (2015). doi:10.1007/978-3-662-47401-3_9
17. Zhou, X., Tao, X., Yong, J., Yang, Z.: Sentiment analysis on tweets for social events. In: Proceedings of the 2013 IEEE 17th International Conference on Computer Supported Cooperative Work in Design (CSCWD), pp. 557–562, June 2013

Hypergraph Drawing by Force-Directed Placement

Naheed Anjum Arafat[✉] and Stéphane Bressan

School of Computing, National University of Singapore, 13 Computing Drive,
Singapore 117417, Singapore
e0001887@u.nus.edu, steph@nus.edu.sg

Abstract. We propose a family of algorithms that transform a hypergraph drawing problem into a graph drawing problem and leverage force-
directed graph drawing algorithms in order to draw hypergraphs. We
propose and discuss a number of criteria to evaluate the quality of the
drawings from the points of view of aesthetics and of visualization and
analytics. We empirically and comparatively evaluate the quality of the
drawings based on these criteria on both synthetic and real data sets.
Experiments reveal that the algorithms are generally effective and the
drawings generated are aesthetically pleasing.

Keywords: Hypergraph · Visualization · Force-directed graph drawing

1 Introduction

We address the problem of drawing hypergraphs. A *Hypergraph* is a generalization of a graph where multi-ary relations exists among the objects including
binary relations. Edges of a hypergraph are called *hyperedges*. Hypergraphs have
found application in many areas of science [5,8,9,12,15] and a key challenge
in visualizing these hypergraphs is the exponential increase in the number of
potential relations with the number of objects [2]. The need to perform visual
analytics on data with multi-ary, complex relations calls for practically effective
and visually pleasing drawings of hypergraphs.

Fruchterman and Reingold, in their 1991 seminal paper [11], presented the
popular force-based graph layout algorithm that bears their names. In this algorithm, the graph is modeled as a physical system where vertices are attracted
and/or repelled according to some force function, eventually resulting in an equi-
librium configuration. The family of algorithms proposed in this paper transform
a hypergraph into a graph and leverage force-directed drawing algorithms to
draw the graph. The final positions of the vertices of the graph are then used to
draw the hyperedges of the hypergraph.

Inspired by the desirable properties of hypergraph drawing proposed by
Mäkinen [13], we propose and discuss some criteria for good hypergraph drawing. We also devise several metrics to empirically and comparatively evaluate
our algorithms based on both synthetic and real data sets. The experiments

© Springer International Publishing AG 2017
D. Benslimane et al. (Eds.): DEXA 2017, Part II, LNCS 10439, pp. 387–394, 2017.
DOI: 10.1007/978-3-319-64471-4_31

reveal that the approach is practical, efficient and generally effective, and, more importantly, the drawings generated are not only aesthetically pleasing but also readable for the purpose of visualization and analytics.

2   Related Work

Two classes of hypergraph drawing have been frequently used in the literature as mentioned by Mäkinen [13] namely, Subset based and Edge based. In both of them, vertices of the hypergraphs are drawn as points in the plane. In Edge based drawings, hyperedges are drawn as smooth curves connecting their vertices. In Subset based drawings, hyperedges are drawn as closed curves enveloping their vertices.

Mäkinen [13] formulates the desirable properties of a subset based hypergraph drawing. Bertault F. and Eades P. [4] demonstrate a method for drawing hypergraph in subset standard. The drawing algorithm constructs a Euclidean Steiner tree from the position of the vertices in a hyperedge, uses force-directed graph drawing algorithm to get the location of the vertices and draws a contour around the edges of the tree.

Since hyperedges are sets, visualizing hypergraphs is closely related to the approaches for visualizing sets. Euler diagram is amongst the earliest set visualization methods. Closed curves such as ellipses, circles or polygons represent sets and curve overlaps represent set relations. Many algorithms exist for drawing Euler diagrams. They differ from each other by their definitions of Euler diagram and drawable instances. Flower and Howse [10] define concrete Euler diagram and propose an algorithm to generate concrete diagrams automatically up to three sets. An extended definition of Euler diagram is given in [18] and the problem of generating diagrams up to eight sets is addressed. It is worth to mention that, the Subset based drawings differ from Euler diagrams in that the former does not impose regional constraints (e.g. not allowing empty zones, allowing exactly two points of intersections between contours representing sets) as the latter [16].

Recently, Simonetto et al. [17] propose an approach for generating Euler-like diagrams which are Euler diagrams with no regional restrictions. The sets are drawn as closed bezier curves. Subset based drawings such as the one presented in this paper are closely related to Euler-like drawings such as the one proposed by Simonetto et al. [17]. Both of these methods draw sets spatially repositioning the set elements and thus do not cater to the cases where preserving the semantics of the layout is important (e.g. scatter plots, geographical maps). Other approaches such as Bubble Sets [6], LineSets [1], Kelp diagrams [7] and KelpFusion [14] are designed for those cases. Simonetto et al. [17] applies a force-directed algorithm that preserves edge-crossing properties on the intersection graph of the hypergraph whereas our algorithm is able to apply any force-directed algorithm on a class of graphs derived from the hypergraph. Simonetto et al. [17] approximates the set boundaries by computing polygons whereas our algorithm computes convex polygons of the set of points. Thus the resulting diagram might be concave [1] violating one of the aesthetics we propose in this paper.

3   Aesthetics

We propose the following criteria of a good hypergraph drawing.

Firstly, The *Concavity* metric refers to the number of non-convex hyperedge drawings drawn by an algorithm. Since convex shapes are visually simpler, minimizing non-convex shapes results in good hypergraph drawing. It is to be noted that, minimizing non-convex shapes i.e. minimizing the *Concavity* metric helps ensure the first criterion as well.

Secondly, The *Planarity* metric is defined as the number of non-adjacent hyperedge crossings of a drawing. Drawing of a hypergraph, ideally, should have no crossing between any pair of non-adjacent hyperedges. In practice, however, having a crossing-free drawing of a hypergraph is rather difficult. Thus it is desirable to have as little non-adjacent hyperedge crossing (*Planarity*) as possible. Drawings with better *Planarity* help avoid clutter and thus ambiguity of the relations being represented.

Fig. 1. Two drawings of the hyperedges $\{a, b, c\}, \{b, f\}, \{d, e, f\}$. The drawing on the left has a concave shape, crossing between a non-adjacent hyperedge pair, poor Coverage and non-uniform distribution of vertices (clutter). The drawing on the right is aesthetically superior to the one on the left as it has no concavity, no crossing, comparatively better Coverage and Regularity.

Thirdly, The *Coverage* metric refers to the ratio of the 'mean area per vertex' of the drawing to the 'mean area per vertex' of the entire drawing canvas. The 'mean area per vertex' of a drawing of a hypergraph $H(X, E)$, Mean-APV$_{drawing}(H)$ is defined as

$$\text{Mean-APV}_{drawing}(H) = \frac{\sum_{i=1}^{|E|} \frac{Area(E_i)}{|E_i|}}{|E|} \tag{1}$$

where $Area(E_i)$ is the area of the shape representing E_i and $|E_i|$ is the number of vertices in the hyperedge E_i (the cardinality of E_i). The 'Mean area per vertex' of the drawing canvas, Mean-APV$_{canvas}$ is computed as -

$$\text{Mean-APV}_{canvas}(H) = \frac{Area_{canvas}}{|X|} \tag{2}$$

where, $Area_{canvas}$ is the area of the drawing canvas. Thus, the *Coverage* of the hypergraph $H(X, E)$ is defined as

$$Coverage(H) = \frac{\text{Mean-APV}_{drawing}(H)}{\text{Mean-APV}_{canvas}(H)}. \tag{3}$$

Maximizing the *Coverage* metric implies that the drawing canvas is utilized properly by the drawing, in other words, the drawing is sparse in some way. A *Coverage* value close to 1 implies almost 100% utilization of the drawing area.

Fourthly, The *Regularity* metric is a measure of uniformity of the vertices over the drawing canvas. Maximizing the *Regularity* criterion helps ensure less clutter of the vertices over the drawing canvas. Since the *Coverage* criteria itself fails to capture uniformity of vertices in cases of drawings with crossings, *Regularity* gives insights into the distribution of vertices over the drawing area in those cases.

Figure 1 illustrates how the proposed metrics encapsulate the aesthetics of a drawing. Interested readers may refer to [3] for implementation of these metrics.

4   Algorithms

The family of algorithms we propose initializes the coordinates of the position of the vertices of the hypergraph in a certain way, transforms the input hypergraph into a graph termed as the *associated graph* of the input hypergraph, draw the graph using force-directed graph layout algorithm to find the coordinates of the vertices of the input hypergraph and envelop the vertices each hyperedge inside a closed curve.

It is well-known that the initial position of vertices influences the performance of the force-based graph layout algorithm. Vertices of the hypergraph can be initialized randomly (*Random initialization*), in a circular fashion or uniformly over the drawing canvas. In *Circular initialization*, for each vertex x_i, its position $x_i.pos$ is initialized to the coordinate of a randomly generated point on a circle of radius $k|E_j|$ where $k \in \mathbb{N}$ and $x_i \in E_j$. In *Grid based initialization*, the drawing canvas is divided into grids, the vertices are associated with grids sequentially and $x_i.pos$ is initialized to a random point inside the grid x_i is placed in.

From Hypergraph to Graph. We propose four different ways of constructing associated graphs of a hypergraph- namely, complete associated graph, star associated graph, cycle associated graph and wheel associated graph. Each of these constructions gives rise to an algorithm.

Consider a hypergraph H denoted by the tuple $(X, E = \{E_1, E_2, \ldots, E_n\})$. For practical purposes, also consider $x_i.pos$ denotes the position vector of an arbitrary vertex $x_i \in X$ on the drawing canvas. Furthermore, $\{x_1, x_2, \ldots, x_{|E_i|}\}$ denotes the set of vertices in the arbitrary hyperedge E_i.

A *complete associated graph* $C(H)$ of the hypergraph H is a graph whose set of vertices is X and for each hyperedge E_i, any pair of distinct vertices in E_i are connected by a unique edge in $C(H)$. To illustrate, the complete associated graph of the hypergraph $H_1 = (\{a, b, c, d\}, \{\{a, b, c, d\}, \{c, d\}, \{a\}\})$ consists of $\{a, b, c, d\}$ as the set of vertices and $\{(a, b), (b, c), (c, d), (a, c), (b, d), (a, d)\}$ as the set of edges. The drawing algorithm which transforms a hypergraph into its complete associated graph is termed as the *Complete algorithm*. The motivation behind the *Complete* algorithm stems explicitly from the underlying principle of the force-directed graph layout algorithms and implicitly from the *Planar* characteristic of good drawing. Intuitively, fewer crossings are expected to occur among a set of hyperedges if the constituent vertices of a hyperedge are spatially closer to each other than to the vertices of the other hyperedges.

A *cycle associated graph* $Cy(H)$ of the hypergraph H is a graph whose set of vertices is X and for each hyperedge E_i, a cycle is formed by adding edges $(x_1, x_2), (x_2, x_3), \ldots, (x_{|E_i|}, x_1)$ in $Cy(H)$. The cycle formed is unique if we consider the vertices in E_i sorted clockwise according to their position in the drawing. The drawing algorithm which transforms a hypergraph into its cycle associated graph is named the *Cycle algorithm*. To illustrate, the cycle associated graph of the hypergraph H_1 mentioned before consists of $\{a, b, c, d\}$ as the set of vertices and $\{(a, b), (b, c), (c, d), (a, d)\}$ as the set of edges. Sometimes attractive forces between the vertices in the complete subgraphs of $C(H)$ are too strong which in turn results in a cluttered drawing. The desire to have a sparse drawing with good *Coverage* and *Regularity* is the driving force of the *Cycle* algorithm since $Cy(H)$ is a subgraph of $C(H)$.

If we allow vertices other than X into our associated graph, transformations of other kinds emerge. Given the position vectors of the vertices in X and an arbitrary hyperedge E_i in the hypergraph H, the barycenter of the vertices in E_i denoted as b_i is the unique vertex located at the position $\sum_{i=1}^{k} \frac{x_i \cdot pos}{k}$. The set of barycenters of H denoted as B is $\{b_1, b_2, \ldots, b_n\}$.

A *star associated graph* $S(H)$ of a hypergraph $H = (X, E)$ is a graph whose set of vertices is $H \cup B$ and for each hyperedge E_i and its barycenter b_i, a star is formed by adding edges $(b_i, x_1), (b_i, x_2), \ldots, (b_i, x_{|E_i|})$. To illustrate, $S(H_1)$ of the hypergraph H_1 mentioned before consists of $\{a, b, c, d, b_1, b_2\}$ as the set of vertices and $\{(b_1, a), (b_1, b), (b_1, c), (b_1, d), (b_2, c), (b_2, d)\}$ as the set of edges. The drawing algorithm which transforms a hypergraph into its star associated graph is named the *Star algorithm*. The design principle of the star algorithm follows from the fact that, spatial nearness among the vertices from the same hyperedge and remoteness among the vertices from distinct hyperedge can be achieved if all the vertices feel the attractive force towards the barycenter.

A *wheel associated graph* $W(H)$ of the hypergraph H is a graph whose set of vertices is $H \cup B$ and for each hyperedge E_i and its barycenter b_i, a wheel is formed by adding edges $(b_i, x_1), (b_i, x_2), \ldots, (b_i, x_{|E_i|}), (x_1, x_2), (x_2, x_3), \ldots,$ $(x_{|E_i|}, x_1)$. To illustrate, $W(H_1)$ of the hypergraph H_1 mentioned above consists of $(\{a, b, c, d, b_1, b_2\}$ as the set of vertices and $\{(b_1, a), (b_1, b), (b_1, c), (b_1, d),$ $(b_2, c), (b_2, d), (a, b), (b, c), (c, d), (a, d)\})$ as the set of edges. The drawing algorithm which transforms a hypergraph into its wheel associated graph is named the *Wheel algorithm*. Note that, the set of hyperedges in the wheel associated graph is the union of the set of hyperedges of the cycle and the star associated graphs i.e. $W(H) = S(H) \cup Cy(H)$.

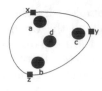

Fig. 2. The vertices of a hyperedge $\{a, b, c, d\}$ after drawing its Complete associated graph.

Fig. 3. Convex hull of the vertices a, b, c, d and its bordering vertices a, b, c.

Fig. 4. Pair-wise outtangents of the bordering vertices and points x, y, z as their intersections.

Fig. 5. The hyperedge drawn as a closed Catmull-Rom Spline going through x, y, z.

Drawing Hypergraph from Associated Graph. The Force-directed graph drawing algorithm applied to an associated graph results in an embedding of the vertices of its corresponding hypergraph. Each hyperedge of that hypergraph is then drawn as a closed curve enveloping its vertices. Figures 2, 3, 4 and 5 illustrate this process in sequence.

5   Experiments, Results, and Analysis

We empirically and comparatively evaluate the effectiveness and scalability of the algorithms on both synthetic and real dataset. We generate random hypergraphs with 2000 vertices and varying number of hyperedges and use them as our synthetic dataset. We use the DBLP [1] co-authorship network as our real dataset. Readers may refer to [3] for details about our experimental results.

The *Complete* and the *Wheel* algorithms have better Planarity than the rest. The reason is the attractive forces among the vertices in the Complete and the Wheel associated graphs are higher than the other associated graphs since a wheel and complete graph has more edges than the star or the cycle graph over the same number of vertices. The *Cycle* and the *Star* algorithm have better Coverage than the others due to the dominance of repulsive forces among the vertices in the associated graphs. In terms of Regularity, the performance varies depending on the granularity parameter in the experiment. We also observe that the Grid based initialization has the same effect on the performance as the circular initialization. Random initialization results in better Regularity than the Circular initialization. In Scalability experiment, the performance of the algorithms on the metrics are consistent and similar as in the effectiveness experiment. Figure 6 illustrates some drawings generated by one of our algorithms.

---

[1] http://dblp.uni-trier.de/xml/.

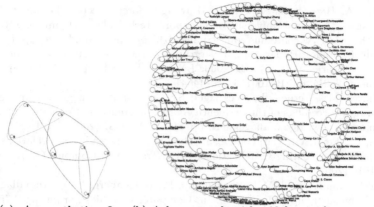

(a) A synthetic 3-uniform hypergraph with 5 hyperedge.

(b) A hypergraph with 100 hyperedges sampled from the DBLP dataset

Fig. 6. Example of some drawings by the *Wheel* algorithm

6 Conclusion

We propose a family of algorithms for drawing hypergraphs. We also propose a set of measurable criteria for evaluating the performance of the algorithms. We empirically evaluate the effectiveness and scalability of our proposed algorithms with different initial positioning of the vertices. The drawings by our algorithms are not only aesthetically pleasing in a qualitative way but also follow a set of quantitative criteria of good drawing. However, the drawings generated have lesser uniformity than expected stemming from scattering of disconnected components by the underlying force-directed graph layout algorithm. In future, we would like to extend Fruchterman-Reingold's algorithm from two-dimensional canvas to higher dimensions by modeling the hyperedges as elastic manifolds with the hope of having better drawings.

Acknowledgement. This work is supported by the National Research Foundation, Prime Minister's Office, Singapore under its Campus for Research Excellence and Technological Enterprise (CREATE) programme.

References

1. Alper, B., Riche, N., Ramos, G., Czerwinski, M.: Design study of linesets, a novel set visualization technique. IEEE Trans. Vis. Comput. Graph. **17**(12), 2259–2267 (2011)
2. Alsallakh, B., Micallef, L., Aigner, W., Hauser, H., Miksch, S., Rodgers, P.: Visualizing sets and set-typed data: State-of-the-art and future challenges. In: Eurographics conference on Visualization (EuroVis)-State of The Art Reports, pp. 1–21 (2014)

3. Arafat, N.A., Bressan, S.: Hypergraph drawing by force-directed placement. School of Computing, National University of Singapore, Technical report number TRC6/17 (2017). http://www.comp.nus.edu.sg/naheed/files/hypergraphdrawing.pdf

4. Bertault, F., Eades, P.: Drawing hypergraphs in the subset standard (short demo paper). In: Marks, J. (ed.) GD 2000. LNCS, vol. 1984, pp. 164–169. Springer, Heidelberg (2001). doi:10.1007/3-540-44541-2_15

5. Brinkmeier, M., Werner, J., Recknagel, S.: Communities in graphs and hypergraphs. In: Proceedings of the Sixteenth ACM Conference on Conference on Information and Knowledge Management, pp. 869–872. ACM (2007)

6. Collins, C., Penn, G., Carpendale, S.: Bubble sets: revealing set relations with isocontours over existing visualizations. IEEE Trans. Vis. Comput. Graph. **15**(6), 1009–1016 (2009)

7. Dinkla, K., van Kreveld, M.J., Speckmann, B., Westenberg, M.A.: Kelp diagrams: point set membership visualization. In: Computer Graphics Forum, pp. 875–884. Wiley Online Library (2012)

8. Eschbach, T., Günther, W., Becker, B.: Orthogonal hypergraph drawing for improved visibility. J. Graph Algorithms Appl. **10**(2), 141–157 (2006)

9. Fagin, R.: Degrees of acyclicity for hypergraphs and relational database schemes. J. ACM (JACM) **30**(3), 514–550 (1983)

10. Flower, J., Howse, J.: Generating Euler diagrams. In: Hegarty, M., Meyer, B., Narayanan, N.H. (eds.) Diagrams 2002. LNCS (LNAI), vol. 2317, pp. 61–75. Springer, Heidelberg (2002). doi:10.1007/3-540-46037-3_6

11. Fruchterman, T.M., Reingold, E.M.: Graph drawing by force-directed placement. Softw. Pract. Exp. **21**(11), 1129–1164 (1991)

12. Lundgren, J.R.: Food webs, competition graphs, competition-common enemy graphs, and niche graphs. In: Roberts, F. (ed.) Applications of Combinatorics and Graph Theory to the Biological and Social Sciences. The IMA Volumes in Mathematics and its Applications, vol. 17, pp. 221–243. Springer, New York (1989)

13. Mäkinen, E.: How to draw a hypergraph. Int. J. Comput. Math. **34**(3–4), 177–185 (1990)

14. Meulemans, W., Riche, N.H., Speckmann, B., Alper, B., Dwyer, T.: Kelpfusion: a hybrid set visualization technique. IEEE Trans. Vis. Comput. Graph. **19**(11), 1846–1858 (2013)

15. Ramadan, E., Tarafdar, A., Pothen, A.: A hypergraph model for the yeast protein complex network. In: Parallel and Distributed Processing Symposium, 2004. Proceedings. 18th International. p. 189. IEEE (2004)

16. Santamaría, R., Therón, R.: Visualization of intersecting groups based on hypergraphs. IEICE Trans. Inf. Syst. **93**(7), 1957–1964 (2010)

17. Simonetto, P., Auber, D., Archambault, D.: Fully automatic visualisation of overlapping sets. In: Computer Graphics Forum, pp. 967–974. Wiley Online Library (2009)

18. Verroust, A., Viaud, M.-L.: Ensuring the drawability of extended Euler diagrams for up to 8 sets. In: Blackwell, A.F., Marriott, K., Shimojima, A. (eds.) Diagrams 2004. LNCS (LNAI), vol. 2980, pp. 128–141. Springer, Heidelberg (2004). doi:10.1007/978-3-540-25931-2_13

A Fast Heuristic for Finding Near-Optimal Groups for Vehicle Platooning in Road Networks

Dietrich Steinmetz[✉], Gerrit Burmester, and Sven Hartmann

Department of Informatics,
Clausthal University of Technology, Clausthal-Zellerfeld, Germany
{dietrich.steinmetz,gerrit.burmester,sven.hartmann}@tu-clausthal.de

Abstract. Grouping vehicles into platoons with short distance can reduce fuel consumption up to 21% [1] and improve capacity of roads. The Vehicle Platooning Problem deals with route planing to form platoons and therefore minimize the overall costs of all considered vehicles. This article is focused on the subject to find groups with most fuel savings potential in acceptable time. We propose a novel spatial grouping algorithm to determine near-optimal groups. Our approach uses fast geometrical heuristics, consisting of direction-comparison, a modified version of a geometric-median-calculation and a comparison of intersections areas between two vehicles respectively. For evaluations is *same-start unlimited ILP* (SSU ILP) used to solves the Vehicle Platooning Problem to get the optimal solution. Driving in found platoons saves round about 2% to 3% fuel in average compared to the sum of particular shortest paths of the vehicles. The algorithm is tested in simulations on randomly created vehicles on different graphs with the size of 5.000, 10.000 40.000 and edges and round about 0.5 times nodes respectively. The performance is evaluated and the results are compared to the total possible amount of savings.

1 Introduction

Optimal grouping of heavy-duty vehicles (HDVs) into Platoons allow to reduce fuel consumption, increase capacity of road network and safety in road traffic. This opimization problem is defined as vehicle platooning problem and belongs to class of NP-Hard problems, shown in [3]. To reduce the complexity, we divide vehicle platooning problem into two phases. In first phase we assign given vehicles into near-optimal groups with high fuel saving potential. In this regard we created a new heuristic to find optimal groups for large number of vehicles in large road networks. In second phase we compute optimal routes for platooning, based on founded groups. For this purpose we use *same-start unlimited ILP* (SSU ILP), defined in [3]. Important factors like safety constraint, traffic condition [2], speed profiles, vehicles deadlines etc. will not be considered in this work.

The structure of the paper is as follows. Section 2 formalizes the problem and gives information about the complexity. Section 3 proposes our novel heuristic for determining near-optimal groups for platooning in road networks. Section 4

© Springer International Publishing AG 2017
D. Benslimane et al. (Eds.): DEXA 2017, Part II, LNCS 10439, pp. 395–405, 2017.
DOI: 10.1007/978-3-319-64471-4_32

presents the experimental evaluation and discusses the results with respect to performance, savings, and quality of groups. Section 5 outlines the state-of-the-art and discusses related work. Finally, Sect. 6 concludes the paper and proposes ideas for future research.

2   Problem Statement

A road network can be represented by a graph $G = (V, E, w, \theta)$. The nodes $\{v_i\} = V$ are three-leg or four-leg interchanges and $E \subseteq V \times V$ represents the roads connecting the nodes. Every edge has a weight $w : E \to \mathbb{R}^+$ and every node has geographical coordinates - latitude and longitude - depicted by $\theta : V \to (\mathbb{R}, \mathbb{R})$. The weight w of an edge e is the length of the road.

Furthermore there are vehicles assumed, which are a tuple defined by a start node and an end node, so that $h_i = (a_h, b_h) \in V \times V | a_h \neq b_h$, with a_h: start node of vehicle h and b_h: end node of vehicle h. Every vehicle has a shortest path from its start node to its end node. A path P is a sequence of nodes $(s, ..., d) \in V \times ... \times V$ and has the length of the sum of weight of the edges connecting the nodes respectively. Vehicles do not have travel time limitations.

Platoon is defined as a subset of vehicles H with $|H| > 2$. Exactly one vehicle is a platoon leader and other vehicles are platoon followers, with descreased fuel cost by factor η on platooned edges.

Fuel-optimal group $f_k = \{h_1 ... h_l\}$ is a subset of the vehicle set $H = \{h_1 ... h_n\}$, with a computed score value between every vehicle h_n and h_l of $\lambda \in]0, 1] \; \forall n, l = 1, ..., l : l > n$. The λ indicates the platooning ability of two vehicles, λ is used to form fuel-optimal groups in Sect. 3.

The vehicle platooning problem is known to be computationally hard. In [3] the decision version of the problem has been shown to be NP-complete for general graphs, as well as for planar graphs.

3   Our Approach for Finding Near-Optimal Platooning Groups

Our approach computes for all vehicle pairs score value $\lambda \in [0, 1]$, which indicates platoon incentive of vehicle pair. This computation include three steps, shown in Algorithm 2. First step compares the directions, second step computes optimal linear platooning and third step computes and compares convex hulls for each vehicle pair. Based on pairwise incentives we compute non-disjoint groups of vehicles, with $\lambda > 0$, presented in Algorithm 3. This produce many conflicted groups, because in most cases one vehicle drive only in one platoon. For each group we compute with SSU ILP optimal platooning routes and determine group savings. We formulated ILP (Eq. 7) to select best disjoint groups, routes in this groups are final platoons. Algorithm 1 shows the full procedure.

Computing Pairwise Incentives. The basic idea of our approach (Algorithm 2) to use simple methods to identify pairwise platoon possibilities and vice versa. In first step we exclude vehicle pairs with very different directions. For each vehicle we compute vehicle-vector $\underline{h}$, shown in (Eq. 1).

$$\underline{h} = \begin{pmatrix} \theta(d_h).lat - \theta(s_h).lat \\ \theta(d_h).lon - \theta(s_h).lon \end{pmatrix} \tag{1}$$

Thus vehicles which drive in opposite ways or at least have a large angle between their vectors will not be investigated further. Thus every combination has to satisfy

$$\frac{\underline{h_l} * \underline{h_n}}{\|\underline{h_l}\| * \|\underline{h_n}\|} > \varphi \tag{2}$$

with $\varphi \in [-1, 1]$.

In second step we consider only vehicle pairs, which were not disqualified in first step. For each remained pair we compute linear platooning, inspired by geometric median approach. The idea to find optimal platoon vector between two vehicle-vectors $\underline{h_l}$ and $\underline{h_n}$. Platoon vector can be depict by two points z_1, z_2 in the euclidean room. First point definded as mergening point, consequently the second point is a splitting point. Cost of the platoon vector are reduced by linear saving factor η'. Overall linear platooning is sum of linear distances between starting point of h_l and merging point, starting point of h_n and merging point, merging point and splitting point with reduced cost, splitting point and ending point of h_l and ending point of h_n. We formulated this optimization problem as following cost function

Min $g(z_1, z_2) = \|\underline{s_n} - \underline{z_1}\| + \|\underline{s_l} - \underline{z_1}\| + \|\underline{d_n} - \underline{z_2}\| + \|\underline{d_l} - \underline{z_2}\| + \|\underline{z_1} - \underline{z_2}\| + \|\underline{z_1} - \underline{z_2}\| * \eta'$

It should be noted that since the addends in g are all convex functions so that, the objective function is also convex. Thus the local minimum of this sub problem is also the global solution for this sub-problem.

The calculated value by cost function g will be compared with sum of linear distances of vehicles h_l and h_n. This comparison allow to derive a continuously value between 0 and 1, which represents platooning probability. Vehicle pairs with 0 value will be disqualified for further computations and vehicle pairs with value 1 can high probably form a platoon.

In third step we compute for each vehicle a shortest path P_h with $A\ star$ algorithm. Based on P_h we compute a platooning area for each vehicle, whereby the area is limited by fuel saving factor $\eta = 0.1$. In other words the detour from P_h to any node in area should be less then $\eta * w(P_h)$. Therefore we construct a polygon around the shortest path to prune unreachable nodes, shown in Fig. 1a. The algorithm selects particular nodes $Y = \{a_h, y_1, y_2, \ldots, y_q, b_h\}$ from subset of nodes included in P_h. For polygon construction we use nodes from Y set.

Initially we compute vector $\underline{w_0}$, this vector is orthogonal to vector $\underline{h}$ with length $\eta * w(P_h)$, depicted in Eq. 5.

$$\underline{w_0} = \begin{pmatrix} 0 & -1 \\ 1 & 0 \end{pmatrix} * \frac{\underline{h}}{\|\underline{h}\|} * w(P_h) * \eta \tag{5}$$

$$f_{con}(y) = \begin{cases} \|\underline{w_0}\| * [1 - \dfrac{w(P_{s,y})}{w(P_h)}], & \dfrac{w(P_{s,y})}{w(P_h)} \leqslant \dfrac{1}{2} \\ \|\underline{w_0}\| * \dfrac{w(P_{s,y})}{w(P_h)}, & \dfrac{w(P_{s,y})}{w(P_h)} > \dfrac{1}{2} \end{cases} \tag{6}$$

For each point in Y we compute alternating vectors with length $f_{con}(y)$ (Eq. 6), Fig. 1a shows an example. End points of computed vectors are used to generate a polygon(grey polygon) and finally we compute convex hull $conv(Y)$ (blue dotted polygon), to to simplify a polygon.

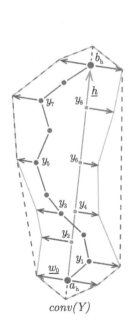

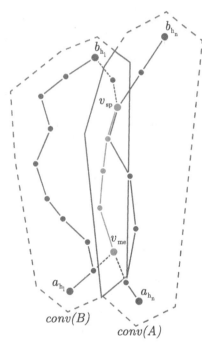

(a) Convex hull $conv(Y)$ for vehicle h.

(b) Example of two convex hulls $conv(A)$ and $conv(B)$ produced from h_l and h_n vehicles. The polygon in red shows intersection area. The green path represents the common road of both vehicles.

Fig. 1. (a) Convex hull $conv(Y)$ for vehicle h. (b) Example of two convex hulls $conv(A)$ and $conv(B)$ produced from h_l and h_n vehicles. The polygon in red shows intersection area. The green path represents the common road of both vehicles. (Color figure online)

Figure 1b shows P_{h_l} (blue solid line) and P_{h_n} (purple solid line) vehicles routes and corresponding convex hulls $conv(A)$ and $conv(B)$. Intersection of convex hulls produces a further convex hull (red polygon), in this area vehicles h_l and h_n can form a platoon. This area includes a platooning route (green line) with margin v_{me} and splitting node v_{sp}.

The size of intersection area, determine continually probability $\in [0,1]$ for platoon forming of vehicles h_l and h_n. Empty instersection area, means it is not possible to form a platoon. Large instersection area (relativ to smallest convex hull) means high incentive to platoon.

The calculated probability values in step 2 and step 3 will be composed to λ value for each vehicle pair. λ value will be stored in R matrix with $|H|$ rows and $|H|$ columns.

Forming Candidate Groups. Next we describe how we form candidate groups, see Algorithm 3. The output of the spatial-grouping-algorithm is a matrix R. For every value $\lambda > 0$ in that matrix a group consisting of two vehicles is build. Please recall that at this stage the groups are not yet required to be mutually disjoint. After this initial group forming every vehicle successively is being attempted to join another group to form even bigger groups until no vehicle can be assigned in an already existing group due to the values in the matrix. Thus all possible groups can be listed and this are the groups which will be examined further.

By using an SSU ILP formulation like [3] the savings for every group can be calculated.

Selecting the Best Mutually Disjoint Groups. The last part of the heuristic is to determinate the best disjoint combination of groups. Therefore the following integer linear problem (Eq. 7) has to be solved.

$$\text{maximize} \sum_{k \in F} x_k \cdot \alpha_k$$
$$\text{subject to} \sum_{\substack{k \in F: \\ h \in f_k}} x_k \leq 1 \forall h \in H \tag{7}$$
$$x_k \in \{0,1\} \forall k \in F$$

Where x_k is the decision variable for group k. α_k are the savings computed by the SSU ILP for every group k while f_k contains a set of vehicles, which are in that group. F is a set of all possible groups f_k.

Algorithm 1. Algorithm for finding near-optimal platooning groups

Data: A road network G and a list of vehicles H with start nodes s and end nodes d.
Result: A list M with near-optimal disjoint platooning groups.

$R = $ **Compute pairwise incentives:** Algorithm 2 (G, H)
$F = $ **Compute candidate groups:** Algorithm 3 (R)
for all f_k in F **do**
    load group relevant subgraph SG from graph database
    compute platooning routes in group f_k with SSU ILP and SG
    add group f_k rating
$M = $ determinate best disjoint groups from F with ILP (Eq. 7)
return M

Algorithm 2. Compute pairwise incentives

Data: A road network G and a list of vehicles H with start nodes s and end nodes d.
Result: A matrix R with values of the incentive to drive together.

calculate all vectors $\underline{h}$
for all $\underline{h_n}$ **do**
    if $\underline{h_l} * \underline{h_n} > 0$ **then**
        if Min $g(z_1, z_2) < \underline{h_l} + \underline{h_n}$ **then**
            if overlap of convex hull exist **then**
                fill matrix R in row h_l and column h_n with calculated value $\lambda \in]0, 1]$

Algorithm 3. Compute candidate groups

Data: A matrix R containing values $\lambda \in [0, 1]$ describing the incentive of two vehicles driving together.
Result: A list F of (potentially non-disjoint) groups of vehicles.

create an empty list of groups $F = \{\}$
for all $\lambda_{h_l, h_n} > 0$ in R **do**
    add a group containing h_l and h_n to F
for all $h_n \in R$ **do**
    for all $f_k \in F$ **do**
        if $h_n \notin f_k$ **then**
            if all $\lambda_{h_l, h_n} > 0 : \forall h_l \in f_k$ **then** duplicate group f_k and add the new
                group to F, then add h_n to new group
        else go to next group

---

4   Experimental Evaluation

The spatial-grouping-algorithm has been implemented with programming language R (version 3.3.1), with all data stored in an underlying Neo4j graph database (version 3.0.3). For experiments we used the following system configuration: Intel Core i7-6700 CPU @3.40 GHz, 32.00 GB DDR3, Windows 10 64-Bit-Version, Samsung SSD 850 PRO 256 GB.

We run the algorithm through various graphs with sized of 3.400, 7.000 and 27.000 nodes with 5.000, 10.000 and 40.000 edges respectively. Every node is at least incidental with 3 edges, so that there are no unnecessary nodes, which could be excluded from the processing. On the different graphs evaluated we generated random vehicles and run tests with sizes of 10, 30, 50 and 70 vehicles. Every size is tested for different graph instance sizes. Since the SSU ILP cannot compute the global solution for more vehicles or bigger graphs in a reasonable time only this instances can be evaluated in relation to the global optimum. The parameters are staying the same for the algorithm. We assume a value of $\eta = 10\%$ even if it could be more due to possible air drag savings [4,5], because this value represents a good average and is also used in related works [3]. Since the performance of the SSU ILP decreases exponential for bigger graphs with higher numbers of vehicles, only the top groups are examined for the graphs with over 40.000 edges. We also evaluated the performance in bigger graphs for 118.000, 374347 and 632524 edges.

4.1 Performance

In the following tables the performance of the heuristic is displayed. The first column indicates the number of vehicles of the test. In the second column the time, which the heuristic needs to initialize, calculate the scores, create the groups and upload the groups is listed. Following that the times for creating the scores and groups is given. 'Time of ILP (Eq. 7) solving groups' is the time the SSU ILP needs to calculate the savings of every created group and the fifth column states the overall time, which the heuristic needs to compute the objective and the group paths. For comparison also the time is given, which a SSU ILP would need to compute all routes without grouping the vehicles first.

Table 1. Performance comparison in seconds between heuristic and optimal solving for 5.000 edges.

Number of vehicles	Heuristic times				SSU ILP
	Heuristic time without ILP	Time to create score and groups	Time of ILP solving groups	Cumulative time	Time for opt. solution
10	2.03	0.56	100.49	102.52	15.40
30	3.51	2.14	91.44	94.95	85.89
50	6.90	5.17	164.02	170.92	195.34
70	14.52	12.80	101.50	116.02	584.19

The heuristics takes much less time for bigger instances and graph sizes, which can be evaluated through Tables 1, 2, and 3. Although the SSU ILP formulation computes faster solutions for the smallest vehicles sizes it's performance grows much more rapidly than the heuristic.

The performance and the results of our heuristic are compared to the Best-Pair Heuristic, and a Hub Heuristic [3], see Sect. 5 for details.

Since the performance of the first heuristic is large for bigger instances we compared only the instance of 10 vehicles to our algorithm and show, that the spatial-grouping-algorithm can even solve big graph instanzes. The results of

Table 2. Performance comparison in seconds between heuristic and optimal solving for 10.000 edges.

Number of vehicles	Heuristic times				SSU ILP
	Heuristic time without ILP	Time to create score and groups	Time of ILP solving groups	Cumulative time	Time for opt. solution
10	0.95	0.33	48.52	49.47	84.56
30	3.73	2.32	178.85	182.58	681.98
50	7.41	6.14	134.52	141.93	2517.60
70	13.63	12.23	135.88	149.51	17009.00

Table 3. Performance comparison in seconds between heuristic and optimal solving for 40.000 edges.

Number of vehicles	Heuristic times				SSU ILP
	Heuristic time without ILP	Time to create score and groups	Time of ILP solving groups	Cumulative time	Time for opt. solution
10	1.35	0.47	868.29	869.64	167.56
30	3.31	2.37	647.17	650.48	4397.60
50	7.256	6.39	418.11	425.37	>200000
70	14.46	12.91	797.18	811.64	>200000

the Best-Pair-Heuristic are quite good and show that a minimum of 50% of the savings can be achieved, while the mean is about 80% to 90% of the optimal solution in a graph of 647 nodes and 1390 edges. Also the Hub-Heuristic shows good results with a mean of 80%, but a bigger standard deviation. Through a subsequent local search algorithm, which is also explained in [3] in almost all tests the actual optimal solution could be accomplished for both algorithms within minutes.

The spatial-grouping-algorithm in combination with an SSU ILP and (Eq. 7) calculated the optimal solution in under 2 min for graph sizes of up to 10.000 edges. Even graph sizes of up to 40.000 edges were solved within minutes and without loss of quality.

4.2   Savings

In the tables below the savings are evaluated. The first column indicates the number of vehicles. After that the optimal result is displayed, which is calculated with a SSU ILP where the vehicles aren't divided into groups. The amount of time needed to calculate these objectives is given in Sect. 4.1. The objectives of the heuristic is stated in the third column. In the fourth and fifth column gap between the optimal solution and the heuristic solution is given. The savings are near the optimal solution and the difference increases for more vehicles. It also increases with the number of vehicles (Tables 4, 5, and 6).

Table 4. Experimental run for 5.000 edges, platooning fuel consumption. Optimal solutions compared with heuristic solutions.

Number of vehicles	Optimal result	Result of heuristic	Difference optimal solution and heuristic	Difference in %
10	380.72	380.81	0.09	0.024
30	1125.10	1134.60	9.50	0.84
50	1825.80	1864.00	38.20	2.05
70	2570.30	2644.70	74.41	2.81

Table 5. Experimental run for 10.000 edges, platooning fuel consumption. Optimal solutions compared with heuristic solutions.

Number of vehicles	Optimal result	Result of heuristic	Difference optimal solution and heuristic	Difference in %
10	1150.80	1151.10	0.28	0.025
30	3092.10	3132.70	40.53	1.29
50	5368.40	5476.50	108.13	1.97
70	7534.10	7778.40	244.26	3.14

Table 6. Experimental run for 40.000 edges, platooning fuel consumption. Optimal solutions compared with heuristic solutions.

Number of vehicles	Optimal result	Result of heuristic	Difference optimal solution and heuristic	Difference in %
10	1882.20	1882.50	0.29	0.015
30	5014.30	5095.90	81.55	1.60
50	8256,73	8432.60	175,87	2.13
70	11359,97	11686.00	326,03	2.87

The heuristics works with a small gap which increases for bigger instances but remains at an acceptable level. It is assumed, that the gap grows with increasing vehicle input sizes which can be adjusted through the parameters of the algorithm at the expense of performance.

4.3 Quality of Groups

The created groups are tested individually with a SSU ILP formulation in each test. Over 95% are successful and the vehicles in the particular group can drive in routes to make savings which are greater than 0. Only some of the tested groups wouldn't lead to a better objective and can't archive a positive amount of savings. This can be lead back to a bad selection of parameters, which could improve the solution.

5 Related Work

The topic of vehicles driving in a platoon is explored since the 1960s [6, 7] however the vehicle platooning problem developed in the last 5 years and has become more and more popular.

In [3] two heuristics are presented which can solve the vehicle platooning problem with a good approximation to the optimal solution.

The two algorithms presented in the paper can solve up to 200 HDVs of the *"same-start-problem"* – which is a relaxation of the problem whereat every vehicle has the same start node – and up to 10 HDVs with different start nodes in a graph containing up to 647 nodes and 1390 edges. The results reveal that 1% to 2% of fuel can be saved within the experiments of 10 vehicles, while the *"same-start-problem"* had savings up to 9%. In [8] platooning rates of 1800 heavy-duty vehicles are analyzed. The data contained the sparse vehicle positions over a one-day-period and map-matching and path-interference algorithms were used to determinate the vehicle paths. The researched objective is the difference between spontaneous platooning, occouring randomly because of same path of vehicles and coordinated platooning of the data by using virtual coordination schemes. The results show that the spontaneous platooning rate is about 0.07% – meaning the vehicles save 0.07% of fuel unlike no vehicle is platooning – and increases up to ten times while using coordination schemes.

Furthermore [9] presents methods to adjust vehicle speeds to form platoons by only knowing the vehicle's position, speed and destination. Small problem instances of 4 or less vehicles can be exactly solved and a heuristic algorithm is proposed which can solve the problem in a reasonable time and finds 90% of the the optimal solutions. However the details of forming platoons will not be part of this work and will be assumed without time or fuel loss.

The presented approaches solve the platooning problem partly in acceptable time with good results, nevertheless they have some weaknesses. The complexity of all approaches depends on the size of the graph, so that for big graph instances the execution time can be worth. [3] runs experiments with only small numbers of vehicles and isn't considered to solve for a bigger vehicle quantity.

6 Conclusion

The spatial-grouping-algorithm has a high performance and computes the groups for vehicles numbers up to 70. The difference of the objective is negligible for small instances and grows with the vehicle numbers, but is yet near the global optimum with a gap of approximately maximum 2% with up to 50 vehicles. Even in graphs with a high number of nodes and edges the spatial-grouping-algorithm performs well and has since the created groups are mostly successful the output will also be near to the optimal solution of the instances.

References

1. Bonnet, C., Fritz, H.: Fuel consumption reduction experienced by two promote-chauffeur trucks in electronic towbar operation. In: 7th World Congress on Intelligent Systems (2000)
2. Wang, D., Cao, W., Xu, M., Li, J.: ETCPS: an effective and scalable traffic condition prediction system. In: Navathe, S.B., Wu, W., Shekhar, S., Du, X., Wang, X.S., Xiong, H. (eds.) DASFAA 2016. LNCS, vol. 9643, pp. 419–436. Springer, Cham (2016). doi:10.1007/978-3-319-32049-6_26
3. Larsson, E., Sennton, G., Larson, J.: The vehicle platooning problem: computational complexity and heuristics. Transp. Res. Part C Emerg. Technol. **60**, 258–277 (2015)
4. Al Alam, A., Gattami, A., Johansson, K.H.: An experimental study on the fuel reduction potential of heavy duty vehicle platooning. In: 13th International IEEE Conference on Intelligent Transportation Systems (ITSC), pp. 306–311. IEEE (2010)
5. Lammert, M.P., Duran, A., Diez, J., Burton, K., Nicholson, A.: Effect of platooning on fuel consumption of class 8 vehicles over a range of speeds, following distances and mass. SAE Int. J. Commer. Veh. **7**(2014–01–2438), 626–639 (2014)
6. Levine, W., Athans, M.: On the optimal error regulation of a string of moving vehicles. IEEE Trans. Autom. Control **11**(3), 355–361 (1966)
7. Melzer, S., Kuo, B.: Optimal regulation of systems described by a countably infinite number of objects. Automatica **7**(3), 359–366 (1971)
8. Liang, K.Y., Martensson, J., Johansson, K.H.: Fuel-saving potentials of platooning evaluated through sparse heavy-duty vehicle position data. In: IEEE Intelligent Vehicles Symposium, pp. 1061–1068. IEEE (2014)
9. Larson, J., Kammer, C., Liang, K.Y., Johansson, K.H.: Coordinated route optimization for heavy-duty vehicle platoons. In: 16th International IEEE Conference on Intelligent Transportation Systems (ITSC), pp. 1196–1202. IEEE (2013)

Semantic Clustering and Data Classification

F-SED: Feature-Centric Social Event Detection

Elio Mansour[1]([⊠]), Gilbert Tekli[3], Philippe Arnould[1], Richard Chbeir[2], and Yudith Cardinale[4]

[1] University Pau & Pays Adour, LIUPPA, EA3000, 40000 Mont De Marsan, France
{elio.mansour,philippe.arnould}@univ-pau.fr
[2] University Pau & Pays Adour, LIUPPA, EA3000, 64600 Anglet, France
richard.chbeir@univ-pau.fr
[3] University of Balamand, Souk El Ghareb, Lebanon
gilbert.tekli@fty.balamand.edu.lb
[4] Dept. de Computación, Universidad Simón Bolívar, Caracas, Venezuela
ycardinale@usb.ve

Abstract. In the context of social media, existent works offer social-event-based organization of multimedia objects (e.g., photos, videos) by mainly considering spatio-temporal data, while neglecting other user-related information (e.g., people, user interests). In this paper we propose an automated, extensible, and incremental Feature-centric Social Event Detection (F-SED) approach, based on Formal Concept Analysis (FCA), to organize shared multimedia objects on social media platforms and sharing applications. F-SED simultaneously considers various event features (e.g., temporal, geographical, social (user related)), and uses the latter to detect different feature-centric events (e.g., user-centric, location-centric). Our experimental results show that detection accuracy is improved when, besides spatio-temporal information, other features, such as social, are considered. We also show that the performance of our prototype is quasi-linear in most cases.

Keywords: Social Event Detection · Social networks · Semantic clustering · Multimedia sharing · Formal Concept Analysis

1 Introduction

With the rapid evolution of social media, more users are now connected to collaboration and information sharing platforms (e.g., Facebook, Google+) and other sharing applications (e.g., Iphotos[1]). These platforms offer users the opportunity to share and manage various multimedia objects (e.g., photos, videos) taken by different users, during life events [12]. Due to excessive sharing on these platforms, the organization of the shared objects has become challenging. As a result, many works, such as the ones on Social Event Detection (SED), have evolved around the organization of shared data on social media platforms and sharing applications. SED works propose an organization of the shared objects by grouping them based on their related social events.

[1] http://www.apple.com/ios/photos

© Springer International Publishing AG 2017
D. Benslimane et al. (Eds.): DEXA 2017, Part II, LNCS 10439, pp. 409–426, 2017.
DOI: 10.1007/978-3-319-64471-4_33

Currently, some online platforms, stand-alone applications, and other works [3, 4, 16, 20, 21] provide an organization of shared objects through event detection, using clustering techniques. Some are based on metadata (e.g., Facebook organizes multimedia content based on publishing timestamps, Iphotos combines photo creation timestamps and locations to organize a user's photo library). Others use the visual attributes of shared objects (e.g., textures, colors) coupled with metadata [11, 13, 15] in order to detect several events.

Even-though these works offer automatic organization, they present limitations related to: (i) no user-centric processing, these methods do not include user-related features (e.g., social, interesting event topics) in the processing; (ii) lack of extensibility, these methods heavily rely on spatio-temporal features and do not allow the integration of various event features; (iii) the processing is based on one data source, neglecting the consideration of multiple data sources (i.e., multiple publishers, users); (iv) incremental processing is only implemented in recent applications, allowing a continuous integration of new data (e.g., photos shared/published on different dates/times) in the set of already processed data. Then, there is a need of providing a better event detection approach, considering user-related features, in a more meaningful way for users.

To answer this need, we propose a Feature-centric Social Event Detection (F-SED) approach, for automatic detection of "feature-centric" social events. Our method considers various features of an event (e.g., time, geo-location, social (e.g., users), topics, etc.), and allows the user to choose at least one central feature. Based on the chosen feature (or set of features) our approach detects the corresponding feature-centric events (e.g., topic-centered events if the topic feature is selected, user-centric events if the social feature is selected). F-SED's clustering technique simultaneously considers the various features, objects shared by different sources (several data publishers) on different dates/times, as well as data from one user (i.e., to organize a user data library). In addition, by integrating the user and his interests (*social and topics* features) in the clustering, the user automatically receives the events that interest him the most. F-SED is defined based on an adaptation of FCA (Formal Concept Analysis) [8, 22], a backbone that provides an extensible and incremental clustering technique, handles high dimensional data, and requires low human intervention. The comparative study on clustering techniques cannot be shown in this paper due to space limitations. We implement F-SED as a desktop-based prototype in order to evaluate the approach in real case scenarios with the ReSEED Dataset [18]. Our experimental results show that the event detection accuracy is improved when the social feature is taken into consideration. In addition, our performance results show quasi-linear behavior in most cases.

The rest of the paper is organized as follows. Section 2 reviews Social Event Detection works. Section 3 introduces FCA. Section 4 details and formally defines the F-SED approach. The implementation and evaluation are discussed in Sect. 5. Finally, Sect. 6 concludes and highlights future perspectives.

2   Related Work

In the literature, several Social Event Detection (SED) approaches have emerged for detecting events. Since most shared objects (e.g., photos, videos) on Social Networks and sharing applications are uploaded randomly without prior knowledge on the occurring events, these approaches mainly use unsupervised clustering techniques [3,13,15,16,21]. Since there are no commonly adopted criteria, we propose the following set of criteria to compare the referenced works:

1. *User-centric processing*: This criterion measures if user related data (e.g., names, interesting event topics) is integrated in the detection process to provide more meaningful and personalized results.
2. *Extensibility*: This criterion states if multiple event features are considered (e.g., *visual, social, topics*) in addition to time and locations for improved event detection.
3. *Multi-source*: This criterion indicates if multiple data sources (other publishers, participants) are considered, since various participants publish/share event related data.
4. *Incremental (continuous) processing*: This criterion considers the possibility of processing newly published data without having to repeat the entire processing, because participants could share event related data on different dates/times.
5. *Level of human intervention*: This criterion measures how frequently users participate in the event detection process; since huge amounts of data are shared, it is important that user interventions become less frequent; we consider low intervention if users provide data input and initial configuration; moderate if users intervene in result correction/optimization; and high intervention when users participate in the whole process.

SED approaches can be grouped into two categories: approaches that rely on the metadata of shared objects [3,16,19,21], denoted metadata-based, and approaches that rely on visual attributes (e.g., colors, shapes) and metadata [6,7,11,13,15], denoted hybrid. We could not find approaches that only rely on visual attributes since grouping visually similar objects does not necessarily mean that the latter belong to the same event.

Metadata-based approaches: In [16], the authors aim to detect social events based on image metadata, using temporal, geo-location, and photo-creator information. They perform a multi-level clustering for these features. A first clustering separates photos into groups by distinct time values. A second one is executed on the first level clusters based on geo-location information. Finally, a third clustering based on the second level clusters is executed using the creator-names. In [3], the authors use time and GPS data to cluster photos into events using the mean-shift algorithm. First, the authors find baseline clusters based on time, then GPS location attributes are integrated. In [19], the authors use time and location information from twitter feeds to detect various events (e.g., earthquakes). In [21], the authors rely on textual tags such as time, geo-location, image title, descriptions, and user supplied tags to cluster photos into events, thus detecting soccer matches that happened in Madrid. These approaches need moderate human intervention. However, they are not incremental, user-centric, nor extensible. Metadata is also used by stand-alone applications for photo management to detect social events in a user's multimedia library. For example, Iphotos is an iOS mobile application that clusters photos and videos based on time and geo-location. These applications require no human intervention, they automatically cluster objects found in a user's library. However, they do not consider other event features (e.g., social), nor other photo sources (photos taken by other participants/collaborators). They mainly focus on time and location, photos taken at the same day and place of an event are merged with the event.

Hybrid approaches: Many hybrid approaches rely on both visual and metadata attributes. In [13,15], the authors combine visual object attributes with temporal information, geo-locations, and user-supplied tags for their clustering procedures. Visual and tag similarity graphs are combined in [13] for the clustering. The tags used are crucial for event detection because they enable the distinction between events and landmarks. The authors consider photos of landmarks to have variant and distant photo capture timestamps because landmarks are photographed at any given date of the year and by many users. They also consider events to have a smaller distance separating timestamps while the number of users (participants) is also lower. They define an event based on time and location. While in [15], although metadata information is used in a different way, it is also crucial for distinguishing objects from events. The authors divide the geographical map of the world into square tiles and then extract the photos of each tile using geo-location metadata. They later use other metadata combined with visual features to detect objects and events. In [7,11], the authors combine temporal metadata with different sets of visual attributes for annotation and event clustering purposes. In [7], they combine time with color, texture, and facial (face detection) information. While in [11], the authors add to the previously mentioned attributes, aperture, exposure time, and focal length. In [6], the author relies more on temporal metadata than visual attributes for correct event detection, since he considers that photos/videos associated with one event are often not visually similar. Hybrid approaches consider different types of object attributes (visual, temporal information, geo-locations, tags, etc.). However, regrouping visually similar objects does not imply that they belong to the same event. Therefore, metadata is required to boost the accuracy of such approaches. Since these methods process visual attributes (e.g., through photo/video processing techniques), they end up having a higher processing cost than the approaches that only process metadata. Some approaches require more human intervention, because they prompt the user to correct/optimize the results.

Table 1 summarizes the evaluations of SED approaches based on the aforementioned criteria. Metadata-based approaches [3,16,21] need low to moderate human intervention and provide good event detection accuracy, since metadata describes data related to the events (e.g., dates, locations, tags). However, these works lack the incremental processing needed to match the flow of publishing/sharing.

Table 1. SED approaches comparison

Criteria	Metadata-based		Hybrid
	[3,16,21]	Stand-alone applications	[6,7,11,13,15]
The level of human intervention	Low - Moderate	Low	Moderate - High
The incremental processing	No	Partially[a]	No
Extensibility	No	No	No
Multi-source	Partially[a]	Partially[a]	Partially[a]
User-centric processing	No	No	No

[a] Partially states that not all approaches of a category are multi-source or incremental

Recently, incremental processing was integrated in some works (e.g., stand-alone applications). Hybrid methods are costly computation-wise and require human intervention thus making continuous processing hard to implement [14]. In contrast, hybrid methods [6,7,11,13,15] offer more event features by combining visual attributes with metadata to improve accuracy. Finally, the two categories of works do not fully consider the social feature in event detection and lack the extensibility and user-centric processing needed to provide more personalized, and therefore interesting, results to the user.

3 FCA Preliminaries and Definitions

After studying various clustering techniques [1,2,10,17], we chose Formal Concept Analysis (FCA) [8,22] as the backbone for our F-SED approach. FCA is incremental and extensible (criteria 4 and 2). It examines data through object/attribute relationships, extracts formal concepts and orders the latter hierarchically in a Concept Lattice which is generated through a four step process [5]:

Step 1: Defining a **Formal Context** (Definition 1) from the input data, based on object/attribute relations represented in a cross-table.

Definition 1. A *Formal Context*: is a triplet $\langle X, Y, I \rangle$ where:

- X is a non-empty set of objects
- Y is a non-empty set of attributes
- I is a binary relation between X and Y mapping objects from X to attributes from Y, i.e., $I \subseteq X \times Y$.

Table 2 shows an example, where photos are objects and photo attributes (locations, photo creator names, and dates) are attributes. The cross-joins represent the mapping of photos to their respective photo attributes, e.g., photo 1 *was taken in Biarritz by John on 17/08/2016.* ∎

Table 2. Formal context example

		Names				Locations			Dates		
		John	Patrick	Dana	Ellen	Biarritz	Munich	Paris	17/08/2016	12/12/2012	02/02/2016
Photos	1	x				x			x		
	2	x				x			x		
	3		x				x			x	
	4			x			x			x	
	5				x			x			x

Step 2: Adopting **Concept Forming Operators** to extract **Formal Concepts** (Definition 2). FCA has two concept forming operators:

- $\uparrow: 2^X \to 2^Y$ (Operator mapping objects to attributes)
- $\downarrow: 2^Y \to 2^X$ (Operator mapping attributes to objects).

For example, from the cross-table shown in Table 2, we have $\{3\}^\uparrow = \{$Patrick, Munich, $12/12/2012\}$ and $\{02/02/2016\}^\downarrow = \{5\}$.

Definition 2. *A **Formal Concept** in $\langle X, Y, I \rangle$ is a pair $\langle A_i, B_i \rangle$ of $A_i \subseteq X$ and $B_i \subseteq Y$ such that: $A_i^\uparrow = B_i \wedge B_i^\downarrow = A_i$.*

Consider the set of photos $A_1 = \{1, 2\}$ and the set of attributes $B_1 = \{$John, Biarritz, $17/08/2016\}$. $A_1^\uparrow = \{$John, Biarritz, $17/08/2016\}$ and $B_1^\downarrow = \{1, 2\}$. Thus, since $A_1^\uparrow = B_1$ and $B_1^\downarrow = A_1$, the pair $\langle A_1, B_1 \rangle$ is a Formal Concept. ∎

Step 3: Extracting a **Subconcept/Superconcept Ordering** relation for **Formal Concept** (cf. Definition 2) ordering by defining the most general concept and the most specific concept for each pair. The ordering relation is denoted $\leq$.

For example, from Table 2, let $A_1 = \{3\}$, $B_1 = \{$Patrick, Munich, $12/12/2012\}$, $A_2 = \{3, 4\}$, and $B_2 = \{$Munich, $12/12/2012\}$. According to Definition 2, $\langle A_1, B_1 \rangle$ and $\langle A_2, B_2 \rangle$ are formal concepts. In addition, $A_1 \subseteq A_2$ therefore, $\langle A_1, B_1 \rangle \leqslant \langle A_2, B_2 \rangle$. This means that formal concept $\langle A_1, B_1 \rangle$ is a subconcept of formal concept $\langle A_2, B_2 \rangle$ (which is the superconcept).

Step 4: Generating the **Concept Lattice**, which represents the concepts from the most general one (top) to the most specific (bottom). The lattice is defined as the ordered set of all formal concepts extracted from the data (based on $\leq$).

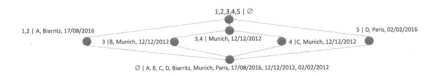

Fig. 1. The Concept/ Galois Lattice

For the example shown in Table 2, Fig. 1 illustrates the Concept Lattice. The next section formally describes our F-SED approach and how these FCA four steps are integrated and adapted for the clustering of shared objects (content).

4   An Approach for Feature-Centric Social Event Detection

In this section, we propose F-SED, an approach to detect feature-centric social events from a set of multimedia objects (e.g., photos, videos) shared by different users. F-SED mainly relies on a modular framework (Fig. 2) that integrates FCA as the backbone clustering technique, to provide an extensible and incremental approach. F-SED aims to detect feature-centric social events which are defined by at least three event features: (i) *temporal* feature (represented by a *time interval*), (ii) *geographical* feature (represented by a *location* value), and (iii) *social* feature (e.g., *photo creator name*). In order to focus more on the user's interests, F-SED also supports additional features such as *topics* (based on *tags* or *annotations*) and various levels of granularities for each feature. In order to organize a set of shared objects according to feature-centric events, F-SED processing is split into three main steps: (i) data pre-processing and extraction (executed by the Pre-processor and Attribute Extractor modules); (ii) lattice construction (executed by the Event Candidates Lattice Builder module); in this step we integrate and adapt the FCA clustering technique (described in Sect. 3) for shared objects clustering; (iii) event detection (carried out by the Feature-Centric Event Detector and Rule Selector modules). The user interacts with the system through the Front End. In the following, we detail each processing step and module.

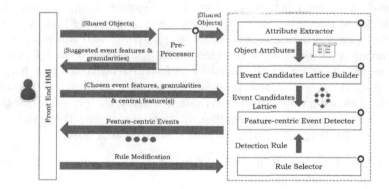

Fig. 2. F-SED Architecture

4.1   Data Pre-processing and Extraction

Through the Front End, one or multiple user(s), when considering a multi-source scenario (criterion 3), send(s) a set of shared objects (Definition 3). The purpose of this step is to extract the attributes of each shared object. An attribute is defined as a value associated with an attribute data type (Definition 4). We define a data type function denoted dt, that returns the attribute data type of a value based on the shared object attributes.

Definition 3. *A **Shared Object** is defined as a 2-tuple, $so : \langle id, V \rangle$, where:*

- id is the unique identifier of a shared object
- V is a set of attribute values according to a given ADT, such that $\forall a_i \in ADT \quad \exists v_i \in V \mid dt(v_i) = a_i$. ∎

Definition 4. ***ADT*** *is a set of attribute data types defined in a metric space, such that $\forall a \in ADT$, $a : \langle l, t, \Omega, d, f \rangle$, where:*

- l is a label
- t denotes the primitive data type of the attribute where
  $t \in \{Integer, Float, Boolean, Date, Time, Character, String\}$
- Ω is the domain (range) of the attribute values
- d is the function that returns the distance between any two values v_i and v_j of attribute data type a
- f is the event feature mapped to the attribute data type. ∎

The *Pre-processor* checks the attributes of each object and based on the data distribution, suggests event features (cf. Definition 5) and granularities to be used in the clustering. For example, the *Pre-processor* suggests the following features: *temporal, geographical, social,* and *topics* based on the availability of *photo creation timestamps, photo locations, photo creator names,* and *tags* or *annotations* respectively. The social feature can identify a single user (e.g., a user name) or a group of users (e.g., family, colleagues, friends), thus allowing the detection of user-centric events, or group-centric events (e.g., family events, work-related events). In addition, considering photo tags and annotations (either added manually by the user or detected by data inference or image processing techniques [9]), it can integrate a fourth event feature related to topics. Regarding granularities, the *Pre-processor* proposes for example to cluster *time* based on *day*, if photo creation values include day values. Finally, the *Pre-processor* detects the user's sharing space (cf. Definition 6), thus regrouping the collaborators, friends, other participants, and users that created/shared objects.

Definition 5. Features ***F*** is a set of event features where $\forall$ f $\in$ F, $f :$ $\langle label, G, interval, gran \rangle$, where:

- *label* is the feature label
- G is a set of granularities associated to the feature
- *interval* is a boolean value indicating if the feature is generated as an interval (true), or not (false)
- *gran* is a function that converts any granularity g_i to another g_j where $g_i, g_j \in$ G.

For example, the *temporal* event feature can be represented by:
$f_0 : (\textit{``Time''}, \{year, month, week, day, hour, minute, second\}, True, gran_{time})$.
$gran_{time}(1\,year) = 12$ months.   ■

Definition 6. A Sharing Space for a user u_0, denoted as $\boldsymbol{SS}_{u_0}$, is defined as a 3-tuple: $SS_{u_0} : \langle U, USO, SO \rangle$, where:

- U is the set of user names in the sharing space of the user $u_0 \mid u_0 \in U$
  (U is used to extract the *social* feature)
- $USO : U \longrightarrow SO$ is an association function mapping users to their respective shared objects such that, $USO(u_i) = \bigcup_{i=1}^{n} so_i$ where so_i is a shared object and n is the number of objects shared by u_i
- SO is a set of all shared objects in $SS_{u_0} \mid SO = \bigcup_{i=1}^{|U|} USO(u_i)$ ■

Once the user(s) validate(s) the features and granularities proposed by the *Pre-processor*, the *Attribute Extractor* extracts the object attributes. It also stores the shared objects, attributes, and values related to the chosen features for the lattice construction. In addition, the user specifies the feature (or set of features) that he would like to consider as central for the feature-centric event detection. No more human intervention is needed (low intervention-criterion 5).

4.2 Lattice Construction

In this step, we process the previously extracted object attributes, and shared objects into lattice attributes and objects, in order to generate one output: the lattice. The *Feature-centric Event Lattice Builder* is the FCA backbone. It integrates the four step process of FCA clustering described in Sect. 3. To do so, we formally define lattice attribute types in Definition 7. These types will be used when defining the lattice attributes (cf. Definition 8). Finally, for object/lattice attribute mapping, we define a binary cross rule denoted BXR (cf. Definition 9).

Definition 7. *lat* is a lattice attribute type representing an interval $[a, b[$ where $lat : \langle a, b, T \rangle$, where:

- a is the lower boundary value
- b is the upper boundary value
- T is a value representing the period having a primitive data type of either integer or float, such that:
  - $dt(a) = dt(b) \in ADT$ and
  - $b = a + T$.   ■

Definition 8. A lattice attribute, denoted $\boldsymbol{la}$, is defined as a 4-tuple $la : \langle f, SS_{u0}, lat, y \rangle$ where:

- $f \in F$ is the event feature mapped to lattice attribute la
- SS_{u_0} is the sharing space in which the detection will take place (cf. Definition 6)
- lat (cf. Definition 7) is the lattice attribute type

– y is a granularity | y ∈ f.G and

$$lat.T = \begin{cases} y & if \ f.interval = True \\ 0 & Otherwise \end{cases}$$

$lat.a = so_i.v_j$, where:

- $so_i \in SS_{u_0}.SO$ and
- $(v_j \in so_i.V) \wedge (dt(v_j).f = f)$. ■

Definition 9. A binary cross rule, denoted as **BXR**, is defined as a function that maps a shared object x to its respective lattice attribute y where $x.v_i \in x.V$:

$$BXR = \begin{cases} 1 & if \ (y.lat.T = 0 \wedge y.lat.a = x.v_i) \vee \\ & (y.lat.T \neq 0 \wedge x.v_i \in [y.lat.a, y.lat.b[) \\ 0 & Otherwise \end{cases}$$ ■

Then the *Feature-centric Event Lattice Builder* constructs the F-SED formal context, denoted **ffc** (cf. Definition 10). Once the **ffc** is created, formal concepts are extracted and a lattice is generated. This process is described in steps 2–4 of Sect. 3. This lattice is called an Event Candidate Lattice, where each node is a potential feature-centric event.

Definition 10. A F-SED Formal Context, denoted **ffc**, is defined as a 6-tuple $ffc : \langle SS_{u0}, F, f_{LAG}, X, Y, I \rangle$, where

- SS_{u0} is the sharing space in which the detection takes place
- F is the set of event features
- f_{LAG} is the function that generates the lattice attributes, described in Algorithm 1
- $X = SS_{u_0}.SO$ is the set of shared objects
- $Y = \bigcup_{i=0}^{|X.V|-1}\{la_i\}$ is the set of lattice attributes | $X.V = \bigcup_{\forall so \in X}\{so.V\}$ is the union of all attribute values from the shared objects in SS_{u0}
- I is a BXR(x,y) where $x \in X \wedge y \in Y$. ■

In Algorithm 1, we detail the lattice attribute generation process. This starts by extracting all object attribute values (lines 5–11). If the value is mapped to a feature that is generated as an interval (e.g., *time*), the algorithm calls the Create-Intervals function (lines 19–23). If not (e.g., *social*), the algorithm generates a lattice attribute type having a null period and creates the corresponding lattice attribute (lines 13–18). This step allows the creation of generic lattice attributes from various features, thus providing extensibility (criterion 2). Algorithm 2 details the Create-Intervals function. This process extracts all values related to the same feature (lines 4–9), orders them (line 10), selects a minimum and a maximum value (lines 11–12), and creates periodic intervals starting from the minimum to the maximum value (lines 14–22). The period is calculated based on the chosen feature granularity (line 15). This makes the detection more user-centric (criterion 1). Finally, the result is added the the output of Algorithm 1.

Algorithm 1. Lattice Attribute Generation (cf. Definition 10 - f_{LAG})

```
1   Input: SS_{u_0}
2   Output: RES                              // List of all lattice attributes
3   VAL = new List()                         // Shared Objects attribute values list
4   PD = new List()                          // Processed event features list
5   foreach so ∈ SS_{u_0}.SO do
6       foreach v ∈ so.V                     // This loop extracts all object attribute values
7       do                                       from all objects in SS_{u_0} and stores them in the
8           if (v ∉ VAL) then                 VAL list
9               VAL←v
10      end
11  end
12  foreach v ∈ VAL do
13      if (not dt(v).f.Interval)            // If the value is not generated as an
14      then                                     interval
15          lat ← LAT(v, lat.a + lat.T, 0)
16          la ← LA(dt(v).f, SS_{u_0}, lat, dt(v).f.g)     // Create la with lat.T=0
17
18          RES ← la
19      else
20          if (dt(v).f ∉ PD) then
21              RES ← (Create-Intervals(VAL,v,PD, SS_{u_0}))     // Call
22                                                                   Create-Intervals
23      end                                                          function
24  end
25  return RES
```

Algorithm 2. Create-Intervals

```
1   Input: VAL, v, PD, SS_{u_0}      // Input provided by Algorithm 1, line 21
2   Output: LAI                       // Generated lattice attributes intervals
3   int i = 0
4   TEMP = new List()                 // Temporary object attribute list
5   foreach val ∈ VAL do
6       if (dt(val).f == dt(v).f)     // Extract all object attribute
7       then                              values having the same feature as
8           TEMP ← val                    v and store them in TEMP
9   end
10  Order_{ascending}(TEMP)           // Order TEMP ascending
11  min ← TEMP.get(0)                 // min is the first element of TEMP
12  max ← TEMP.get(|TEMP| − 1)        // max is the last element of TEMP
13  lat ← LAT()
14  while (lat.b < max) do
15      lat ← LAT(min, lat.a + (i+1) × lat.T, dt(v).f.g)
16      if (lat.b > max)              // This loop creates
17      then                              intervals of period
18          lat.b ← max                   lat.T = f.g (feature
19      la ← LA(dt(v).f, SS_{u0}, lat, dt(v).f.g)     granularity)
20      LAI ← la
21      i++
22  end
23  PD ← dt(v).f      // Add feature to the list of processed features
24  return LAI
```

Feature	Condition
Time	Time Range
Geo	Geo Range
Social (central)	Specific value
Topic	Topic Range

(a)

Feature	Condition
Time	Time Range
Geo (central)	Specific Value
Social	Social Range
Topic	Topic Range

(b)

Feature	Condition
Time	Time Range
Geo	Geo Range
Social	Social Range
Topic (central)	Specific Value

(c)

Feature	Condition
Time (central)	Specific Value
Geo	Geo Range
Social	Social Range
Topic	Topic Range

(d)

Fig. 3. Default detection rule

4.3 Event Detection

The *Feature-centric Event Detector* module uses the previously generated lattice, an event detection rule, and the central features chosen by the user in order to detect feature-centric events (cf. Definition 11). We define a default detection rule, as a set of lattice attributes that comply with the two conditions mentioned in Definition 11. The rule is extensible, thus allowing the integration of multiple event features (e.g., *Time, Geo-location, Social, Topic*), each represented by the corresponding lattice attribute. This rule uses the selected central features in order to target the related feature-centric events. For example, the rules illustrated in Fig. 3(a), (b), (c), and (d) detect user, geo, topic, and time-centric events respectively. Finally, for testing purposes, users can change/add detection rules using the *Rule Selector* module. Since the lattice is not affected by the rule change, only the event detection step is repeated based on the new detection rule.

Definition 11. A feature-centric Event, denoted fce, is a Formal Concept defined as a 4-tuple $fce : \langle ffc, central_F, A, B \rangle$, where:

- ffc is a F-SED Formal Context (Definition 10)
- $central_F$ is the set of central features selected by the user $|central_F \subseteq ffc.F$
- A is a set of shared objects $| A \subseteq ffc.X$
- B is a set of lattice attributes $| B \subseteq ffc.Y$ where $\forall b_i, b_j \in B \wedge i \neq j$:
  - **Condition 1:** $b_i.f \neq b_j.f$
  - **Condition 2:** if $b_i.f.label = c_f.label | \forall c_f \in central_F$, then $d(b_i.lat.a, so_j.v_k) = 0 | \forall so_j \in A \wedge \forall v_k \in so_j.V, \quad dt(b_i.lat.a) = dt(so_j.v_k)$. ∎

5   Implementation and Evaluation

In order to validate our approach, we developed a Java desktop prototype and used the Colibri-java library for lattice generation. We evaluated the algorithm's performance based on execution time and memory consumption, and the quality of our detection process by measuring its accuracy. The objective of the experimentation is to show that the approach is generic and accurate when given optimal features/granularities. We do not aim at comparing accuracy results with other works.

ReSEED Dataset: To evaluate the detection results, we used the ReSEED Dataset, generated during the Social Event Detection of MediaEval 2013 [18]. It contains real photos crawled from Flickr, that were captured during real social events which are heterogeneous in size (cf. Fig. 4) and in topics (e.g., birthdays, weddings). The dataset contains 437370 photos assigned to 21169 events. In our evaluation, we used three event features: *time, location, and social*, since ReSEED photos have *time, geo, and social* attributes. In ReSEED, 98.3% of photos contain *capture time*, while only 45.9% of the photos have a *location*. We had to select photos having these attributes from the dataset. This left us with 60434 photos from the entire dataset. In ReSEED, the ground truth used for result verification assigns photos to social events. Since, our approach is focused on feature-centric events (in this experimentation, user-centric events), we modified the ground truth to split the social events into their corresponding user-centric events. Since the splitting is based on the event features, we need to specify the feature granularities during the process. The latter are not specified in ReSEED, therefore we chose the lowest granularity values: *day* for *time*, *street* for *geo*, and *photo creator name* for *social*. The ground truth refactoring process is described in Fig. 5. First, we extracted the photos of each event in the ground truth. Second, we used the *timestamps of photo capture* to group photos by *day*. Third, we split the resulting clusters into distinct groups based on *street* values. Finally, the result was further split based on distinct *photo creators*.

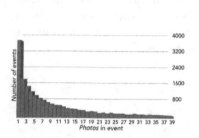

Fig. 4. ReSEED photo distribution

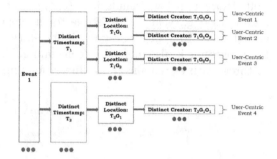

Fig. 5. Refactoring ReSEED ground truth

Performance Evaluation: The performance tests were conducted on a machine equipped with an Intel i7 2.60 GHZ processor and 16 GB of RAM. The aim was to test the performance of our F-SED algorithm. We considered two criteria for this task: (i) total execution time and (ii) memory overhead.

Use Cases: The performance is highly affected by the number of photos, generated attributes, and clusters. We noticed that granularities *day* for *time* and *street* for *geo* generate more clusters and attributes than any other granularity combination. Therefore, we used *day* and *street* to test the prototype's performance in three worst case scenarios:

- Case 1: We selected the biggest event (1400 photos) as input. We varied the number of photos progressively from 1 to 1400. Since all photos are related to one event, the number of detected clusters should be one.
- Case 2: We extracted 400 events each having exactly one photo. We varied the number of photos from 100, 200, 300 to 400. The number of generated clusters for each iteration should be 100, 200, 300, and 400 respectively.
- Case 3: The goal is to test with as many photos as possible related to different events. We varied the number of photos from 15000, 30000, 45000 to 60434. Since thousands of events contain only one or two photos per event (worst case scenario), this case will generate the most clusters.

Results and Discussion: In Cases 1 and 2 (Fig. 6a and b), where the number of photos does not exceed 1400 and 400 respectively, the total execution time is quasi-linear. However, in Case 3 (Fig. 6c), we clustered the entire dataset (60434 photos). The total execution time tends to be exponential, in accordance with the time complexity of FCA. When considering RAM usage, we noticed a linear evolution for the three cases (Fig. 6d, e and f). RAM consumption is significantly higher in Case 2, where we generated 400 clusters, than in Case 1, where we generated one cluster. In Case 3, RAM consumption is the highest because both the number of photos at the input, and the number of generated clusters (detected events) were the highest. Other tests were conducted, Fig. 7 (left) shows that low granularities (e.g., day) consume more execution time than high ones (e.g., year). This is due to the generation of more lattice attributes and clusters. In addition, Fig. 7 (right), shows that considering more features in the processing is also more time consuming. Nonetheless, the evolution from one to three features remains quasi-linear, making the process extensible.

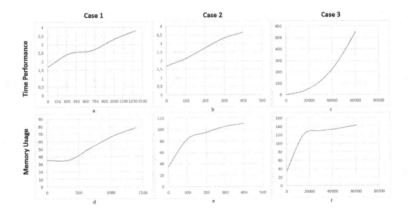

Fig. 6. Performance results

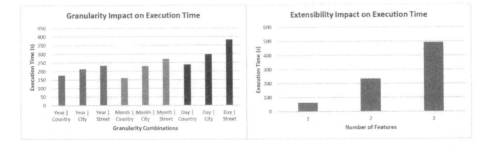

Fig. 7. Granularity and extensibility impact

Table 3. Detection rule

Combination	Number of features	Features considered in the detection rule
1	3	Time, Geo, Social
2	2	Time, Geo
3		Time, Social
4		Geo, Social
5	1	Time
6	1	Geo
7	1	Social

Table 4. Granularity combinations

Combination	Granularities: Time / Geo
1	Year / Country
2	Year / City
3	Year / Street
4	Month / Country
5	Month / City
6	Month / Street
7	Day / Country
8	Day / City
9	Day / Street

Accuracy Evaluation: We chose to consider the criteria proposed by MediaEval for clustering quality evaluation. We calculated the F-score, based on the Recall (R) and Precision (PR), and the Normalized Mutual Information (NMI) using ReSEED's evaluation tool. These criteria are commonly adopted in information retrieval and social event detection. A high F-score indicates a high quality of photo to user-centric event assignment while NMI will be used to measure the information overlap between our clustering result and the ground truth data. Therefore, a high NMI indicates accurate clustering result.

Use Cases: Since we considered the *time*, *geo*, and *social* features, we identified all possible combinations of the detection rule (see Table 3). In order to test granularity impacts, Table 4 sums up the different granularity combinations. When applying detection rules to granularity combinations, we get 63 use cases. We measured for each one the NMI and F-Score.

Results and Discussion: Results shown in Table 5, highlight the following:

(i) *Detection rule/features impact:* Our detection rule (based on *time*, *geo*, and *social* features) generates the highest NMI and F-score (NMI: 0.9999 and F-Score: 0.9995). It also exceeds all other detection rules (e.g., the one including solely *time* and *geo* features) in every granularity combination. This underlines the importance of the *social* feature in the detection task. Moreover, it highlights F-SED's extensibility, which allows the integration of additional features and the accurate detection of user-centric events. Nonetheless, accuracy can still be improved, few photos were assigned to the wrong clusters, due to the closeness in time and space of the latter.

(ii) *Granularity impact:* The results improve, when the clustering is based on granularities closer to the ones used in the ground truth. For example, in the case of granularities *year, country*, the F-Score achieved based on *time* and *geo* features is 0.1911, but for the detection rule that considers only the *social* feature the F-Score is higher: 0.5376. This is because the granularities for *time* and *geo* are the most general (*year and country*). Therefore, the

Table 5. Clustering results

Detection rule	Measure	Granularities								
		Year			Month			Day		
		Country	City	Street	Country	City	Street	Country	City	Street
Time geo social	F-score	0.6399	0.8180	0.8662	0.7964	0.8619	0.8948	0.9535	0.9742	**0.9995**
	NMI	0.9181	0.9602	0.9729	0.9549	0.9703	0.9789	0.9880	0.9938	**0.9999**
Time geo	F-score	0.1911	0.7678	0.8473	0.4943	0.8367	0.8821	0.8854	0.9542	0.9892
	NMI	0.7113	0.9475	0.9684	0.8707	0.9637	0.9759	0.9729	0.9894	0.9977
Time social	F-score	0.6245	0.6245	0.6245	0.7939	0.7939	0.7939	0.9534	0.9534	0.9534
	NMI	0.9143	0.9143	0.9143	0.9544	0.9544	0.9544	0.9879	0.9879	0.9879
Geo social	F-score	0.5085	0.7718	0.8357	0.5085	0.7718	0.8357	0.5085	0.7718	0.8357
	NMI	0.8742	0.9470	0.9653	0.8742	0.9470	0.9653	0.8742	0.9470	0.9653
Time	F-score	0.0220	0.0220	0.0220	0.1399	0.1399	0.1399	0.7278	0.7278	0.7278
	NMI	0.3971	0.3971	0.3971	0.7069	0.7069	0.7069	0.9392	0.9392	0.9392
Geo	F-score	0.0559	0.6958	0.8343	0.0559	0.6958	0.8343	0.0559	0.6958	0.8343
	NMI	0.5084	0.9241	0.9646	0.5084	0.9241	0.9646	0.5084	0.9241	0.9646
Social	F-score	0.5376	0.5376	0.5376	0.5376	0.5376	0.5376	0.5376	0.5376	0.5376
	NMI	0.8755	0.8755	0.8755	0.8755	0.8755	0.8755	0.8755	0.8755	0.8755

impact factor of granularities is more important than that of the number of features considered in the detection rule. Some rules can exceed others for specific granularity combinations (e.g., *Time Geo* exceeds *Time Social* and *Geo Social* for granularities Year/Month/Day-Street while Time Social exceeds the other two rules for Year/Month/Day-Country). The best result can be achieved by considering the maximal number of features having correct granularities. This indicates that the granularities should not be fixed for all scenarios. When given the best granularities, our approach detects the user-centric events very accurately. With these results, we find it interesting to study the user data distribution to deduce the granularities that boost his detection results the most or the possibility of disregarding certain features with non optimal (result-wise) granularities. This can become an advantage for reducing processing costs. In addition, we would also like to investigate inter user-centric event distances for *time* and *geo* essentially.

6   Conclusion and Future Work

Social Event Detection (SED) has become essential towards automatic, semantic organization of photo collections and other multimedia objects on the web. In this paper, we propose a generic framework for Feature-centric Social Event Detection (F-SED) based on Formal Concept Analysis (FCA). Our approach aims at integrating various features (e.g., *social, topics*), making the detection task more user-centric, extensible, incremental, and as automated as possible, thus reducing human intervention. We developed a prototype for testing purposes. The results show that our approach achieved high accuracy in most cases, especially when the social feature is considered. As future work, we are investigating the detection of optimal granularities based on the user's data distribution and sharing frequency and the impact of other features. We would also like to improve our accuracy by considering spatio-temporal distances between clusters and noise handling techniques. In addition, we aim at testing the incremental processing of our algorithm. Finally, we want to extend our work in order to detect social events based on existent ties between feature-centric events.

References

1. Berkhin, P.: A survey of clustering data mining techniques. In: Kogan, J., Nicholas, C., Teboulle, M. (eds.) Grouping Multidimensional Data, pp. 25–71. Springer, Heidelberg (2006). doi:10.1007/3-540-28349-8_2
2. Burmeister, P.: Formal concept analysis with ConImp: Introduction to the basic features. Technische Universität Darmstadt, Fachbereich Mathematik (2003)
3. Cao, L., et al.: Image annotation within the context of personal photo collections using hierarchical event and scene models. IEEE Trans. Multimedia 11(2), 208–219 (2009)
4. Chen, L., Roy, A.: Event detection from flickr data through wavelet-based spatial analysis. In: Conference on Information and Knowledge Management, pp. 523–532 (2009)
5. Choi, V.: Faster algorithms for constructing a concept (Galois) lattice. CoRR abs/cs/0602069 (2006). http://www.bibsonomy.org/bibtex/2a6eb1dc7b4615fc635dfd633fa950cd8/dblp
6. Cooper, M., et al.: Temporal event clustering for digital photo collections. ACM Trans. Multimedia Comput. Commun. Appl. 1(3), 269–288 (2005)
7. Cui, J., et al.: EasyAlbum: an interactive photo annotation system based on face clustering and re-ranking. In: Conference on Human Factors in Computing Systems, pp. 367–376. ACM (2007)
8. Ganter, B., Wille, R.: Formal Concept Analysis: Mathematical Foundations. Springer Science & Business Media, Heidelberg (2012)
9. Hanbury, A.: A survey of methods for image annotation. J. Vis. Lang. Comput. 19(5), 617–627 (2008)
10. Jain, A.K.: Data clustering: 50 years beyond k-means. Pattern Recogn. Lett. 31(8), 651–666 (2010)
11. Mei, T., et al.: Probabilistic multimodality fusion for event based home photo clustering. In: International Conference on Multimedia and Expo, pp. 1757–1760 (2006)
12. Oeldorf-Hirsch, A., Sundar, S.S.: Social and technological motivations for online photo sharing. J. Broadcast. Electron. Media 60(4), 624–642 (2016)
13. Papadopoulos, S., et al.: Cluster-based landmark and event detection for tagged photo collections. IEEE MultiMedia 18(1), 52–63 (2011)
14. Park, S.C., Park, M.K., Kang, M.G.: Super-resolution image reconstruction: a technical overview. IEEE Sig. Process. Mag. 20(3), 21–36 (2003)
15. Quack, T., Leibe, B., Van Gool, L.: World-scale mining of objects and events from community photo collections. In: International Conference on Content-Based Image and Video Retrieval, pp. 47–56 (2008)

16. Raad, E.J., Chbeir, R.: Foto2events: from photos to event discovery and linking in online social networks. In: International Conference on Big Data and Cloud Computing, pp. 508–515. IEEE (2014)

17. Rehman, S.U., et al.: DBSCAN: past, present and future. In: International Conference on Applications of Digital Information and Web Technologies, pp. 232–238 (2014)

18. Reuter, T., et al.: ReSEED: social event detection dataset. In: Conference on Multimedia Systems, pp. 35–40. ACM (2014)

19. Sakaki, T., Okazaki, M., Matsuo, Y.: Earthquake shakes Twitter users: real-time event detection by social sensors. In: Proceedings of International Conference on WWW, pp. 851–860. ACM (2010)

20. Sayyadi, H., et al.: Event detection and tracking in social streams. In: ICWSM (2009)

21. Sheba, S., Ramadoss, B., Balasundaram, S.R.: Event detection refinement using external tags for flickr collections. In: Mohapatra, D.P., Patnaik, S. (eds.) Intelligent Computing, Networking, and Informatics. AISC, vol. 243, pp. 369–375. Springer, New Delhi (2014). doi:10.1007/978-81-322-1665-0_35

22. Wille, R.: Restructuring lattice theory: an approach based on hierarchies of concepts. In: Rival, I. (ed.) Ordered Sets. NATO Advanced Study Institutes Series, vol. 83, pp. 445–470. Springer, Dordrecht (1982). doi:10.1007/978-94-009-7798-3_15

Generating Fake but Realistic Headlines Using Deep Neural Networks

Ashish Dandekar[1]([⊠]), Remmy A.M. Zen[2], and Stéphane Bressan[1]

[1] National University of Singapore, Singapore, Singapore
ashishdandekar@u.nus.edu, steph@nus.edu.sg
[2] Universitas Indonesia, Jakarta, Indonesia
remmy.augusta@ui.ac.id

Abstract. Social media platforms such as Twitter and Facebook implement filters to detect fake news as they foresee their transition from social media platform to primary sources of news. The robustness of such filters lies in the variety and the quality of the data used to train them. There is, therefore, a need for a tool that automatically generates fake but realistic news.

In this paper, we propose a deep learning model that automatically generates news headlines. The model is trained with a corpus of existing headlines from different topics. Once trained, the model generates a fake but realistic headline given a seed and a topic. For example, given the seed "Kim Jong Un" and the topic "Business", the model generates the headline "kim jong un says climate change is already making money".

In order to better capture and leverage the syntactic structure of the headlines for the task of synthetic headline generation, we extend the architecture - Contextual Long Short Term Memory, proposed by Ghosh et al. - to also learn a part-of-speech model. We empirically and comparatively evaluate the performance of the proposed model on a real corpora of headlines. We compare our proposed approach and its variants using Long Short Term Memory and Gated Recurrent Units as the building blocks. We evaluate and compare the topical coherence of the generated headlines using a state-of-the-art classifier. We, also, evaluate the quality of the generated headline using a machine translation quality metric and its novelty using a metric we propose for this purpose. We show that the proposed model is practical and competitively efficient and effective.

Keywords: Deep learning · Natural language generation · Text classification

1 Introduction

In the Digital News Report 2016[1], Reuters Institute for the Study of Journalism claims that 51% of the people in their study indicate the use of social media

---

[1] http://www.digitalnewsreport.org/survey/2016/overview-key-findings-2016/.

© Springer International Publishing AG 2017
D. Benslimane et al. (Eds.): DEXA 2017, Part II, LNCS 10439, pp. 427–440, 2017.
DOI: 10.1007/978-3-319-64471-4_34

platforms as their primary source of news. This transition of social media platforms to news sources further accentuates the issue of the trustworthiness of the news which is published on the social media platforms. In order to address this, social media platform like Facebook has already started working with five fact-checking organizations to implement a filter which can flag fake news on the platform[2].

Starting from the traditional problem of spam filtering to a more sophisticated problem of anomaly detection, machine learning techniques provide a toolbox to solve such a spectrum of problems. Machine learning techniques require a good quality training data for the filters to be robust and effective. To train fake news filters, they need a large amount of fake but realistic news. Fake news, which are generated by a juxtaposition of a couple of news without any context, do not lead to robust filtering. Therefore, there is a need of a tool which automatically generates a large amount of good quality fake but realistic news.

In this paper, we propose a deep learning model that automatically generates news headlines given a seed and the context. For instance, for a seed "obama says that", typical news headlines generated under *technology* context reads "obama says that google is having new surface pro with retina display design" whereas the headline generated under *business* context reads "obama says that facebook is going to drop on q1 profit". For the same seed with *medicine* and *entertainment* as the topics, typical generated headlines are "obama says that study says west africa ebola outbreak has killed million" and "obama says that he was called out of kim kardashian kanye west wedding" respectively.

We expect that the news headlines generated by the model should not only adhere to the provided context but also to conform to the structure of the sentence. In order to catch the attention of the readers, news headlines follow the structure which deviates from the conventional grammar to a certain extent. We extend the architecture of Contextual Long Short Term Memory (CLSTM), proposed by Ghosh et al. [9], to learn the part-of-speech model for news headlines. We compare Recurrent Neural Networks (RNNs) variants towards the effectiveness of generating news headlines. We qualitatively and quantitatively compare the topical coherence and the syntactic quality of the generated headlines and show that the proposed model is competitively efficient and effective.

Section 2 presents the related work. Section 3 delineates the proposed model along with some prerequisites in the neural network. We present experiments and evaluation in Sect. 4. Section 5 concludes the work by discussing the insights and the work underway.

2   Related Work

In the last four-five years, with the advancement in the computing powers, neural networks have taken a rebirth. Neural networks with multiple hidden layers, dubbed as "Deep Neural Networks", have been applied in many fields starting

---

[2] https://www.theguardian.com/technology/2016/dec/15/facebook-flag-fake-news-fact-check.

from classical fields like multimedia and text analysis [11,18,28,29] to more applied fields [7,32]. Different categories of neural networks have been shown to be effective and specific to different kinds of tasks. For instance, Restricted Boltzmann Machines are widely used for unsupervised learning as well as for dimensionality reduction [13] whereas Convolutional Neural Networks are widely used for image classification task [18].

Recurrent Neural Networks [28] (RNNs) are used learn the patterns in the sequence data due to their ability to capture interdependence among the observations [10,12]. In [5], Chung et al. show that the extensions of RNN, namely Long Short Term Memory (LSTM) [14] and Gated Recurrent Unit (GRU) [3], are more effective than simple RNNs at capturing longer trends in the sequence data. However, they do not conclude which of these gated recurrent model is better than the other. Readers are advised to refer to [22] for an extensive survey of RNNs and their successors.

Recurrent neural networks and their extensions are widely used by researchers in the domain of text analysis and language modeling. Sutskever et al. [29] have used multiplicative RNN to generate text. In [10], Graves has used LSTM to generate text data as well as images with cursive script corresponding to the input text. Autoencoder [13] is a class of neural networks which researchers have widely used for finding latent patterns in the data. Li et al. [19] have used LSTM-autoencoder to generates text preserving the multi-sentence structure in the paragraphs. They give entire paragraph as the input to the system that outputs the text which is both semantically and syntactically closer to the input paragraph. Tomas et al. [24,25] have proposed RNN based language models which have shown to outperform classical probabilistic language models. In [26], Tomas et al. provide a context along with the text as an input to RNN and later predict the next word given the context of preceding text. They use LDA [2] to find topics in the text and propose a technique to compute topical features of the input which are fed to RNN along with the input. Ghosh et al. [9] have extended idea in [26] by using LSTM instead of RNN. They use the language model at the level of a word as well as at the level of a sentence and perform experiments to predict next word as well as next sentence given the input concatenated with the topic. There have been evidences of LSTM outperforming GRU for the task of language modeling [15,16]. Nevertheless, we compare our proposed model using both of these gated recurrent building blocks. We use the simple RNN as our baseline for the comparison.

Despite these applications of deep neural networks on the textual data, there are few caveats in these applications. For instance, although in [9] authors develop CLSTM which is able to generate text, they evaluate its predictive properties purely using objective metric like perplexity. The model is not truly evaluated to see how effective it is towards generating the data. In this paper, our aim is to use deep neural networks to generate the text and hence evaluate the quality of synthetically generated text against its topical coherence as well as grammatical coherence.

3   Methodology

3.1   Background: Recurrent Neural Network

Recurrent Neural Network (RNN) is an adaptation of the standard feedforward neural network wherein connections between hidden layers form a loop. Simple RNN architecture consists of an input layer (x), a hidden layer (h), and an output layer (y). Unlike the standard feedforward networks, the hidden layer of RNN receives an additional input from the previous hidden layer. These recurrent connections give RNN the power to learn sequential patterns in the input. We use the many-to-many variant of RNN architecture which outputs n-gram given the previous n-gram as the input. For instance, given { *(hello, how, are)* } trigram as the input, RNN outputs { *(how, are, you)* } as the preceding trigram.

Bengio et al. [1] show that learning the long-term dependencies using gradient descent becomes difficult because the gradients eventually either vanish or explode. The gated recurrent models, LSTM [14] and GRU [3], alleviate these problems by adding gates and memory cells (in the case of LSTM) in the hidden layer to control the information flow. LSTM introduces three gates namely forget gate (f), input gate (i), and output gate (o). Forget gate filters the amount of information to retain from the previous step, whereas input and output gate defines the amount of information to store in the memory cell and the amount of information to transfer to the next step, respectively. Equation 1 shows the formula to calculate the forget gate activations at a certain step t. For given layers or gates m and n, W_{mn} denotes the weight matrix and b_m is the bias vector for the respective gate. h is the activation vector for the hidden state and $\sigma(\cdot)$ denotes the sigmoid function. Readers are advised to refer to [14] for the complete formulae of each gate and layer in LSTM.

$$f_t = \sigma(W_{fx}x_t + W_{fh}h_{t-1} + b_f) \tag{1}$$

GRU simplifies LSTM by merging the memory cell and the hidden state, so there is only one output in GRU. It uses two gates which are update and reset gate. Update gate unifies the input gate and the forget gate in LSTM to control the amount of information from the previous hidden state. The reset gate combines the input with the previous hidden state to generate the current hidden state.

3.2   Proposed Syntacto-Contextual Architecture

Simple RNNs predict the next word solely based on the word dependencies which are learnt during the training phase. Given a certain text as a seed, the seed may give rise to different texts depending on the context. Refer to the Sect. 1 for an illustration. [9] extends the standard LSTM to Contextual Long Short Term Memory (CLSTM) model which accepts the context as an input along with the text. For example, an input pair { *(where, is, your), (technology)* } generates an output like { *(is, your, phone)* }. CLSTM is a special case of the architecture shown in Fig. 1a using LSTM as the gated recurrent model.

In order to use the model for the purpose of text generation, contextual information is not sufficient to obtain a good quality output. A good quality text is coherent not only in terms of its semantics but also in terms of its syntax. By providing the syntactic information along with the text, we extend the contextual model to Syntacto-Contextual (SC) models. Figure 1b shows the general architecture of the proposed model. We encode the patterns in the syntactic meta information and input text using the gated recurrent units and, later, merge them with the context. The proposed model not only outputs text but also corresponding syntactic information. For instance, an input {(where, is, your), (adverb, verb, pronoun), (technology)} generates output like {(is, your, phone), (verb, pronoun, noun)}.

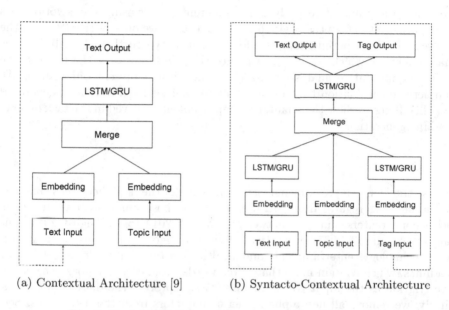

(a) Contextual Architecture [9]   (b) Syntacto-Contextual Architecture

Fig. 1. Contextual and syntacto-contextual architectures

Mathematically, the addition of context and syntactic information amounts to learning a few extra weight parameters. Specifically, in case of LSTM, Eq. 1 will be modified to Eqs. 2 and 3, for CLSTM and SCLSTM respectively. In Eqs. 2 and 3, p represents topic embedding and s represents embedding of the syntactic information.

$$f_t = \sigma(W_{fx}x_t + W_{fh}h_{t-1} + b_f + \mathbf{W_{pf}p_t}) \tag{2}$$
$$f_t = \sigma(W_{fx}x_t + W_{fh}h_{t-1} + b_f + \mathbf{W_{pf}p_t} + \mathbf{W_{sf}s_t}) \tag{3}$$

For the current study, we annotate the text input with the part-of-speech tags using Penn Treebank tagset [23]. We learn the parameters of the model using stochastic gradient descent by minimizing the loss for both output text and output tags. We, also, work on a variation of the contextual architecture which does not accept topic as an input and uses conventional RNN instead of LSTM. This model is treated as the baseline against which all of the models will be compared.

For each of the model, we embed the input in a vector space. We merge the inputs by column wise concatenation of the vectors. We perform experiments using both LSTM and GRU as the gated recurrent units. The output layer is a softmax layer that represents the probability of each word or tag. We sample from that probability to get the next word and tag output.

4   Experimentation and Results

We conduct a comparative study on five different models using a real-world News Aggregator Dataset. In the beginning of this section, we present the details of the dataset and the experimental setup for the study. We, further, describe various quality metrics which we use to evaluate the effectiveness of the models. We perform quantitative analysis using these metric and present our results. We complete the evaluation by presenting micro-analysis for a sample of generated news headlines to show the qualitative improvement observed in the task of news headline generation.

4.1   Dataset

We use the News Aggregator dataset[3] consisting of the news headlines collected by the news aggregator from 11,242 online news hostnames, such as time.com, forbes.com, reuters.com, etc. between 10 March 2014 to 10 August 2014. The dataset contains 422,937 news articles divided into four categories, namely business, technology, entertainment, and health. We randomly select 45000 news headlines, which contain more than three words, from each category because we give trigram as the input to the models. We preprocess the data in two steps. Firstly, we remove all non alpha-numeric characters from the news titles. Secondly, we convert all the text into lower case. After the preprocessing, the data contains 4,274,380 unique trigrams and 39,461 unique words.

4.2   Experimental Setup

All programs are run on Linux machine with quad core 2.40 GHz Intel® Core i7™ processor with 64 GB memory. The machine is equipped with two Nvidia GTX 1080 GPUs. Python® 2.7.6 is used as the scripting language. We use a high-level neural network Python library, *Keras* [4] which runs on top of *Theano* [30]. We use categorical cross entropy as our loss function and use ADAM [17] as an optimizer to automatically adjust the learning rate.

We conduct experiments and comparatively evaluate five models. We refer to those models as, **baseline** - a simple RNN model, **CLSTM** - contextual

---

[3] https://archive.ics.uci.edu/ml/datasets/News+Aggregator.

architecture with LSTM as the gated recurrent model, **CGRU** - contextual architecture with GRU as the gated recurrent model, **SCLSTM** - syntacto-contextual architecture with LSTM as the gated recurrent model, **SCGRU** - syntacto-contextual architecture with GRU as the gated recurrent model, in the rest of the evaluation. All inputs are embedded into a 200-dimensional vector space. We use recurrent layers each with 512 hidden units with 0.5 dropout rate to prevent overfitting. To control the randomness of the prediction, we set the temperature parameter in our output softmax layer to 0.4. We use the batch size of 32 to train the model until the validation error stops decreasing.

4.3   Evaluation Metrics

In this section, we present different evaluation metrics that we use for the quantitative analysis. Along with purely objective quantitative metrics such as perplexity, machine translation quality metric, and topical precision, we use metrics like grammatical correctness, n-gram repetition for a finer effectiveness analysis. Additionally, we devise a novelty metric to qualitatively analyse the current use case of news headline generation.

Perplexity is commonly used as the performance measure [9,10,15,16] to evaluate the predictive power of a language model. Given N test data with w_t as the target outputs, the perplexity is calculated by using Eq. 4, where $p_{w_t}^i$ is the probability of the target output of sample i. A good language model assigns a higher probability to the word that actually occurs in the test data. Thus, a *language model with lower perplexity is a better model.*

$$perplexity = 2^{-\frac{1}{N} \sum_{i=1}^{N} \log p_{w_t}^i} \tag{4}$$

As it happens, the exponent in the Eq. 4 is the approximation of cross-entropy[4], which is the loss function we minimize to train the model, given a sequence of fixed length.

Although the task under consideration of the presented work is not of a word or a topic prediction, we simply use perplexity as a purely objective baseline metric. We complement it by using various application specific measures in order to evaluate the effectiveness of the quality of the generated text.

Topical coherence refers to the extent to which the generated text adheres to the desired topic. In order to evaluate the topical coherence, one requires a faithful classifier which predicts the topic of generated text. We treat the topics predicted by the classifier as the ground truth to quantitatively evaluate the topical coherence. The proposed method generates a news headline given a seed and a topic of the news. People have widely used Multinomial naive Bayes classifier to deal with text data due to independence among the words given a certain class[5]. We train a Multinomial naive Bayes classifier with Laplace smoothing on the news dataset consisting of 45000 news from each of the four

---

[4] http://cs224d.stanford.edu/lecture_notes/notes4.pdf.
[5] https://www.kaggle.com/uciml/news-aggregator-dataset.

categories. We hold out 20% of the data for validation. By proper tuning of the smoothing parameter, we achieve 89% validation accuracy on the news dataset. We do not use this metric for the baseline model.

Taking existing text as a reference, a **quality** metric evaluates the effectiveness of the generated text in correspondence to the reference. Such a metric measures the *closeness* of the generated text to the reference text. Metrics such as BLEU [27], Rouge [8], NIST [21] are widely used to evaluate the quality of machine translation. All of these metrics use "gold standard", which is either the original text or the text written by the domain experts, to check the quality of the generated text. We use BLEU as the metric to evaluate the quality of generated text. For a generated news headline, we calculate its BLEU score by taking all the sentences in the respective topic from the dataset as the reference. Interested readers should refer to [33] for a detailed qualitative and quantitative interpretation of BLEU scores.

With the motivation of the current work presented in the Sect. 1, we want the generated text from our model to be as *novel* as possible. So as to have a robust fake news filter, the fake news, which is used to train the model, should not be a juxtaposition of few existing news headlines. More the patterns it learns from the training data to generate a single headline, more novel is the generated headline. We define **novelty** of the generated output as the number of unique patterns the model learns from the training data in order to generate that output. We realize this metric by calculating longest common sentence common to the generated headline and each of the headline in the dataset. Each of these sentences stands as a pattern that the model has learned to generate the text. *Novelty* of a generated headline is taken as the number of unique longest common sentences.

The good quality generated text should be both *novel* and grammatically correct. **Grammatical correctness** refers the judgment on whether the generated text adheres to the set of grammatical rules defined by a certain language. Researchers either employ experts for evaluation or use advanced grammatical evaluation tools which require the gold standard reference for the evaluation [6]. We use an open-source grammar and spell checker software called LanguageTool[6] to check the grammatical correctness of our generated headlines. LanguageTool uses NLP based 1516 English grammar rules to detect syntactical errors. Aside from NLP based rules, it used English specific spelling rules to detect spelling errors in the text. To evaluate grammatical correctness, we calculate the percentage of grammatically correct sentences as predicted by the LanguageTool.

We find that LanguageTool only recognizes word repetition as an error. Consider a generated headline *beverly hills hotel for the first in the first in the world* as an example. In this headline, there is a trigram repetition - *the first in* - that passes LanguageTool grammatical test. Such headlines are not said to be good quality headlines. We add new rules with a regular expression to detect such repetitions. We count **n-gram repetitions** within a sentence for values of n greater than two.

---

[6] API and Java package available at https://languagetool.org.

4.4   Results

To generate the output, we need an initial trigram as a seed. We randomly pick the initial seed from the set of news headlines from the specified topic. We use windowing technique to generate the next output. We remove the first word and append the output to the back of the seed to generate the next output. The process stop when specified sentence length is generated. We generate 100 sentences for each topic in which each sentence contains 3 seed words and 10 generated words.

Quantitative Evaluation. Table 1 summarizes the quantitative evaluation of all the models using metrics described in Sect. 4.3. Scores in bold numbers denote the best value for each metric. We can see that for Contextual architecture, GRU is a better gated recurrent model. Conversely, LSTM is better for Syntacto-Contextual architecture.

For Syntacto-Contextual architecture, we only consider the perplexity of the text output to make a fair comparison with the Contextual architecture. We analyze that our Syntacto-Contextual architecture has a higher perplexity score because the model jointly minimizes both text and syntactical output losses. On the other hand, the baseline model has a low perplexity score because it simply predicts the next trigram with control on neither the context nor the syntax.

A high score on classification precision substantiates that all of these models generate headlines which are coherent with the topic label with which they are generated. We observe that all of the models achieve a competitive BLEU score. Although Contextual architecture performs slightly better in terms of BLEU score, Syntacto-Contextual architecture achieves a higher novelty score. In the qualitative evaluation, we present a more detailed comparative analysis of BLEU scores and novelty scores.

We observe that the news headlines generated by Syntacto-Contextual architecture are more grammatically correct than other models. Figure 2 shows the histogram of n-gram repetitions in the generated news headline. We see that the Syntacto-Contextual architecture gives rise to news headlines with less number of n-gram repetitions.

Lastly, we have empirically evaluated, but not presented here, the time taken by different models for one epoch. CLSTM takes 2000 s for one epoch whereas SCLSTM takes 2131 s for one epoch. Despite the Syntacto-Contextual architecture being a more complex architecture than Contextual architecture, it shows that it is competitively efficient.

Qualitative Evaluation. Table 2 presents the samples of generated news from CLSTM proposed by [9] and SCLSTM, which outweighs the rest of the models in the quantitative analysis.

In Table 1, we see that the Contextual architecture models receive a higher BLEU score than the proposed architecture models. BLEU score is calculated using n-gram precisions with the news headlines as the reference. It is not always

Table 1. Quantitative evaluation.

	Baseline	CLSTM [9]	CGRU	SCLSTM	SCGRU
Perplexity	108.383	119.10	**92.22**	146.93	175.83
Topical coherence (%)	-	84.25	77.25	**94.75**	87.50
Quality (BLEU)	0.613	0.637	**0.655**	0.633	0.625
Novelty	21.605	24.67	25.21	**26.57**	25.65
Grammatical correctness (%)	28.25	49.75	50.75	**75.25**	69.00
n-gram Repetitions	11	30	12	**5**	8

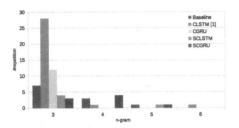

Fig. 2. n-gram repetition analysis   **Fig. 3.** Boxplot for novelty metric

necessary that the higher BLEU score leads towards a good quality text generation. Qualitative analysis of generated headlines shows that the higher BLEU score, in the most cases, is the result of the juxtaposition of the existing news headlines. For instance, consider a headline generated by CLSTM model as an example - "justin bieber apologizes for racist joke in new york city to take on" - which receives a BLEU score of 0.92. When we search for the same news in the dataset, we find that this generated news is a combination of two patterns from the following two headlines, "justin bieber apologizes for racist joke" and "uber temporarily cuts fares in new york city to take on city cabs". Whereas the headline generated by SCLSTM with the same seed is quite a novel headline. In the training dataset there is mention of neither Justin Bieber joking on Twitter nor joke for gay fans. Similar observation can be made with the news related to Fukushima. In the training data set there is no news headline which links Fukushima with climate change. Additionally, there is no training data which links higher growth risk to climate change as well. Thus, we observe that the headlines generated using SCLSTM are qualitatively better than CLSTM.

All of the models presented in the work are probabilistic models. Text generation being a probabilistic event, on the one hand it is possible that contextual architecture generates a good quality headline at a certain occasion. For instance, we see that CLSTM also generates some good quality news headlines such as "the fault in our stars trailer for the hunger games mockingjay part teaser". On the other hand, it is possible that Syntacto-Contextual architecture generates some news headline with poor quality or repetitions, such as "obama warns

Table 2. Generated news headlines

Category	CLSTM [9]	SCLSTM
Medicine	first case chikungunya virus in saudi arabia reports new mers cov in the	first case chikungunya virus found in florida state health care system draws attention
	mosquitoes test positive for west africa in guinea in june to be the	mosquitoes test positive for west nile virus found in storage room at home
	ticks and lyme disease in the us in the us are the best	ticks and lyme disease rates double in autism risk in china in west
Business	us accuses china and russia to pay billion in billion in us in	us accuses china of using internet explorer bug leaves users to change passwords
	wake of massive data breach of possible to buy stake in us	wake of massive recall of million vehicles for ignition switch problem in china
	japan fukushima nuclear plant in kansas for first time in the first time	japan fukushima nuclear plant linked to higher growth risk of climate change ipcc
Entertainment	justin bieber apologizes for racist joke in new york city to take on	justin bieber apologizes for racist joke on twitter for gay fans have
	the fault in our stars trailer for the hunger games mockingjay part teaser	the fault in our stars star chris hemsworth and elsa pataky reveal his
	giant practical spaceship interiors for joint venture in mexico production of star wars	giant practical spaceship hits the hollywood walk of fame induction ceremony in
Technology	first android wear watches and google be to be available in the uk	first android wear watch google play edition is now available on xbox one
	samsung sues newspaper for anti vaccine and other devices may be the best	samsung sues newspaper over facebook experiment on users with new profile feature is
	obama warns google glass to be forgotten on the us government issues recall	obama warns google apple to make android support for mobile for mobile

google apple to make android support for mobile for mobile". In order to qualitatively analyse the novelty of generated sentence, we need to observe how likely
such events occur. Figure 3 shows the boxplot of novelty numbers we calculate
for each of 400 generated news headlines using different models. As discussed
earlier, we want our model to generate novel news headlines. So, we prefer higher

novelty scores. Although the mean novelty of all of the models lie around 24, we see that SCLSTM is more likely to generate the novel headlines. Additionally, we observe that contextual and Syntacto-Contextual architectures performs better than the baseline model.

As mentioned in the quantitative evaluation, Contextual architecture gives rise to news headlines with a large number of n-gram repetitions. In an extreme case, CLSTM model generates the following headline, "lorillard *inc nyse wmt wal mart stores* inc nyse wmt wal mart stores", that contains 6-gram repetition. The news headline generated by CLSTM - "Samsung sues newspaper for anti vaccine and other devices may be the best"- exemplifies the smaller topical coherence observed for the Contextual architecture models.

In order to garner the opinion of real-world users, we use CrowdFlower[7] to conduct a crowdsource based study. In this study, we generate two news headlines using CLSTM and SCLSTM using the same seed and ask the workers to choose a more realistic headline between two. We generate such a pair of headlines for 200 different seeds. Each pair is evaluated by three workers and majority vote is used to choose the right answer. At the end of the study, 66% workers agree that SCLSTM generates more realistic headlines than CLSTM.

5   Discussion and Future Work

In [9], Ghosh et al. proposed a deep learning model to predict the next word or sentence given the context of the input text. In this work, we adapted and extended their model towards automatic generation of news headlines. The contribution of the proposed work is two-fold. Firstly, in order to generate news headlines which are not only topically coherent but also syntactically sensible, we proposed an architecture that learns part-of-speech model along with the context of the textual input. Secondly, we performed thorough qualitative and quantitative analysis to assess the quality of the generated news headlines using existing metrics as well as a novelty metric proposed for the current application. We comparatively evaluated the proposed models with [9] and a baseline. To this end, we show that the proposed approach is competitively better and generates good quality news headlines given a seed and the topic of the interest.

Through this work, we direct our methodology for data-driven text generation towards a "constraint and generate" paradigm from a more brute-force way of "generate and test". Quality assessment of the generated data using generative model remains an open problem in the literature [31]. We use the measure of quality, which in our case is the grammatical correctness, as an additional constraint for the model in order to generate the good quality data. The usage of POS tags as the syntactic element is mere a special case in this application. We can think of more sophisticated meta information to enrich the quality of text generation. Ontological categories can be an alternative option.

---

[7] https://www.crowdflower.com/.

Acknowledgement. This work is partially supported by the National Research Foundation, Prime Ministers Office, Singapore under its Campus for Research Excellence and Technological Enterprise (CREATE) programme and by the National University of Singapore under a grant from Singapore Ministry of Education for research project number T1 251RES1607. We kindly acknowledged UCI for maintaining repository of publicly available Machine Learning dataset [20].

References

1. Bengio, Y., Simard, P., Frasconi, P.: Learning long-term dependencies with gradient descent is difficult. IEEE Trans. Neural Netw. **5**(2), 157–166 (1994)
2. Blei, D.M., Ng, A.Y., Jordan, M.I.: Latent dirichlet allocation. J. Mach. Learn. Res. **3**, 993–1022 (2003)
3. Cho, K., Van Merriënboer, B., Gulcehre, C., Bahdanau, D., Bougares, F., Schwenk, H., Bengio, Y.: Learning phrase representations using RNN encoder-decoder for statistical machine translation. arXiv preprint arXiv:1406.1078 (2014)
4. Chollet, F.: Keras (2015). https://github.com/fchollet/keras
5. Chung, J., Gulcehre, C., Cho, K., Bengio, Y.: Empirical evaluation of gated recurrent neural networks on sequence modeling. arXiv preprint arXiv:1412.3555 (2014)
6. Dahlmeier, D., Ng, H.T.: Better evaluation for grammatical error correction. In: Proceedings of the 2012 Conference of the North American Chapter of the Association for Computational Linguistics: Human Language Technologies, pp. 568–572. Association for Computational Linguistics (2012)
7. Deng, L., Yu, D., et al.: Deep learning: methods and applications. Found. Trends ® Sig. Process. **7**(3–4), 197–387 (2014)
8. Doddington, G.: Automatic evaluation of machine translation quality using n-gram co-occurrence statistics. In: Proceedings of the Second International Conference on Human Language Technology Research, pp. 138–145. Morgan Kaufmann Publishers Inc. (2002)
9. Ghosh, S., Vinyals, O., Strope, B., Roy, S., Dean, T., Heck, L.: Contextual LSTM (CLSTM) models for large scale NLP tasks. arXiv preprint arXiv:1602.06291 (2016)
10. Graves, A.: Generating sequences with recurrent neural networks. arXiv preprint arXiv:1308.0850 (2013)
11. Graves, A., Jaitly, N., Mohamed, A.R.: Hybrid speech recognition with deep bidirectional LSTM. In: 2013 IEEE Workshop on Automatic Speech Recognition and Understanding (ASRU), pp. 273–278. IEEE (2013)
12. Graves, A., Mohamed, A.R., Hinton, G.: Speech recognition with deep recurrent neural networks. In: 2013 IEEE International Conference on Acoustics, Speech and Signal Processing (ICASSP), pp. 6645–6649. IEEE (2013)
13. Hinton, G.E., Salakhutdinov, R.R.: Reducing the dimensionality of data with neural networks. Science **313**(5786), 504–507 (2006)
14. Hochreiter, S., Schmidhuber, J.: Long short-term memory. Neural Comput. **9**(8), 1735–1780 (1997)
15. Irie, K., Tüske, Z., Alkhouli, T., Schlüter, R., Ney, H.: LSTM, GRU, highway and a bit of attention: an empirical overview for language modeling in speech recognition. In: INTERSPEECH (2016)
16. Jozefowicz, R., Zaremba, W., Sutskever, I.: An empirical exploration of recurrent network architectures. In: Proceedings of the 32nd ICML, pp. 2342–2350 (2015)

17. Kingma, D., Ba, J.: Adam: A method for stochastic optimization. arXiv preprint arXiv:1412.6980 (2014)
18. Krizhevsky, A., Sutskever, I., Hinton, G.E.: Imagenet classification with deep convolutional neural networks. In: Advances in Neural Information Processing Systems, pp. 1097–1105 (2012)
19. Li, J., Luong, M.T., Jurafsky, D.: A hierarchical neural autoencoder for paragraphs and documents. arXiv preprint arXiv:1506.01057 (2015)
20. Lichman, M.: UCI machine learning repository (2013). http://archive.ics.uci.edu/ml
21. Lin, C.Y.: Rouge: a package for automatic evaluation of summaries. In: Proceedings of the ACL-04 Workshop, vol. 8 (2004)
22. Lipton, Z.C., Berkowitz, J., Elkan, C.: A critical review of recurrent neural networks for sequence learning. arXiv preprint arXiv:1506.00019 (2015)
23. Marcus, M.P., Marcinkiewicz, M.A., Santorini, B.: Building a large annotated corpus of English: the penn treebank. Comput. Linguist. **19**(2), 313–330 (1993)
24. Mikolov, T., Karafiát, M., Burget, L., Cernocký, J., Khudanpur, S.: Recurrent neural network based language model. In: Interspeech, vol. 2, p. 3 (2010)
25. Mikolov, T., Kombrink, S., Burget, L., Černocký, J., Khudanpur, S.: Extensions of recurrent neural network language model. In: 2011 IEEE International Conference on Acoustics, Speech and Signal Processing (ICASSP), pp. 5528–5531. IEEE (2011)
26. Mikolov, T., Zweig, G.: Context dependent recurrent neural network language model. SLT **12**, 234–239 (2012)
27. Papineni, K., Roukos, S., Ward, T., Zhu, W.J.: BLEU: a method for automatic evaluation of machine translation. In: Proceedings of the 40th Annual Meeting on Association for Computational Linguistics, pp. 311–318. Association for Computational Linguistics (2002)
28. Sutskever, I.: Training recurrent neural networks. Ph.D. thesis, University of Toronto (2013)
29. Sutskever, I., Martens, J., Hinton, G.E.: Generating text with recurrent neural networks. In: Proceedings of the 28th International Conference on Machine Learning (ICML-2011), pp. 1017–1024 (2011)
30. Theano Development Team: Theano: A Python framework for fast computation of mathematical expressions. arXiv e-prints abs/1605.02688, May 2016. http://arxiv.org/abs/1605.02688
31. Theis, L., Oord, A.V.D., Bethge, M.: A note on the evaluation of generative models. arXiv preprint arXiv:1511.01844 (2015)
32. Wang, H., Wang, N., Yeung, D.Y.: Collaborative deep learning for recommender systems. In: Proceedings of the 21th ACM SIGKDD International Conference on Knowledge Discovery and Data Mining, pp. 1235–1244. ACM (2015)
33. Zhang, Y., Vogel, S., Waibel, A.: Interpreting BLEU/NIST scores: how much improvement do we need to have a better system?. In: Proceedings of the Fourth International Conference on Language Resources and Evaluation (LREC), pp. 2051–2054 (2004)

Qualitative AHP Method Based on Multiple Criteria Levels Under Group of Experts

Amel Ennaceur[1,2($\boxtimes$)], Zied Elouedi[1], and Eric Lefevre[3]

[1] Université de Tunis, Institut Supérieur de Gestion, LARODEC, Tunis, Tunisia
amel_naceur@yahoo.fr, zied.elouedi@gmx.fr
[2] Université de Jendouba, Faculté des Sciences Juridiques, Economiques et de
Gestion de Jendouba, Jendouba, Tunisie
[3] Univ. Artois, EA 3926, LGI2A, 62400 Béthune, France
eric.lefevre@univ-artois.fr

Abstract. In this paper, we consider a multi-criteria group decision making problem. We propose a novel version of the Analytic Hierarchy Process under the belief function theory. The presented approach uses groups of experts to express their assessments regarding the evaluation criteria and the evaluation alternatives. It considers also more complex multi-criteria decision problems that have multiple criteria levels. The presented method is illustrated with some examples.

1 Introduction

The multi-criteria decision-making (MCDM) can be defined as a field which refers to making decisions in the presence of multiple and conflicting criteria [11]. In this research, we focus on one of the most popular MCDM approach, namely the Analytic Hierarchy Process (AHP) [6]. Within the AHP context, many extensions were introduced [10]. Their main objective is to handle uncertainty under the expert assessments [4]. For instance, in some cases, decision maker is unable to express his judgements using crisp numbers and to provide a complete pair-wise comparison procedure. Qualitative AHP is then one of the very useful tools to tackle this drawback [3].

Even though selecting single expert-based alternatives according to conflicting criteria has received significant attention [11], handling such problems with group of experts and based on multiple criteria levels is still an open subject. Consequently, many extensions, under the belief function framework, have been introduced [1,10]. Regarding these methods, the expert may be unable to provide quantitative numbers to describe his opinions. For that reason, we propose to extend the Qualitative AHP into a group decision making context based on multiple criteria levels. In this work, we present additional usefulness of Ennaceur et al.'s method [3] for handling more complex MCDM problems. In fact, the Qualitative AHP method combines the standard AHP and the belief function theory to adequately model uncertain human judgments and to represent the expert assessment easily. However, in many decision problems, the expert is

© Springer International Publishing AG 2017
D. Benslimane et al. (Eds.): DEXA 2017, Part II, LNCS 10439, pp. 441–449, 2017.
DOI: 10.1007/978-3-319-64471-4_35

able to decompose the problem into: goal, criteria, sub-criteria and alternatives. Therefore, an improved version of the Qualitative AHP is proposed to take into account the fact that many multi-criteria problems might be modeled under multiple criteria levels. Moreover, this new approach is extended into a group decision making environment.

The rest of the paper is organized as follows: In Sect. 2, we introduce the belief function theory. Next, Sect. 3 describes the Qualitative AHP method. In Sects. 4 and 5, we detail an improved version of our Qualitative AHP under multiple criteria levels and based on a group decision-maker context. Section 6 concludes this paper.

2 Belief Function Theory

For brevity, we will not consider in detail what this model is. The interested reader should refer to [7]. We present the basic concepts as interpreted by the Transferable Belief Model (TBM). Let Θ be a finite set of elementary hypotheses, called a frame of discernment. Let 2^{Θ} be all the subsets of Θ [7]. The basic belief assignment (bba) is a function m, that represents the portion of belief committed exactly to the event A. The belief function theory offers many interesting tools. The discounting technique allows to take in consideration the reliability of the information source that generates the bba m [9]. Also, to combine beliefs induced by distinct pieces of evidence, we can use the conjunctive rule of combination [8]. Moreover, It is necessary when making a decision, to select the most likely hypothesis. One of the most used solutions within the belief function theory is the pignistic probability. More details can be found in [7].

3 AHP Method with Belief Preference Relations

In this section, we consider a revised version of the AHP model, namely Qualitative AHP. To describe the approach, we present its different steps:

Step 1: Model the problem as a hierarchy based on three levels. At the highest level, we find the main objective. Then, in the middle, the sets of criteria $\Omega = \{c_1, \ldots, c_n\}$ for evaluating the sets of alternatives $\Theta = \{a_1, \ldots, a_m\}$, which will be in the lowest level. Then, we define the subsets of criteria and alternatives. As presented in [3], we put together criteria (or alternatives) that have the same degree of preference.

Step 2: Establish priorities among the elements of the hierarchy. Each element is paired and compared. In this context, Ennaceur et al. model [2] is used to transform preferences relations into constraints of an optimization problem. Its resolution, according to an uncertainty measure (UM) [5] generates the most uncertain belief functions. For instance, we consider the criterion based on pairwise comparison matrix, we get:

$$\text{Max } H(m^{\Omega}) = m^{\Omega}(C_1) * log_2(|C_1|/m^{\Omega}(C_1)) + m^{\Omega}(C_2)log_2(|C_2|/m^{\Omega}(C_2))$$
$$+ ... + m^{\Omega}(C_n) * log_2(|C_n|/m^{\Omega}(C_n)) + m^{\Omega}(\Omega) * log_2(|\Omega|/m^{\Omega}(\Omega));$$
$$s.t.$$
$$bel^{\Omega}(C_i) - bel^{\Omega}(C_j) \geq \gamma \quad \forall (C_i, C_j), C_i \succ C_j$$
$$bel^{\Omega}(C_i) - bel^{\Omega}(C_j) \leq \gamma \quad \forall (C_i, C_j), C_i \succeq C_j$$
$$bel^{\Omega}(C_i) - bel^{\Omega}(C_j) \geq \varepsilon \quad \forall (C_i, C_j), C_i \succeq C_j$$
$$bel^{\Omega}(C_i) - bel^{\Omega}(C_j) \geq -\varepsilon \quad \forall (C_i, C_j), C_i \sim C_j$$
$$bel^{\Omega}(C_i) - bel^{\Omega}(C_j) \leq \varepsilon \quad \forall (C_i, C_j), C_i \sim C_j$$
$$\sum_{C_i \in \mathcal{F}(m^{\Omega})} m^{\Omega}(C_i) = 1, m^{\Omega}(A) \geq 0, \forall A \subseteq \Omega; m^{\Omega}(\emptyset) = 0$$

where H is the uncertainty measure. The preference relation is represented by the first constraint. Next, the weak preference relation is illustrated by the second and third constraints. The indifference relation corresponds to the fourth and fifth constraints. The expert has to specify the indifference threshold ε and the preference threshold γ as two constants.

Step 3: Assume that criteria weights and alternatives scores are described by a bba defined on the possible responses. Thus, m^{Ω} denotes the criterion bba and $m_{c_k}^{\Theta}$ denotes the alternative bba, according to c_k.

Step 4: Use the pignistic probabilities. At the level of criteria, the bba m^{Ω} is transformed into pignistic probabilities as follows:

$$BetP^{\Omega}(c_i) = \sum_{C_j \subseteq \Omega} \frac{|c_i \cap C_j|}{|C_j|} \frac{m^{\Omega}(C_j)}{(1 - m^{\Omega}(\emptyset))}, \quad \forall c_i \in \Omega \quad (1)$$

Step 5: Consider each pignistic probability $(BetP^{\Omega}(c_i))$ as a measure of reliability. For each specific criterion c_i, β_i is its corresponding measure of reliability. For each $i, k = 1, \ldots, n$:

$$\beta_i = \frac{BetP^{\Omega}(c_i)}{max_k BetP^{\Omega}(c_k)} \quad (2)$$

Step 6: Synthesize the overall judgment. We have to update the alternatives priorities with their corresponding criteria weight. The obtained bba's are discounted such as:

$$^{\alpha}m_{c_k}^{\Theta}(A_j) = \beta_k . m_{c_k}^{\Theta}(A_j), \quad \forall A_j \subset \Theta \quad (3)$$

$$^{\alpha}m_{c_k}^{\Theta}(\Theta) = (1 - \beta_k) + \beta_k . m_{c_k}^{\Theta}(\Theta) \quad (4)$$

Step 7: Combine the overall bba's to get a single representation by using the conjunctive rule ($m^{\Theta} = \bigcap {}^{\alpha}m_{c_k}^{\Theta}$) and choose the best alternatives by computing its pignistic probabilities.

Example. Let us consider a problem of buying a car. This case study involves four criteria: $\Omega = \{\text{Style } (c_1), \text{ Price } (c_2), \text{ Fuel } (c_3), \text{ Reliability } (c_4)\}$ and three selected alternatives: $\Theta = \{\text{Peugeot } (p), \text{Renault } (r), \text{Ford } (f)\}$. The expert has identified three subsets of criteria: $\{c_1\}$, $\{c_4\}$ and $\{c_2, c_3\}$. Along with the qualitative pair-wise comparison, the preference relations defined in Table 1 was obtained. After deriving the criteria weight, the corresponding bba is transformed into pignistic probabilities as presented in Table 1. At the level of alternatives, the same steps is repeated.

Table 1. Preference relations matrix

Criteria	$\{c_1\}$	$\{c_4\}$	$\{c_2, c_3\}$
$\{c_1\}$	-	$\succ$	$\succ$
$\{c_4\}$	-	-	$\succ$
$\{c_2, c_3\}$	-	-	-

Subsets of criteria	$\{c_1\}$	$\{c_4\}$	$\{c_2, c_3\}$	Ω
m^{Ω}	0.228	0.218	0.208	0.346
Criteria	$\{c_1\}$	$\{c_4\}$	$\{c_3\}$	$\{c_2\}$
$BetP^{\Omega}$	0.315	0.305	0.190	0.190

We move now to the next stage to calculate the reliability measure. We obtain $\beta_{\{c_1\}} = 1$, $\beta_{\{c_4\}} = 0.96$, $\beta_{\{c_3\}} = 0.6$ and $\beta_{\{c_2\}} = 0.6$.

Next, we need to update the alternatives priorities by the criteria weight. The discounted bba's are defined in Table 2.

Table 2. The discounted bba's

	$\alpha_1 m_{c_1}^{\Theta}(.)$		$\alpha_4 m_{c_4}^{\Theta}(.)$		$\alpha_3 m_{c_3}^{\Theta}(.)$		$\alpha_2 m_{c_2}^{\Theta}(.)$
$\{p\}$	0.505	$\{p\}$	0.306	$\{r\}$	0.303	$\{f\}$	0.303
$\{p, r, f\}$	0.495	$\{r, f\}$	0.514	$\{p, r, f\}$	0.697	$\{p, r, f\}$	0.697
		$\{p, r, f\}$	0.180				

The next step is to combine the overall evidences and the alternative ranking is, finally, obtained according to the pignistic transformation. We obtain $BetP^{\Theta}(p) = 0.410$, $BetP^{\Theta}(r) = 0.295$ and $BetP^{\Theta}(f) = 0.295$.

4   Qualitative AHP Method Under Multiple Criteria Levels

At this stage, our main aim is to propose an extension of Qualitative AHP model under multiple levels of criteria. The originality of this work is to introduce a new hierarchy level, namely sub-criteria, in order to handle more complex multi-criteria problem. Let us consider a case when there are two criteria levels. Our MCDM problem is defined as follows: $\Theta = \{a_1, \ldots, a_m\}$ represents the set of alternatives, $\Omega = \{c_1, \ldots, c_n\}$ is a set of criteria. For each criterion, we have Ω_l, the set of k^l sub-criteria (denoted by sc_j^l with $j = 1, \ldots, k^l$) corresponding to the criterion c_l. For example, the second criterion c_2 has three sub-criteria denoted by sc_1^2, sc_2^2 and sc_3^2. We start by computing their relative scores. At each hierarchy level, the expert has to assess the relative importance of each criterion regarding the main objective. His evaluations have to be modeled using preference relations in order to convert them into an optimization problem. A basic belief assignment is computed. The resulting basic belief mass, denoted by $m^{\Omega}(C_j)$ (where C_j is a single or a group of criteria), has to satisfy the following relation: $\sum_{C_j \subseteq \Omega} m^{\Omega}(C_j) = 1$.

The same process is repeated at the sub-criterion level to get the local priorities. Suppose that the j-th criterion is selected by the expert to evaluate its corresponding sub-criteria. The relative bba should satisfy: $\sum_{SC_i^j \subseteq \Omega_j} m^{\Omega_j}(SC_i^j) = 1$. where $i = 1, \ldots, k^j$ with k^j is the number of sub-criterion according to the j-th criterion and $j = 1, \ldots, n$.

We move to the alternative level to get $m^{\Theta}(A_k)$ (which represents the belief about the subset of alternatives A_k regarding each sub-criterion).

Now, we calculate the result and synthesize the solution by aggregating all the obtained bba's. Therefore, we must start by computing the global priority of each sub-criterion. Since we know how much the priority of each sub-criterion contributes to the priority of its parent, we can now calculate their global priorities. That will show us the priority of each sub-criterion regarding the goal.

As with Qualitative AHP method, we start by the criterion level. We transform m^{Ω} to its relative pignistic probabilities $BetP^{\Omega}$, since our beliefs are defined on groups of criteria. The same technique is repeated on the sub-criterion level to get the pignistic probabilities. We convert m^{Ω_j} into $BetP^{\Omega_j}$.

For each sub-criterion, we can now compute its global priority by: $GP(sc_i^j) = BetP^{\Omega}(c_j) * BetP^{\Omega_j}(sc_i^j)$. Then, to compute the alternatives ranking, we must apply the Qualitative AHP model from step 5 in Sect. 3, to compute the reliability measures, to update the alternatives priorities and to combine the overall bba's to select the highest alternatives.

Example. Let us continue with the same problem. We consider that the criterion c_1 has two sub-criteria sc_1^1 and sc_2^1. The necessary calculations have already been presented in the previous example. We obtain the following local priorities presented in Table 3. After computing the corresponding $BetP^{\Omega_1}$, we can deduce the global priority as defined in Table 3.

Table 3. The local priorities assigned to the criteria and sub-criteria

Criteria	$BetP^{\Omega}$	Sub-criteria	$BetP^{\Omega_1}$	GP
c_1	0.315	sc_1^1	0.310	0.098
		sc_2^1	0.690	0.217
c_4	0.305			0.305
c_2	0.190			0.190
c_3	0.190			0.190

In this case, one more step need to be made. In fact, our aim here is to quantify the priority of each criterion regarding the main objective. The global priorities throughout the hierarchy will add up to 1 (see Table 3).

5   Belief AHP Method Under Multiple Criteria Levels Based on Group of Experts

In many complex problems, collective decisions are made based on a group of expert. Accordingly, we propose to extend the Qualitative AHP under a group decision-making environment. The objective is then to aggregate evidence from a member of group, to select the most appropriate alternative.

Step 1: Expert weights. The main aim of this step is to give weights to experts to quantify their importance. In this case, pair-wise comparison is suggested to generate their importances. We consider the experts as the set of alternatives that are compared regarding the reliability criterion. As presented previously, the belief pair-wise comparison generates a bba, that represents the part of belief committed exactly to each expert. This bba is transformed into a reliability measure, denoted by β_i corresponding to expert i.

Step 2: Expert elicitation. Each expert starts by identifying all the focal elements. Then, he compares all the identified elements through binary relations. The next step is to transform the obtained assessments into an optimization problem and to generate the least informative bba. A bba (m_i) is generated corresponding to each expert i.

Step 3: Aggregation process. We proceed to the aggregation of the obtained bba's. First, we start by the discounting technique. Indeed, the idea is to measure most heavily the bba evaluated according to the most importance expert and conversely for the less important ones. So, each obtained bba is discounted by its corresponding measure of reliability, as follows:

$$P_i(A) = m_i(A) * \beta_i, \forall A \subseteq \Theta \tag{5}$$

$$P_i(\Theta) = (1 - \beta_i) + \beta_i.m_i(\Theta) \tag{6}$$

where $i = 1, \ldots, \eta$ with η is the number of experts.

Then, we move to compute the final decision. An intuitive definition of the strategy to fuse these bba's will be through the conjunctive rule of combination and the ranking of alternatives is obtained using the pignistic probabilities.

Example. Let us consider the previous problem with four decision makers. The first step is to assign a degree of importance to each expert. We have $\beta_1 = 1$, $\beta_2 = 0.9$, $\beta_3 = 0.2$ and $\beta_4 = 0.1$. After applying the Qualitative AHP method, we obtain Table 4.

Next, each bba corresponding to each expert is discounted according to its measure of reliability. Then, we combine the different bba's in order to generate the collective decision. We obtain: $m_{\text{global}}(\{p\}) = 0.070$, $m_{\text{global}}(\{r\}) =$

Table 4. The final result using the Qualitative AHP approach regarding each decision maker

Alternatives	$\{p\}$	$\{r\}$	$\{f\}$	$\{p,r\}$	$\{p,f\}$	$\{r,f\}$	$\emptyset$	Θ
m_1	0.193	0.073	0.073			0.124	0.495	0.042
m_2	0.154	0.063	0.080	0.106		0.100	0.395	0.102
m_3	0.090	0.108	0.034	0.171			0.455	0.142
m_4	0.073	0.073	0.144			0.102	0.592	0.016

0.042, $m_{\text{global}}(\{f\}) = 0.032$, $m_{\text{global}}(\{f,r\}) = 0.032$, $m_{\text{global}}(\{p,r\}) = 0.002$, $m_{\text{global}}(\emptyset) = 0.815$ and $m_{\text{global}}(\Theta) = 0.009$. The obtained ranking is illustrated as follows: $BetP_{\text{global}}(p) = 0.395$, $BetP_{\text{global}}(r) = 0.334$ and $BetP_{\text{global}}(f) = 0.271$.

Now, we analyze the different results. At this step, we propose to combine the obtained priority except one, in order to study the influence of the decision made by this expert on the decision of the group. This process is repeated for all the expert priorities. The aim of the process is to identify experts who provide preferences that are significantly different from the group, and to provide these experts with the opportunity to update these preferences to be closer to the majority.

The main objective is to show that the best alternative is supported by the majority of expert and there is not a contradictory alternative. We can notice that experts 1 and 2 are considered as reliable sources of information, since they

Table 5. The final result using the Qualitative AHP approach

Alternatives	$P_1 \cap P_2 \cap P_3$	$BetP_{P_1 \cap P_2 \cap P_3}$	$P_1 \cap P_2 \cap P_4$	$BetP_{P_1 \cap P_2 \cap P_4}$
$\emptyset$	0.799		0.790	
$\{p\}$	0.076	0.400	0.079	0.392
$\{r\}$	0.046	0.334	0.045	0.324
$\{f\}$	0.034	0.266	0.038	0.284
$\{f,r\}$	0.033		0.037	
$\{p,r\}$	0.002		0.002	
Θ	0.010		0.009	

Alternatives	$P_1 \cap P_3 \cap P_4$	$BetP_{P_1 \cap P_3 \cap P_4}$	$P_2 \cap P_3 \cap P_4$	$BetP_{P_2 \cap P_3 \cap P_4}$
$\emptyset$	0.590		0.468	
$\{p\}$	0.155	0.406	0.117	0.377
$\{r\}$	0.067	0.307	0.059	0.334
$\{f\}$	0.059	0.287	0.067	0.289
$\{f,r\}$	0.094		0.062	
$\{p,r\}$	0.001		0.069	
Θ	0.034		0.158	

have almost the highest reliability measure. However, experts 3 and 4 are treated as not fully reliable. As we can see, we have the P_1 and P_2 support the same alternative and they consider that the best one is p and the worst is f. However, P_3 and P_4 present a contradictory information.

Consequently, we can notice that by combining the preferences of experts 1, 2 and 3 or experts 1, 2 and 4, have generated almost the same ranking of alternatives (see Table 5). We find that the most preferred alternative is p even if experts 3 and 4 prefer the alternatives r and f respectively. This is because experts 3 and 4 are not considered reliable. Besides, as shown in Table 5, when combining evidences $P_1 \odot P_3 \odot P_4$ or $P_2 \odot P_3 \odot P_4$, we know that experts 3 and 4 are not fully reliable. Therefore, we can consider only the preference induced from expert 1 and 2 respectively. So, we can conclude that the most preferred alternative is p.

6   Conclusion

In this paper, we have formulated qualitative AHP method in an environment characterized by imperfection. Our approach deals with qualitative reasoning to model the uncertainty related to experts assessment. The advantage of this newly proposed model is its ability to handle multi-criteria level problem. It is also able to manage more complex problem by solving a multi-criteria group decision making problem. A future research idea is to study the effect of changing the weight of the groups of experts by a sensitivity analysis. In fact, the idea of assigning importance to a group of expert has been investigated, with arguments given as to the need for and against its utilization.

References

1. Beynon, M.: A method of aggregation in DS/AHP for group decision-making with the non-equivalent importance of individuals in the group. Comput. Oper. Res. **32**, 1881–1896 (2005)
2. Ennaceur, A., Elouedi, Z., Lefevre, É.: Modeling qualitative assessments under the belief function framework. In: Cuzzolin, F. (ed.) BELIEF 2014. LNCS, vol. 8764, pp. 171–179. Springer, Cham (2014). doi:10.1007/978-3-319-11191-9_19
3. Ennaceur, A., Elouedi, Z., Lefevre, E.: Multi-criteria decision making method with belief preference relations. Int. J. Uncertain. Fuzziness Knowl. Based Syst. **22**(4), 573–590 (2014)
4. Holder, R.D.: Some comments on Saaty's AHP. Manage. Sci. **41**, 1091–1095 (1995)
5. Pal, N., Bezdek, J., Hemasinha, R.: Uncertainty measures for evidential reasoning I: a review. Int. J. Approx. Reason. **7**, 165–183 (1992)
6. Saaty, T.: The Analytic Hierarchy Process. McGraw-Hill, New York (1980)
7. Shafer, G.: A Mathematical Theory of Evidence. Princeton University Press, Princeton (1976)
8. Smets, P.: The combination of evidence in the transferable belief model. IEEE Pattern Anal. Mach. Intell. **12**, 447–458 (1990)
9. Smets, P.: Transferable belief model for expert judgments and reliability problems. Reliab. Eng. Syst. Saf. **38**, 59–66 (1992)

10. Utkin, L.V., Simanova, N.V.: The DS/AHP method under partial information about criteria and alternatives by several levels of criteria. Int. J. Inf. Technol. Decis. Making **11**(2), 307–326 (2012)
11. Zeleny, M.: Multiple Criteria Decision Making. McGraw-Hill Book Company, New York (1982)

Exploit Label Embeddings for Enhancing Network Classification

Yiqi Chen[1], Tieyun Qian[1(✉)], Ming Zhong[1], and Xuhui Li[2]

[1] State Key Laboratory of Software Engineering, Wuhan University, Wuhan, China
{fyiqic16,qty,clock}@whu.edu.cn
[2] School of Information Management, Wuhan University, Wuhan, China
lixuhuig@whu.edu.cn

Abstract. Learning representations for network has aroused great research interests in recent years. Existing approaches embed vertices into a low dimensional continuous space which encodes local or global network structures. While these methods show improvements over traditional representations, they do not utilize the label information until the learned embeddings are used for training classifier. That is, the process of representation learning is *separated* from the labels and thus is unsupervised. In this paper, we propose a novel method which learns the embeddings for vertices under the supervision of labels. The key idea is to regard the label as the context of a vertex. More specifically, we attach a true or virtual label node for each training or test sample, and update the embeddings for vertices and labels to maximize the probability of both the neighbors and their labels in the context. We conduct extensive experiments on three real datasets. Results demonstrate that our method outperforms the state-of-the-art approaches by a large margin.

Keywords: Deep learning · Representation learning · Network classification

1 Introduction

Networked data, such as users in social networks, authors or documents in publication networks, are becoming pervasive nowadays. Network classification is a very important problem for many applications like recommendation or targeted advertisement. However, the data sparsity severely deteriorates the performance of network classification algorithms. Graph representation learning, which aims to embed the sparse network into a low-dimensional dense space, has long been a fundamental task and attracted great research interests.

Traditionally, networks or graphs are represented as adjacency, Laplacian, or affinity matrices, and the spectral properties of matrix representations are exploited for learning low dimensional vectors as features for graphs. Typical methods include IsoMap [3], Laplacian EigenMap [4], locally linear embedding (LLE) [12], and spectral clustering [16]. These methods rely on the solving of leading eigenvectors and are computationally expensive. For example, Isomap

D. Benslimane et al. (Eds.): DEXA 2017, Part II, LNCS 10439, pp. 450–458, 2017.
DOI: 10.1007/978-3-319-64471-4_36

and Laplacian EigenMap have a quadratic time complexity to the number of the nodes. Real world social networks like Facebook or Twitter contain millions of nodes and billions of edges. Hence it is extremely hard for classic approaches to scale to real social networks.

Recently, distributed representations of words and documents in a low dimensional vector space have shown great success in many natural language processing tasks. In particular, Mikolov [9] introduced an efficient neural network based model, SkipGram, to learn distributed representations of words. SkipGram is based on the hypothesis that words in similar context tend to have similar meanings. It has been proven that SkipGram can learn very expressive representations for words.

Inspired by the SkipGram model, researchers learned representations for vertices based on an analogy between a network and a document [7,11,13]. The basic idea is to sample sequences of nodes from the network and to convert a network into an ordered sequence of nodes. These methods differ mainly in the sampling procedure to generate sequence. DeepWalk [11] combined skip-gram with random walk. LINE [13] exploited the first-order and second-order proximity, i.e., the local and global structure, into the process. Node2Vec [7] offered a flexible sampling strategy by smoothly interpolating between the breadth-first search and depth-first search.

While the above three network representation methods show improvements over traditional techniques, they are all designed for a general purpose rather than the network classification. In this paper, we investigate the problem of learning representation for network classification. Our goal is to leverage the label information for enhancing the performance of classification. To this end, we propose a novel *label enhanced network embedding* (LENE) model which combines the label into the objective function by regarding it as the context of a node. We first attach a true or virtual label node for each training or test instance. We then update the embeddings for vertices and labels to maximize the probability of both the neighbors and their labels in the context. We conduct extensive experiments on three real datasets. Results demonstrate that our method significantly outperforms three state-of-the-art approaches.

2   Label Enhanced Network Embedding

This section we introduce our label enhanced network embedding (LENE) model.

2.1   Label as Context

The goal of existing graph embedding approaches is to learn the low dimensional representations reflecting the network structure. In the case of network classification, some nodes share the same labels with other nodes, and such relationships should convey useful information. Figure 1(a) shows an example network containing five nodes v_1, v_2, v_3, v_4, v_5 and two labels l_1, l_2, where the solid lines

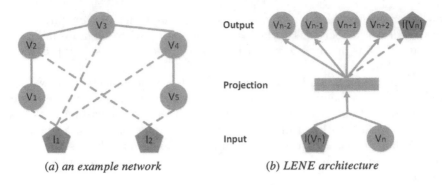

(a) an example network   (b) LENE architecture

Fig. 1. Illustration of LENE model

denote the edges between nodes in the graph and the dashed lines denote the links between nodes and their corresponding labels.

For a possible walk $v_1v_2v_3v_4v_5$ in Fig. 1(a), we can infer that v_2 and v_4 should be close to each other in the embedding space since they have similar network structures. However, if we take the label node into consideration, we can find that l_2 will "pull" v_2 and v_5 closely and "push" them away from the remaining nodes (including v_4).

In order to capture such information, we regard the label as the context of nodes and generate a *label walk* by replacing one node each time in the original *node walk* with its corresponding label. For example, we have the following label walks for the node walk $v_1v_2v_3v_4v_5$ in Fig. 1(a).

$$l_1v_2v_3v_4v_5, \; v_1l_2v_3v_4v_5, \; v_1v_2l_1v_4v_5, \; v_1v_2v_3l_1v_5, \; v_1v_2v_3v_4l_2$$

2.2   Incorporating Label Context into Objective

Given a graph $G = (V, E, L)$, we consider the problem of embedding the vertices and the labels in G into a low dimensional continuous space: $\Phi_1 : v \in V \mapsto R^{|V| \times d}$ and $\Phi_2 : l \in L \mapsto R^{|L| \times d}$. The objective is to maximize the conditional probability $Pr(N_{v_i} | v_i, l_i)$ and $Pr(l_i | v_i)$ over given data, where v_i is a vertex in V, l_i the label of v_i, and N_{v_i} the context nodes in either a node walk or a label walk of v_i. More formally, the objective function can be defined as follows.

$$J(\Phi_1, \Phi_2) = - \sum_{v_i \in V} logPr(N(v_i) | v_i, l_i, \Phi_1, \Phi_2) \tag{1}$$

We solve the optimization problem using the stochastic gradient descent (SGD) [5] with the negative sampling [10] over the model parameters.

In practice, it is expensive to label the nodes and most of the vertices are unlabeled. To fully utilize the label walks, we further assign a virtual label for test nodes. This is done simply by examining the neighbors' labels in the walks a test node involved and selecting the most popular one.

2.3   Label Enhanced Network Embedding

We now present our label enhanced network embedding (LENE) framework by utilizing the above defined node walks, label walks, and the objective function. Figure 1(b) illustrates the architecture of LENE. The *input, projection*, and *output* in Fig. 1(b) indicates the input layer, hidden layer, and output layer of a neural network model, respectively. Our LENE approach learns the label embedding in the same way as DeepWalk learns the node embedding. The difference is that we incorporate the label node into the architecture to capture the relationships between nodes sharing the same labels.

We illustrate the procedure in Algorithm 1. The LENE algorithm consists of two components, i.e., a node walk and label walk generator and an update procedure. The core of LENE is shown in lines 2–9 in Algorithm 1. We iterate over all the nodes in G to generate walks (lines 4–8) and corresponding label walks (line 6). This procedure will repeat r times to start walks for each node(line 2).

Algorithm 1. LENE(G, t, r, w)

Input: Graph $G = (V, E, L)$,
the walk length t,
the number of walks r,
and the window size w
the learning rate η
Output: matrix of node embedding $\Phi_1 \in R^{|V| \times d}$
matrix of label embedding $\Phi_2 \in R^{|L| \times d}$

1. Initialization: Sample Φ_1 and Φ_2 from $U^{|V| \times d}$ and $U^{|L| \times d}$
2. **for** $i = 0$ to r **do**
3. $\mathcal{O} = \text{Shuffle}(V)$
4.   **for each** $v_i \in \mathcal{O}$ **do**
5. $NW_{v_i} = NodeWalk(G, v_i, t)$
6. $LW_{v_i} = LabelWalk(G, l_i, t)$
7. $\text{SkipNL}(\Phi_1, \Phi_2, NW_{v_i}, LW_{v_i}, w, \eta)$
8.   **end for**
9. **end for**

We adapt the SkipGram algorithm [9] to updating the parameters in accordance with our objective function $J(\Phi_1, \Phi_2)$ defined in Eq. 1. We call the adapted version as SkipNL and show it in Algorithm 2.

Algorithm 2. SkipNL(Φ_1, Φ_2, NW_{v_i}, LW_{v_i}, w, η)

1. **for each** $v_j \in NW_{v_i}$ **do**
2.   **for each** $u_k \in NW_{v_i}[j - w : j + w]$ **do**
3. $J(\Phi_1, \Phi_2) = -logPr(\boldsymbol{u_k}|\boldsymbol{v_j}, \Phi_1, \Phi_2)$
4. $\Phi_1 = \Phi_1 - \eta \frac{\partial J(\Phi_1, \Phi_2)}{\partial \Phi_1}$
5.   **end for**
6. **end for**
7. **for each** $l_j \in LW_{v_i}$ **do**
8.   **for each** $u_k \in LW_{v_i}[j - w : j + w]$ **do**
9. $J(\Phi_1, \Phi_2) = -logPr(\boldsymbol{u_k}|\boldsymbol{l_j}, \Phi_1, \Phi_2)$
10. $\Phi_2 = \Phi_2 - \eta \frac{\partial J(\Phi_1, \Phi_2)}{\partial \Phi_2}$
11.   **end for**
12. **end for**

In the original SkipGram, a window is sliding from the beginning to the end of the sentence to access the context of each word. For example, for a given sentence "We really like this book" and a window size of 2, the central word "like" has four contexts "we really" and "this book". In the analog of a node walk $v_1v_2v_3v_4v_5$ to the sentence, the embedding of central node $\boldsymbol{v_3}$ would be used to update embeddings for context nodes $\boldsymbol{v_1}$, $\boldsymbol{v_2}$, $\boldsymbol{v_4}$, $\boldsymbol{v_5}$. This is exactly what DeepWalk does, as shown in lines 1–6 in Algorithm 2. Our SkipNL algorithm further generalizes it by using the embedding of the label node $\boldsymbol{l_3}$ to update $\boldsymbol{v_1}$, $\boldsymbol{v_2}$, $\boldsymbol{v_4}$, $\boldsymbol{v_5}$ (lines 7–12).

3 Experimental Evaluation

In this section, we first provide an overview of the datasets in our experiment. We then introduce the baselines and experimental settings. Finally we report and analyze the experimental results.

3.1 Experimental Setup

Datasets. We conduct experiments on three well known and publicly available datasets.

DBLP [2] includes approximately 16000 scientific publications chosen from the DBLP database including three categories: "Database", "Machine Learning", and "Theory". The nodes are papers and links are co-authorship information between documents.

BlogCatalog [15] is a network of social relations among the authors of the bloggers. There are 39 classes (labels) representing the topic categories provided by the authors. Two of our baselines, DeepWalk [11] and Node2Vec [7], conducts experiments on the same dataset.

The statistics for three datasets are summarized in Table 1. By evaluating on such datasets with varied characteristics, we seed to investigate the applicability of our model to different types of data.

Table 1. The statistics for datasets

	BlogCatalog	DBLP
# of labels	39	3
# of nodes	10,312	16,808
# of edges	333,983	359,192
avg. degree	64.78	21.37

Baselines. We conduct extensive experiments to compare our methods with three the state-of-the-art baselines.

DeepWalk [11] learns the node representations based on the combination of skip-gram model and random walks.

LINE [13] exploits the first-order and second-order proximity, i.e., the local and global structure, into the learning process.

Node2Vec [7] extends DeepWalk by using two parameters to control the sampling strategies of random walks for generating neighborhood of nodes.

We do not compare our method with traditional approaches because Deep-Walk, LINE, and Node2Vec have already shown significant improvements over a number of baselines like Graph Factorization [1], Spectral Clustering [16], Modularity [14], EdgeCluster [15], and weighted-vote Relational Neighbor [8]. We thus only show our improvements over these three baselines.

Settings. We train a one *vs.* rest the logistic regression classifier implemented by Liblinear [6] for each class and select the class with maximum scores as the label. We take representations of vertices as features to train classifiers, and evaluate classification performance using 10-fold cross-validation with different training ratios.

For a fair comparison with the relevant baselines, we use the typical settings in DeepWalk [11] and Node2Vec [7]. Specifically, we set the dimension $d = 128$ and the number of walks = 10, walk length = 80, window size = 10, the number of negative samples = 5 and learning rate = 0.025 on all datasets. For LINE, we set the total number of samples = 10 million.

3.2   Results

We report the average micro-F_1 and macro-F_1 as the evaluation metrics. The scores in bold represent the highest performance among all methods.

Comparison Results on DBLP. We first report classification results on DBLP dataset in Tabel 2. It is clear that LENE trained with only 10% of the training data has already outperforms the baselines when they are provided with the 90% of the training samples. This strongly demonstrates that our approach benefits a lot from label contexts even if there are only a small fraction of labeled nodes. The performance of LENE becomes much more attractive with more training data. For example, LENE achieves at least 8% and 7% improvements over other methods on 90% training data.

Table 2. Average Macro-F_1 and Micro-F_1 score on DBLP

macroF$_1$(%)	10	20	30	40	50	60	70	80	90
LENE	**81.75**	**83.67**	**84.70**	**85.47**	**85.89**	**86.70**	**86.93**	**87.06**	**87.71**
DeepWalk	79.39	80.47	80.74	80.67	80.60	80.51	80.77	80.84	80.94
LINE	73.44	74.27	74.39	74.44	74.73	74.98	74.96	74.94	75.07
Node2Vec	79.93	80.34	80.58	80.82	80.85	81.09	80.92	80.86	80.89
microF$_1$(%)	10	20	30	40	50	60	70	80	90
LENE	**82.25**	**84.07**	**85.06**	**85.85**	**86.28**	**87.04**	**87.27**	**87.42**	**88.02**
DeepWalk	80.17	81.10	81.36	81.32	81.26	81.16	81.39	81.48	81.56
LINE	74.60	75.36	75.46	75.50	75.74	75.91	75.93	75.90	76.03
Node2Vec	80.61	80.94	81.21	81.41	81.48	81.71	81.53	81.49	81.54

Comparison Results on BlogCatalog. We finally list classification results on BlogCatalog in Table 3. We can see that LENE consistently outperforms all other baselines from Table 3. However, the improvements of LENE over other methods on BlogCatalog is not as significant as those on DBLP. The reasons can be two fold. First, the classification task on BlogCatalog is hard since it has much more classes than that on DBLP while having similar node numbers. Second, nodes in BlogCatalog can be multi-labeled, which brings in ambiguous in classification. Nevertheless, the enhanced performance of LENE indicates that it is robust to both simple and difficult classification problems.

3.3   Sensitivity

Like three baselines, our LENE algorithm involves a number of parameters including the walk length t, the number of walks r, and the window size w. For the learning rate η, we just set it to the default value 0.025, as the baselines do.

Table 3. Average Macro-F_1 and Micro-F_1 score on BlogCatalog

macroF$_1$(%)	10	20	30	40	50	60	70	80	90
LENE	**21.05**	**22.98**	**24.30**	**25.04**	**25.67**	**26.25**	**26.83**	**27.74**	**27.84**
DeepWalk	18.64	20.28	21.07	21.64	22.08	22.17	22.46	22.72	23.25
LINE	15.67	17.35	18.67	18.89	19.51	19.94	20.46	20.53	21.02
Node2Vec	19.70	21.69	23.00	23.21	23.83	24.06	24.47	25.30	25.57
microF$_1$(%)	10	20	30	40	50	60	70	80	90
LENE	**35.56**	**36.99**	**38.03**	**38.52**	**39.14**	**39.75**	**39.76**	**40.17**	**40.48**
DeepWalk	34.56	35.58	36.25	36.54	37.00	37.22	37.40	37.56	37.67
LINE	32.38	33.59	34.47	34.83	35.14	35.35	35.74	35.81	36.09
Node2Vec	35.06	36.15	37.13	37.37	37.78	38.10	38.50	38.78	38.88

We examine how the different choices of the parameters affect the performance of LENE on the BlogCatalog dataset using a 50-50 split between training and test data. Except for the parameter being tested, we use default values for all other parameters. We report the macro-F1 score as a function of parameters t, r, and w in Fig. 2.

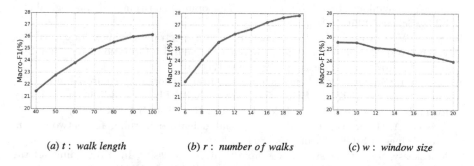

(a) t : *walk length*   (b) r : *number of walks*   (c) w : *window size*

Fig. 2. Parameter sensitivity study on BlogCatalog

We observe that increasing the number and length of walks improves performance in Fig. 2(a) and (b). However, the number of walks r has a higher impact on the performance than the walk length t. This is reasonable because a large number of walks are beneficial to construct homogeneous neighborhoods. In contrast, a long walk denotes a depth first search in the network bringing in more diversified nodes. This happens to the window size w as well which is more directly affected by the neighbor nodes and we see a decrease of the curve in Fig. 2(c).

4 Conclusion

We propose a novel label enhanced network embedding (LENE) approach. The main contribution is to incorporate the label information into the learning process. In particular, we generate label walks by replacing the node with its corresponding label node, and enhance the network representation by sharing the label contexts. We conduct extensive experiments on real world datasets and results demonstrate that our algorithm performs significantly better than the state-of-the-art baselines. In the future, we plan to investigate how the improved embeddings can be used to other tasks like link prediction. In addition, we would like to extend LENE to more complicated networks such as weighted or directed graphs.

Acknowledgment. The work described in this paper has been supported in part by the NSFC project (61572376), and the 111 project(B07037).

References

1. Ahmed, A., Shervashidze, N., Narayanamurthy, S., Josifovski, V., Smola, A.J.: Distributed large-scale natural graph factorization. In: Proceedings of WWW, pp. 37–48 (2013)
2. Angelova, R., Weikum, G.: Graph-based text classification: learn from your neighbors. In: Proceeding of SIGIR, pp. 485–492 (2006)
3. Balasubramanian, M., Schwartz, E.L.: The isomap algorithm and topological stability. Science **295**(5552), 7 (2002)
4. Belkin, M., Niyogi, P.: Laplacian eigenmaps and spectral techniques for embedding and clustering. In: International Conference on Neural Information Processing Systems: Natural and Synthetic, pp. 585–591 (2001)
5. Bottou, L.: Stochastic gradient learning in neural networks. In: Neuro-Nîmes (1991)
6. Fan, R.E., Chang, K.W., Hsieh, C.J., Wang, X.R., Lin, C.J.: Liblinear: a library for large linear classification. J. Mach. Learn. Res. **9**, 1871–1874 (2008)
7. Grover, A., Leskovec, J.: node2vec: Scalable feature learning for networks. In: Proceedings of SIGKDD, pp. 855–864 (2016)
8. Macskassy, S.A., Provost, F.: A simple relational classier. In: Proceedings of the Second Workshop on Multi-Relational Data Mining (MRDM) at KDD-2003, pp. 64–76 (2003)
9. Mikolov, T., Chen, K., Corrado, G., Dean, J.: Efficient estimation of word representations in vector space. CoRR abs/1301.3781 (2013)
10. Mikolov, T., Sutskever, I., Chen, K., Corrado, G., Dean., J.: Distributed representations of words and phrases and their compositionality. In: Proceedings of NIPS, pp. 3111–3119 (2013)
11. Perozzi, B., Al-Rfou, R., Skiena., S.: Deepwalk: online learning of social representations. In: Proceedings of SIGKDD, pp. 701–710 (2014)
12. Roweis, S.T., Saul, L.K.: Nonlinear dimensionality reduction by locally linear embedding. Science **290**(5500), 2323–2326 (2000)
13. Tang, J., Qu, M., Wang, M., Zhang, M., Yan, J., Mei, Q.: Line: large-scale information network embedding. In: Proceedings of WWW, pp. 1067–1077 (2015)
14. Tang, L., Liu, H.: Relational learning via latent social dimensions. In: ACM SIGKDD International Conference on Knowledge Discovery and Data Mining, Paris, France, pp. 817–826, 28 June–July 2009
15. Tang, L., Liu, H.: Scalable learning of collective behavior based on sparse social dimensions. In: Proceeding of ACM Conference on Information and Knowledge Management (CIKM), pp. 1107–1116 (2009)
16. Tang, L., Liu, H.: Leveraging social media networks for classification. Data Min. Knowl. Disc. **23**(3), 447–478 (2011)

Author Index

Printed in the United States
By Bookmasters